RICE DISEASES

This and other publications of the
Commonwealth Agricultural Bureaux
can be obtained through any major bookseller or
direct from
Commonwealth Agricultural Bureaux,
Central Sales, Farnham Royal, Slough SL2 3BN, England

RICE DISEASES

BY

S. H. OU, Ph.D.

Plant Pathologist, The International Rice Research Institute, Los Baños, Laguna, Philippines

COMMONWEALTH MYCOLOGICAL INSTITUTE
KEW, SURREY, ENGLAND
1972

First published in 1972 by the
Commonwealth Mycological Institute
under the authority of the
Executive Council, Commonwealth Agricultural Bureaux
Farnham Royal, Slough, England

SBN 85198 217 4

PRINTED IN GREAT BRITAIN BY THE EASTERN PRESS LTD., LONDON AND READING

Dedicated to:

MY PARENTS

and two inspiring teachers

J. C. WALKER

S. C. TENG

FOREWORD

THE rice plant is attacked by many diseases, the more serious ones causing substantial losses in crop yield.

This book represents the most complete analysis of the diseases of the rice plant that has ever been assembled in one volume, and it is the only thing of its kind that has appeared during the past 20 years.

The information is derived from research and observations in both the temperate zones and the tropics, and the literature review is intended to be complete as far as recent findings are concerned.

The writer, Dr. S. H. Ou, has spent more than 25 years as a rice pathologist working first in mainland China and later in Taiwan. He accepted assignments with the Food and Agriculture Organization of the United Nations in Iraq and Thailand, and has headed the plant pathology department of the International Rice Research Institute since its inception in 1962. Dr. Ou is recognized as an eminent authority on rice diseases, and particularly on the rice blast disease. His treatment of this, the most serious and ubiquitous of all rice diseases, is exceedingly thorough.

His complete and up-to-date discussion of the virus diseases of the rice plant is of particular value because so much new information on the virus diseases of tropical rice has come to light during the past few years.

This book is designed to serve as a reference work for all students of phytopathology and, of course, for all scientists who seek information on the history, occurrence, nature and control of all known diseases of the rice plant.

ROBERT F. CHANDLER, Jr., Director
The International Rice Research Institute

CONTENTS

Part III. Fungus Diseases—Foliage Diseases

Part IV. Fungus Diseases—Diseases of Stem, Leaf Sheath and Root

Part V. Fungus Diseases—Seedling Diseases

Part VI. Fungus Diseases—Diseases of Grain and Inflorescence

LIST OF ILLUSTRATIONS

PREFACE

THE widening gap between the slow rate of increase in rice production and the high rate of increase in the rice-eating population has been one of the most acute food problems in the world in recent years.

In an attempt to increase rice production, many improved varieties have been developed in several countries of Southeast Asia in recent years. Many of them, such as the varieties Khao-tah-haeng 17 and Leuang-awn 29 of Thailand, Malinja and Mashsuri of Malaysia, and Tjeremas of the Philippines are, however, susceptible to blast disease and have been discarded after a brief period of use. New virus diseases causing heavy losses have been reported in many countries and bacterial blight has been very destructive in the tropics. Several other diseases may also cause serious losses in various areas. Furthermore, disease problems will be increasingly important in the future, as the necessity for using more and more fertilizers for higher production tends to aggravate many diseases, and as fewer varieties are planted in large areas the probability of greater losses due to disease epidemics increases.

In spite of the fact that most rice is grown in the tropics, relatively little work has in the past been done on rice disease problems in tropical areas. The recent acute nature of the food problem has, however, stimulated an increasing number of people in the rice-growing countries in the tropics to conduct research and develop control measures.

The literature on rice diseases, unfortunately, has been very scattered and much of it has been unavailable to many tropical research workers. Working on rice diseases in India, Padwick (1950) once said: " I soon realised how much of the relevant literature was unavailable to me, and how much was not only unavailable to other plant pathologists in the rice-growing countries, but was also written in languages which many could not read." The rapid accumulation of literature on rice diseases in more and more journals during the last 20 years has greatly worsened the situation. Padwick's *Manual of Rice Diseases* has been out of print for several years. We have often been requested by eager young workers to supply ' all relevant literature ' on a common disease such as blast or bacterial blight. Requests of this type are often difficult to comply with. Many rice workers in several tropical countries cannot identify such a disease as bacterial blight or distinguish leaf spot caused by blast from brown spot. We feel there is at the present a great need for a comprehensive reference book on rice diseases, written in a commonly used language, for students in both research and extension work in rice pathology.

This book is aimed primarily at presenting the status of knowledge at this moment on each of the rice diseases known. The survey of literature has been made as complete as possible. Various views on a given subject are presented even though some of them are controversial and a few are of doubtful accuracy. Extensive relevant literature is cited in which researchers may look for further specific details.

Descriptions and illustrations are presented to assist in the proper diagnosis of rice diseases. Sufficient details are also given to enable workers to identify the causal organisms and to prepare specific media with which to culture them. Some emphasis is given to the discussion of varietal resistance and breeding for disease resistance. I believe that this is one of the fundamental approaches to the solution of rice disease problems and it has not yet been fully explored.

Physiological diseases are treated very briefly. Such problems as nitrogen deficiency, iron toxicity and other nutritional relations are discussed in publications on plant physiology or soil chemistry. An article entitled ' Nutritional disorders of rice plant in Asia ' by Tanaka and Yoshida has recently been published as a technical bulletin of the IRRI. References in the literature to problems arising from cold injury or from air or water pollution are very few, although damage from these causes has occurred. A few diseases, such as ' akiochi ', ' akagare ', bronzing and straighthead, which are well known as physiological diseases of rice, are briefly discussed.

S. H. Ou
The International Rice Research Institute
Los Baños, Laguna, Philippines

ACKNOWLEDGMENTS

THE task of compiling an extensive work such as this would have been almost impossible for me without the opportunity, given by the International Rice Research Institute (IRRI), of a year of study leave, during which the preliminary draft was done.

The faithful assistance of collecting many of the references and checking all the literature citations by the library of IRRI is greatly appreciated. The use of references in the libraries of the Department of Plant Pathology and College of Agriculture and Life Sciences of the University of Wisconsin and in the National Agricultural Library, U.S. Department of Agriculture, is also appreciated. The accommodation and secretarial assistance provided by the Department of Plant Pathology, University of Wisconsin, in which I spent the study leave as a visiting professor, is gratefully acknowledged. A great deal of editorial work rendered by the publisher, the Commonwealth Mycological Institute, is much appreciated.

Personally, I wish to thank Dr. R. F. Chandler, Jr. and Dr. A. C. McClung for their interest and encouragement from the planning stage to the publication of the book; Dr. J. C. Walker who kindly read the entire manuscript; Dr. P. R. Jennings for his continuing interest and discussion during the preparation of the manuscript; Miss J. S. Galang and Mrs. M. K. Samuels for typing the final manuscripts and other colleagues for their assistance in one way or another.

The reviews of Padwick (1950, *Manual of rice diseases*), Diehl (1954, *A list of references to diseases of rice*), Wei (1957, *Manual of rice pathogens*), and Hara (1959, *Monograph of rice diseases*), and the proceedings of the symposia on rice blast disease (1965, Johns Hopkins Press) and on virus diseases of rice (1969, Johns Hopkins Press) have facilitated the preparation of the manuscript. Illustrations and data by various authors used in the manuscript are acknowledged in the text.

PART I

VIRUS DISEASES

Twelve distinct virus diseases are now known to occur on rice. Some of them have been known since the beginning of the century, and rice dwarf virus was, in fact, the first plant virus demonstrated to be transmitted by an insect. It was also the first case to be found of a virus passing from one generation of the vector to the next through the eggs. Other rice virus diseases have, however, been identified only recently.

All these diseases are transmitted by leafhoppers or planthoppers (Fig. I–1), except yellow mottle, mosaic and necrosis mosaic, and each occurs in a specific geographical region. A brief summary of the rice virus diseases reported, their vectors and geographical distribution is given in Table I–1:

TABLE I–1

SUMMARY OF RICE VIRUS DISEASES

Disease	Vector/transmission	Geographical distribution
Dwarf	*Nephotettix cincticeps, N. apicalis, Recilia (Inazuma) dorsalis*	Japan, ? Korea, ? China
Stripe	*Laodelphax striatellus, Unkanodes sapporonus, Ribautodelphax albifascia*	Japan, Korea
Yellow dwarf	*Nephotettix impicticeps, N. cincticeps, N. apicalis*	Throughout Asia
Black-streaked dwarf	*Laodelphax striatellus, Unkanodes sapporonus, Ribautodelphax albifascia*	Japan
Hoja blanca	*Sogatodes oryzicola, S. cubanus*	North, Central and South America
Tungro	*Nephotettix impicticeps, N. apicalis, Recilia (Inazuma) dorsalis*	Philippines
Penyakit merah	*Nephotettix impicticeps*	Malaysia
Yellow-orange leaf	*Nephotettix impicticeps, N. apicalis*	Thailand
Mentek	*Nephotettix impicticeps*	Indonesia
Leaf yellowing	*Nephotettix impicticeps*	India
Transitory yellowing	*Nephotettix apicalis, N. cincticeps*	Taiwan
Orange leaf	*Recilia (Inazuma) dorsalis*	Philippines, Thailand, Ceylon
Grassy stunt	*Nilaparvata lugens*	Philippines, Ceylon, ? India
Yellow mottle	*Sesselia pusilla* and mechanical	Kenya
Mosaic	Mechanical	Philippines
Necrosis mosaic	Soil	Japan

As may be seen from Table I–1, a single virus may be transmitted by more than one vector and one insect species may be a vector of more than one virus.

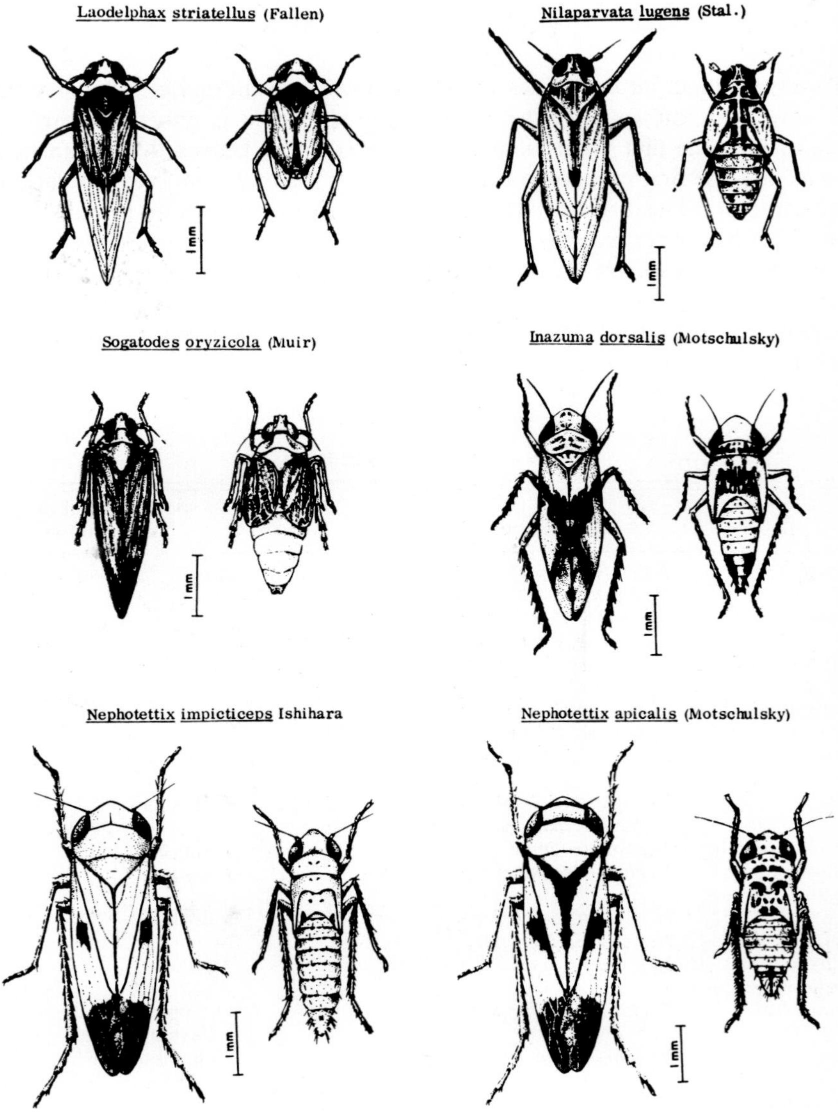

Fig. I–1. Common vectors of rice virus diseases.

Laodelphax striatellus—left, macropterous male; right, brachypterous male. *Nilaparvata lugens*—left, macropterous male; right, nymph. *Sogatodes oryzicola*—left, macropterous male; right, brachypterous female. *Inazuma* (*Recilia*) *dorsalis*—left, male adult; right, nymph. *Nephotettix impicticeps*—left, male adult; right, nymph. *Nephotettix apicalis*—left, male adult; right, nymph. (Drawings after Ling, 1968b).

The tungro, penyakit merah, yellow-orange leaf, mentek and leaf yellowing viruses were identified only recently. Available information indicates that they are similar or perhaps identical. Penyakit merah and mentek have been known for several decades but were previously thought to be physiological diseases. This group includes the most destructive virus diseases of rice known in South-east Asia. The dwarf and stripe viruses are important in Japan and Korea. Hoja blanca is widespread in Latin America and also occurs in U.S.A.

While dwarf, stripe and hoja blanca viruses are transovarial, i.e. the viruses pass through eggs from one generation of the insect vector to the next, the tungro virus is nonpersistent, i.e. the vector retains the virus only for a maximum of 5 or 6 days after each acquisition feeding. This nonpersistent virus-vector relationship is unique among the leafhopper-borne viruses. The other viruses are of the persistent type, i.e. once the vector has become infective the virus is retained throughout the life of the insect. Thus, in the transovarial type, once

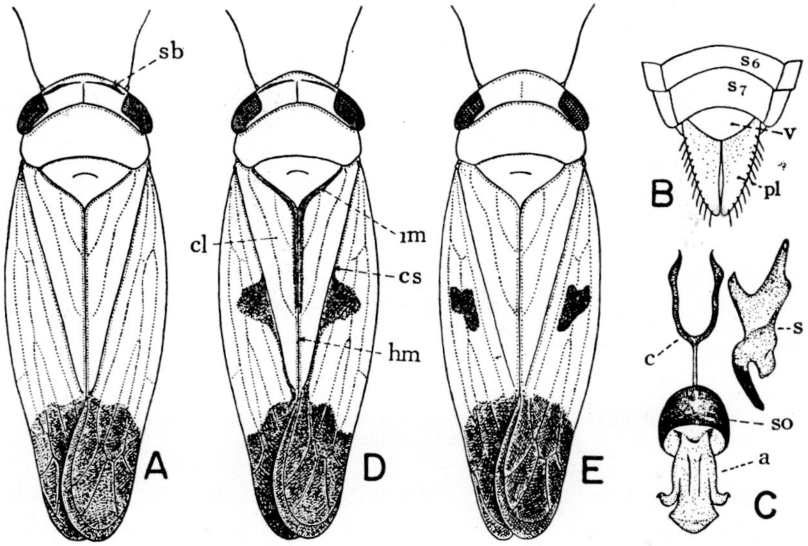

Fig. I–2. A to C, *Nephotettix cincticeps* (Uhler) A, typical male form. sb, submarginal band; B, male genital segments in ventral view. pl, plate; s, sternite; v, valve; C, internal characters of male genitalia in ventral view. a, aedeagus, c, connective, s. stype, so, socle; D, *N. apicalis* (Motschulsky), typical male form. cl, clavus; cs, claval suture; hm, hind margin of tegmen; im, inner margin of tegmen; E, *N. impicticeps* Ishihara, typical male form. (From Ishihara, 1965.)

the viruses have been acquired, the insects remain infective for many generations through the female individuals. In the persistent type, the virus has to be acquired once in the life of the insect. In the nonpersistent type, the insect has to acquire the virus repeatedly to remain continuously infective. Also in the transovarial type, virus continuity may be accomplished by the insect alone, whereas in the persistent and nonpersistent types diseased host plants are required to complete the cycle under natural conditions.

Nephotettix, the rice green leafhopper, is the most important genus in the transmission of several rice viruses. The taxonomy of the genus is confusing and the identification of species, particularly in tropical Asia, is often difficult. Ishihara (1964, 1965) (for these and the following literature citations, see under Dwarf, pp. 13–14), recognized three species, *N. cincticeps*, *N. apicalis* and *N. impicticeps* (Fig. I–2). Nasu (1963, 1969), however, found intermediate and

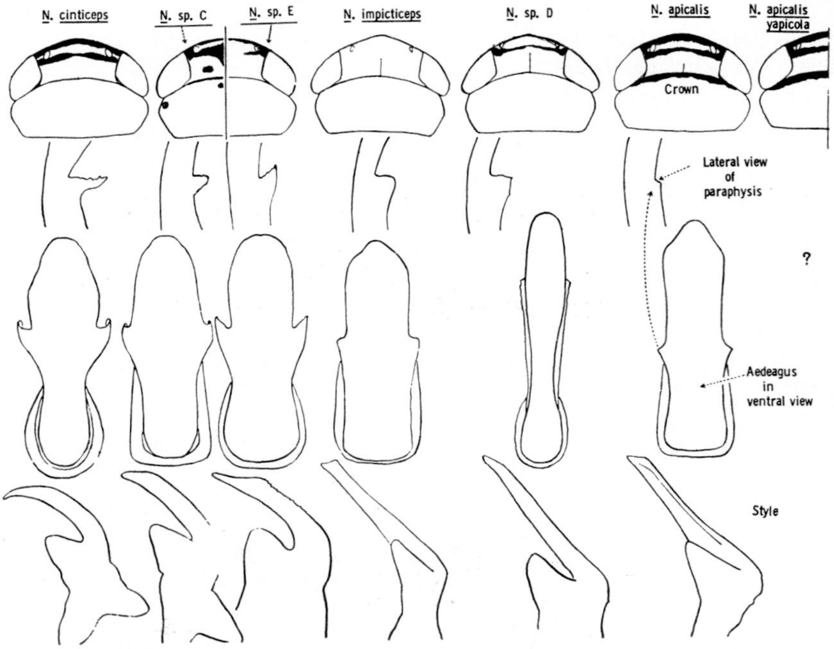

Fig. I–3. Morphological characters of rice green leafhopper species and intermediate forms. (Nasu, 1969.)

other forms (Fig. I–3). Recently, hybridization between *N. apicalis* and *N. impicticeps* has been observed (IRRI, 1968; Ling, 1968a). If such hybridization occurs in nature, classification of species in the genus will be further complicated. Nasu (1969) also prepared a preliminary distribution map of the three species in Asia (Fig. I–4). Besides investigations into morphology, physiological and ecological studies on these leafhoppers in tropical Asia are needed.

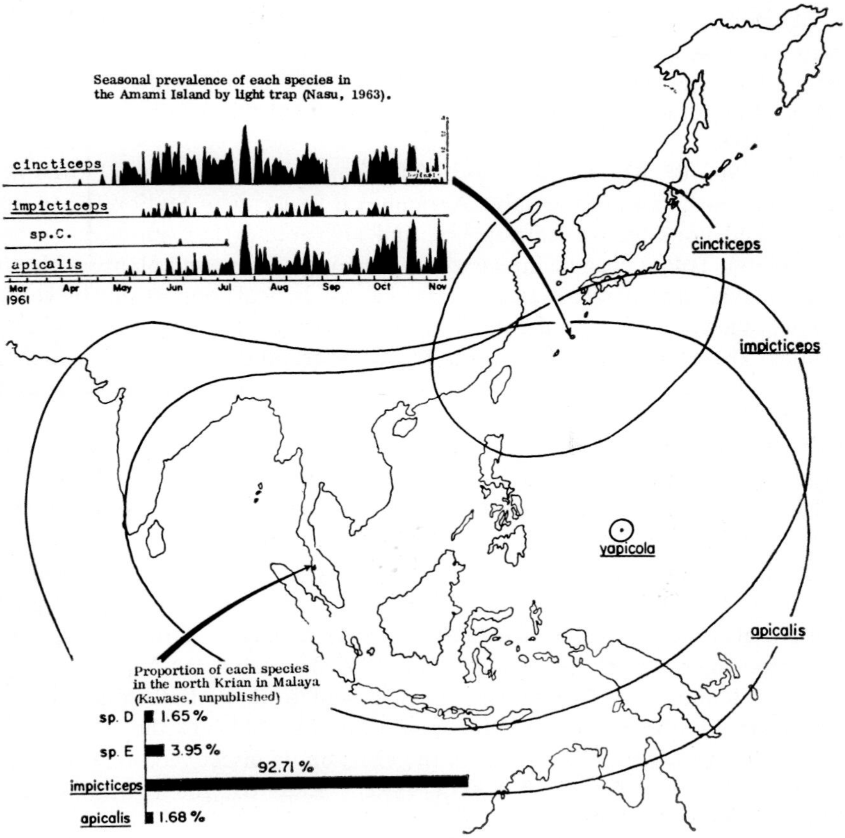

Fig. I–4. Distribution of rice green leafhoppers in Asia. (Nasu, 1969.)

DWARF

History

According to Ishikawa (1928), dwarf disease was first noticed in Shiga Prefecture of Japan in 1883 by Hashimoto, a rice grower, who suspected that the disease was in some way related to leafhoppers. In 1894, he experimentally proved the causal relationship of leafhoppers to rice dwarf, but did not publish the results and the leafhopper he dealt with was not identified. Takata in 1895 first reported the relationship between the insect and the disease and the leafhopper involved was identified as *Inazuma* (now *Recilia*) *dorsalis* (Fukushi, 1965). In 1900, Shiga Agricultural Experiment Station, after several years of study, published a report to the effect that *Nephotettix cincticeps* was the true cause of dwarf, while other species including *Inazuma dorsalis* had no connection with the disease. For some time *N. cincticeps* was considered the only insect related to dwarf disease.

During 1902–08 Ando and Onuki of the Imperial Agricultural Experiment Station and Nishizawa of the Shiga Agricultural Experiment Station found both infective and noninfective leafhoppers, and also found that noninfective leafhoppers from Tokyo became infective after feeding on diseased plants. This led to the conclusion that rice dwarf was not caused by leafhoppers but by certain unknown agents carried by them. The Shiga Agricultural Experiment Station continued the work on transmission for 20 years and definitely reached the conclusion that the unknown agent is a virus. It was thus established that dwarf is a virus disease transmitted by leafhoppers (Fukushi, 1965).

Fukushi studied the disease in detail. Among other findings, he reported the transmission of the virus through the eggs of the vectors (Fukushi, 1933, 1939). He (Fukushi, 1937) also found that *Inazuma dorsalis* was a vector, substantiating the early claim of Takata.

Fukushi & Kimura (1959) succeeded in transmitting the virus to virus-free leafhoppers by injecting fluid from viruliferous leafhoppers or diseased plants using fine glass capillaries. Fukushi *et al.* (1960) first revealed the virus particles by electron microscopy. Yoshii & Kiso (1959b) isolated RNA from diseased tissue and Kimura (1962) extracted nucleic acid from partially purified virus preparations; both proved to be infective when injected into the abdomens of nonviruliferous insects. The effect of the virus on its insect vectors was studied by Nasu (1963) and Yoshii & Kiso (1959a). Recently, the virus was highly purified by phospholipase treatment (Toyada *et al.*, 1965).

For control of the disease by elimination of its vectors, new insecticides have been tested and used. In addition, ecological studies of the insect vectors have been made so that by forecasting outbreaks the chemicals may be used more efficiently. Thus it can be seen that dwarf is one of the most extensively studied rice virus diseases in Japan.

Dwarf disease has also been called rice stunt disease by Katsura (1936) and others, and these names have been used interchangeably.

Distribution and Losses

The disease is known to occur in most parts of Japan except Hokkaido.

About 100,000 hectares have been damaged annually in recent years and the losses in yield are estimated at about 15,000 tons (Iida, 1969).

Outside Japan, it is probably present in Korea (Iida, 1969) and China (Siang, 1952; Wei, 1957). The identity of the ' dwarf ' and ' stunt ' diseases reported from the Philippines (Agati *et al.*, 1941; Reyes, 1957; Reyes *et al.*, 1959; Serrano, 1957) is uncertain but the records probably refer to tungro and other diseases (Ou & Ling, 1966).

Symptoms

The characteristic symptoms of dwarf disease are marked stunting of the entire plant and the presence of chlorotic or whitish specks on the leaves (Plate I). The specks vary in size and often fuse together to form interrupted streaks along the veins. Specks may also be seen on the leaf sheath. The colour of other parts of the leaves is often darker green than normal. The older leaves of infected plants sometimes show diffuse yellowing on the distal parts, and along the margin.

Infected plants usually produce many diminutive tillers. However, when infection occurs at a very early stage of growth, the number of tillers may be reduced. Root growth is much arrested, the small roots extending horizontally.

Diseased plants usually survive until harvest time, remaining more or less green. Heads, if produced, are poor, carrying small numbers of grains, many of which are unfilled. The grains are often covered with dark brown blotches.

Vectors and Transmission

As mentioned above (under history), *Nephotettix cincticeps* (Uhler) and *Recilia dorsalis* (Motsch.) have long been known as vectors of the virus. More recently, *N. apicalis* (Motsch.) was also found able to transmit the virus (Nasu, 1963).

Transmission of rice dwarf virus was studied in detail by Fukushi (1934, 1940) and more recently by Shinkai (1962), Nasu (1963) and others.

' Active ' and ' inactive ' individuals. Fukushi (1934) first demonstrated that in a population of vector species, some are potential transmitters and others are nontransmitters. Percentages of these active or inactive individuals in a population vary according to the locality. In the case of *N. cincticeps*, 0% to 69%, and of *R. dorsalis*, 2% to 43% active individuals have been found in different localities of Japan (Shinkai, 1962). In Northern Japan, where the disease is not present, the population showed no active individuals. This ability or inability was shown to be genetically controlled and inheritable (Fukushi, 1940; Shinkai, 1962).

Acquisition and incubation. The majority of active individuals of *N. cincticeps* acquire the virus by feeding for 1 day on diseased plants. Young nymphs, 1st or 2nd instar, seem to be more efficient, and may sometimes acquire the virus merely by feeding for 1 to 3 minutes. About half the active individuals of *R. dorsalis* may acquire the virus by feeding for 1 day. The minimum time required was found to be 30 minutes (Fukushi, 1934; Shinkai, 1962).

Plate I

Upper left—*Dwarf* virus infected plants in field showing stunting and excessive tillering.

Upper right—*Dwarf* virus infected plants in glasshouse showing chlorotic specks forming streaks on leaf blades.

Lower left—*Yellow Dwarf* infected plants in field showing yellowing, stunting and excessive tillering.

Lower right—*Grassy Stunt* infected plant in field, showing stunting, yellow-green leaves and excessive tillering.

PLATE I

In *N. cincticeps*, the incubation period for the newly acquired virus varies from 4 to 58 days, 12 to 35 days for most individuals. The incubation period in *R. dorsalis* is 9 to 42 days, most commonly 10 to 15 days. Usually the incubation periods are shorter at higher temperatures and longer at lower temperatures. The incubation period often becomes indefinite when the temperature is lower than 13°C or infection does not occur even if the leafhoppers have acquired the virus. The virus concentration in the leafhoppers was shown to increase most rapidly 15–20 days after acquisition (Kimura, 1962).

Infection feeding and latent period. About half of the viruliferous individuals of both vector species can transmit the virus to healthy rice seedlings after feeding for 1 hour, and feeding for several hours to a day is usually enough for most individuals to become capable of transmitting the virus. The minimum feeding time necessary for successful infection was reported to be 3 minutes for *N. cincticeps* and 10 minutes for *R. dorsalis* (Shinkai, 1962).

The latent period, the interval between infection and symptom appearance, is 8–10 days in the early stages of growth of the rice plant (up to the 10th leaf stage). At the 13th leaf stage the latent period is 27 days. Beyond the 13th leaf stage, typical symptoms may not appear (Shinkai, 1952).

Effect of age and sex. Younger nymphs are usually more efficient in acquiring virus than older nymphs or adults; there seems to be no appreciable difference between male and female. Infectivity declines as the adults age. In daily transfer of individual infective leafhoppers to healthy seedlings, some individuals may transmit the virus almost every day, while others show intermittent patterns. Nymphs and young adults show more continuous patterns than older adults.

It was found that the percentage of transmitting individuals of *N. cincticeps* which have acquired virus in autumn decreases almost by half after overwintering. Congenitally infective individuals, however, mostly remain infective after overwintering.

Transovarial passage. Passage of virus through the eggs to the progeny of infective females was first found in this virus by Fukushi and has been demonstrated several times by others (Fukushi, 1933, 1940; Shinkai, 1962; Nasu, 1963). It occurs in each of the known vector species when the female is infective, but the rate of passage is higher in *N. cincticeps* and *N. apicalis* than in *R. dorsalis*. Tests have shown that when both parents are infective, about 85% of the offspring are viruliferous, as compared with about 60% when the male is noninfective. In *N. cincticeps*, the virus could be passed transovarially to 6 or more succeeding generations, but in *R. dorsalis*, not beyond the 4th generation (Shinkai, 1962).

Congenitally infective nymphs usually begin transmitting the virus a few days after hatching, but occasionally infection may occur on the day of hatching. These congenitally infective individuals show high infectivity, and usually remain infective almost until they die. There are instances, however, where a female from a congenitally infective clone fails to transmit the virus throughout its life, and yet produces progeny with high infectivity (Fukushi, 1940; Shinkai, 1962).

Transmission of virus by injection. To prove further the transovarial passage of the virus, Fukushi & Kimura (1959) ground 200 eggs from infective female leafhoppers, diluted the material with phosphate buffer 1/100 or 1/1000, and injected it into the abdomen of noninfective leafhoppers by fine glass capillaries. About 10% of each group of about 50 became infective. When the virus was carried through several serial passages in this manner, it was again very evident that the virus multiplies in the insect body. It was calculated that the virus would have undergone a dilution of 10^{-11} at the 4th passage, while the dilution end point of the fluid from an infective insect was 10^{-4} (Kimura, 1962). The multiplication of virus in the insect vector was demonstrated earlier by Fukushi (1935) in transovarial virus passages. The injection technique has been very useful in assaying the infectivity of virus preparations.

Host Range

According to the work of Fukushi (1934) and Shinkai (1962), the following plants have been shown to be susceptible: *Alopecurus aequalis* Sobol., *A. japonicus* Steud., *Avena sativa* L., *Echinochloa crusgalli* var. *frumentosa* Wight, *E. crusgalli* var. *oryzicola* Ohwi, *Glyceria acutiflora* Torr., *Hordeum sativum* var. *hexastichon* Hack., *Oryza sativa* L., *O. cubensis* Ekman, *Panicum miliaceum* L., *Paspalum thunbergii* Kunth., *Poa annua* L., *P. pratensis* L., *Secale cereale* L. and *Triticum aestivum* L.

Wheat and barley have never been found infected in the field. Among the weeds, naturally infected specimens of *E. crusgalli* and *P. thunbergii* have been observed, with symptoms similar to those on rice.

Transmission Cycle

N. cincticeps overwinters as adults and 3rd and 4th instar nymphs in Japan. The adults from the overwintered population oviposit on gramineous weeds and sometimes on early planted rice seedlings in seedbeds; very few individuals breed in wheat and barley fields.

Since the virus is transmitted transovarially, the percentage of viruliferous individuals is maintained at a certain level throughout the year and is not much affected by external sources of virus. Disease incidence depends upon the population of the early generations, which synchronize with the attractive and infective stage of the rice plants. In recent years, the early and regular planting of rice in Japan has extended the period of infection. The adults of the year's first generation and nymphs arising from it have been found to be mostly responsible for the transmission of the disease.

Properties of the Virus

The first electron micrographs of rice dwarf virus were published by Fukushi *et al.* (1960). The partially purified virus particles have a hexagonal outline and are 70mμ in diameter, with a surrounding envelope. They possess a central dark area approximately 40–50mμ in diameter. Later (Fukushi *et al.*, 1962) such particles were found in infected plants and insect cells and were frequently seen to be closely packed in crystalline-like array. The fine structure of the particle was later further revealed (Fukushi & Shikata, 1963), and Kimura & Shikata (1968) gave a structural model of the virus particles. The virus may be isolated by alternating cycles of centrifugation at 20,000 rpm for 60 minutes and 8000 rpm for 30 minutes (Shikata, 1966).

The virus survives for 48 hours in extracted leafhopper fluids stored at 0–4°C; infectivity is maintained for up to a year at −30 to −35°C. The thermal inactivation point is 40–45°C and dilution end point 10^{-3} (plant juice) and 10^{-4} (insect fluid) (Kimura & Fukushi, 1960).

The virus was further purified by removing the enveloping material of the particle with phospholipase prepared from snake venom or from pancreatin (Toyoda *et al.*, 1965) (Figs. I–5, 6). Assay by means of the insect vector indicated that the purified particles retained their infectivity.

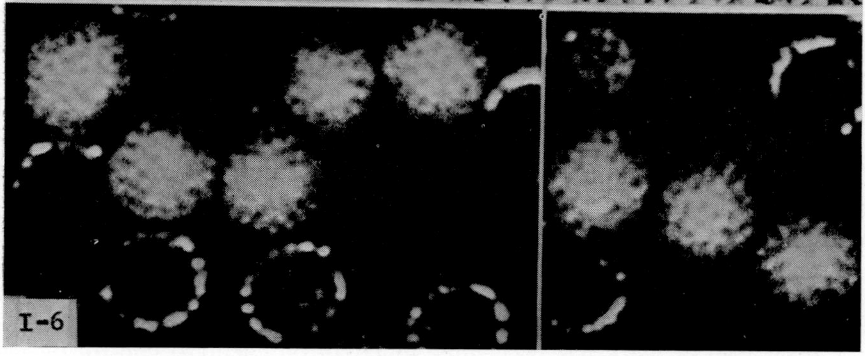

Fig. I–5. Electron micrograph of rice dwarf virus, fraction C shadowed with chromium.
Fig. I–6. The same stained with phosphotungstate, showing the presence of capsomeres on each particle and the contamination of empty shells after suspension in distilled water. (Both from Toyoda *et al.*, 1965.)

Yoshii & Kiso (1959b) isolated RNA from affected rice plants and injected it into leafhoppers which then became infective. Kimura (1962) extracted nucleic acid from a partially purified preparation; the nucleic acid was found infective by injecting it into the vector. Suzuki *et al.* (1965) believed that the RNA of the virus is double-stranded, or of a double-helical structure. Kimura & Suzuki (1965) developed a fluorescent antibody technique for the detection of virus antigen.

Effect of the Virus on the Insect

The deleterious effect of rice dwarf virus on its vectors was studied by Nasu (1963). In congenitally viruliferous leafhoppers, the virus caused high mortality, and impaired the fecundity of the insects. Infective females laid fewer eggs, 28–68% less in *N. cincticeps*, 37–45% less in *N. apicalis* and 41% less in *R. dorsalis*; some individuals laid no eggs. Yoshii & Kiso (1959a) described metabolic abnormalities and Nasu (1963) noted cytological changes in the infective leafhoppers.

Nasu (1965) demonstrated, by electron microscopy, the possible route of the transovarial passage of the virus in the insect vector. Mitsuhashi (1965) successfully cultured the viruliferous embryonic tissues of leafhoppers.

Varietal Resistance

Until comparatively recently there had been few studies made of varietal resistance. Yasuo *et al.* (1960) tested a number of varieties in the field and found that Japanese varieties were susceptible while foreign varieties showed considerable differences in their degree of resistance. Varieties considered highly resistant were Hyahunichi-to, Pe Bi Hun, Te-tep, Loktjan, Kaladumai and Dahrial.

Kimura and Nishizawa experimented with seedling resistance tests. Among the methods tried, they recommended two. (1) Viruliferous green rice leafhoppers are reared in a cage in the greenhouse and are allowed to feed on diseased as well as healthy seedlings. A large population may be kept throughout the year and a level of 10–20% viruliferous individuals may be maintained. Rice seedlings are brought into the room for inoculation for 7 to 12 days. (2) About 30 seeds of each variety are sown in a pot. At the 2-leaf or 3-leaf stage, they are covered with a cage into which are introduced 10 adult male insects, taken from a population containing 40% or more viruliferous individuals. The tests again showed that Japanese varieties were susceptible while among the foreign varieties tested C203-1, Te-tep, Loktjan, Peta, Depi, Intan, Kaeu N525, Bluebonnet, Karalath, Chiem Chank, and 5 other Chinese varieties were resistant (Sakurai, 1969).

Chemical Control of the Vectors

A great deal of work has been done in Japan on chemical control of leafhoppers and planthoppers as vectors of rice viruses. Many new chemicals have been tested, including fenitrothion (Sumithion), diazinon, carbaryl (Sevin), disulfoton (Disyston), vamidothion and malathion, in the form of dusts, emulsions and granules. The effectiveness of these chemicals against the various species of vectors varies. Carbaryl has been considered effective against *N. cincticeps*. In areas where malathion has been used repeatedly, *N. cincticeps* has begun to show tolerance of the chemical. Usually when the insect population is low, chemical control of the vector reduces virus disease incidence effectively, but when the population is very high, the use of insecticides may not give any control of the disease (Kisimoto, 1969).

LITERATURE CITED

AGATI, J. A., SISON, P. L. & ABALOS, P. 1941. A progress report on the rice maladies recently observed in Central Luzon with special reference to the 'stunt or dwarf' disease: 1. *Philipp. J. Agric.* **12**:197–210.

FUKUSHI, T. 1933. Transmission of virus through the eggs of an insect vector. *Proc. imp. Acad. Japan* **9**:457–460.

FUKUSHI, T. 1934. Studies on the dwarf disease of rice plant. *J. Fac. Agric. Hokkaido Univ.* **37**:41–164.

FUKUSHI, T. 1935. Multiplication of virus in its insect vector. *Proc. imp. Acad. Japan* **11**:301–303.

FUKUSHI, T. 1937. An insect vector of the dwarf disease of rice plant. *Proc. imp. Acad. Japan* **13**:328–331.

FUKUSHI, T. 1939. Retention of virus by its insect vector through several generations. *Proc. imp. Acad. Japan* **15**:142–145.

FUKUSHI, T. 1940. Further studies on the dwarf disease of rice plant. *J. Fac. Agric. Hokkaido Univ.* **45**:83–154.

FUKUSHI, T. 1965. Relationships between leafhoppers and rice viruses in Japan. Papers presented at Conference on relationships between arthropods and plant-pathogenic viruses: 1–21. Tokyo. (Mimeographed.)

FUKUSHI, T. & KIMURA, I. 1959. On some properties of the rice dwarf virus. *Proc. Japan Acad.* **35**:482–484.

FUKUSHI, T. & SHIKATA, E. 1963. Fine structure of rice dwarf virus. *Virology* **21**:500–503.

FUKUSHI, T., SHIKATA, E. & KIMURA, I. 1962. Some morphological characters of rice dwarf virus. *Virology* **18**:192–205.

FUKUSHI, T., SHIKATA, E., KIMURA, I. & NEMOTO, M. 1960. Electron microscopic studies on the rice dwarf virus. *Proc. Japan Acad.* **36**:352–357.

IIDA, T. T. 1969. Dwarf, yellow dwarf, stripe, and black-streaked dwarf diseases of rice. In *The virus diseases of the rice plant: Proc. Symp. at IRRI, April, 1967*: 3–11. Baltimore, Maryland, Johns Hopkins Press.

IRRI (INTERNATIONAL RICE RESEARCH INSTITUTE). Annual Report for 1967.

ISHIHARA, T. 1964. Revision of the genus *Nephotettix* (Hemiptera: Deltocephalidae). *Trans. Shikoku ent. Soc.* **8**:39–44.

ISHIHARA, T. 1965. Taxonomic position of some leafhoppers known as virus vectors. Paper presented at Conference on relationships between arthropods and plant-pathogenic viruses. Tokyo. (Also published by Entom. Lab., Coll. Agric., Ehime Univ., Japan.)

ISHIHAWA, R. 1928. The merit of Hatsuzo Hashimoto, the earliest investigator of dwarf disease of rice plant. *J. Pl. Prot., Tokyo* **15**:218–222. [Jap.]

KATSURA, S. 1936. The stunt disease of Japanese rice, the first plant virosis shown to be transmitted by an insect vector. *Phytopathology* **26**:887–895.

KIMURA, I. 1962. Further studies on the rice dwarf virus. I; II. *Ann. phytopath. Soc. Japan* **27**:197–203; 204–213. [Jap. Engl. summ.]

KIMURA, I. & FUKUSHI, T. 1960. Studies on the rice dwarf virus. *Ann. phytopath. Soc. Japan* **25**:131–135. [Jap. Engl. summ.]

KIMURA, I. & SHIKATA, E. 1968. Structural model of rice dwarf virus. *Proc. Japan Acad.* **44**:538–543.

KIMURA, I. & SUZUKI, N. 1965. Purification of rice dwarf virus and the application of immunofluorescent antibody technique for the detection of RDV antigen. *J. Pl. Prot., Tokyo* **19**:137–140. [Jap.]

KISIMOTO, R. 1969. Ecology of insect vectors, forecasting, and chemical control. In *The virus diseases of the rice plant*: *Proc. Symp. at IRRI, April, 1967*: 243–255. Baltimore, Maryland, Johns Hopkins Press.

LING, K. C. 1968a. Hybrids of *Nephotettix impicticeps* Ish. and *N. apicalis* (Metsch.) and their ability to transmit the tungro virus of rice. *Bull. ent. Res.* **58**:393–398.

LING, K. C. 1968b. Virus diseases of the rice plant. Pamphlet, Int. Rice Res. Inst., Los Banos, Laguna, Philippines.

MITSUHASHI, J. 1965. Multiplication of plant virus in the leafhopper vector cells grown *in vitro*. Paper presented at Conference on relationships between arthropods and plant-pathogenic viruses. Tokyo. (Mimeographed.)

NASU, S. 1963. Studies on some leafhoppers and planthoppers which transmit virus diseases of rice plant in Japan. *Bull. Kyushu agric. Exp. Stn* **8**:153–349. [Jap.]

NASU, S. 1965. Electron microscopic observations on transovarial passage of rice dwarf virus. Paper presented at Conference on relationships between arthropods and plant-pathogenic viruses. Tokyo. (Mimeographed.)

NASU, S. 1969. Vectors of rice viruses in Asia. In *The virus diseases of the rice plant*: *Proc. Symp. at IRRI, April*, 1967: 93–109. Baltimore, Maryland, Johns Hopkins Press.

OU, S. H. & LING, K. C. 1966. Virus diseases of rice in the South Pacific. *Pl. Prot. Bull. F.A.O.* **14**:113–121.

REYES, G. M. 1957. Rice dwarf disease in the Philippines. *Pl. Prot. Bull. F.A.O.* **6**:17–19.

REYES, G. M., LEGASPI, B. M. & MORALES, M. T. 1959. Progress of studies on the dwarf or stunt (virus) disease of rice in the Philippines. *Philipp. J. Agric.* **24**:27–43.

SAKURAI, Y. 1969. Varietal resistance to stripe, dwarf, yellow dwarf, and black-streaked dwarf. In *The virus diseases of the rice plant*: *Proc. Symp. at IRRI, April, 1967*: 257–275. Baltimore, Maryland, Johns Hopkins Press.

SERRANO, F. B. 1957. Rice ' accep na pula ' or stunt disease—a serious menace to the Philippine rice industry. *Philipp. J. Sci.* **86**:203–230.

SHIKATA, E. 1966. Electron microscopic studies on plant viruses. *J. Fac. Agric. Hokkaido Univ.* **55**:1–110.

SHINKAI, A. 1962. Studies on insect transmission of rice virus diseases in Japan. *Bull. natn. Inst. Agric. Sci., Tokyo*, Ser. C **14**:1–112. [Jap. Engl. summ.]

SIANG, W. N. 1952. Host index to non-fungus diseases of plants in China. *Pl. Dis. Reptr*, Suppl. 215:165–186.

SUZUKI, N., KIMURA, I. & MIURA, K. 1965. Purification of rice dwarf virus and double helical structure of its RNA. Paper presented at Conference on relationships between arthropods and plant-pathogenic viruses. Tokyo. (Mimeographed.)

TOYODA, S., KIMURA, I. & SUZUKI, N. 1965. Purification of rice dwarf virus. *Ann. phytopath. Soc. Japan* **30**:225–230.

WEI, C. T. 1957. Manual of rice pathogens. Peiping, Science Press. [Chin.]

YASUO, S., YAMAGUCHI, T. & ISHII, M. 1960. Experimental results in plant diseases. Studies on the stripe and dwarf of rice plant. Central agric. Exp. Stn, Japan. (Mimeographed.)

YOSHII, H. & KISO, A. 1959a. Studies on the nature of insect transmission in plant viruses. V. On the abnormal metabolism of the virus transmitting green rice leafhopper, *Nephotettix bipunctatus cincticeps* affected with the rice stunt virus. *Virus* **9**:415–422. [Jap.]

YOSHII, H. & KISO, A. 1959b. Studies on the nature of insect transmission in plant viruses. IX. On the biological activity of ribonucleic acid from stunt virus-infected rice plant. *Virus* **9**:582–589.

STRIPE

HISTORY, DISTRIBUTION AND IMPORTANCE

Stripe disease has been known since the 1890s in Kanto-Tozan district, eastern Japan, where it has caused severe damage. In 1931, Kuribayashi demonstrated that it was due to a virus transmitted by the smaller brown planthopper, *Laodelphax* (*Delphacodes*) *striatellus*. The disease was later found in other localities, but only in limited areas. More recently it has become increasingly prevalent, especially in the warmer southwestern districts, the increase coinciding with the extension of early planting of the rice crop, which promotes the multiplication of the vector insect. Stripe is now considered to be the most important rice virus disease in Japan. It is still absent from North Japan, Hokkaido and parts of Tohoku, despite the wide presence of the vector. Annual occurrence in recent years has been about 200,000 hectares, with a loss in yield of 40,000 tons. Detailed transmission studies were made by Yamada & Yamamoto (1955, 1956) and Shinkai (1962). Both proved the transovarial passage of the virus.

Insecticidal control has been found effective in certain instances. Resistant varieties, incorporating resistance from *indica* selections, have been developed in Japan.

In Korea, the disease was noticed and its transmission studied in the early 1940s (Lee, 1969). It was present all over South Korea, but losses were higher in the south, where 15–35% of the hills were infected during recent years. It appears that it also occurs in mainland China (Chen, 1964).

Symptoms

The characteristic symptoms are the presence of wide, chlorotic stripes or general chlorosis on the leaves, and lack of vigour. The emerging leaves fail to unfold properly, elongate, and become twisted and drooping. Other leaves which emerge during later growth may unfold more or less normally, but show irregular, chlorotic mottlings which also appear in a stripe pattern. Grey, necrotic streaks frequently appear on the chlorotic areas, enlarge and cause prompt death of the leaf (Plate II).

When infection occurs early, the entire plant may die prematurely, or the result may be considerable stunting. Later infection causes only slight stunting. Tillers of infected plants are usually much reduced in number.

Diseased plants produce only a few poor heads, with malformed spikelets, or none at all. The heads also show difficulty in emerging from the leaf sheath (Plate II). Sometimes these are the only symptoms to be seen, if infection takes place at a late stage in the growth of the plant.

In more resistant varieties, only mottling or chlorotic mosaic may be seen on the leaves.

Vectors and Transmission

Kuribayashi (1931) experimentally proved the transmission of the virus by *Laodelphax striatellus* (Fall.). This insect was believed to be the sole vector until a few years ago, when *Unkanodes sapporonus* (Mats.) was also found to be a vector (Shinkai, 1966). Because rice is not a favoured host of *U. sapporonus*, *L. striatellus* is still the major vector in rice fields. More recently Shinkai (1967) found another vector, *Ribautodelphax albifascia* (Mats.).

Kuribayashi (1931) first reported that 14–54% of the adult individuals transmitted stripe disease. More intensive study by Shinkai (1962) indicated that about 20% were active transmitters. Kisimoto (1967) reported that genetically there are two types of low acquisitive ability, dominant and incompletely dominant.

Shinkai (1962) found the minimum acquisition feeding time to be 15 minutes, most of the active individuals acquiring the virus by feeding for 1 day. The minimum incubation period in the insect was 5 days, maximum 21 days, the

PLATE II

Upper left—*Stripe* virus infected plants in field showing chlorosis, and drooping and dying leaves.

Upper right—*Stripe* virus infected plants showing abnormal emergence of panicles.

Lower left—*Hoja blanca* virus infected field.

Lower right—*Hoja blanca* virus infected plants showing chlorotic leaves.

PLATE II

period for most individuals being 5–10 days. The insects are highly viruliferous 1–2 weeks following the incubation period.

The minimum inoculation feeding time is 3 minutes; after feeding for 1 hour, half the individuals were able to infect test rice plants. The latent period in rice is 10–25 days, depending on the age of the plant. It is longer in older plants, being 25 days in plants inoculated at the 13-leaf stage.

Ability to transmit the virus decreases markedly as the insects become old. Females are more efficient transmitters than males.

In transmission studies, Kuribayashi (1931) found that some progeny of infective females produced stripe disease on healthy plants, but he thought that this was due to experimental errors, the young nymphs from the eggs perhaps having fed on unknown infected plants. More than 20 years later, Yamada & Yamamoto (1955, 1956) and Shinkai (1962) proved that the virus was passed through the eggs to a very high percentage of the progeny for 23 (Yamada & Yamamoto) and 40 (Shinkai) consecutive generations. After 6 years, 95% of the individuals of the 40th generation proved to be infective, showing that there is no progressive decline of the virus. This provides further evidence of virus multiplication in the insect vector. Ovarial transmission was only possible through viruliferous females. Nonviruliferous females are unable to transmit the virus even if they mate with a viruliferous male. Most individuals which received the virus through the egg began to transmit the virus to plants from the day of hatching. They remained highly infective throughout the nymph stage and also for 2–3 weeks of the early adult stage.

Individuals which receive the virus through the egg decrease markedly in infectivity after overwintering. High percentages of these overwintered individuals do, however, pass the virus to their progeny through the egg.

Transmission of the virus to insect vectors may also be accomplished experimentally by injecting virus preparation into the abdomen of the insects, as in the case of rice dwarf virus. Okuyama & Asuyama (1959) tried another method, in which the lower part of a cut leaf sheath, about 5 cm long, is immersed in the virus preparation. Insects feed on the upper part for 48 hours. Two out of 48 insects thus tested transmitted the virus. Saito *et al.* (1964), using leaf blades, succeeded in securing transmission by 3 out of 19 insects.

Harmful effects of this virus on the insect vectors were reported by Nasu (1963). Individuals from eggs laid by infective planthoppers died prematurely, if they hatched. The mortality of the nymphs was also high, particularly in the first and second instars.

Host Range

The works of Amano (1937), Shinkai (1955, 1956), Yamada & Yamamoto (1956) and Sugiyama (1966) have proved that many members of the Gramineae are hosts of the virus. The known species are: *Agrostis alba* L., *Alopecurus aequalis* Sobol., *A. japonicus* Steud., *Avena sativa* L., *A. fatua* L., *Beckmannia syzigachne* (Steud.) Fernald, *Briza minor* L., *Bromus catharticus* Vahl, *Cynodon dactylon* Pers., *Cynosurus cristatus* L., *Dactylis glomerata* L., *Digitaria adscendens* (H.B.K.) Henr., *D. violascens* Link, *Echinochloa crusgalli* (L.) Beauv.,

E. crusgalli var. *frumentacea* Wight, *Eragrostis multicaulis* Steud., *Glyceria acutiflora* Torr., *Hordeum sativum* var. *hexastichon* Hack., *H. sativum* var. *vulgare* Hack., *Leersia oryzoides* var. *japonica* Hack., *Lolium multicaulis* Steud., *L. perenne* L., *Oryza sativa* L., *Panicum miliaceum* L., *Pennisetum alopecuroides* (L.) Spreng., *Phleum pratense* L., *Poa annua* L., *Saccharum koenigii* Retz., *Secale cereale* L., *Setaria italica* Beauv., *S. viridis* (L.) Beauv., *Sorghum halepensis* Pers., *S. sudanense* Stapf, *Trisetum bifidum* Ohwi, *Triticum aestivum* L., *Zea mays* L. and *Zoysia japonica* Steud. (Iida, 1969).

On most of these plants only mild symptoms, consisting of chlorotic streaks or mosaic mottling, are seen. *Setaria italica, S. viridis, Digitaria adscendens, D. violascens* and *Eragrostis multiflorum* are frequently infected in the field. *Zoysia japonica* is not considered to be a host by Shinkai (1962).

Transmission Cycle

The main habitats of the overwintered first generation of *L. striatellus* are wheat and barley fields and fields with gramineous weeds. Large numbers of nymphs are often collected in May. The harvest time of wheat and barley markedly affects the survival rate of the nymphs. The adults migrate to the rice fields and this is the important transmission period. The migration is accomplished by the macropterous forms; following migration brachypterous forms are produced. When the population density becomes high the macropterous form increases. The migrant density has been studied by the use of sticky traps (Yoshimeki, 1966), water-pan traps (Kisimoto, 1966, 1968) and sweeping. Kisimoto (1966) found that water-pans painted yellow attracted the greatest number of insects.

In the first and second generations, the insects are infective from the transovarial passage of the virus. In the later generations, they also acquire the virus by feeding on infected plants.

Properties of the Virus

Purification of rice stripe virus was first attempted by Okuyama & Asuyama (1959) by differential centrifugation. They found the highest percentage of particles in the range 30–50mμ. By an insect transmission technique, they deduced that these were the stripe virus particles. Methods of purification were improved by Saito *et al.* (1964) and Kitani & Kiso (1965). Saito *et al.* reported the particles as spherical, with an average diameter of 29mμ. Kitani & Kiso (1967) found the diameter to be 30–35mμ; they also found the dilution end point to be 10^{-4} for fluid from viruliferous insects and 10^{-3} for sap from diseased leaves. The thermal inactivation point was found to be 55°C for 3 minutes. At −20°C, the extracts remained infective for 8 months.

Okuyama *et al.* (1968) studied the possible site of virus multiplication and also the pathogenic effects of the virus on the vector. On the 5th to 15th day after injection, the virus was estimated to have multiplied 5 times. It was recovered from the eggs, male and female reproductive organs, and blood and fat bodies, but not from the alimentary canal. Head, thorax and abdomen were found to contain virus 13 days after incubation but not before incubation. There were no consistent or conspicuous cytological changes in blood corpuscles, reproductive cells, salivary glands, alimentary canal, dermis or fat body of viruliferous planthoppers but glycogen and polysaccharides in the fat body and mycetome were relatively less. The dilution end point was found to be between 10^{-2} and 10^{-3}.

Saito & Iwata (1964) applied the haemagglutination test for detection of viruliferous planthoppers and Kitani & Kiso (1967) employed the fluorescent antibody technique for detecting the virus in insects and rice plants. These methods make possible the detection of the virus in small quantities and in specific positions.

Varietal Resistance

In recent years, varietal resistance to stripe has been intensively studied and breeding for resistance has been successful. Early tests limited to Japanese paddy rice revealed no resistant varieties, but Yamaguchi *et al.* (1965) found that Japanese upland and most *indica* type foreign varieties were highly resistant.

In field tests, Suzuki *et al.* (1960) noticed that early planting is favourable for disease development. Yamaguchi *et al.* (1965) found it easy to separate out highly resistant and highly susceptible varieties, but it was found difficult to distinguish varieties of intermediate resistance from susceptible ones.

Sakurai *et al.* (1963, 1964) developed a seedling test method which facilitates resistance testing and breeding. The method is essentially as follows: (1) about 30 seeds of each variety are planted in Petri dishes filled with soil; (2) at the 1·5 leaf stage, the seedlings are covered by a glass cylinder with a screen at the top, the lower end fitting into one Petri dish; (3) a volume (40–70mm²) of 2nd or 3rd instar nymphs are introduced into the glass cylinder at 27°C under artificial light, left for 2 days, and then removed; (4) the inoculated seedlings are transplanted into a seedling box.

Readings are taken every 3 or 4 days. Infected seedlings are classified, 3–4 weeks after transplanting, according to the following symptoms:

Highly susceptible, grade A—Seedlings severely stunted; yellowish white stripe running through the leaves which roll and finally die.

Highly susceptible, grade B—Resemble A, but leaves do not die.

Moderately susceptible, grade Bt—Resemble B, but growth of seedlings is fairly good.

Moderately resistant, grade Cr—Growth of seedlings good. Leaves sparsely dotted with yellowish white lesions and slightly rolled.

Highly resistant, grade C—Resemble Cr, but leaves not rolled.

Highly resistant, grade D—Symptoms masked, growth vigorous.

A disease rating index is calculated as follows:

$$\frac{100\text{ A} + 80\text{ B} + 600\text{ Bt} + 40\text{ Cr} + 20\text{ C} + 5\text{ D}}{\text{Total number of seedlings examined}}$$

To minimize the effect on the disease index of age of insects, percentage of viruliferous individuals and other factors, a standard variety To-to is included in each test, so that a ‘ ratio of disease rating index ’ may be calculated.

All 123 Japanese paddy rice varieties tested were susceptible. Forty-six of the 53 Japanese upland varieties were resistant, including Chata-wase, Tamasari, Hakaburi, Kuronbo and Sensho. Various degrees of resistance or susceptibility were found among foreign varieties, and many of them were resistant, including Charnack, Bason Takakal, Hatadavi, Loktjan, Te-tep, Tadukan and Kannonsen. All ponlai varieties were susceptible (Sakurai, 1969).

A resistant strain ' St No. 1 ' was selected for commercial planting from a cross of Modan × Norin No. 8. Another sister line ' Chugoku No. 31 ' was also selected (Sakurai, 1969; Toriyama *et al.*, 1966). They were also reported to be blast resistant, producing only small numbers of lesions (Sakurai & Toriyama, 1967).

Okamoto & Inoue (1967) reported some varieties to be resistant to the insect vector but susceptible to the virus. Others are susceptible to the insect but resistant to the virus.

Inheritance of Resistance

Toriyama (1966, 1969) studied the inheritance of resistance to stripe virus and concluded that resistance in Japanese upland rice varieties is controlled by two pairs of complementary dominant genes, St_1 and St_2, and resistance in *indica* varieties is completely dominant and controlled by one major gene, St_3. The gene action of St_3 varies with varieties, and the gene for a higher degree of resistance shows complete dominance over the gene for a lower degree. St_1 has complementary dominant action not only with St_2 but also with St_3, and St_3 is an allelomorph of St_2. Modifying genes may also be present which influence the degree of resistance. Washio *et al.* (1968) reported that among the crosses between the susceptible variety Kibiyoshi and 5 resistant varieties, resistance in Zenith is controlled by 2 pairs of completely dominant genes, while that in Surjumkhi, Charnack, Russia 35, and Ketan-Nangka is controlled by the incompletely dominant gene St_3.

Yamaguchi *et al.* (1965) reported that resistance to stripe was due to the combination of 3 factors: (1) preference of the vectors for certain plants; (2) resistance to disease development; and (3) sensitivity of plants to disease development at different ages. They also found that the resistance of F_1 plants from crosses between resistant and susceptible parents was influenced by the female parents.

Chemical Control

Of various insecticides tested, diazinon has been most commonly recommended. The effectiveness of chemical control varies with the density of the vector migrant populations which transmit the virus. It has been reported that in years of heavy migration of *L. striatellus*, even daily application of malathion emulsion was not effective, whereas in years of light migration a single application of BHC granules yielded good results. Some attempts have been made to apply chemicals over large areas (Kisimoto, 1969).

Laboratories in Japan have also been looking for compounds which inactivate the virus. Blasticidin-S, primarily used for blast disease control, was found to reduce the transmission ability of *L. striatellus* if administered orally (Kitani & Kiso, 1966; Harai *et al.*, 1968).

LITERATURE CITED

AMANO, E. 1937. Relation of rice stripe with several gramineous plants. *J. Pl. Prot., Tokyo* **24**:774–780. [Jap.]

CHEN, L. 1964. A preliminary survey of stripe disease of paddy rice in Yuyang Hsian. *Chekiang Agric. Sci.* **3**:123–127. [Chin.]

HIRAI, T., SAITO, T., ONDA, H., KITANI, K. & KISO, A. 1968. Inhibition by blasticidin-S of the ability of leafhoppers to transmit rice stripe virus. *Phytopathology* **58**:602–604.

IIDA, T. T. 1969. Dwarf, yellow dwarf, stripe, and black-streaked dwarf diseases of rice. In *The virus diseases of the rice plant: Proc. Symp. at IRRI, April, 1967*: 3–11. Baltimore, Maryland, Johns Hopkins Press.

KISIMOTO, R. 1966. Ecology of *Laodelphax striatellus* Fallen and its control. *J. Pl. Prot., Tokyo* **20**:126–130. [Jap.]

KISIMOTO, R. 1967. Genetic variation in the ability of a planthopper, *Laodelphax striatellus* (Fallen) to acquire the rice stripe virus. *Virology* **32**:144–152.

KISIMOTO, R. 1968. Yellow pan water trap for sampling the small brown planthopper *Laodelphax striatellus* (Fallen), a vector of the rice stripe virus. *Appl. Ent. Zool.* **3**:37–48.

KISIMOTO, R. 1969. Ecology of insect vectors, forecasting, and chemical control. In *The virus diseases of the rice plant: Proc. Symp. at IRRI, April, 1967*: 243–255. Baltimore, Maryland, Johns Hopkins Press.

KITANI, K. & KISO, A. 1965. Studies on rice stripe disease. I. Purification of rice stripe virus. *Ann. phytopath. Soc. Japan* **30**:84. [Abs., Jap.]

KITANI, K. & KISO, A. 1966. Studies on the rice stripe virus disease. IV. Effect of blasticidin-S on the transmitting ability of infective small brown planthopper, *Laodelphax striatellus* (Fallen), transmitting the rice stripe virus. *Ann. phytopath. Soc. Japan* **32**:89. [Abs., Jap.]

KITANI, K. & KISO, A. 1967. Localization of rice stripe virus antigen in tissues of diseased rice and wheat leaves and viruliferous planthoppers, *Laodelphax striatellus*, by fluorescent antibody technique. *Ann. phytopath. Soc. Japan* **33**:108. [Abs., Jap.]

KURIBAYASHI, K. 1931. On the relation between rice stripe disease and *Delphacodes striatellus* Fall. *J. Pl. Prot., Tokyo* **18**:565–571; 636–640. [Jap.]

LEE, S. C. 1969. Rice stripe disease in Korea. In *The virus diseases of the rice plant: Proc. Symp. at IRRI, April, 1967*: 67–73. Baltimore, Maryland, Johns Hopkins Press.

NASU, S. 1963. Studies on some leafhoppers and planthoppers which transmit virus diseases of rice plant in Japan. *Bull. Kyushu agric. Exp. Stn* **8**:153–349. [Jap.]

OKAMOTO, D. & INOUE, H. 1967. Studies on the smaller brown planthopper, *Laodelphax striatellus* (Fallen) as a vector of rice stripe virus. 2. Varietal resistance of rice to the smaller brown planthopper. *Bull. Chugoku agric. exp. Stn*, E, **1**:115–136. [Jap. Engl. summ.]

OKUYAMA, S. & ASUYAMA, H. 1959. Isolation of rice stripe virus from diseased rice leaves. *Ann. phytopath. Soc. Japan* **24**:35. [Abs., Jap.]

OKUYAMA, S., YORA, K. & ASUYAMA, H. 1968. Multiplication of the rice stripe virus in its insect vector, *Laodelphax striatellus* Fall. *Ann. phytopath. Soc. Japan* **34**:255–262. [Jap. Engl. summ.]

SAITO, Y., INABA, T. & TAKAHASHI, K. 1964. Purification and morphology of rice stripe virus. *Ann. phytopath. Soc. Japan* **29**:286. [Abs., Jap.]

SAITO, Y. & IWATA, Y. 1964. Hemagglutination test for titration of plant virus. *Virology* **22**:426–428.

SAKURAI, Y. 1969. Varietal resistance to stripe, dwarf, yellow dwarf and black-streaked dwarf. In *The virus diseases of the rice plant: Proc. Symp. at IRRI, April, 1967*: 257–275. Baltimore, Maryland, Johns Hopkins Press.

SAKURAI, Y. & EZUKA, A. 1964. The seedling test method of varietal resistance of rice plant to stripe virus disease. 2. The resistance of various varieties and strains of rice plant by the method of seedling test. *Bull. Chugoku agric. Exp. Stn*, A, **10**:51–70.

SAKURAI, Y., EZUKA, A. & OKAMOTO, H. 1963. The seedling test method of varietal resistance of rice plant to stripe virus disease (Part I). *Bull. Chugoku agric. Exp. Stn*, A, **9**:113–125.

SAKURAI, Y. & TORIYAMA, K. 1967. Field resistance of rice plant to *Piricularia oryzae* and its testing method. In *Proc. symp. rice diseases and their control by growing resistant varieties and other measures.* Agric. For. Fish. Res. Council, Tokyo, 1967.

SHINKAI, A. 1955; 1956. Host range of rice stripe disease. *Ann. phytopath. Soc. Japan* **20**:100; **21**:47. [Abs., Jap.]
SHINKAI, A. 1962. Studies on insect transmission of rice virus diseases in Japan. *Bull. natn. Inst. Agric. Sci., Tokyo*, Ser. C, **14**:1–112. [Jap. Engl. summ.]
SHINKAI, A. 1966. Methods for studying virus transmission by *Delphacidae* and *Cicadellidae*. *J. Pl. Prot., Tokyo* **20**:137–140. [Jap.]
SHINKAI, A. 1967. Transmission of stripe and black-streaked dwarf viruses by a new planthopper. *Ann. phytopath. Soc. Japan* **33**:318. [Abs., Jap.]
SUGIYAMA, M. 1966. Studies on methods of forecasting rice stripe disease. I. Kinds of indicator plants. *Ann. phytopath. Soc. Japan* **32**:83. [Abs., Jap.]
SUZUKI, H., KATO, T., KAWAGUCHI, K. & SASAMURA, H. 1960. On the testing method of resistance of rice varieties to rice stripe, *Oryza* virus 2 Kuribayashi, in the frequently affected paddy field. *Bull. Tochigi pref. agric. exp. Stn* **4**:1–13. [Jap. Engl. summ.]
TORIYAMA, K. 1966. Breeding of rice varieties for direct seeding, especially for resistance to stripe disease. *Recent Advan. Breed.* **7**:60–66. [Jap.]
TORIYAMA, K. 1969. Genetics of and breeding for resistance to rice virus diseases. In *The virus diseases of the rice plant: Proc. Symp. at IRRI, April, 1967*: 313–334. Baltimore, Maryland, Johns Hopkins Press.
TORIYAMA, K., SAKURAI, Y., WASHIO, O. & EZUKA, A. 1966. A newly bred rice line, Chugoku No. 31, with stripe disease resistance transferred from an indica variety. *Bull. Chugoku agric. exp. Stn*, A, **13**:41–54.
WASHIO, O., TORIYAMA, K., EZUKA, A. & SAKURAI, Y. 1968. Studies on the breeding of rice varieties resistant to stripe disease. III. Genetic studies on resistance to stripe in foreign varieties. *Jap. J. Breed.* **18**:167–172.
YAMADA, W. & YAMAMOTO, H. 1955; 1956. Studies on the stripe disease of rice plant. I. On the virus transmission by insect, *Delphacodes striatella* Fallen. III. Host plants, incubation period in rice plant and retention and overwintering of the virus in the insect, *Delphacodes striatella* Fallen. *Spec. Bull. Okayama pref. Agric. Exp. Stn* **52**:93–112; **55**:35–56. [Jap. Engl. summ.]
YAMAGUCHI, T., YASUO, S. & ISHII, M. 1965. Studies on rice stripe disease. II. Study on varietal resistance to stripe disease of rice plant. *J. cent. agric. Exp. Stn* **8**:109–160. [Jap. Engl. summ.]
YOSHIMEKI, M. 1966. Ecological and physiological studies on the dynamics of the migratory population of rice planthoppers and leafhoppers. *Bull. Kyushu agric. Exp. Stn* **12**:1–78.

YELLOW DWARF

HISTORY AND DISTRIBUTION

The disease was first recognized in the warm, southwestern region of Japan soon after 1910. In 1943, workers at Kochi Agriculture Experiment Station suggested that the leafhopper *Nephotettix cincticeps* transmitted the virus. This was later confirmed by Iida & Shinkai (1950); further studies were later made by Shinkai (1962) and others. In recent years, the disease has spread from the southern coastal regions to the northern inland areas of Japan (Iida, 1969).

Outside Japan, the disease has been known since 1932 in Taiwan, where it has caused considerable damage (Kurosawa, 1940; Chiu *et al.*, 1964) and in Hainan Island (Hashioka, 1952). Recently, it was confirmed by transmission experiments that it is present in the Philippines (Rivera & Ou, 1964; Palomar & Rivera, 1968), in Thailand (Wathanakul & Weerapat, 1969), in India (Raychaudhuri *et al.*, 1967) and in Malaysia (Lim & Goh, 1968). Diseased plants have also been observed in Indonesia and Ceylon. This indicates the wide distribution and common occurrence of the disease in the tropics of Asia.

Recently, the disease was strongly suspected by Japanese workers to be caused by a mycoplasma-like organism together with diseases of other crop plants with the ‘ yellowing ’ type of symptoms.

Symptoms

The characteristic symptoms are general chlorosis, pronounced stunting and profuse tillering. The chlorotic leaves are uniformly pale green or pale yellow. Discoloration first appears on the newly emerging young leaves and all the succeeding leaves show chlorosis. On resistant varieties, only faint mottling may appear (Plate I).

Plants infected early may die prematurely, but most often they survive until the crop matures. However, they produce no heads or very poor ones. Plants infected late may not show any symptoms, but characteristic symptoms appear on the ratoon plants grown from the cut stubbles.

Vectors and Transmission

Besides *Nephotettix cincticeps*, mentioned above, *N. impicticeps* (Shinkai, 1959, 1962) and *N. apicalis* (Ouchi & Suenaga, 1963) have been found to be vectors. The disease was transmitted by *N. apicalis* in the Philippines (Rivera & Ou, 1964), by *N.* sp. in Thailand (Wathanakul & Weerapat, 1969) and by *N. impicticeps* in Malaysia (Lim & Goh, 1968).

Most of the leafhoppers acquire the causal agent by feeding on diseased plants for 1–3 hours. The minimum acquisition time was 10 min for *N. cincticeps* and 30 min for *N. impicticeps*. There is a long incubation period, 20–39 days, usually 25–30 days, in the insect. The minimum inoculation feeding time was reported to be 1 min for *N. cincticeps* and 3 min for *N. impicticeps*. About half of the infective individuals of both species cause infection with 1 hour of feeding. The latent period on rice plants is also long. It varies with the temperature from less than 1 month in warm seasons to 3 months in cool weather. Once infective, the insects can transmit the agent to new healthy plants daily throughout their lives (10–60 days). The agent was found not to be transmitted transovarially (Shinkai, 1962). Populations of *N. cincticeps* collected from areas in which the disease is prevalent or not prevalent are equally capable of transmission.

Host Range

Only 2 species were found to be alternative hosts of yellow dwarf disease by Shinkai (1951, 1960, 1962): *Alopecurus aequalis* Sobol. and *Glyceria acutiflora.* Torr. The symptoms on these plants are similar to those on rice. Only *A. aequalis* was found to be naturally infected.

Transmission Cycle

The causal organism overwinters in leafhoppers and also in the wild grass *Alopecurus aequalis*. It was found that the earlier in autumn leafhoppers acquire the organism, and the higher the winter temperature, the higher is the percentage of infectious individuals after overwintering.

Because of the relatively long incubation and latent periods, the cycle of transmission is relatively slow. In epidemic fields, the percentage of infective leafhoppers late in the autumn can, however, be extremely high; 80% infection has been recorded in Japan (Shinkai, 1962). In the tropics, ratoon plants growing from the stubble were often found to be heavily infected, thus accounting for many of the late infections. In the first and second generations, the percentages of infective individuals are usually very low.

Properties of the Causal Agent

The partially purified virus was reported (Takahashi, 1964) as polyhedral particles about 55mμ in diameter. Recently, however, workers in Japan have found reason to suspect that this disease is due to a mycoplasma-like organism, and that earlier reports of virus particles were erroneous (personal communications).

Varietal Resistance

Rice plants are susceptible to yellow dwarf at all stages of growth, though with age they become more resistant to other viruses. The effects on plant growth and yield reduction due to infection are, however, less when infection is late (Mori *et al.*, 1963; IRRI, 1966; Palomar & Rivera, 1968).

In field tests in Taiwan, Hashioka (1952) found considerable differences in resistance or susceptibility among the more than 300 varieties tested. Komori & Takano (1964) found a glutinous variety, Saitama Mochi No. 10, and several foreign varieties (Kaladumai, Loktjan, Pe Bi Hun and Te-tep) to be resistant in Japan. In these field tests, new ratoon growth was examined for determining disease incidence because disease symptoms may not show up in the original crop.

A seedling test has been experimented with by Morinaka & Sakurai (Sakurai, 1969). Because the virus is not transovarial and has a long incubation period, young nymphs have to be used on diseased plants to acquire the virus. The inoculation method is somewhat the same as that for stripe virus. To determine the incidence of the disease, plants are cut back after several weeks' growth, after which new leaves will show clear symptoms. The results of such tests indicated that such a method is generally too severe in comparison with field tests. Varieties found to be resistant are Saitama Mochi No. 10, Kagura Mochi, Mangetsu Mochi, Shinano Mochi No. 3 and Te-tep. All are glutinous except Te-tep.

Preliminary experiments showed that F_1 plants of a cross between Saitama Mochi No. 10 (resistant) × Manryo (susceptible) were moderately resistant Sakurai, 1969).

Chemical Control

Attempts at control of the vectors by aerial application of insecticides over large areas have been made in Japan. Malathion or carbaryl dust was used, usually soon after transplanting, but sometimes in the autumn to kill the overwintering insects. Good results have been obtained in several instances.

LITERATURE CITED

Chiu, R. J. 1964. Virus diseases of rice in Taiwan. Paper presented at 10th meeting of FAO-IRC Working Party on Rice Production and Protection, Manila.

HASHIOKA, Y. 1952. Varietal resistance of rice to the brown spot and yellow dwarf. (Studies on pathological breeding of rice. VI). *Jap. J. Breed.* **2**:14–16.

IIDA, T. T. 1969. Dwarf, yellow dwarf, stripe, and black-streaked dwarf diseases of rice. In *The virus diseases of the rice plant: Proc. Symp. at IRRI, April, 1967*: 3–11. Baltimore, Maryland, Johns Hopkins Press.

IIDA, T. T. & SHINKAI, A. 1950. Transmission of rice yellow dwarf by *Nephotettix cincticeps. Ann. phytopath. Soc. Japan* **14**:113–114. [Abs., Jap.]

IRRI (INTERNATIONAL RICE RESEARCH INSTITUTE). 1966 Annual Report.

KOMORI, N. & TAKANO, S. 1964. Varietal resistance of rice plant to rice yellow dwarf in the field. *Proc. Kanto-Tosan Pl. Prot. Soc.* **11**:22. [Jap.]

KUROSAWA, E. 1940. On the rice yellow dwarf occurring in Taiwan. *J. Pl. Prot., Tokyo* **27**:156–161. [Jap.]

LIM, G. S. & GOH, K. G. 1968. Leafhopper transmission of a virus disease of rice locally known as 'Padi Jantan' in Krian, Malaysia. *Malay. agric. J.* **46**:435–450.

MORI, K., MAKINO, A. & OSAWA, T. 1963. Process of disease development and injury in rice yellow dwarf. *Ann. phytopath. Soc. Japan* **28**:83. [Abs., Jap.]

OUCHI, Y. & SUENAGA, H. 1963. On the transmissibility by the leafhopper *Nephotettix apicalis* of rice yellow dwarf virus. *Proc. Ass. Pl. Prot., Kyushu* **9**:60–61. [Jap.]

PALOMAR, M. K. & RIVERA, C. T. 1967. Yellow dwarf of rice in the Philippines. *Philipp. Phytopath.* **3**:27–34.

RAYCHAUDHURI, S. P., MISHRA, M. D. & GHOSH, A. 1967. Preliminary note on the occurrence and transmission of rice yellow dwarf virus in India. *Pl. Dis. Reptr* **51**:1040–1041.

RIVERA, C. T. & OU, S. H. 1964. Studies on the virus diseases of rice in the Philippines. Int. Rice Res. Inst., Los Banos, Laguna, Philippines. (Mimeographed.)

SAKURAI, Y. 1969. Varietal resistance to stripe, dwarf, yellow dwarf, and black-streaked dwarf. In *The virus diseases of the rice plant: Proc. Symp. at IRRI, April, 1967*: 257–275. Baltimore, Maryland, Johns Hopkins Press.

SHINKAI, A. 1951. Host range and problem of virus transmission through seeds of rice yellow dwarf. *Ann. phytopath. Soc. Japan* **15**:176. (Abs., Jap.]

SHINKAI, A. 1959. Transmission of rice yellow dwarf by *Nephotettix bipunctatus* Fab. *Ann. phytopath. Soc. Japan* **24**:36. [Abs., Jap.]

SHINKAI, A. 1960. Virus diseases of rice. In HIDAKA, J. *et al., Plant Viruses*: 246–263. [Jap.]

SHINKAI, A. 1962. Studies on insect transmissions of rice virus diseases in Japan. *Bull. natn. Inst. agric. Sci., Tokyo*, Ser. C **14**:1–112. [Jap. Engl. summ.]

TAKAHASHI, Y. 1964. Isolation and purification of the rice yellow dwarf virus. *Ann. phytopath. Soc. Japan* **29**:73. [Abs., Jap.]

WATHANAKUL, L. & WEERAPAT, P. 1969. Virus diseases of rice in Thailand. In *The virus diseases of the rice plant: Proc. Symp. at IRRI, April, 1967*: 79–85. Baltimore, Maryland, Johns Hopkins Press.

BLACK-STREAKED DWARF

HISTORY AND DISTRIBUTION

The disease was first recognized as being distinct from dwarf in 1952 (Kuribayashi & Shinkai, 1952) though it had occurred for many years in the eastern and southern parts of Japan; it was also reported to be transmitted by *Laodelphax striatellus*. Local epidemics have caused losses in yield of rice as well as wheat, barley and maize. The disease is not known outside Japan.

SYMPTOMS

Diseased plants show pronounced stunting and also darkening of the foliage. The most characteristic symptom is the presence of galls which are elongated swellings extending along the major veins on the lower side of leaf blades, on the outside of leaf sheaths, and on culms. These swellings are parenchymatous proliferations in the phloem which have broken through the epidermis. Inclusion bodies were found in these cells (Kashiwagi, 1966). The swellings form grey or dark brown streaks of varying lengths; they usually appear late and are not numerous. Black-streaked dwarf is essentially a phloem gall disease, similar to Fiji disease of sugarcane. It is not to be confused with a malady of similar nature called leaf gall of rice, which was found to be the result of insect injury (Maramorosch *et al.*, 1961).

Diseased plants usually survive until harvest time. The leaf blades are often twisted, particularly in their proximal parts. No heads or very poor ones are formed and the heads tend to remain only halfway emerged from the leaf sheath. Grains are often marked with dark brown blotches.

Vectors and Transmission

Laodelphax striatellus (Fallen) was demonstrated to be the vector by Kuribayashi & Shinkai (1952). More recently *Unkanodes sapporonus* (Mats.) and *Ribautodelphax albifascia* (Mats.) were also found to be vectors (Shinkai, 1966, 1967). Both insects are in addition vectors of stripe disease. There is considerable variation among individual insects in their ability to transmit the virus.

Most active individuals of *L. striatellus* acquire the virus by feeding on diseased plants for 1 day; the minimum acquisition feeding time is 30 min. Young nymphs acquire the virus more easily than old ones, percentage acquisition being 83–88 for the 1st instar and 19–27 for the 5th. The incubation period in the insect is 7–35 days, most commonly 7–12 days.

The minimum infection feeding time reported is 5 minutes and about half of the individuals can infect rice plants after feeding for 1–3 hours (Shinkai, 1962). A longer period of infection feeding is required to secure 100% infection. The latent period in rice plants is 14–24 days.

Most individuals remain viruliferous until quite old. No transovarial virus passage has been demonstrated (Shinkai, 1962).

Host Range

The following species of plants have been found to be susceptible (Shinkai, 1957); *Alopecurus aequalis* Sobol., *A. japonicus* Steud., *Avena sativa* L., *Beckmannia syzigachne* (Steud.) Fernald, *Cynosurus cristatus* L., *Digitaria adscendens* (H.B.K.) Henr., *D. violascens* Link, *Echinochloa crusgalli* (L.) Beauv., *E. crusgalli* var. *frumentacea* Wight, *E. crusgalli* var. *oryzicola* Ohwi, *Eragrostis multicaulis* Steud., *Glyceria acutiflora* Torr., *Hordeum sativum* var. *hexastichon* Hack., *H. sativum* var. *vulgare* Hack., *Lolium multiflorum* Lam., *L. perenne* L., *Oryza sativa* L., *Panicum miliaceum* L., *Phleum pratense* L., *Poa annua* L., *Secale cereale* L., *Setaria italica* Beauv., *S. viridis* (L.) Beauv., *Triticum aestivum* L., *Trisetum bifidum* (Thunb.) Ohwi and *Zea mays* L. Several of the cereals are also favoured food of the smaller brown planthopper. Among the weeds, *A. aequalis* was most commonly infected in the field.

Transmission Cycle

The virus is perpetuated from one season to the next in overwintering planthoppers, the percentage of viruliferous individuals in the overwintering population being high. Virus transmission begins in the early spring, and hoppers remain viruliferous until early summer. In the second generation, which appears in midsummer, the percentage of viruliferous individuals is low, but builds up again in the autumn (Shinkai, 1962).

Properties of the Virus

The virus was partially purified by Shikata *et al.* (1967). Differential centrifugation and density gradient sedimentation in sucrose solution reveal spherical particles, 80–120mμ in diameter. Electron microscopy of ultrathin sections of diseased plants reveals two kinds of particles, one 50–55mμ and the other 75–85mμ in diameter. However, as the larger particles were found in both infected plants and insects, they are believed to be the virus particles.

The virus was also purified by Kitagawa & Shikata (1969a) from an extract of diseased leaves, and was shown to consist of spherical particles 60mμ in diameter. Kitagawa & Shikata (1969b) found the thermal inactivation point in expressed sap to be between 50 and 60°C (10 minutes). In extracts of infected plants and insects the virus survived for 6 days at 4°C. At -30 to -35°C a high degree of infectivity was found after 232 days' storage. The dilution end point was 10^{-4}—10^{-5} in plant sap and 10^{-5}—10^{-6} in viruliferous insects. The virus was stable in several buffer solutions but was inactivated when diseased plant sap was treated with chloroform or a mixture of chloroform and *n*-butanol. It was affected by treatment with EDTA, sodium deoxycholate, carbon tetrachloride and other chemicals.

Varietal Resistance

Rice plants are susceptible to the virus at an early stage of growth, becoming somewhat resistant after maximum tillering has occurred. Early infection causes more damage to the plants than does late infection.

Since the disease rarely becomes epidemic in the field, seedling inoculation has been tried for varietal resistance screening by Morinaka & Sakurai (1967). As the same vectors transmit both stripe and black-streaked dwarf viruses, the seedling test used was similar to that for stripe. The insects, however, have to acquire the virus by feeding on diseased plants at the young nymph stage. Four weeks after inoculation, the percentage of diseased plants and the degree of stunting and vein enation were determined. Varieties found to be resistant included Loktjan, Amareriyo, Tadukan, Te-tep, Pusur, Modan, So-do-so, KA-35, and some others.

Morinaka *et al.* (1969) found that when the resistant variety Te-tep was crossed with two susceptible varieties, resistance was controlled by one major gene Bs.

LITERATURE CITED

Kashiwagi, Y. 1966. Staining of inclusion bodies in tumour cells of rice infected with rice black-streaked dwarf virus. *Ann. phytopath. Soc. Japan* **32**:168–170.

Kitagawa, Y. & Shikata, E. 1969a. Purification of rice black-streaked dwarf virus. *Mem. Fac. Agric. Hokkaido Univ.* **6**:446–451. [Jap. Engl. summ.]

Kitagawa, Y. & Shikata, E. 1969b. On some properties of rice black-streaked dwarf virus. *Mem. Fac. Agric. Hokkaido Univ.* **6**:439–445. [Jap. Engl. summ.]

Kuribayashi, K. & Shinkai, A. 1952. On the new disease of rice, black-streaked dwarf. *Ann. phytopath. Soc. Japan* **16**:41. [Abs., Jap.]

MARAMOROSCH, K., CALICA, C. A., AGATI, J. A. & PABLEO, G. 1961. Further studies on the maize and rice leaf galls induced by *Cicadulina bipunctella*. *Entomologia exp. appl.* **4**:86–89.

MORINAKA, T. & SAKURAI, Y. 1966. Varietal resistance by seedling test method to black-streaked dwarf of rice plant. *Ann. phytopath. Soc. Japan* **32**:89–90. [Abs., Jap.]

MORINAKA, T. & SAKURAI, Y. 1967. Studies on the varietal resistance to black-streaked dwarf of rice plant. 1. Varietal resistance in field and seedling tests. *Bull. Chugoku agric. Exp. Stn*, E, **1**:25–42. [Jap. Engl. summ.]

MORINAKA, T., TORIYAMA, K. & SAKURAI, Y. 1969. Inheritance of resistance to black-streaked dwarf virus in rice. *Jap. J. Breed.* **19**:74–78. [Jap. Engl. summ.]

SHIKATA, E., LO, Y., MATSUMOTO, T. & YAMADA, K. 1967. Electron microscopic studies on rice black-streaked dwarf virus. *Ann. phytopath. Soc. Japan* **22**:34. [Abs., Jap.]

SHINKAI, A. 1957. Host range and problems concerning transmission of rice black-streaked dwarf. *Ann. phytopath. Soc. Japan* **22**:34. [Abs., Jap.]

SHINKAI, A. 1962. Studies on insect transmission of rice virus diseases in Japan. *Bull. natn. Inst. agric. Sci., Tokyo*, Ser. C **14**:1–112. [Jap. Engl. summ.]

SHINKAI, A. 1966. Methods for studying virus transmission by Delphacidae and Cicadellidae. *J. Pl. Prot., Tokyo* **20**:137–140. [Jap.]

SHINKAI, A. 1967. Transmission of stripe and black-streaked dwarf viruses by a new plant-hopper. *Ann. phytopath. Soc. Japan* **33**:318. [Abs., Jap.]

HOJA BLANCA

HISTORY AND DISTRIBUTION

Hoja blanca (white leaf) has been recognized as a new rice disease in Latin America since 1957 (Anon., 1957; Adair & Ingram, 1957). It was, however, noticed and described in Colombia as early as 1935 (Galvez, 1969). It was first observed in Panama in 1952, in Cuba in 1954, and in Venezuela in 1956, where a severe outbreak of the disease caused yield losses of 25–50% (Atkins & Adair, 1957). Since then the disease has been reported from almost all rice-growing countries in Latin America and also from U.S.A. (CMI Distribution Map 359). The limited and sporadic occurrence in early years of the disease, which caused minor damage prior to 1957, was perhaps due to the use of local unimproved but resistant varieties. Since the extensive cultivation of improved varieties, such as Bluebonnet 50 and Century Patna 231, the disease has become widespread (Galvez, 1969).

Malaguti *et al.* (1957) demonstrated transmission by mixed groups of plant-hoppers and leafhoppers (*Sogatodes oryzicola* and *Hortensia similis*). Acuna *et al.* (1958) identified *Sogatodes oryzicola* as the vector. Testing for varietal resistance soon showed varieties of the *japonica* type to be resistant.

The symptoms of hoja blanca are similar to those of stripe disease in Japan. This led to the suspicion that hoja blanca might be caused by a virus (Mukoo & Iida, 1957) and to a co-operative study of varietal reaction to the two diseases which led to the conclusion that they are caused by different viruses (Atkins *et al.*, 1961).

Recently, under the title ' A preliminary investigation on the transmission and hosts of the hoja blanca disease of rice ' (Wang, Chen & Pai, 1964) there appeared an article in a review journal which suggests that the disease may be present in Mainland China, but this remains very doubtful.

SYMPTOMS

The major field symptoms are the appearance of white or chlorotic stripes on the leaves or the occurrence of completely white leaves, stunting of the whole plant and poor or partial filling of the grains at maturity (Plate II).

On artificially inoculated seedlings, the first symptoms to appear are small, chlorotic spots at the base of the leaf immediately above the one inoculated. On the next younger leaf, one or more longitudinal white stripes appear. The succeeding leaves are either almost entirely white or mottled. Diseased plants are stunted. If infected very young, the seedlings are killed.

The panicles of diseased plants are often only partially extruded and are small and deformed. The floral parts are absent or the spikelets are sterile; because of this, few or no seeds are found on the panicles. Root size and number are reduced and many roots become brown and die.

Symptoms on several other hosts are similar. Chlorosis and necrosis are more severe on barley, oats, rye and wheat. These plants are killed rapidly, and usually only 2 or 3 symptom-bearing leaves emerge before death (Galvez, 1969).

These symptoms differ from those of stripe disease. In the latter the central young leaves fail to unfold and have a tendency to bend downward whereas in the case of hoja blanca the leaves develop normally, apart from being chlorotic (Mukoo & Iida, 1957).

VECTORS AND TRANSMISSION

Hoja blanca is transmitted by *Sogatodes* (*Sogata*) *oryzicola* (Muir) which is the principal vector and also by *S. cubanus* (Crawf.) which is a minor vector. Galvez *et al.* (1960) transmitted the virus, using *S. cubanus*, from rice to *Echinochloa colonum* and from *Echinochloa* to *Echinochloa* but not from rice to rice or *Echinochloa* to rice. Recently such transmissions have been reported to be possible by forced feeding with highly active individuals of *S. cubanus* (Galvez, 1968). Rice is the preferred feeding host of *S. oryzicola* whereas *Echinochloa* is preferred by *S. cubanus*. Many other Fulgorids and Cicadellids have been tested as vectors with negative results.

The proportion of active individuals in a natural population of *S. oryzicola* is about 7–12% (Acuna *et al.*, 1958). McMillian reported that the percentage of active individuals could be increased by selective breeding and a scheme was worked out (McMillian *et al.*, 1962). However, the insects selected had only a few days of their life span remaining in which they could act as vectors. Hendrick *et al.* (1965) developed a technique that increased the proportion of active vectors from the original 5% to 50–75% and then to 80–100% by continuous selective breeding of active individuals as parents. The method has also been used in the hoja blanca laboratory in Louisiana (Everett, 1969). When selection ceases, however, the population reverts to 15–30% active individuals in 2–3 generations. Galvez (1968) obtained a 95–100% active colony which has been maintained for 10 generations at the highly active level. Once infective the insect remains capable of transmitting the virus until it is old or even until it dies.

McMillian *et al.* (1961, 1962) reported that after acquisition feeding times of 1, 6 and 12 hours, 20, 80 and 100% respectively of the potential vectors became infective. Galvez found the minimum acquisition feeding time to be 15 minutes, and within 1 hour of feeding all the insects of this highly active colony became infective.

Early observations (Acuna *et al.*, 1958; McMillian *et al.*, 1961, 1962; Galvez *et al.*, 1960, 1961) indicated wide variability in the incubation period of the virus in the vector, varying from 6 to 37 days. With a highly active virus-free colony Galvez (1968) demonstrated the incubation period to be 30–36 days. The early reports of relatively short incubation periods are believed to have been due to transovarial virus passage.

Infection feeding of 1, 7, 15, 24 and 48 hours resulted in 40, 90, 90, 100 and 100% transmission respectively in experiments by McMillian *et al.* (1961, 1962) and feeding for 0·5, 1, 3 and 6 hours resulted in 35, 65, 100 and 100% transmission respectively in tests made by Galvez (1968).

The incubation period in the plant varies greatly with the age of the plant, the site of inoculation, varietal resistance etc. McMillian *et al.* (1961, 1962) found the period to be 6–34 days, Galvez *et al.* (1961), 5–21 days. On susceptible varieties at the first leaf stage, symptoms become visible 3 or 4 days after inoculation, as soon as the new leaf emerges. The symptoms never show up on the inoculated leaf.

Transovarial transmission was first demonstrated by Acuna & Ramos (1959) who found 60–80% transmission through the eggs of transmitting females. Galvez (1968) showed that 96% of 500 eggs tested transmitted the virus, apparently without loss of concentration after 10 generations.

The virus is not transmitted mechanically, through soil, or through seeds.

Host Range

Galvez *et al.* (1960, 1961) transmitted the virus from rice to *Echinochloa colonum* (L.) Link, *Leptochloa* sp., *Digitaria* sp., *Triticum sativum* L., *Hordeum vulgare* L. and *Avena sativa* L., but not to sugarcane, maize or sorghum. Lamey *et al.* (1960, 1961) also transmitted the virus to *Secale cereale* L., barley, oats and wheat. Gibler *et al.* (1961) reported the natural infection by the virus of wheat and oats in fields adjacent to infected rice areas.

Infection Cycle

In view of the recent observation that there is a relatively long incubation period in the vector, most of the male insects will not be able to become vectors before they die even if they acquire the virus at an early stage in their development. The female insects, having a longer life span, may act as vectors in late adult life when they acquire the virus at the early nymph stage. Most of the vectors acquire the virus through the eggs. Maximum disease incidence occurs when high insect population peaks coincide with the most susceptible stage of the rice, i.e. during the first 45 days.

Properties of the Virus

Herold *et al.* (1968) demonstrated spherical particles approximately 42mμ in diameter both in dipping preparations and purified suspensions of the virus from diseased rice leaves. Shikata & Galvez (1969), however, found thread-like particles in the vector and infected rice plants.

Varietal Resistance

While differences in varietal resistance were noted by earlier workers, systematic screening programmes initiated by the U.S. Department of Agriculture in 1957 (Atkins & Adair, 1957) showed that all the major rice varieties from U.S.A. grown commercially in Latin America were susceptible. Several thousand lines were planted in infected areas of Cuba and Venezuela and later in Colombia. To assure optimum conditions for good disease development, planting was done at the correct time and a low seeding rate was used; in addition susceptible varieties were planted as a check (Lamey *et al.*, 1961). A visual disease rating of 0 to 9 was used (Atkins & Adair, 1957). Jennings (1963) found a direct relationship between the scale reading and yield loss, there being about 10% yield loss for each unit. Most commercial U.S. varieties were susceptible while those of *japonica* type were resistant. Most *indica* types were susceptible, a few exceptions being Mass, Peta, Bengawan, Co. 10, N.10-B, S-67, Salak, Tjahai, Pandhori No. 4 etc. Later, breeders' materials were tested more or less in the same way. Uniform hoja blanca nurseries and virus × strain nurseries were established in several Latin American countries (Lamey, 1969).

Although field tests have been successful, greenhouse tests by artificial inoculation have also been developed and perfected (Lamey *et al.*, 1961, 1964; Lamey, 1969) to save time and space and to reduce the number of plants to be handled. This became possible when methods were found of increasing the percentage of active transmitters and when other techniques were developed. During these studies, it was found that the age of the plant has profound effects on its susceptibility (Lamey *et al.*, 1965). The 1–3 leaf stage was the most susceptible and many of the field resistant varieties were susceptible when inoculated at the 1 leaf stage. Artificial testing for resistance was therefore done with older plants (Galvez, 1968).

The major U.S. resistant varieties and sources of resistance for breeding programmes are Gulfrose (A Brunimissie selection (R) × Zenith) and Lacrosse (Colusa-Blue Rose × Shoemed (R)-Fortuna (R)). Lacrosse has been used in the development of Northrose (Lacrosse × Arkrose), Nova (Lacrosse × Zenith-Nira) and Nova 66, a selection from Nova (Johnston, 1958; Bollich, Scott & Beachell, 1965; Johnston & Henry, 1965; Johnston, Templeton & Atkins, 1965; Johnston *et al.*, 1966). Tainan-iku No. 487 has also been used extensively as a source of resistance and Sadri, Pandhori No. 4 and Arkrose have been used to a limited extent.

Recently, in Colombia, several highly resistant lines, the ICA lines, have been developed, and some are in commercial use. These lines are also tolerant of the vectors.

Resistance appears to be dominant to susceptibility and is relatively simply inherited (Beachell & Jennings, 1961). Acuna & Ramos (1959) observed that a field-resistant variety had thicker cell walls, longer protective trichomes and larger stomata. Van Hoof *et al.* (1962) pointed out that variety SML 81b was possibly field resistant because of its erect leaves which result in lower humidity among the rice plants. The effect of erect leaves on humidity is probably operative on the insect vectors, since they thrive under highly humid conditions.

Chemical Control

Several workers (Galvez, 1969) have tried to control hoja blanca through chemical control of the vectors. This has been partially successful with varieties which show some resistance but not with susceptible varieties. Chemical control helps to reduce the population and, consequently, the damage caused by the virus.

LITERATURE CITED

Acuna-Gale, J., Ramos-Ledon, L. & Lopez-Cardet, Y. 1958. *Sogata oryzicola* Muir vector de la enfermedad virosa hoja blanca del Arroz en Cuba. *Agrotecnia 1958*: 23–34.

Acuea-Gale, J. & Ramos-Ledon, L. 1959. Informes de interes general en relacion con el Arroz. *Bull. Admin. Estab. Arroz, Cuba* 11.

Adair, C. R. & Ingram, J. W. 1957. Plans for the study of hoja blanca, a new rice disease. *Rice J.* **60** (4):12.

Anon. 1957. Hoja blanca, a threat to U.S. rice. *Rice J.* **60** (4):14, 48.

Atkins, J. G. & Adair, C. R. 1957. Recent discovery of hoja blanca, a new rice disease in Florida, and varietal resistance tests in Cuba and Venezuela. *Pl. Dis. Reptr* **41**:911–915.

Atkins, J. G., Goto, K. & Yasuo, S. 1961. Comparative reactions of rice varieties to the stripe and hoja blanca virus diseases. *Int. Rice Commn Newsl.* **10** (4):5–8.

Beachell, H. M. & Jennings, P. R. 1961. Mode of inheritance of hoja blanca resistance in rice. Abs. in *Proc. Rice Tech. Working Group, 1960.*

Everett, T. R. 1969. Vectors of hoja blanca virus. In *The virus diseases of the rice plant: Proc. Symp. at IRRI, April, 1967*: 111–121. Baltimore, Maryland, Johns Hopkins Press.

Galvez E., G. E. 1968. Transmission studies of the hoja blanca virus with highly active virus-free colonies of *Sogatodes oryzicola. Phytopathology* **58**:818–821.

Galvez E., G. E. 1969. Hoja blanca disease of rice. In *The virus diseases of the rice plant: Proc. Symp. at IRRI, April, 1967*: 35–49. Baltimore, Maryland, Johns Hopkins Press.

Galvez E., G. E., Thurston, H. D. & Jennings, P. R. 1960. Transmission of hoja blanca of rice by the planthopper, *Sogata cubana. Pl. Dis. Reptr* **44**:394.

Galvez E., G. E., Thurston, H. D. & Jennings, P. R. 1961. Host range and insect transmission of the hoja blanca disease of rice. *Pl. Dis. Reptr* **45**:949–953.

Gibler, J. W., Jennings, P. R. & Krull, C. F. 1961. Natural occurrence of hoja blanca on wheat and oats. *Pl. Dis. Reptr* **45**: 334.

Hendrick, R. D., Everett, T. R., Lamey, H. A. & Showers, W. B. 1965. An improved method of selecting and breeding for active vectors of hoja blanca virus. *J. econ. Ent.* **58**:538–542.

Herold, F., Trujillo, G. & Munz, K. 1968. Virus-like particles related to hoja blanca disease of rice. *Phytopathology* **58**: 546–547.

Jennings, P. R. 1963. Estimating yield loss in rice caused by hoja blanca. *Phytopathology* **53**:492.

Johnston, T. H. 1958. Registration of rice varieties. *Agron. J.* **50**:694–700.

Johnston, T. H. & Henry, S. E. 1965. Northrose rice (Reg. No. 23). *Crop Sci.* **5**:285.

Johnston, T. H., Templeton, G. E. & Atkins, J. G. 1965. Nova rice (Reg. No. 24). *Crop Sci.* **5**:285–286.

Lamey, H. A. 1969. Varietal resistance to hoja blanca. In *The virus diseases of the rice plant: Proc. Symp. at IRRI, April, 1967*: 293–311. Baltimore, Maryland, Johns Hopkins Press.

Lamey, H. A. & McMillian, W. W. 1960. Hosts besides rice for virus and vector of hoja blanca. Abs. in *Phytopathology* **50**:643.

Lamey, H. A., Lindberg, G. D. & Brister, C. D. 1964. A greenhouse testing method to determine hoja blanca reaction of rice selections. *Pl. Dis. Reptr* **48**:176–179.

Lamey, H. A., McMillian, W. W. & McGuire, J. U. 1961. Transmission and host range studies on hoja blanca. Abs. in *Proc. Rice Tech. Working Group, 1960.*

LAMEY, H. A., SHOWERS, W. B. & EVERETT, T. R. 1965. Developmental stage of rice plant affects susceptibility to hoja blanca virus. Abs. in *Phytopathology* **55**:1065.
MALAGUTI, G., DIAZ C., H. & ANGELES, N. 1957. The virus hoja blanca in rice. *Agron. trop.* **6**:157–163.
MCMILLIAN, W. W., MCGUIRE, J. U. & LAMEY, H. A. 1961. Hoja blanca studies at Camaguey, Cuba. Abs. in *Proc. Rice Tech. Working Group, 1960.*
MCMILLIAN, W. W., MCGUIRE, J. U. & LAMEY, H. A. 1962. Hoja blanca transmission studies on rice. *J. econ. Ent.* **55**:796–797.
MUKOO, H. & IIDA, T. 1957. Informe sobre la investigacion de la hoja blanca del arroz en Cuba. *Bull. Admin. Estab. Arroz, Cuba* **1**:5–12.
SHIKATA, E. & GALVEZ E., G. E. 1969. Fine flexuous threadlike particles in cells of plant and insect hosts infected with rice hoja blanca virus. *Virology* **39**:635–641.
VAN HOOF, H. A., STUBBS, R. W. & WOUTERS, L. 1962. Beschouwingen over hoja blanca en zijn overbrenger *Sogata oryzicola* Muir. *Surin. Landb.* **10**:3–18.
WANG, F. M., CHEN, Y. L. & PAI, K. C. 1964. A preliminary investigation on the transmission and hosts of the hoja blanca disease of rice. *Pl. Prot.* (Mainland China) **2**:9–10. [Chin.]

TRANSITORY YELLOWING

HISTORY AND DISTRIBUTION

This disease was first seen in epidemic form in 1960 in the southern part of Taiwan. It was at first considered to be a physiological disorder called ‘ suffocation disease ’ (Chang, 1961; Takahashi, 1961), but in 1963 it was demonstrated to be a virus disease transmitted by *Nephotettix apicalis* (Chiu *et al.*, 1965). The disease is only known to be present in Taiwan where, during the last few years, a few to many thousands of hectares of rice have been infected each year, mostly in the second crop season.

SYMPTOMS

The major symptoms are leaf discoloration, reduced tillering and stunting. Two to three weeks after transplanting, a typical diseased plant has one or two lower leaves becoming distinctly yellow, then turning bright yellow or orange-buff according to the severity of the disease. Finally the diseased leaves wither. The leaf discoloration usually starts from the tip of a leaf and from the lower leaves upwards. Rusty-coloured flecks of various sizes are often observed on these discoloured leaves.

The reduction in tillering and the stunting are obvious on susceptible varieties. Infected plants bear poor panicles or none. Plants infected early show severe symptoms, but no symptoms may appear when they are infected at the later stages of growth. In the most acute cases, plants die before flowering.

The diseased plants frequently show a greater degree of recovery under glasshouse conditions. Following the typical appearance of leaf yellowing, infected plants seem to recover gradually and produce no symptoms during the later stages of growth. Frequently, healthy-looking tillers grow from a diseased stalk. For this reason the name ‘ transitory yellowing ’ was given to the disease.

Large, round inclusion bodies were found internally in the parenchyma cells, sieve tubes etc., around vascular bundles (Su & Huang, 1965). Starch accumulation can be detected in diseased leaves (Hsieh, 1966) as in the case of tungro virus. These features help in distinguishing this disease from the truly physiological ‘ suffocation ’ disease. Chen & Shikata (1968) found bullet-shaped particles associated with diseased plants, the particles measuring 96 × 120–140mμ (negative stained preparation of clarified leaf sap) or 94 × 180–210mμ (ultrathin sections).

Vectors and Transmission

Chiu *et al.* (1965) first identified *Nephotettix apicalis* (Motsch.) as a vector; *N. cincticeps* also was later found to transmit the virus (Chiu & Jean, 1969; Chiu *et al.*, 1968).

In different colonies, 41 to 62% of the individuals are active transmitters. A majority of active insects acquire the virus by a few hours' feeding on diseased plants, 90% by 1-day feeding. The minimum period is 5 to 10 minutes. The incubation period in the insects varies from a minimum of 3 days to a maximum of 34 days, but is usually 9–16 days. Infection feeding of 5–10 minutes, 5–10 hours and 24 hours resulted in successful transmission by 28%, 45% and 88% of the insects respectively. The latent period in the plants is 2–4 weeks (Chiu *et al.*, 1965; Chiu & Jean, 1969).

Once infective, the insects remain viruliferous for life. Low temperature seems to reduce the rate of transmission (Chiu & Jean, 1969).

Varietal Resistance and Control

Field tests during the last few years have revealed a number of resistant varieties (Chiu, 1964). They include Wu-ku-chin-yu, Chung-lin-chung, Chu-tze, Hy-lu-tuen and Kaohsiung 22.

Insecticides such as carbaryl have been tested in attempts to reduce the population of the vectors and thus the damage by the virus to the rice crop.

LITERATURE CITED

Chang, S. C. 1961. Control of suffocating disease of rice plant in Taiwan. *Soils Fertil. Taiwan, 1961*: 1–4.

Chen, M. J. & Shikata, E. 1968. Electron microscopy of virus-like particles associated with transitory yellowing virus-infected rice plants in Taiwan. *Pl. Prot. Bull., Taiwan* **10** (2): 19–28.

Chiu, R. J. 1964. Virus diseases of rice in Taiwan—a general review. Paper presented at 10th meeting of the FAO-IRC Working Party on Rice Production and Protection, Manila. 10pp.

Chiu, R. J. & Jean, J. H. 1969. Leafhopper transmission of transitory yellowing of rice. In *The virus diseases of the rice plant*: *Proc. Symp. at IRRI, April, 1967*: 131–137. Baltimore, Maryland, Johns Hopkins Press.

Chiu, R. J., Jean, J. H., Chen, M. H. & Lo, T. C. 1968. Transmission of transitory yellowing virus of rice by two leafhoppers. *Phytopathology* **58**:740–745.

Chiu, R. J., Lo, T. C., Pi, C. L. & Chen, M. H. 1965. Transitory yellowing of rice and its transmission by the leafhopper, *Nephotettix apicalis apicalis* (Motsch.). *Bot. Bull. Acad. sin., Taipei* **6**:1–18.

Hsieh, S. P. Y. 1966. Accumulation of starch in rice leaves infected with transitory yellowing and its application to differentiate transitory yellowing from suffocating disease. *Pl. Prot. Bull., Taiwan* **8**:205–210.

Su, H. J. & Huang, J. H. 1965. Intracellular inclusion bodies in the rice plants affected with transitory yellowing. *Bot. Bull. Acad. sin., Taipei* **6**:170–181.

Takahashi, J. 1961. Physiological disease of rice in Taiwan. *Soils Fertil. Taiwan, 1961*: 10–14.

TUNGRO AND SIMILAR DISEASES

History and Distribution

Tungro (meaning degenerated growth) was first observed in the Philippines at the experimental farm of the International Rice Research Institute (IRRI) in 1963. It was identified as a distinct virus transmitted by *Nephotettix impicticeps* (IRRI, 1963; Rivera & Ou, 1965) and was found to be one of the most widespread and destructive rice diseases in the country.

Since the early 1940s, rice dwarf or stunt disease (Agati *et al.*, 1941; Reyes, 1957; Reyes *et al.*, 1959), ' accep na pula ' (red disease) or stunt disease (Serrano, 1957) and rice ' cadang-cadang ' (yellowing) (Agati & Peralta, 1939; Peralta & Agati, 1939) have been reported from the Philippines. They, or at least most of them, are now believed to have been tungro disease (Ou & Ling, 1966).

' Penyakit merah ' (red disease) has been known to occur in Malaysia since 1938. It was for a long time considered to be a physiological disorder and it was suspected to be caused by metabolic deficiency of nitrogen (Lockard, 1959) or by hydrogen sulphide or organic acids originating from the anaerobic decomposition of weeds. Recent studies (Ou *et al.*, 1965; Ou & Goh, 1966) have shown that penyakit merah is due to a virus transmitted by *Nephotettix impicticeps*. The symptoms, the nonpersistent manner in which the virus is transmitted by the insect vector and the varietal reaction of those varieties which have been tested are very similar to those of tungro virus, and indeed no distinction between the two viruses has been found.

The ' mentek ' (midget) disease of Indonesia has been known since 1859. In connection with it many studies on soil chemistry and physiology have been made, as it was thought that it was a physiological disorder. Nematodes associated with the disease were also studied (van der Vecht, 1953). The similarities in symptoms, varietal reactions and other circumstantial evidence have led workers to suspect that mentek is similar to tungro and penyakit merah (Ou, 1965). Quite recently some experimental evidence has been obtained to substantiate the suspicion that a tungro-like virus is present in Indonesia which may have been the cause of mentek disease (Rivera *et al.*, 1968). The virus was found to be transmitted by *Nephotettix impicticeps* and to be nonpersistent in the leafhopper. This nonpersistent type of virus-insect relationship strongly suggests that mentek virus is closely related or identical to tungro virus. Mentek is not a serious problem at present because of the extensive cultivation of resistant varieties.

A new virus disease was reported from Thailand in 1964 (Wathanakul, 1965) and was called yellow-orange leaf (Wathanakul & Weerapat, 1969). It was transmitted by *N. impicticeps* and the virus was also found to be nonpersistent. Similarities in symptoms and in varietal reactions in general, together with virus-vector interactions and other factors suggest that the virus is the same as or a strain of tungro virus.

Another virus resembling tungro has been reported from India (Raychaudhuri *et al.*, 1967). It has similar symptoms to tungro and is also transmitted by *N. impicticeps*. As not much information is yet available, the relation of the Indian virus to tungro is yet to be determined.

Baldacci *et al.* (1970) reported a yellowing and stunting disease of rice in Italy resembling tungro by its symptoms. However, they found mycoplasma-like organisms in the infected tissues. The means of transmission are not yet known.

Very recently tungro disease has been found to be very prevalent in East Pakistan and has also been found in West Bengal, India, in areas bordering East Pakistan.

It is of interest to note from the historical point of view that mentek was known in Indonesia in 1859, penyakit merah in Malaysia in 1938, tungro in the Philippines (referred to as dwarf or stunt) in 1941 and yellow-orange leaf in Thailand in 1964. If these diseases prove to be the same, the indication will be that the virus has spread considerably.

IMPORTANCE

According to Serrano (1957), the accep na pula or stunt disease (believed to be the same as tungro) was very destructive all over the major rice regions of the Philippines and caused 30% loss, equivalent to 1·4 million tons, each year in the 1940s. It is still one of the most widespread and destructive diseases in the Philippines.

Penyakit merah has long been endemic in the Krian district of Perak State in West Malaysia. Disease incidence varies from year to year, but many thousands of hectares of rice are affected every year.

During 1934–36, 30,000 to 50,000 hectares were affected by mentek in Indonesia (van der Vecht, 1953) and it was considered one of the major problems in rice production. Resistant varieties, such as Bengawan, have been grown in recent years and the disease is now of no immediate importance, but the virus is still present in old susceptible varieties and so is a potential threat (Rivera, Ou & Tantere, 1968).

Yellow-orange leaf disease has spread widely in Thailand since 1964. In 1966, the estimate was 660,000 hectares infected, of which about half were severely diseased (Lamey *et al.*, 1967; Wathanakul & Weerapat, 1969).

This group of viruses is most important in the countries of Southeast Asia, where most of the world's rice is produced.

SYMPTOMS

The major symptoms on tungro-affected rice plants are stunting and discoloration of the leaves, the colours ranging from various shades of yellow to orange (Plate III). The degree of stunting and discoloration varies considerably with the rice variety, environmental conditions, the age of the plants, as well as the strain of the virus.

The discoloration starts from the tip of the leaf and may or may not extend to the lower part of the leaf blade; often only the upper portion is discoloured. Young leaves may have a mottled appearance and old leaves show rusty-coloured specks of various sizes. The leaf discoloration in various shades of yellow is common in *japonica* varieties and in shades of orange in *indica* varieties.

Stunting is severe in susceptible varieties, but slight in those which are more resistant. The number of tillers produced by infected plants is usually slightly reduced but does not differ greatly from the number produced by healthy plants.

In resistant or tolerant varieties, discoloration of leaves is usually only partial and young leaves which develop later may not show any discoloration. The plants are usually slightly stunted. In moderately resistant varieties, the discoloration of infected plants, conspicuous at one stage, may gradually disappear at later stages. This is often referred to as ' recovery ' from the disease. This perhaps led earlier workers to think of the disease as being physiological in origin. In susceptible varieties, however, the stunting and leaf discoloration usually persist throughout the life of the plant, severely diseased plants may die at an early or a later stage of growth.

In the same variety, the degree of stunting due to the virus decreases as the age of the plant increases. When infection occurs late, no symptoms appear (IRRI, 1965).

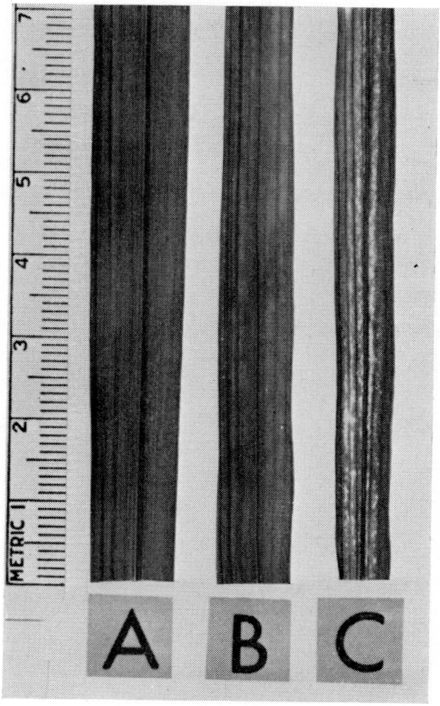

Fig. I–7. Symptoms of " S " and " M " strains of tungro virus on variety FK-135. A, healthy; B, mottling symptoms caused by " M " strain; C, striping symptoms caused by " S " strain.

The tungro virus is now known to have two strains, which have been named S and M (IRRI, 1964; Rivera & Ou, 1967). The S strain produces conspicuous interveinal chlorosis on some varieties, such as FK-135 and Acheh, giving an appearance of yellow or whitish stripes or sometimes irregular chlorotic specks.

PLATE III

Upper picture—*Tungro* virus infected field showing stunting and discoloration of leaves.

Lower left—*Tungro* virus infected seedlings from artificial inoculation. Plants in pot at right are healthy.

Lower right—*Orange leaf* virus infected plants in field showing discoloration and rolling of leaves and dying of plants.

PLATE III

These symptoms are similar to those of the dwarf disease of Japan. The M strain produces a mottled appearance on these varieties (Fig. I–7).

Shading and a high level of nitrogen in the soil seem partially to mask the leaf discoloration.

Diseased leaves have been shown to contain relatively large quantities of starch, and turn black or dark brown when stained with iodine (IRRI, 1966).

VECTORS AND TRANSMISSION

The main vector of tungro virus is *Nephotettix impicticeps* Ishihara, previously known as *N. bipunctatus* (Ishihara, 1964). Recently (IRRI, 1968, 1969) *N. apicalis* and *Recilia* (*Inazuma*) *dorsalis* were also found to transmit the virus, though in lower percentages.

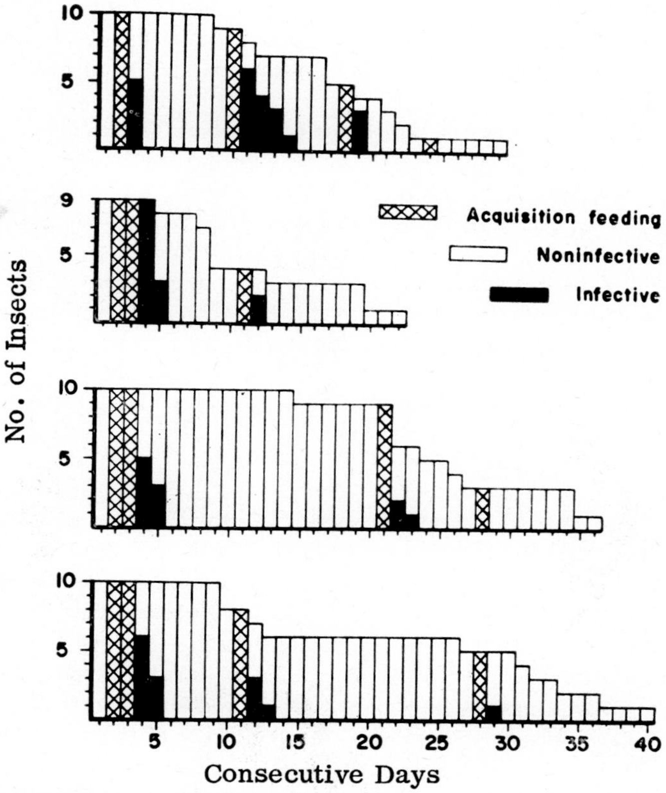

Fig. I–8. Nonpersistence of rice tungro virus in *Nephotettix impicticeps* and effect of reacquisition feeding on the infectivity of the vector in transmitting the virus. (From Ling, 1966.)

About 83% of populations of *N. impicticeps* are active transmitters. The minimum acquisition and inoculation feeding periods are 30 minutes and 15 minutes, respectively. The latent period in the plants is 6–9 days (Rivera & Ou, 1965). Ling (1966) found very unusual virus-vector interactions. There is no apparent incubation period in the insect, which may transmit the virus in 2 hours, including acquisition and inoculation feeding. The insects, however, only retain the virus for not more than 5 or 6 days. After that, insects are not infective unless they acquire the virus again, and this behaviour can be demonstrated repeatedly during the life of the insects. This nonpersistence of the virus in the vector has been studied by Ling (1966) (Fig. I–8). The decrease in infectivity may be measured hourly, about 40–50% reduction taking place 24 hours after termination of the acquisition feeding. Nymphs are as good transmitters as the adults but viruliferous nymphs lose their infectivity after moulting.

In the case of *N. apicalis*, the percentage of active transmitters varies from 0 to 45% in different colonies. This explains why this species was not reported as a vector until active colonies were found recently. Other aspects of transmission ability are also lower than in the case of *N. impicticeps* (IRRI, 1968; Ling, 1970).

Only 4–8% of *R. dorsalis* are able to transmit the virus. Nine other species of leafhoppers and planthoppers were found incapable of transmitting the virus (IRRI, 1968).

Yellow-orange leaf virus is also reported to be transmitted by *Nephotettix apicalis*, although not as efficiently as by *N. impicticeps* (Wathanakul & Weerapat, 1969).

Transmission of penyakit merah and mentek by *N. impicticeps* is much the same as in tungro (Ou *et al.*, 1965; Rivera *et al.*, 1968).

In the case of the leaf yellowing disease of India, the vector ' required acquisition feeding and incubation periods of not less than 5 days ', indicating some difference in the virus-vector interaction of this virus compared with that of tungro (Raychaudhuri *et al.*, 1967).

The virus is not transmitted through the eggs, nor by seed, soil, or mechanical means so far tested.

Host Range

Wathanakul (1964) inoculated 29 species of grass weeds and found *Eleusine indica* (L.) Gaertn., *Echinochloa colonum* (L.) Link and *Echinochloa crusgalli* Beauv. to be alternative hosts of tungro virus. Infected *E. indica* shows stunting and some leaf discoloration. No symptoms were observed on the other hosts but virus was recovered from the inoculated plants.

A recent study (IRRI, 1968) on host range included 63 species in 26 genera and 8 tribes of wild grasses and many species and strains of wild rice. It was found that many species of wild rice are good hosts of the virus. Among the grasses, *Ischaemum rugosum* Salisb., *Dactyloctenium aegyptium* (L.) Beauv. and others are sporadically infected. Often a few plants were infected in one experiment but not at all in subsequent tests.

Infection Cycle

Since the virus is nonpersistent, the insects have to acquire the virus from diseased plants repeatedly to be continuously infective. It is conceivable that the percentage of viruliferous populations may fluctuate greatly depending upon the amount of virus reservoir present. Roguing of diseased plants in the field may therefore reduce the sources of virus.

The virus can only cause serious damage in the tropics where host plants grow and the insect vector multiplies the year round. In areas where rice is grown only once a year and where there is a long, dry, hot period when no rice is grown, such as Central Thailand, the alternative hosts, especially wild rice, are probably important sources of inoculum.

Properties of the Virus

Antiserum to the tungro virus has been obtained by injecting partially purified materials into rabbits. The antiserum reacts, however, not only to tungro virus but also to orange leaf, yellow dwarf and grassy stunt viruses (John, 1965). Apparently the virus was not purified enough to obtain specific antiserum.

An attempt at further purification was made by Galves (1967). The particles are reported to be polyhedral and 30–33mμ. Shikata (personal communication) indicated earlier that the virus particles in the insect were about 30–35mμ in diameter, but they had not been found in plant tissues.

Varietal Resistance

Methods of test. Varietal resistance to tungro has been tested in the field, in small mylar cages in the glasshouse and more recently at IRRI under a system of continuous mass screening which permits the handling of large numbers of varieties (IRRI, 1964, 1965, 1966; Ling, 1969).

The mass screening system is essentially as follows: (1) About 3000 adult insects are produced every 2 or 3 days from eggs through a series of large cages. (2) These insects become highly viruliferous after feeding for 2–4 days on diseased plants before inoculation. They are used for inoculation for about 8–9 hours each day and are then fed again on diseased plants for 15–16 hours to reacquire the virus; they are used again for inoculation the next day. This is repeated as long as they survive. (3) Plants are inoculated at the 2-3 leaf stage (11–13 days) with about 2–3 insects per seedling under a large cage containing 16 pots with a total of 464 seedlings. (4) Seeds are soaked in water and transplanted into pots as soon as they germinate. For each variety, 2 pots each of 29 seedlings are tested in duplicate inoculation sets. (5) The reaction of varieties is checked 12 days after inoculation. By this system, 16 varieties can be tested in duplicate each day.

Classification of resistance. The reaction of varieties to seedling infection varies from about zero to 100%. The percentage of infection has, therefore, been used directly to indicate the degree of susceptibility. Varieties are further classified arbitrarily into 3 groups: (1) resistant, those with 30% or less infected seedlings; (2) intermediate, 30–60% infected; and (3) susceptible, more than 60% infected. Field infection has always been less, e.g. those varieties with 30% or less infection in seedling tests may show only a very small proportion of infected plants in the field.

It was noticed earlier at IRRI that some varieties appearing resistant in the field gave a susceptible reaction in the seedling test. It was also observed that some plants recovered from the disease during the later stages of growth. This tolerance of, or recovery from, the disease is considered as a second criterion of resistance to tungro. For ease of assessment, the degree of reduction in plant height due to the disease, as compared to the height of healthy plants, is used as an index. The reading is made about 6 weeks after inoculation. The four scale units used are:

Scale unit	*Description*
S_0	No reduction in height
S_1	25% reduction in height
S_2	50% reduction in height
S_3	75% or more reduction in height

Varietal reaction. Many thousands of varieties and hybrid lines have been tested for resistance to tungro (IRRI, 1966; Ling, 1969), and several resistant varieties have been identified.

Among them, Pankhari 203 is most resistant, very few seedlings having shown symptoms among many thousands tested. This variety was found to be highly resistant to the vector (IRRI, 1967; Pathak *et al.*, 1969), as are other varieties such as IR8. Resistance to virus and vector is controlled by independent genes, and varieties resistant to the virus may not be resistant to the vector and *vice versa* (IRRI, 1969).

Most interesting is the fact that many of the commercial varieties grown in Indonesia and the Philippines, e.g. Tjeremas, Bengawan, Peta, Intan and Sigadis, are resistant. They are sister selections from a cross between Latisail and Tjina made by van der Muelen (1950) in Indonesia and they are known to be resistant to mentek disease even though the real cause of mentek was not known at the time when they were selected. The same group of varieties is also resistant to penyakit merah in Malaysia (Ou *et al.*, 1965) and to yellow-orange leaf disease in Thailand. This may be considered as another indication that these diseases are similar or closely related.

Many thousands of rice varieties have been tested in the field for yellow-orange leaf disease resistance in Thailand, and a number of them have proved to be resistant (Wathanakul & Weerapat, 1969). Varieties Pankhari 203, Latisail, Ambemohar 102, Kataribhog and Kamod 25-3 were found resistant in India (All-India Co-ord. Rice Improvement Project, 1969).

Inheritance of resistance. Preliminary studies on several crosses between resistant and susceptible varieties have shown that the F_1 population is generally resistant, particularly when a highly resistant parent is involved. The F_2 has a population segregating approximately in a 9 : 7 ratio. These data indicate that resistance is dominant (IRRI, 1966). Recent observations (IRRI, 1968) have shown that among the progeny of crosses between resistant and susceptible varieties, when inoculated at the young seedling stage, many of the infected seedlings recover from the disease and are healthy at a later stage of growth. On the other hand, many healthy-looking seedlings became diseased as they grow older. This has led to the suspicion that resistance at the seedling stage and resistance in the adult plant are inherited separately. Further experiments are needed to confirm this suspicion.

LITERATURE CITED

AGATI, J. A. & DE PERALTA, F. 1939. Rice cadang-cadang in Albay Province. II. Fertilizer treatments. *Philipp. J. Agric.* **10**:271–283.

AGATI, J. A., SISON, P. L. & ABALOS, R. 1941. A progress report on the rice maladies recently observed in Central Luzon with special reference to the stunt or dwarf disease. *Philipp. J. Agric.* **12**:197–210.

BALDACCI, E., AMICI, A., BELLI, G. & CORBETTA, G. 1970. Research on symptomatology, epidemiology and etiology of rice ' giallume ' (yellow). *Riso* **19**:3–9. [Ital. Engl. summ.]

GALVEZ E., G. E. 1967. The purification of virus-like particles from rice tungro virus-infected plants. *Virology* **33**:357–359.

IRRI. (INTERNATIONAL RICE RESEARCH INSTITUTE). Annual Reports for 1963–1969.

ISHIHARA, T. 1964. Revision of the genus *Nephotettix* (*Hemiptera*: *Deltocephalidae*). *Trans. Shikoku ent. Soc.* **8**:39–44.

JOHN, V. T. 1965. On the antigenicity of virus causing ' tungro ' disease of rice. *Pl. Dis. Reptr* **49**:305–306.

LAMEY, H. A., SURIN, P., DISTHAPORN, S. & WATHANAKUL, L. 1967. The epiphytotic of yellow-orange leaf disease of rice in 1966 in Thailand. *Pl. Prot. Bull. F.A.O.* **15**:67–69.

LING, K. C. 1966. Nonpersistence of the tungro virus of rice in its leafhopper vector *Nephotettix impicticeps*. *Phytopathology* **56**:1252–1256.

LING, K. C. 1969. Testing rice varieties for resistance to tungro disease. In *The virus diseases of the rice plant*: *Proc. Symp. at IRRI, April, 1967*: 277–291. Baltimore, Maryland, Johns Hopkins Press.

LING, K. C. 1970. Ability of *Nephotettix apicalis* to transmit the rice tungro virus. *J. econ. Ent.* **63**:582–586.

LOCKARD, R. G. 1959. Mineral nutrition of the rice plant in Malaya, with special reference to penyakit merah. *Bull. Dep. Agric. Malaya* 108, 148pp.

OU, S. H. 1965. Rice diseases of obscure nature in Tropical Asia with special reference to ' Mentek ' disease in Indonesia. *Int. Rice Commn Newsl.* **14** (2):4–10.

OU, S. H. & GOH, K. G. 1966. Further experiment on ' penyakit merah ' disease of rice in Malaysia. *Int. Rice Commn Newsl.* **15** (2):31–32.

OU, S. H. & LING, K. C. 1966. Virus diseases of rice in the South Pacific. *Pl. Prot. Bull. F.A.O.* **14**:113–121.

OU, S. H., RIVERA, C. T., NAVARATNAM, S. J. & GOH, K. G. 1965. Virus nature of ' penyakit merah ' disease of rice in Malaysia. *Pl. Dis. Reptr* **49**:778–782.

PATHAK, M. D., CHENG, C. H. & FORTUNO, M. E. 1969. Resistance to *Nephotettix impicticeps* and *Nilaparvata lugens* in varieties of rice. *Nature, Lond.* **223**:502–504.

PERALTA, F. DE & AGATI, J. A. 1939. The rice cadang-cadang in Albay Province. I. Its probable cause. *Philipp. J. Agric.* **10**:153–171.

RAYCHAUDHURI, S. P., MISHRA, M. D. & GHOSH, A. 1967. Preliminary note on transmission of virus disease resembling tungro of rice in India and other virus-like symptoms. *Pl. Dis. Reptr* **51**:300–301.

REYES, G. M. 1957. Rice dwarf disease in the Philippines. *Pl. Prot. Bull. F.A.O.* **6**:17–19.

REYES, G. M., LEGASPI, B. M. & MORALES, M. T. 1959. Progress of studies on the dwarf or stunt (virus) disease of rice in the Philippines. *Philipp. J. Agric.* **24**:27–43.

RIVERA, C. T. & OU, S. H. 1965. Leafhopper transmission of ' tungro ' disease of rice. *Pl. Dis. Reptr* **49**:127–131.

RIVERA, C. T. & OU, S. H. 1967. Transmission studies of the two strains of rice tungro virus. *Pl. Dis. Reptr* **51**:877–881.

RIVERA, C. T., OU, S. H. & TANTERE, D. M. 1968. Tungro disease of rice in Indonesia. *Pl. Dis. Reptr* **52**:122–124.

SERRANO, F. B. 1957. Rice ' accep na pula ' or stunt disease—a serious menace to the Philippine rice industry. *Philipp. J. Sci.* **86**:203–230.

VAN DER MEULEN, J. G. J. 1951. Rice improvement by hybridization and results obtained. *Contr. gen. agric. Res. Stn Bogor* 116.

VAN DER VECHT, J. 1953. The problem of the ' mentek ' disease of rice in Java. *Contr. gen. agric. Res. Stn Bogor* 137, 88pp.

WATHANAKUL, L. 1964. A study on the host range of tungro and orange leaf viruses of rice. M.S. Thesis, Coll. Agric., Univ. Philippines, 35pp.

WATHANAKUL, L. 1965. Occurrence of a new virus disease of rice in Thailand. Fourth natn. Conf. Agric. and Biology. Kasetsart Univ., Bangkok, Thailand. (Mimeographed.)

WATHANAKUL, L. & WEERAPAT, P. 1969. Virus diseases of rice in Thailand. In *The virus diseases of the rice plant*: *Proc. Symp. at IRRI, April, 1967*: 79–85. Baltimore, Maryland, Johns Hopkins Press.

GRASSY STUNT

History and Distribution

The disease was first noticed in 1962 in the Philippines at the International Rice Research Institute and was also observed at the same time in commercial fields. The transmission of the disease by *Nilaparvata lugens* (Stal.) was demonstrated in 1964 (IRRI, 1964; Rivera *et al.*, 1966). It has also been reported from Ceylon (Abeygunawardena, 1969) and has been observed in Thailand (Wathanakul & Weerapat, 1969) and India (Raychaudhuri *et al.*, 1967). Rice rosette, reported from the Philippines by Bergonia *et al.* (1966), is apparently the same disease.

The disease was at first assumed to be caused by a virus because of the type of disease syndrome and its transmission by a planthopper. Recently, electron microscope observations made by Dr. Shikata of Hokkaido University (personal communication) have revealed the presence of mycoplasma-like bodies in diseased tissues. It is therefore now suspected that the disease may be caused by the mycoplasma-like organisms, as in the case of yellow dwarf disease of rice.

Symptoms

Diseased plants are characterized by severe stunting, excessive tillering and an abnormally erect growth habit. The leaves are short and narrow, yellowish-green and covered with numerous rusty spots often extending into irregular blotches (Plate I). The disease thus to some extent resembles yellow dwarf, but leaves of plants affected by grassy stunt retain more green colour than in the case of yellow dwarf, particularly when the plants are supplied with adequate nitrogenous fertilizer. In addition, the rusty spots are usually not present in the case of yellow dwarf. Diseased plants produce numerous small tillers which give a fan-like or grassy appearance. They live until the crop is mature, but produce very poor panicles, or none at all.

Vector and Transmission

Transmission studies (Rivera *et al.*, 1966; IRRI, 1968) show that about 20–40% of the individuals in a population of *N. lugens* are able to transmit the virus. The male and female, long- and short-winged, light and dark-brown forms have an equal ability to transmit the virus. The percentage of active transmitters may be built up by selective breeding within a population but declines gradually when the selection ceases. Insects may become infective by feeding for 30 minutes on diseased plants. The incubation period in the insects is usually about 10–11 days, but varies from 5 to 28 days. Infection feeding for 5–15 minutes was found to induce a small percentage of infection in healthy plants; increasing the time increased the percentage of infection, which reached a maximum at 24 hours. The incubation period in the plant is 10–20 days. Viruliferous insects retain the virus for life, although they mostly do not transmit the disease every day, but usually on about two out of three days in an intermittent manner.

Experiments have shown that the virus is not transmitted through the seed, nor is it transmitted transovarially. The average life span of viruliferous insects is shorter than that of nonviruliferous ones.

Transmission Cycle

The development of the disease in the field depends to a great extent on the source of virus or infected plants. Experiments (IRRI, 1968) have shown that the percentage of infective individuals in field populations is very low or none at all when they are collected from areas where no disease is present, but is high from infected areas, though in both cases the insects are potential transmitters. The brown planthopper multiplies quickly under favourable conditions, so disease incidence may also increase rapidly.

The macropterous or long-winged forms, which can move over longer distances, are more important than the brachypterous forms in the dissemination of the disease.

Varietal Resistance

For testing the resistance of a large number of varieties, a mass screening method similar to that used for tungro disease has been developed (IRRI, 1968). The method consists of the following: (1) the insects are reared in large cages so that a large number of insects is constantly available; (2) diseased plants are provided for virus acquisition feeding at the nymph stage and the insects are used for inoculation 10–11 days after acquisition; (3) seedlings are inoculated at the 2–3-leaf stage (11–13 days), in small pots, 19 seedlings in each pot; (4) 16 pots (a total of 304 seedlings) are placed in each cage and viruliferous insects are introduced at the rate of 4–6 insects per seedling for 24 hours. They are gently disturbed to ensure even distribution on all the seedlings; (5) after inoculation, the seedlings are taken out of the cages and checked for symptom development, which takes 2 weeks. The remaining insects are used again for the next inoculation, after the number has been adjusted by adding new viruliferous insects. Because of the intermittent manner in which the vector transmits the virus, the results are often not as consistent as in the case of tungro.

Many thousand varieties and hybrid lines and species of wild rice have been tested but none has been found to be highly resistant, except a strain of wild rice, *Oryza nivara*, recently found to be highly resistant at IRRI. Some of the less susceptible varieties are Pagey-tenyagen-dayket-likwet, Mitao, Gendjah Banten, and Gowai 84 (IRRI, 1968). However, some varieties, such as Mudgo, were found to be resistant to the vector (IRRI, 1968; Pathak *et al.*, 1969) and they may prove useful in developing resistant varieties.

Chemical Control

Since the disease is suspected to be due to a mycoplasma-like organism, preliminary pot experiments on treating diseased plants with antibiotics have been made. None of the four tetracyclines so far tried, however, was effective in controlling or minimizing the symptoms (IRRI, 1968).

LITERATURE CITED

ABEYGUNAWARDENA, D. V. W. 1969. The present status of virus diseases of rice in Ceylon. In *The virus diseases of the rice plant*: *Proc. Symp. at IRRI, April, 1967*: 53–57. Baltimore, Maryland, Johns Hopkins Press.

BERGONIA, H. T., CAPULE, N. M., NOVERO, E. P. & CALICA, C. 1966. Rice rosette, a new disease in the Philippines. *Philipp. Jnl Pl. Ind.* **31**:45–52.

IRRI (INTERNATIONAL RICE RESEARCH INSTITUTE). Annual Report for 1964, 1968.

PATHAK, M. C., CHENG, C. H. & FORTUNO, M. E. 1969. Resistance to *Nephotettix impicticeps* and *Nilaparvata lugens* in varieties of rice. *Nature, Lond.* **223**:502–504.

RAYCHAUDHURI, S. P., MISHRA, M. D. & GHOSH, A. 1967. Preliminary note on transmission of virus disease resembling tungro of rice in India and other virus-like symptoms. *Pl. Dis. Reptr* **51**:300–301.

RIVERA, C. T., OU, S. H. & IIDA, T. T. 1966. Grassy stunt disease of rice and its transmission by the planthopper *Nilaparvata lugens* (Stal.). *Pl. Dis. Reptr* **50**:453–456.

WATHANAKUL, L. & WEERAPAT, P. 1969. Virus diseases of rice in Thailand. In *The virus diseases of the rice plant*: *Proc. Symp. at IRRI, April, 1967*: 79–85. Baltimore, Maryland, Johns Hopkins Press.

ORANGE LEAF

Orange leaf disease was first observed in Northern Thailand in 1960 (Ou, 1963). It was again found in the Philippines, on both upland and lowland rice (Rivera *et al.*, 1963) and its presence in Ceylon was recently confirmed by Abeygunawardena (1969). Plants with similar symptoms have been observed in Malaysia and other countries of Southeast Asia, but that a virus is implicated has not been confirmed by transmission experiments.

The distinctive symptoms of the disease are: (1) orange discoloration occurs on the leaves of affected plants, beginning with the lower leaves and starting from the leaf tip; (2) longitudinal rolling of these affected leaves occurs as the disease advances; (3) the number of tillers is reduced but there is no conspicuous stunting; (4) there is rapid death of seedlings, especially when infected young; even plants infected when 60 days old eventually die. High temperature (30°C) favours disease development and plants die sooner at high than at low temperature (20°C) (Plate III).

Preliminary transmission studies (Rivera *et al.*, 1963) showed that the leafhopper *Recilia* (*Inazuma*) *dorsalis* (Motsch.) is the vector. Six other species of insects were tested but were found not to be vectors. Seed from diseased plants failed to transmit the disease, and transmission through the soil or by mechanical inoculation failed. About 14% of the natural insect population of *R. dorsalis* was found to be able to transmit the virus. A minimum feeding period of 5 hours appears to be required for the insect to acquire the virus and 6 hours for inoculation feeding. Once acquired, the virus is retained throughout the life of the insect. The latent period in the insect is about 2–6 days and the incubation period in the plant is about 13–15 days.

Preliminary glasshouse trials with a limited number of varieties indicated that Peta, FB-121 and others are resistant, while susceptible varieties include Tjeremas and BPI-76 (Rivera & Ou, 1964).

LITERATURE CITED

ABEYGUNAWARDENA, D. V. W. 1969. The present status of virus diseases of rice in Ceylon. In *The virus diseases of the rice plant*: *Proc. Symp. at IRRI, April, 1967*: 53–57. Baltimore, Maryland, Johns Hopkins Press.

Ou, S. H. 1963. Report to the Government of Thailand on blast and other diseases of rice. *F.A.O. expand. tech. Assist. Rep.* 1673, 28pp.
Rivera, C. T., Ou, S. H. & Pathak, M. D. 1963. Transmission studies on the orange leaf disease of rice. *Pl. Dis. Reptr* **47**:1045–1048.
Rivera, C. T. & Ou, S. H. 1964. Studies on the virus diseases of rice in the Philippines, 15pp. (Mimeographed.)

YELLOW MOTTLE

This disease has recently been reported from fields around Lake Victoria, Kenya (Bakker, 1970). It presents some unusual features, e.g. it is mechanically transmissible, and the virus is present in guttation fluid and in the irrigation water of heavily infected fields.

Symptoms

The disease causes stunting and reduced tillering, mottling and yellowish streaking of the leaves, malformation and incomplete emergence of the panicles, and sterility. Infection, even at a late stage of growth, results in considerable reduction in yield. In severe cases affected plants may die.

In the field diseased plants may be observed 3–4 weeks after transplanting and are easily noticeable because of their yellowish appearance. The youngest leaves show mottling or a mild yellow-green streaking. In mechanically inoculated plants, the first symptoms are a few yellow-green spots on the youngest leaves. These spots enlarge parallel to the veins and give rise to the characteristic symptoms.

Transmission

The virus is transmissible to healthy seedlings by mechanical inoculation of leaf sap from diseased plants, with the aid of carborundum. Symptoms appear about 7 days after inoculation.

The vector is *Sesselia pusilla* Gerstaeker (Chrysomelidae, Galerucinae). In limited tests this beetle was found to be able to transmit the virus for 1–5 consecutive days in daily transfers, after feeding for 3–4 days on diseased plants.

The virus was recovered from sap obtained from the roots of diseased plants, from guttation fluid and from irrigation water in heavily infected fields.

Seed and soil transmission tests gave negative results.

Host Range

Several varieties of *Oryza sativa* L. were infected by the virus by inoculation. *O. barthii* A. Chev. and *O. punctata* Kotscky ex Bteud. were also infected but symptoms appeared 14 days after inoculation, not 7 days as in *O. sativa.*

In inoculation tests the virus did not infect *O. eichingeri* Peter (forest rice), *Avena sativa* L., *Eleusine coracana* (L.) Azch. & Br., *Hordeum vulgare* L., *Pennisetum typhoides* (Burm.) Stapf & Hubbard, *Saccharum officinarum* L., *Triticum sativum* L., *T. durum* Deaf., *Zea mays* L., 20 other grasses and 9 dicotyledonous plant species.

PROPERTIES OF THE VIRUS

Sap from young rice leaves, 2–3 weeks after inoculation, was still infective at a dilution of 10^{-10}. With sap from plants inoculated 4–5 weeks earlier, a dilution of 10^{-6} was the most infective. The thermal inactivation point using sap from the leaf blade was above 80°C. The virus remained infective *in vitro* for 33 days at room temperature (16–25°C), but not for 51 days. Sap stored at 9°C was still infective after 71 days. The virus was not adversely affected by treatment with organic solvents (chloroform, chloroform + butanol, carbon tetrachloride + ether). It may be purified by repeated centrifugation. The purified preparation contained many spherical particles *c.* 32mμ in diameter.

An antiserum with a dilution of up to 1/256 gave a reaction against crude sap and purified virus as determined by the agar gel diffusion test.

LITERATURE CITED

BAKKER, W. 1970. Rice yellow mottle, a mechanically transmissible virus disease of rice in Kenya. *Neth. J. Pl. Path.* **76**:53–63.

MOSAIC

Mosaic was reported on rice in the Philippines in 1960 (Martinez *et al.*, 1960) Infected plants in the field can be identified by the presence of foliar mottling which is irregular in shape and varies in size from greenish dots to elongated yellowish green lesions which may coalesce to form chlorotic streaks (Fig. I–9).

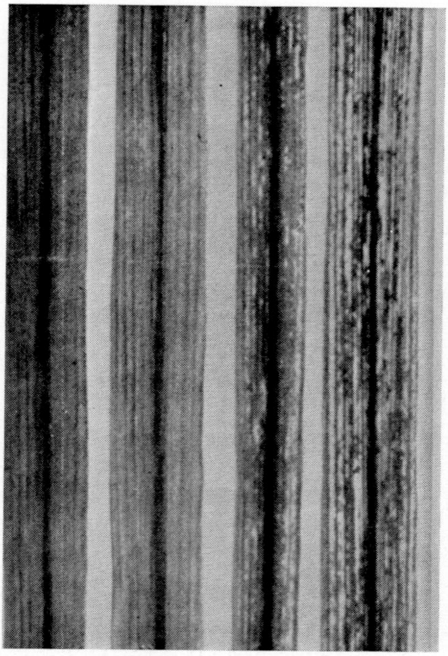

Fig. I–9. Mosaic disease of rice showing two healthy leaves on the left, two infected leaves on the right. (Martinez *et al.*, 1960.)

Mottling is also observed on the leaf sheath. Severely infected plants are stunted and, in the later stages of growth, the leaves turn yellowish brown and eventually wither. The number of tillers is also reduced. The disease has not been observed recently and is of little importance.

Mosaic was found to be transmissible from rice to maize by mechanical means such as carborundum rubbing and pin-pricking with expressed sap. Transmission from rice to rice has not been reported.

It was suspected that the disease is caused by one of the mosaic viruses affecting grasses (Martinez *et al.*, 1960). Rice is also a host of sugarcane mosaic, barley stripe mosaic, barley yellow dwarf, brome mosaic and ryegrass mosaic viruses (Kahn & Dickerson, 1957; Anzalone, 1963; Anzalone & Lamey, 1968; Slykhuis, 1962). Little else is known about the disease.

LITERATURE CITED

ANZALONE, L. 1963. Susceptibility of rice to a strain of the sugarcane mosaic virus. *Pl. Dis. Reptr* **47**:583–584.

ANZALONE, L. & LAMEY, H. A. 1968. Possible differential reaction of certain rice varieties to sugarcane mosaic virus. *Pl. Dis. Reptr* **52**:775–777.

KAHN, R. P. & DICKERSON, O. J. 1957. Susceptibility of rice to systemic infection by three common cereal viruses. Abs. in *Phytopathology* **47**:526.

MARTINEZ, A. L., BERGONIA, H. T., ESCOBER, J. T. & CASTILLO, B. S. 1960. Mosaic of rice in the Philippines. *Pl. Prot. Bull. F.A.O.* **8**:77–78.

SLYKHUIS, J. T. 1962. An international survey for virus diseases of grasses. *Pl. Prot. Bull. F.A.O.* **10**:1–16.

NECROSIS MOSAIC

Necrosis mosaic of rice is a soil-borne virus disease recently described by Fujii *et al.* (1967). It had been observed earlier (Fujii *et al.*, 1966a, b), when it was called rice dwarf disease.

Infected plants are slightly stunted, have a reduced number of tillers, and instead of being normally erect, they are spreading and assume a decumbent growth habit. Mosaic symptoms, characterized by elongated, yellowish spots, which are wavy at the margin, develop first on the lower leaves and then on the upper ones. Necrotic lesions appear in the xylem and parenchyma tissues of the stem, first at the base and later in the upper nodes and leaf sheaths.

Inouye (1968) found rod-shaped particles in dip preparation from roots, sheaths, and leaves of naturally infected seedlings grown in infected soil in the glasshouse. Two peaks of particle-length distribution, 275 and 550mμ, were observed. The diameter of the particles was *c.* 13–14mμ.

LITERATURE CITED

FUJII, S., OKAMOTO, Y., YAMAMOTO, H. & INOUYE, T. 1966a. Rice dwarf disease (Preliminary report). *Ann. phytopath. Soc. Japan* **32**:82. [Abs., Jap.]

FUJII, S., OKAMOTO, Y., IDE, S. & INOUYE, T. 1966b. On the mosaic symptoms observed on rice dwarf disease. *Ann. phytopath. Soc. Japan* **32**:325. [Abs., Jap.]

FUJII, S., OKAMOTO, Y., IDE, S., SHIOMI, M., INOUYE, T., INOUYE, S., ASAYA, M. & MITSUHATA, K. 1967. Necrosis mosaic, a new virus disease of rice. *Ann. phytopath. Soc. Japan* **33**:105. [Abs., Jap.]

INOUYE, T. 1968. Rod-shaped particles associated with necrosis mosaic of rice. *Ann. phytopath. Soc. Japan* **34**:301–304. [Jap. Engl. summ.]

PART II

BACTERIAL DISEASES

BACTERIAL LEAF BLIGHT

HISTORY AND DISTRIBUTION

Bacterial leaf blight is said to have been first seen by farmers in the Fukuoka area of Japan in 1884. During 1908–10 it was commonly observed in the southwest area of Japan and since 1926 it has been recorded in northeastern Japan. The disease increased markedly after 1950, and by 1960 it was known to occur in all parts of Japan except the northern island of Hokkaido (Tagami & Mizukami, 1962).

The study of the disease in Japan commenced in 1901. It was then believed to be physiological in origin, and to be due to acid soil, because dew drops on the infected leaves were acidic in reaction. In 1908 Takaishi found bacterial masses in the dew drops, isolated the organism and successfully inoculated it on to rice leaves; but he did not name the organism. Bokura in 1911 also isolated a bacterium, assumed to be identical with that of Takaishi. He carried out inoculation experiments and concluded that the disease was due to the bacterium and not to the acid soil. After a study of its morphology and physiology, the bacterium was named *Bacillus oryzae* Hori & Bokura. Ishiyama (1922) studied the disease further and renamed the bacterium *Pseudomonas oryzae* Uyeda & Ishiyama according to Migula's system. It was later (1927) transferred to *Bacterium oryzae* (Uyeda & Ishiyama) Nakata according to E. F. Smith's concept, and subsequently to *Xanthomonas oryzae* (Uyeda & Ishiyama) Dowson according to the present generally accepted classification of bacteria.

Since the 1940s, many aspects of the disease have been studied in Japan, including epidemiology, varietal resistance, chemical control etc. Muko & Yoshida (1951) introduced the pricking inoculation method which allowed more accurate measurement of disease development to be made. The method has since been used extensively. Yoshii *et al.* (1953) first isolated bacteriophage, which has been employed for estimating the bacterial population, classifying strains of the bacterium based on phage reaction, and ecological and disease forecasting studies.

Several references (Padwick, 1950; Dickson, 1956; Pordesimo, 1958; Tagami & Mizukami, 1962) have been made to the report of Reinking (1918) that the disease occurred in the Philippines. Actually Reinking reported a 'bacterial leaf stripe' disease and described the symptoms as follows: 'In the young stages the stripes are from 0·5 to 1 millimeter wide and from 3 to 5 millimeters long, run lengthwise, and have a watery, dark green translucent appearance. In this stage, the disease is usually confined to the portion between the larger veins'. Obviously, his was a different disease which is now known as bacterial leaf streak. Since leaf streak was distinguished from leaf blight only after 1957 (Fang *et al.*, 1957), earlier workers have often confused the two diseases.

Reitsma & Schure (1950) studied a disease called ' kresek ' in Indonesia. A bacterium isolated was successfully inoculated to rice seedlings by spraying, needle pricking or submerging the seedlings in a bacterial suspension. The causal organism was called *Xanthomonas kresek* Schure (1953). This disease has now been shown to be a severe form of bacterial leaf blight which is found in various parts of the tropics (Goto, M., 1964).

Bacterial leaf blight was first reported from India by Sreenivasan *et al.* (1959) and by Bhapkar *et al.* (1960) who observed that the disease had been known since 1951 in the Bombay area. Recent studies have shown the disease to be present in most of the rice-growing states of India (Mizukami, 1964; Srivastava & Rao, 1964a; Srivastava, 1967). Since the introduction and cultivation over a large acreage of new high yielding but susceptible rice varieties in recent years, the disease has become one of the most serious problems on rice in India. It has also been reported from Ceylon, East, Central, South and West China, Taiwan, Korea, Thailand, Vietnam and the Philippines, and is now considered to be one of the most destructive diseases of rice in Asia. It has not been found in North or South America nor in Europe, with the possible exception of the U.S.S.R. (Vzoroff, 1938). It has been reported from Malagasy (verbal communication) but is not known to occur on the mainland of Africa.

Symptoms

In the temperate regions the disease usually becomes noticeable in the field at the heading stage. In severe cases it may appear earlier but it is rare in the seedbed.

On seedlings in the seedbed it first appears as tiny water-soaked spots at the margin of fully developed lower leaves. As the spots enlarge, the leaves turn yellow, dry rapidly and wither.

On leaf blades, lesions usually begin at the margin, a few cm from the tip, as water-soaked stripes. The lesions enlarge both in length and width, have a wavy margin, and turn yellow within a few days. The region adjoining the healthy part shows water-soaking. Lesions may start at one or both edges of the leaves. As the disease advances, the lesions cover the entire blade, turn white and later become greyish from the growth of various saprophytic fungi (Plate IV). On susceptible varieties, the lesions extend to the leaf sheath, where they may reach the lower end. Although often beginning from the leaf margins, lesions may also start at any point on the blade if it is injured. In this case also infection begins with stripes, which later may cover most or all of the leaf blade.

On more resistant varieties or under certain conditions, a yellow stripe appears just inside the margin of the leaf blade, with no formation of necrotic lesions for some time. The stripes may eventually turn yellow and necrotic. On susceptible varieties, infected blades wilt and roll as the diseased portion enlarges while the leaves are still green. The entire blade may soon be involved and may then dry up.

On the surface of young lesions, milky or opaque dew drops may be observed in the early morning. They dry up to form small, yellowish, spherical beads which are easily shaken off by wind and drop into the field water. These bacterial masses are seldom found on old lesions which have become white.

In severely diseased fields, grains may also be infected, the disease appearing on the glumes as discoloured spots surrounded by a watersoaked margin. The spots are conspicuous while the grain is young and green. At maturity, they are grey or yellowish white.

In the tropics, two additional types of symptoms are found: ' kresek ' or the withering of leaves and entire young plants; and the production of pale yellow leaves at a later stage of growth (IRRI, 1964; Goto, 1964).

The ' kresek ' symptoms were first described as a separate disease by Reitsma & Schure (1950) in Indonesia (Plate IV). They may be observed 1 or 2 weeks after transplanting, when infected leaves become greyish green and begin to fold up and roll along the midrib. In the transplanting of rice seedlings in tropical countries, the tips of the leaves are often cut off. Usually these cut leaves were found to be attacked first. The earliest evidence of the presence of disease is a green water-soaked spot, just beneath the cut surface, which soon turns greyish green. Rolling and withering of the entire leaf, including the leaf sheath, follow. The bacterium spreads through the xylem vessels to the growing point of the young plant and infects the base of other leaves. As a result, the entire young plant dies. In the earlier stages, when only a few older leaves had withered and could be seen floating on the water, the disease in Java was known as ' kresek '. The final stage, the complete death of the entire plant, is known as ' hama lodoh ' (Reitsma & Schure, 1950). For simplicity, all these symptoms are here referred to as ' kresek ' symptoms.

The kresek symptoms have been found to be common in many tropical countries; they sometimes look like the damage caused by rice stem borers. When infected young plants survive, the growth of tillers is arrested, and they are stunted and yellowish green. The entire field shows uneven growth and whole hills may be missing.

Reitsma and Schure pointed out the importance of leaf cutting in providing entry points for the bacterium. In addition to this, the broken roots which result from the seedlings being pulled from the seedbed also serve as points of entry (IRRI, 1966).

Another type of symptom in the tropics is the production of pale yellow leaves (Plate IV). In the field, such leaves are found in mature plants. While the older leaves are normal and green, the youngest leaf is uniformly pale yellow or has a yellow or greenish yellow broad stripe on the blade. The mechanism of the appearance of such pale yellow leaves has not been studied in detail. No bacteria can be detected in the yellow leaves, but they are copious in the crown of the stem and in the internodes immediately below infected leaves. Apparently the gradual build up of the bacterial population in the lower portion of the stem reaches a point at which little nutrient is available to the young leaves, which therefore become pale yellow. When 3-week-old seedlings are artificially inoculated, pale yellow leaf symptoms are produced 20–30 days after inoculation.

The symptoms on the leaves are sometimes difficult to distinguish from those

PLATE IV

Upper left—*Bacterial Blight* showing infected leaves.

Upper right—*Bacterial Blight* showing pale yellow leaves on the right. Plant at left is healthy.

Lower left—*Bacterial Blight* showing kresek symptoms in field.

Lower right—*Bacterial Blight* showing kresek symptoms on susceptible plants at right (variety JC-70). Plants in pot at left are resistant (variety Zenith).

PLATE IV

of various other leaf diseases, both physiological and parasitic, and the kresek symptoms are not easily separated from rice stem borer damage. A few simple methods may, however, help to identify the disease. One is to cut out a small piece of a leaf lesion, a few mm long, together with adjacent healthy tissue, and examine it under the microscope in a drop of water. Bacterial ooze may be observed emerging from the leaf veins in a few minutes. It should be noted that if the diseased tissue has been dead for a long time, there will be no ooze. One may also cut a piece of leaf 3–4 cm long near one end of a lesion and keep the cut piece between two slides in a moist Petri dish. In a few hours, bacterial droplets may be observed at the cut ends. In the field, one may collect a few diseased leaves, cut them near the lower end of the lesions and put the cut ends into water in a test tube or glass. Bacterial streams may ooze out from the cut end into the water in a few minutes. They may be more easily seen if the container is held up to the light. After 1–2 hours the water becomes turbid. To distinguish kresek symptoms from stem borer damage, one may cut the lower end of a young plant and squeeze it between the fingers. Yellowish bacterial ooze may be seen at the ends if kresek is present.

Damage

In Japan, 300,000 to 400,000 hectares have been affected annually by the disease in recent years. Yield losses in severely infected fields range from 20 to 30%, and may on rare occasions be up to 50%. In the tropics, the disease has been very destructive in the Philippines and Indonesia; losses are higher than in Japan, but few figures are available. Bacterial leaf blight is one of the major diseases of rice in India, where perhaps millions of hectares are severely infected. Losses in yield vary from 6 to 60% in some states (Srivastava, 1967).

Where, as in Japan, the disease appears on older plants, usually at or after maximum tillering or at the rooting stage, it does not affect the number of ears or spikelets. It does, however, cause poor development and lowered quality of grain, and increases the number of under-developed grains, reduces weight and results in poor maturing and a high proportion of broken rice. The earlier the appearance of the disease, the worse are its effects. It also affects the chemical composition of the grains; soluble nitrogenous substances decrease and crude protein increases because the grains do not completely mature (Tagami & Mizukami, 1962).

Many scales and formulae have been developed in Japan for measuring the severity of disease and for assessing losses. The number of diseased leaves alone was found not to be a good index since it is not proportional to the actual losses. In addition, therefore, leaf area affected by the disease is estimated. The following (Inoue & Tsuda, 1959) is one of the formulae used:

$$Y (\%) = 1I + 3II + 4III + 5IV + 7V$$

Y = % yield reduction;

1, 3, 4, 5, 7 = indices of damage;

I =< 20% damaged leaf area, II = 30–40%, III = 50%, IV = 60%, V => 70%

A chart (Fig. II–1) has also been used (Kyushu Agricultural Experiment Station, after Tagami & Mizukami, 1962).

The damage is usually more severe in the tropics since the kresek type of attack kills young plants completely or nearly so, and the lesions on leaves are often large and progress rapidly. Results of an artificial inoculation experiment at IRRI (1967) using two varieties, IR8, susceptible, and Tainan 8, moderately resistant, showed about 75% and 47% yield losses, respectively.

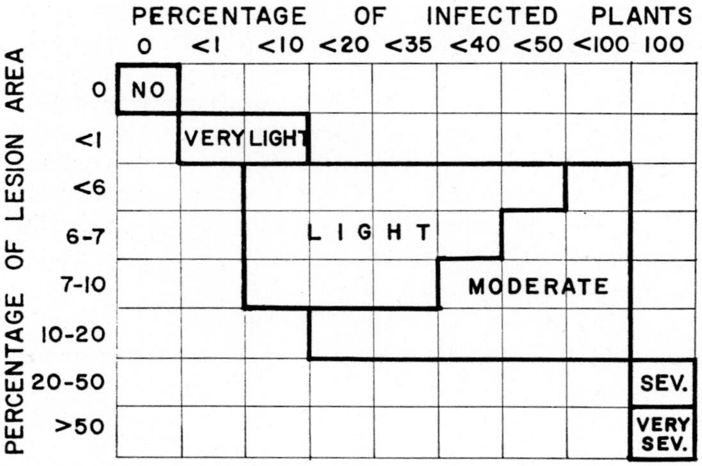

Fig. II–1. Chart used for estimating damage caused by bacterial blight. (After Tagami & Mizukami, 1962.)

Causal Organism

Morphology. Short rods with round ends, 1–2 × 0·8–1μ, monotrichous flagellum of 6–8μ; Gram-negative and non-spore-forming (Ishiyama, 1922). Bacterial cells are surrounded by mucous capsules and joined to form an aggregated mass which is relatively stable even in water. Colonies yellowish on artificial media; the yellow pigments are insoluble in water.

In recent observations under the electron microscope, the size of the bacterial cells was determined as 0·55–0·75 × 1·35–2·17μ from culture media, and 0·45–0·60 × 0·65–1·40μ from host tissue. The flagellum is 8·75μ × 30mμ (Yoshimura & Tahara, 1960) (Fig. II–2).

The mucous capsule is soluble in water and precipitated by acetone. It seems to protect the cells under dry or unfavourable conditions. In chemical composition the capsule is assumed to be a heteropolysaccharide (Mizukami, 1961).

Kuo *et al.* (1970a) treated the bacterium with glycine, tysozyme and penicillin and observed the formation of spheroplasts without cell walls, resembling the L-form of bacteria. These spheroplasts lost pathogenicity and changed the phage adsorption properties.

Physiological characters. Aerobic, does not liquefy gelatin, does not use nitrates, does not produce ammonia, produces H_2S slightly. It produces no indol; it ferments but does not coagulate milk; litmus milk is turned red; it does not produce gas or acid from sugar.

Recent experiments on physiology have shown ammonia and H_2S production, gelatin liquefaction, acid production from sugar and slight starch fermentation. Gelatin liquefaction differs among strains, being greater in virulent strains. There are also other variations among strains (Tagami & Mizukami, 1962; Muko & Isaka, 1964; Shekhawat & Srivastava, 1968).

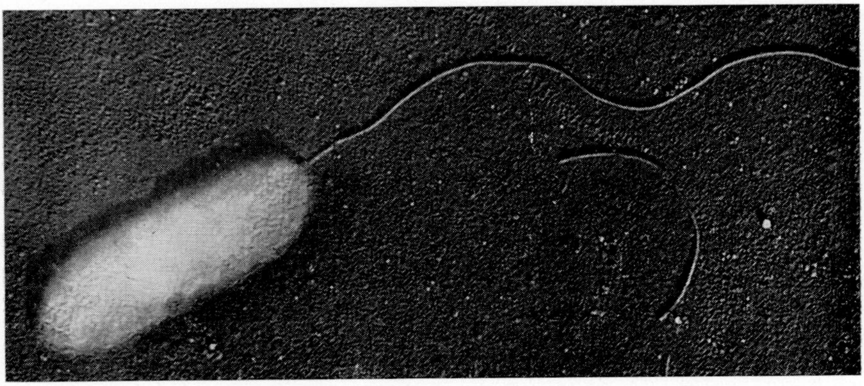

Fig. II–2. *Xanthomonas oryzae* × 24,500. (Courtesy of Dr. S. Wakimoto.)

Physiology. Sucrose is the most favourable carbon source, but other carbon sources, except fructose, are also utilized to some extent (Fang *et al.*, 1957; Watanabe, 1963, 1966). Fang *et al.* (1957) reported inhibition by glucose in potato broth medium. This is probably due to toxin produced when glucose in the medium is autoclaved. With respect to nitrogen, inorganic sources are not utilized except ammonium sulphate to a slight extent. Most favourable sources of nitrogen are glutamic acid and cystine. Asparagine and glutamine are used slightly (Muko & Watanabe, 1958; Watanabe, 1963, 1966). Tanaka's study (1963, 1964a) found glucose and sucrose to be the best carbon sources, and glutamic acid, aspartic acid, methionine, cystine and asparagine to be good nitrogen sources.

Using 6 tropical isolates, Hsu (1966) reported that glucose, galactose and sucrose are the most favourable carbon sources. Mannose and maltose also support good growth in a synthetic medium, but no growth was observed with fructose, dextrin and citric acid. L-glutamic acid and L-aspartic acid are the best sources of nitrogen. Cystine supports some growth at low concentration (5 to 50 ppm).

Culture media generally used are: (1) Wakimoto's potato semi-synthetic medium: potato, 300g; $Ca(NO_3)_2 \cdot 4H_2O$,0·5g; $Na_2HPO_4 \cdot 12H_2O$,2g; peptone, 5g; sucrose, 20g; agar, 15g; water, 1 litre; pH 6·8–7. (2) Several synthetic liquid media have been developed and are used (Table II–1). (3) Nutrient agar, consisting of: peptone, 5g; beef extract, 3g; NaCl, 5g; agar, 15g; water, 1 litre. Small quantities of inorganic salts such as $MgSO_4$, a mixture of $FeSO_4$ and

TABLE II–1

SYNTHETIC MEDIA USED FOR CULTURING *Xanthomonas oryzae*.

Natl. Inst. Agric. Sciences, Japan		Suwa (1962)		Tanaka (1964a)		Watanabe (1963, 1966)	
Na-glutamate	2 g	Na-glutamate	2 g	Sucrose	1.5 g	L-glutamic acid	1 g
KH_2PO_4	2 g	$MgCl_2 \cdot 6H_2O$	1 g	Na-glutamate	1 g	L-cystine	0.5 g
$FeSO_4 \cdot 7H_2O$	0.1 g	KH_2PO_4	0.1 g	KH_2PO_4	0.2 g	$(NH_4)_2HPO_4$	3 g
$MgCl_2 \cdot 6H_2O$	0.05 g	Fe (as EDTA)	1 mg	$MgSO_4 \cdot 7H_2O$	0.02 g	Sucrose	5 g
Sucrose	20 g	Sucrose	5 g	H_3BO_3	0.5 μg	$FeSO_4 \cdot 7H_2O$	0.01 g
Water	1000 ml	Water	1000 ml	$CaCO_3$	10 μg	$MgCl_2 \cdot 6H_2O$	1 g
pH	6.8–7	pH	7	$CuSO_4 \cdot 5H_2O$	1 μg	$MnSO_4 \cdot 4H_2O$	0.01 g
				$FeSO_4(NH_4)_2SO_4 \cdot 6H_2O$	10–50 μg	pH	6.5–7
				KI	0.1 μg		
				$MnSO_4 \cdot 4H_2O$	1–2 μg		
				MoO_3	1 μg		
				$ZnSO_4 \cdot 7H_2O$	5 μg		
				Water	1000 ml		
				pH	6.8		

$MgCl_2$ (0·01–0·05%), and 0·2% K_2HPO_4 promote growth in these media. Growth on nutrient agar is usually slow.

Silva & Buddenhagen (personal communication) found that the following medium promoted better growth: yeast extract, 5g; peptone, 5g; dextrose, 5g; glutamine, 1g; agar, 20g; water, 1 litre. We found that Wakimoto's medium without potato but with the addition of 0·5g $FeSO_4 \cdot 7H_2O$ gave improved growth. A more favourable medium is still to be formulated.

The bacterium does not require vitamins as indispensable growth factors but small amounts of riboflavin, thiamin, calcium panthothenate, nicotine or pyridoxin give some stimulating effect (Watanabe, 1966).

The pH range for the growth of the organism is 4–8·8. The optimum pH values reported are 6–6·5 (by Fang *et al.*, 1957), and 6·2–6·4, 6·8 or 7 in Japan. The interpretation of the term ' optimum pH ' and differences in the strains and media used by various workers may account for the discrepancies.

The minimum temperature for growth is 5–10°C, optimum 26–30°C and maximum 40°C. The best temperature for initial growth of a very dilute suspension is 20°C (Mizukami, 1961).

The bacterium cannot be preserved for a long time in sterile distilled water as can other xanthomonads. It may, however, be kept in phosphate buffer at pH 7·0 and peptone water for a considerable length of time. The best method of maintaining the organism is in a clay suspension. The finer the clay particles (7000 rpm, 20 minutes) the better. A high percentage of viable bacteria may be recovered even after 400 days (Goto, 1969).

Watanabe (1966) isolated polysaccharides from eight strains of the bacterium and found that they were toxic but non-specific, and not correlated with virulence. After hydrolysis, the main constituent was mannose, plus a little galactose, maltose etc. Kuo *et al.* (1970b) also purified the phytotoxic polysaccharides, found them non-specific, and was able to separate them into four fractions with different molecular weights.

Single cell culture. Single cells of the organism are difficult to grow on culture media. Suwa (1960) found that when single cells and the culture medium were pretreated with a 40 ppm solution of $MgCl_2$, 50–80% of the single cells grew successfully. The culture should be incubated at 20°C for initial growth. Suwa (1962) developed a synthetic medium containing EDTA-Fe (ethylenediamine-tetraacetic acid tetrasodium salt-Fe) for culturing single cells.

We tried several media including Suwa's medium to grow single cells. Wakimoto's medium without potato and with the addition of $FeSO_4 \cdot 7H_2O$ mentioned above gave the largest number of colonies on the plate. Suspending the bacterial cells in 1% peptone or fine clay particles also gave larger number of colonies than with water.

The lack of suitable media for single cell growth and the lack of a selective medium have hampered some of the studies of the disease. The bacterial cell population is believed to have been underestimated by most workers because many single cells did not grow. The lack of selective medium renders the direct detection of the bacterium in soils, field water or plant parts very difficult or impossible.

Streptomycin-resistant strains. The strains have been obtained by adding streptomycin to culture media (Tabei *et al.*, 1957; Yamamoto & Kusaka, 1965; Eamchit *et al.*, 1969; IRRI, 1969). Great differences were found in susceptibility to streptomycin but little to other antibiotics. These strains are useful in tracing the bacterium through inoculated plants and in other kinds of experiments when the special media can be used without the need for sterilization. Strains resistant to streptomycin have also been found in nature (Wakimoto & Muko, 1963).

Phage. Bacteriophage was first isolated by Yoshii *et al.* (1953). Wakimoto (1954a) made detailed studies on the biological and physical characteristics of the phage and named it *Xanthomonas oryzae* OP_1 bacteriophage. Based on these studies, methods for detecting the presence and quantitative determination of

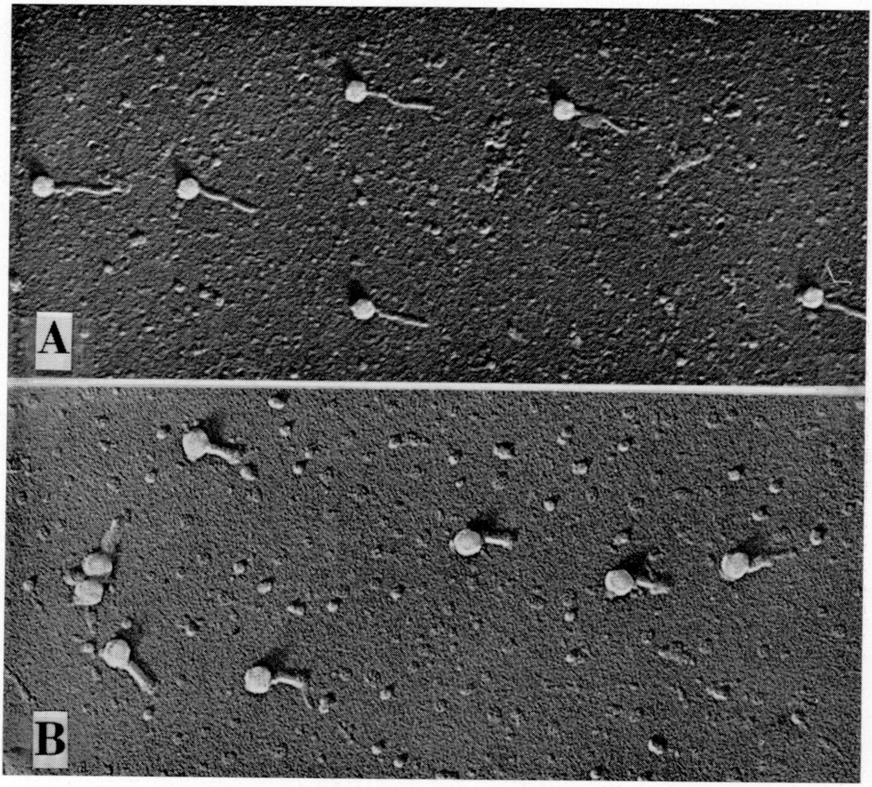

Fig. II–3. *Xanthomonas oryzae* phage. A, Op1 × 43,000. B, Op2 × 52,000. (Courtesy of Dr. S. Wakimoto.)

bacterial populations were developed (Wakimoto, 1954b; Wakimoto & Yoshii, 1955; Wakimoto, 1957). This contributed greatly to subsequent studies on the disease. Three other strains of the phage were identified in Japan, differing in morphological, physical or serological characteristics and host range.

The phages are usually tadpole-shaped, with a more or less polyhedral head and a tail (Fig. II–3). Kuo *et al.* (1967) reported a filamentous phage, Xf, measuring 858×6mμ.

Wakimoto (1960) classified the strains of *X. oryzae* in Japan according to their sensitivity to the phages, as shown in Table II–2.

TABLE II–2

CLASSIFICATION OF BACTERIAL STRAINS OF *X. oryzae* IN JAPAN BASED UPON SENSITIVITY TO PHAGES. (Wakimoto, 1960).

Bacterial strain	Bacteriophage			
	OP_1	OP_1h	OP_1h_2	OP_2
A	+	—	+	+
B	—	+	+	+
C	—	—	—	—
D	—	—	+	+
E	—	—	—	+

The phages are generally considered specific and stable. However, Goto & Okabe (1967b, c) reported that phage Bp1 was changed to Bp2 by repeated passage through the bacterial isolate B20 which originally was not lysed by phage Bp1 but was lysed by phage Bp2. They called this host-controlled modification. They also found phage-resistant mutants in single cell cultures of both *X. oryzae* and *X. translucens* f. sp. *oryzicola* and this type of mutation may be the origin of the lysotypes of the two organisms found in nature. Phage-resistant strains were also noticed earlier by Tanaka (1964b).

In the tropics, Goto (IRRI, 1964; Goto, 1965a) identified 7 phages, called them Bp1 to Bp7, differentiated 15 lysotypes of the bacterial isolates, and studied their morphology, serology and other characteristics (Goto & Okabe, 1965). These phages are quite different from those of Japan (Fig. II–4A). Eight phages,

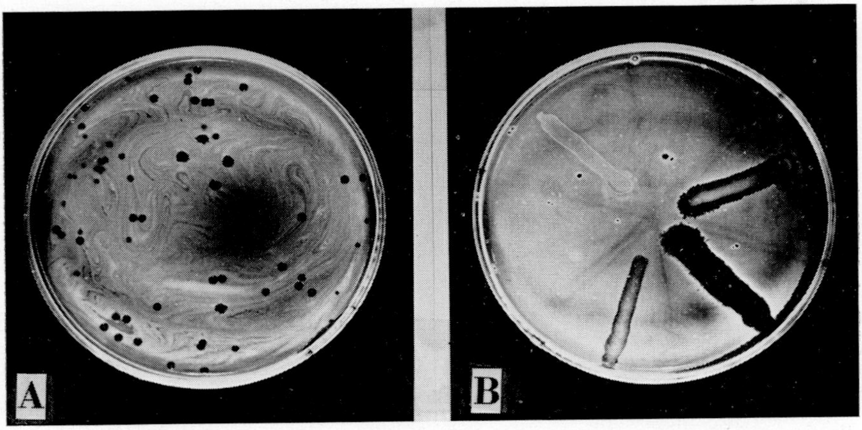

Fig. II–4. A, Plaques of phage Bp1 on bacterial blight isolate B–8. B, Phage reaction of some lysotypes of bacterium; three produced transparent plaques, appearing black in the picture on black background.

Xop1 to Xop8 and 20 lysotypes were reported from India (All-India Coord. Rice Improvement Project, 1968).

Wakimoto (1967) studied the phage sensitivity of some strains from India and showed that all of them belonged to the E type of Japan, whereas most of the Japanese strains belong to the A type.

All studies undertaken have shown that the sensitivity of the bacterial strains to phages and their serological reactions have no relation to one another, nor to the virulence of the strains.

The phages may be isolated from diseased leaves, irrigation water and soil. The sensitivity of bacterial strains to the phages may be tested by streaking the phages on to culture plates of the bacterium (Fig. II–4B).

The bacteriophages have been used for ecological studies (Wakimoto, 1954b, 1957; Wakimoto & Yoshii, 1955; Mizukami, 1966; Mizukami & Wakimoto, 1969) to estimate the population of the bacterium. Recent studies (Goto, 1969) seem to show that: (1) Only when the bacterial cells reach 10^4/ml or more can they be detected by phage, and (2) the phages in paddy water are quickly inactivated by solar radiation. IRRI (1969) reported that the phages survive

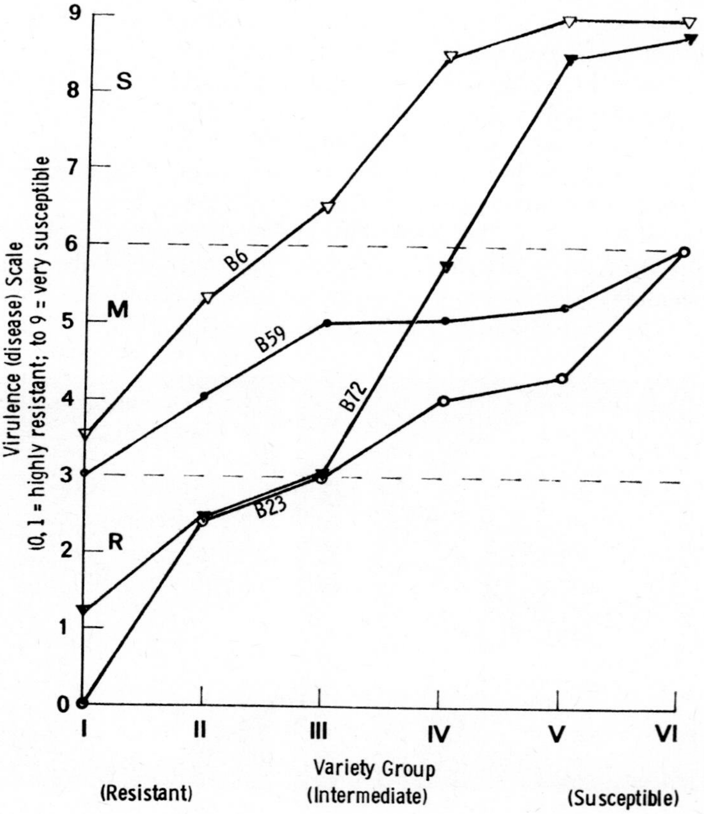

Fig. II–5. Patterns of pathogenicity of *Xanthomonas oryzae* on resistant, intermediate and susceptible rice varieties.

for much longer than the bacterial cells, particularly at higher temperatures. The phage technique for estimating bacterial population has to be used with care.

Virulence. In 1957, a previously resistant variety Asakaze was severely affected by bacterial leaf blight in Japan. The existence of virulent strains of the causal organism caught the attention of several workers (Kuhara *et al.*, 1958). Several studies were made in an attempt to classify the strains on the basis of virulence (Tagami & Mizukami, 1962). In more recent studies (Kuhara *et al.*, 1965; Kusaba *et al.*, 1966; Watanabe, 1966) on some 14–19 resistant, intermediate and susceptible varieties, the strains were divided into two groups, based upon the size of the lesions resulting from inoculation. Group I (or A) strains attacked all the varieties tested, while Group II (or B) strains produced large lesions only on intermediate or susceptible varieties. Each group may further be divided into different types.

Washio *et al.* (1966) in studying the breeding of resistant varieties, separated the bacterial strains into three groups, A, B and C, and the varieties into four groups, I to IV. Group A attacks only Group IV varieties, Group B attacks both III and IV, while Group C attacks II, III and IV. Group I varieties are resistant to all strains.

Reviewing the work done in Japan, Wakimoto (1967) listed 19 resistant, intermediate and susceptible varieties to be used for testing virulence and classifying bacterial strains according to the scheme shown in Table II–3. The 19 varieties, in order from resistant to susceptible, are: Koganemaru, Akashinriki, Norin 27, Asakaze, Kogyoku, Shinzeki 1, Norin 12, Tozan 38, Oita-mii 120, Norin 44, Sasahigure, Nakasengoku, Shinsen, Norin 29, Zensho 17, Taisyoako 66, Jukkoku, Aichi-asahi and Kinmaze.

Fifty strains studied in the Philippines (IRRI, 1966) show the pathogenicity patterns illustrated in Figure II–5.

TABLE II–3

CLASSIFICATION OF STRAINS OF *X. oryzae* IN JAPAN ACCORDING TO THEIR VIRULENCE. (After Wakimoto, 1967).

Strain classification			Reaction of different varieties
Group	A	I	Severely infecting all varieties
	A	II	Slightly infecting all varieties
	A	III	Slightly infecting most varieties except those in resistant group
Group	B	I	Severely infecting varieties which are more susceptible than Norin 12
	B	II	Slightly infecting varieties which are more susceptible than Norin 12
	B	III	Slightly infecting susceptible varieties only

Pattern 1, represented by isolate B6, indicates high virulence, causing moderate reaction in resistant varieties, severe reaction in intermediates and complete killing of susceptible varieties. Pattern 2, represented by isolate B23, indicates the lowest virulence, causing an immune reaction in resistant varieties and only moderate infection in susceptible varieties. Pattern 3, represented by B72, indicates more distinct reactions among resistant, intermediate and susceptible varieties. Pattern 4, represented by B59, contrary to pattern 3, indicates little difference in reaction among the test varieties. Other patterns have also been found, e.g. both resistant and intermediate varieties may show similar reactions but the reaction of susceptible varieties may be very severe. Occasionally a few virulent strains may cause moderately susceptible or susceptible reactions in a few of the resistant varieties.

Strains of the bacterium, as shown above, differ considerably in virulence. However, in general susceptible varieties are more susceptible to all strains and resistant ones more resistant. Physiologic races with distinctly contrary reactions do not seem to exist among the strains. Further studies with more isolates from different countries are needed.

Wakimoto & Yoshii (1954b) reported that strains become more virulent after repeated passage through a resistant variety, but virulence did not change or decrease when strains were passed through susceptible varieties. The effect is apparently a result of the selective action of the host environment. Recently, Goto & Okabe (1967a) reported mutants, in culture, with smaller, translucent, deep yellow colonies and very weak pathogenicity. Preliminary studies at IRRI (1967) have shown single colony subcultures from two isolates to vary considerably in their pathogenicity to rice varieties. Continuous selection of the more virulent single colonies developed very virulent strains; conversely, very weak strains were developed when weak colonies were selected each time (IRRI, 1969).

Strains in the tropics are more virulent than most of the virulent strains in Japan (Anon., 1964; Goto, 1965b, Wakimoto, 1967). Little information is, however, available as to the relative virulence of the bacterial strains present in the different tropical countries of Asia. Wakimoto (1967) studied strains from the tropics and reported that Indian strains are virulent; 6 out of 7 strains from the Philippines were found to be relatively less virulent but the remaining strain was very virulent. Very recently, comparative studies have been made in Hawaii (personal communication) where about 150 isolates from 10 countries in Asia (Burma, Ceylon, India, Indonesia, Japan, Malaysia, Pakistan (East), Philippines, Taiwan and Thailand) and from Australia were tested against nine selected varieties. Results showed that in general isolates from Japan and Australia had low virulence while those from Burma, Ceylon, India, Indonesia and Pakistan had high virulence. Isolates from each country differed, however, in virulence. For example, isolates from India showed variable virulence from low to high. This is one of the many important aspects of the disease which require further study, especially to assist the breeding of resistant rice varieties.

Lin *et al.* (1969) reported that virulent and weak strains are serologically different and could be easily distinguished by the gel diffusion test.

Disease Cycle

Overwintering and source of primary inoculum. The overwintering of the pathogen in a temperate climate has been carefully studied in Japan (Tagami &

Mizukami, 1962; Mizukami, 1966). The organism may live in soil for one to three months, the time depending on soil moisture and acidity. The soil is not considered to be an important source of infection, as this short survival period means that the organism dies before the next year's crop is planted.

In stored seed, the organism can be detected easily until the next crop season. However, although a possible source of infection, seed is not believed to be an important source, as the bacteria rapidly decrease in June (Mizukami, 1961) and in the course of seed soaking for sowing they generally die in a few days (Tagami *et al.*, 1963). Fang *et al.* (1956) found the organism not only on the inside of the glumes but occasionally on the endosperm and considered that the seed is a source of infection.

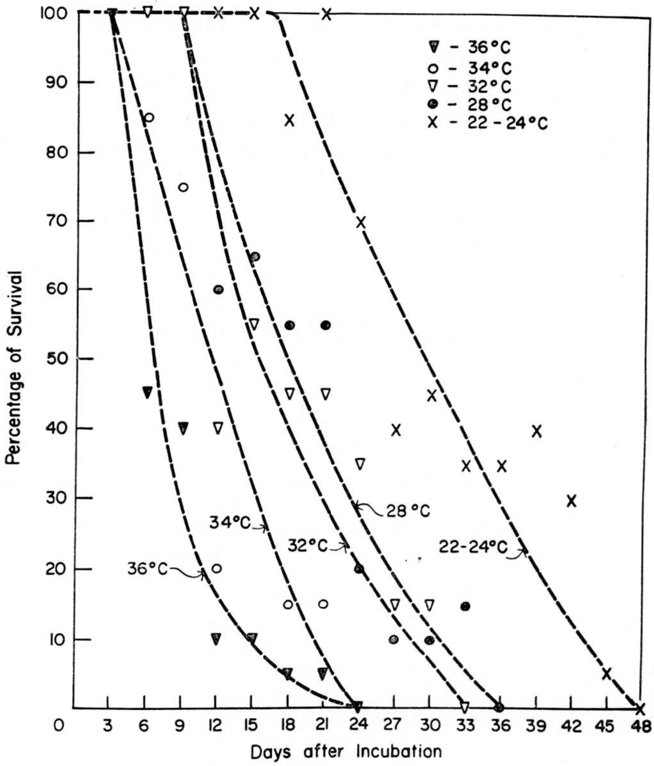

Fig. II–6. Viability of *Xanthomonas oryzae* on rice hulls at different temperatures.

In the tropics, Srivastava & Rao (1964b) found a high percentage of seed infection in India. Recent studies in the Philippines (IRRI, 1968; Eamchit *et al.*, 1969) have shown, however, that the organism survives on seed and infected leaves for only a few weeks at high temperatures (Fig. II–6), although it can survive for several months to one year at low temperatures (4°C) and for two years or more at −30°C. It was reported from India (All-India Coord. Rice Improvement Project, 1969) that the bacterium may be detected on the seeds at harvest but not after three months. No disease symptoms were observed on plants grown from infected seeds. Seed transmission is not considered likely in normal conditions.

Weed hosts, particularly *Leersia sayanuka*, are considered as one of the most important sources of primary inoculum in Japan. The organism survives in the rhizosphere and multiplies early in the spring (Goto *et al.*, 1953; Inoue *et al.*, 1957; Yoshimura *et al.*, 1959). The lesions on *L. sayanuka* develop much earlier than on rice. The bacterium may be found on many kinds of plants near rice fields but does not survive on them through winter.

The organism also survives easily on diseased straw in outdoor piles or indoors (Isaka, 1962). This source of inoculum may be an important one where little or no *Leersia* occurs.

Rice stubble which survives the winter has been found to harbour the organism in the base of the stem and the roots until the following spring.

Wakimoto (1956) reported two forms of the bacterium, the ordinary growth form and a dry form which produces an aggregated mass in the vessels and xylem parenchyma. The latter is smaller than the ordinary form and survives longer under adverse conditions.

The epidemiology of the disease in the tropics has not yet been studied in any detail, but the prevailing high temperatures permit the bacterium to grow at any time of the year. There are plenty of weeds on which the bacterium might survive and secondary growth from rice stubbles may remain alive from one crop season to the next. Some studies at IRRI (unpublished) seem to indicate a high

TABLE II–4

PLAQUE COUNTS OF *Xanthomonas oryzae* AND *X. translucens* F. SP. *oryzicola* PER ML OF WATER SAMPLES FROM CENTRAL LUZON IN MAY TO JUNE 1967 (IRRI, 1967).

Place	Source	Blight	Streak
Capihan, San Rafael, Bulacan	creek water from Angat mountain	0	13
Garlang, San Ildefonso, Bulacan	creek	10	13
Lomboy, Talavera, Nueva Ecija	creek water from Dibabuyan creek	0	25
Maligaya Experiment Station, Munoz, Nueva Ecija	deep well water	169	7
Ilog, Baliuag, Sto. Domingo Nueva Ecija	water from irrigation canal	240	1
Linek, Cuyapo, Nueva Ecija	water from drainage creek	156	12
Hda. Moreno, Sta. Clara, Cuyapo, Nueva Ecija	deep well and mountain water	181	5
San Nicolas, Villasis, Pangasinan	creek from San Manuel river	7	3
San Jose, Urdaneta, Pangasinan	Agno river	120	3
Bamban, Tarlac	Tarlac river, water from Layak mountain	0	40
Mabiga, Mabalacat, Pampanga	natural spring	5	23
Sampaloc, Apalit, Pampanga	Pampanga river	9	50

bacterial population all the year round in irrigation water, in canals, and in rice fields. These factors may be part of the reason for the prevalence of the disease in the tropics. In certain regions of the tropics there is a period of high temperature and drought during which conditions would appear to be unfavourable for the bacterial population. However, bacteriophage was detected in water reservoirs and streams in Central Luzon, Philippines, after several months of dry weather and when no rice was present (IRRI, 1967) (Table II–4).

Preliminary laboratory tests at IRRI (unpublished) showed that phages live longer than the bacterium at temperatures between 28 and 36°C. The presence of phage may not, therefore, indicate the presence of the bacterium, particularly when the bacterial population has declined at the end of the growing season; but it would be a good indicator when the bacterial population is increasing.

Infection and growth of the bacterium. Water pores on hydathodes on the leaf blade, growth cracks caused by the emergence of new roots at the base of the leaf sheath and wounds are points where the pathogen invades. It is not, however, known to enter through stomata.

Wounds on rice leaves are common and are favourable avenues for entry; new wounds are more conducive to infection than old. Kiryu *et al.* (1954) found the percentage of infection of inoculated leaves to be related to the time between injury and inoculation, as follows: 5 minutes after injury, 99·9%; 20 min, 95·7%; 80 min, 62·7%; 320 min, 32·9%; 21·3 hr, 0·4%. A few cells of the bacterium placed on wounds do not grow, whereas abundant inoculum tends to result in high percentages of infection. The minimum concentration required is about 10,000 cells in each ml (Mizukami, 1961). After a lag phase of 1–2 days, the bacterium attains logarithmic growth in susceptible varieties, and soon reaches the vascular tissues through which the pathogen spreads inside the plant.

Water pores are found on the hydathodes along the upper surface of the leaf near the edges (Fig. II–7). The bacterium enters the water pore, and multiplies in the epitheme, into which the vessel opens. When sufficient bacterial multiplication has taken place in the epitheme, some of the bacteria invade the vascular system and some ooze out from the water pore (Fig. II–8) (Tabei & Muko, 1960). The number of hydathodes differs according to variety and age of leaf, upper leaves and those of susceptible varieties often having more.

In the tropics, the leaf tips of rice seedlings are often cut before transplanting. In addition, roots are broken when seedlings are pulled from the seedbed. The bacterium is attracted by the broken roots and swarms into them (Mizukami, 1957), which facilitates infection. These are important courts of infection and are partly responsible for the severe kresek symptoms found in the tropics.

Seasonal development. While this disease is usually observed in Japan at the maximum tillering stage or later, the causal bacterium is often detected during the latter part of the seedbed stage. The bacterium is brought into the irrigation water from overwintering sources. After the rice is transplanted, the organism increases, first in the lower and then in the upper leaves, long before any visible symptoms appear. The bacterial population varies according to environmental

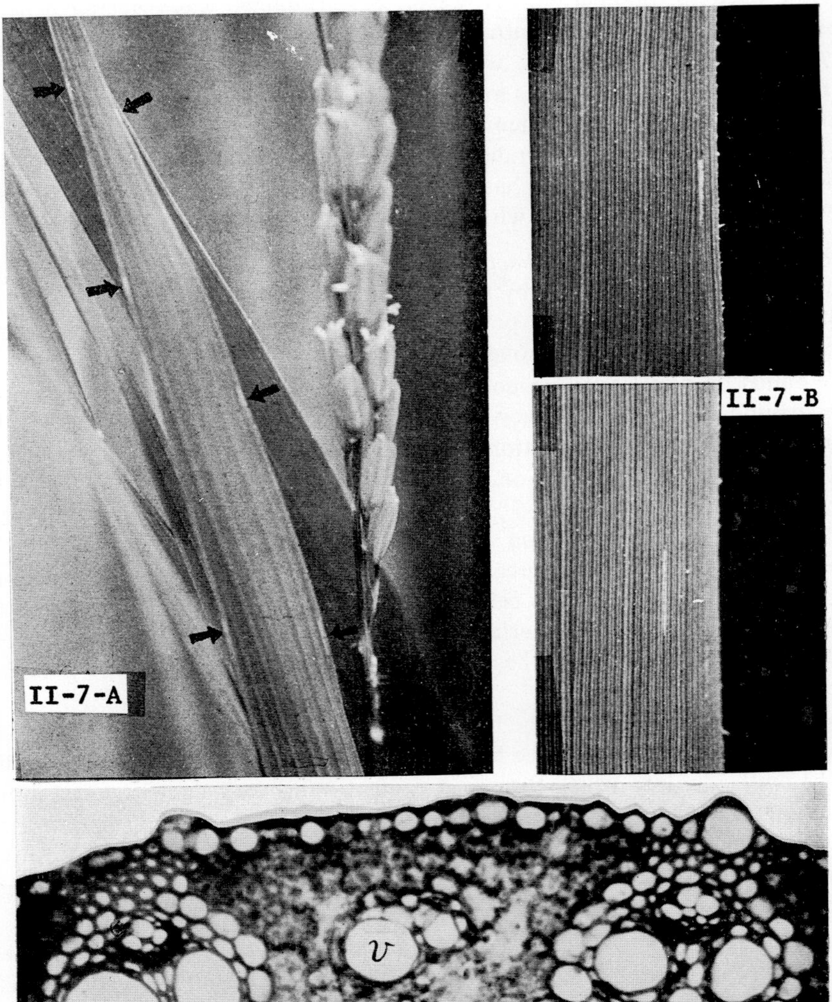

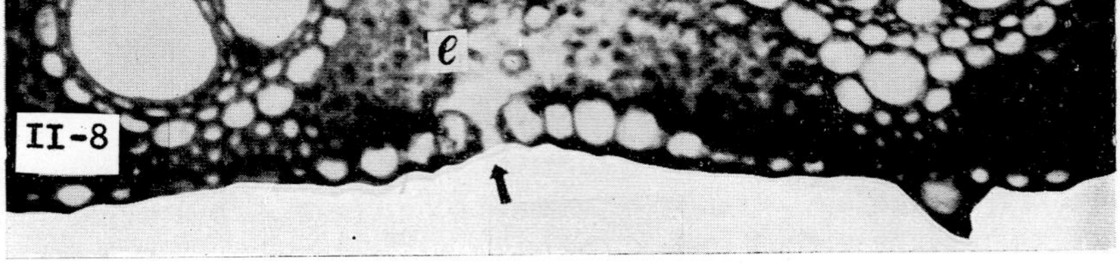

Fig. II–7. A, Water exudation system of rice leaf. B, Enlarged. (Rearranged from Tabei & Mukoo, 1960.)

Fig. II–8. Cross-section of a portion of rice leaf showing open water pore (↑), epitheme (*e*), and vessel (*v*). (× 450). (From Tabei & Mukoo, 1960.)

conditions and this affects disease development (Tagami *et al.*, 1964). Small numbers of bacteria may be detected by washing or macerating leaves or other materials, filtering through cotton, centrifuging, and inoculating to a susceptible variety or testing with bacteriophage.

The disease sometimes starts in small patches in a field and spreads from such small centres to other parts of the field. In Southeast Asia, the frequent occurrences of typhoons and rainstorms are major factors in the dissemination of the organism. The storms not only effectively spread the pathogen, but also cause wounding of the rice. Because of its frequent relationship with bad weather, bacterial leaf blight is often referred to as post-typhoon disease or rain-borne disease.

In dry weather the bacterial exudates, in the form of very small beads, drop into the irrigation water. The bacterial population in the water, as estimated by bacteriophage plaques, frequently reaches 100,000 per ml of water from diseased fields in the Philippines. The contaminated water is channelled or floods from field to field during the rainy season and this results in widespread dissemination of the pathogen.

Effect of environmental conditions. There are certain areas in Japan where the disease is endemic. Many studies have been made and these have revealed that such areas are usually along rivers, near marshes, or in mountain basins. Such locations are low, wet and poorly drained, are easily flooded during the seedbed or growing stages, are often subject to mists, and frequently have adjacent areas of *Leersia sayanuka.*

In connection with forecasting of bacterial blight, climatic factors have been subject to analysis. Incidence of the disease has been found to be correlated with total precipitation, the occurrence of heavy rain, flooding and the presence of deep irrigation water, and severe wind. Relatively high temperatures during the growth of the rice crop increase incidence, but severe summer heat and drought suppress the disease. By inoculation experiments, Muko *et al.* (1957) found that high temperature (25–30°C) is more favourable for disease development than low (21°C); at 17°C there was almost no development. Watanabe (1966) obtained similar results. Kresek symptoms may be observed in less than 20 days after inoculation at 31°C whereas symptom development took 40 days or more at 21°C in tests at IRRI (1967).

Many experiments have been done to investigate the effects of fertilizers on the incidence of the disease (Tagami & Mizukami, 1962). Excess nitrogen, especially in organic form and as a late top dressing, increases the disease, as do phosphate and potassium deficiency and excess silicate and magnesium. The effect of nitrogen on the disease is perhaps mainly due to the enhanced vegetative growth of the plants, which influences the humidity and dissemination of the pathogen. Nitrogen does not seem to affect the enlargement of individual lesions (IRRI, 1966).

Host Range

The weed hosts of *X. oryzae* in Japan were first found by artificial inoculation (Goto *et al.*, 1953); infected weeds were later found in nature. Among them,

Leersia sayanuka Ohwi is the most important as it serves as a common over-wintering host. *L. oryzoides* (L.) Sw., *L. japonica* and *Zizania latifolia* are also naturally infected (Tagami & Mizukami, 1962).

In the tropics, *Leptochloa chinensis* (L.) Nees, *L. filiformis* and *L. panacea* have been found as weed hosts in the Philippines; large lesions were produced on these plants by artificial inoculation with needles (IRRI, 1967). *Cyperus rotundus* L. and *C. defformis* are also reported from India as being alternative hosts found infected in nature (Chattopadhyay & Mukherjee, 1968), but this was not confirmed by others (All-India Coord. Rice Improvement Project, 1968).

Varietal Resistance

Methods of testing. Varietal resistance may be tested in the field under natural conditions. However, several seasons are required for such tests because incidence of the disease in nature varies from season to season. Reitsma & Schure (1950) used three methods (spraying with a bacterial suspension, pricking the leaves with a needle, and immersing the seedlings in a bacterial suspension) to prove the infectivity of kresek disease in Indonesia. Disease development when the spraying method is employed is usually slower than with the needle pricking method, and it may not give uniform results on the inoculated plants. Muko & Yoshida (1951) in Japan studied the needle inoculation method, which was found to provide efficient inoculation and to give an accurate measurement of disease development. The results of needle inoculation and natural infection were proved to be identical (Muko *et al.*, 1952). Recently, the immersion method, i.e. submerging the seedlings in or flooding the nursery bed with

Fig. II–9. Inoculation needle and pad for field inoculation of bacterial leaf blight.

bacterial suspension, has been developed and used on certain occasions (Fang *et al.*, 1956; Yoshimura & Iwata, 1965). Clipping off the leaf or root tips of seedlings and then submerging the seedlings in a bacterial suspension also offers a good method of inoculation (IRRI, 1966). Injection of the bacterial suspension into the leaf veins by means of a micropipette (0·5mm diameter) has also been used (Goto *et al.*, 1953).

Several modifications of the needle inoculation method may be used. For example, the number of needles may vary from 1 to 20, 1 or a few being used for young and small leaves, a greater number for large leaves. Usually, 4–6 needles are sufficient. The needles are mounted on a rubber bottle stopper or some other suitable object. A cotton pad soaked in bacterial suspension provides the inoculum. The needles may be arranged in any convenient way so that the leaves are punctured and at the same time inoculated in one operation. One of the arrangements used at IRRI is shown in Fig. II–9.

Devadath (1970) reported that a mixture of virulent and weak strains in the inoculum suppressed the symptom development of the virulent strain. Another report (All-India Coord. Rice Improvement Project, 1969) indicated, however, that a mixture of virulent and weak strains in the ratio of 1 : 1, when the bacterial cell concentration was 10^8/ml, did not inhibit lesion development, whereas it did with a large proportion (19 : 1) of the weak strain.

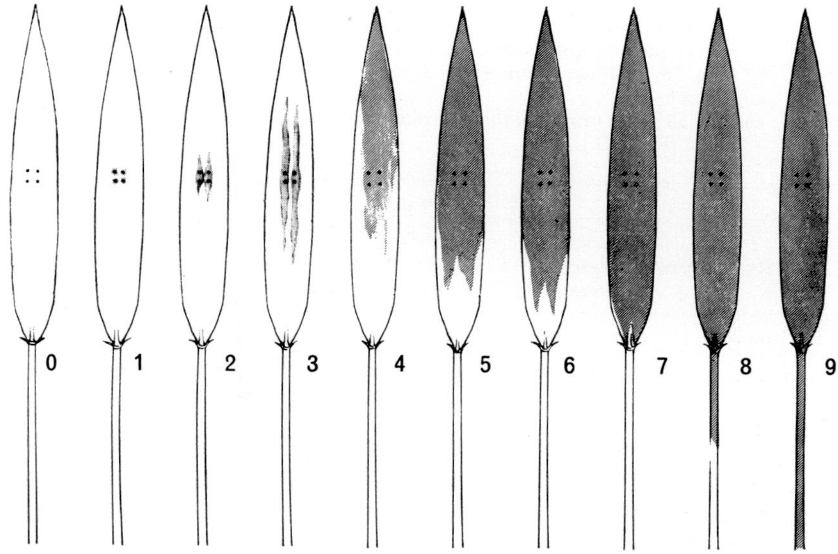

Fig. II–10. Diagram showing varietal reaction (scale) to bacterial leaf blight on flag leaves.

Standard scales for measuring degree of resistance. In Japan, diseased areas of the inoculated leaves have been measured by index units of 0–4, or a transparent scale board may be used on which 1mm squares are marked (Tagami & Mizukami, 1962). In the tropics, because of the much more extensive lesions which occur on inoculated leaves, a standard scale of 10 units has been used. (Table II–5, Fig. II–10) (IRRI, 1965).

TABLE II–5

STANDARD OF SCALES FOR INDICATING DEGREE OF RESISTANCE TO BACTERIAL LEAF BLIGHT. (IRRI, 1965).

Scale	Seedling Stage	Flowering stage—flag leaf
0	No lesion observable (immune).	Same.
1	Lesions restricted to 1–2 mm around the points of inoculation.	Same.
2	Lesions more or less elliptical not more than 2–3 cm long.	Same.
3	Lesions elongated, extend to about ½ length of leaf blade or less.	Lesions elongated, less than ½ leaf length.
4	Lesions extend both in length and width destroying ¾ of leaf blade.	Lesions broad and coalesce, the upper portion of the leaves often dead, lesions extend to about ¼ of the lower half of leaf surface below the points of inoculation.
5	Lesions extend to and destroy the entire leaf blade.	Lesions coalesce and upper portion of leaves dead, lesions extend to about ½ of the lower half of the leaf surface.
6	Lesions extend to and destroy the entire leaf blade and less than ½ of leaf sheath.	Lesions extend to about ¾ of the lower half of the leaf blade.
7	Entire leaf blade and more than ½ leaf sheath destroyed and a few pale yellow symptoms appear.	Lesions extend to the base and destroy the entire leaf blade.
8	Leaf blade and leaf sheath destroyed, many pale yellow plants, and less than 25 seedlings completely killed (Kresek).	Lesions destroy the entire leaf blade and extend to about ½ of leaf sheath.
9	About 50% or more seedlings entirely killed (Kresek).	Lesions completely destroy the leaf blade and sheath.

TABLE II–6

REPRESENTATIVE RESISTANT AND SUSCEPTIBLE VARIETIES OF JAPAN AND THEIR PARENTAGE. (After Tagami & Mizukami, 1962.)

Resistance	Variety	Parentage
Highly resistant	Kogyoku (Kidama)	Shirosenbon × Shobei
	Zensho 26	Shoyii × Shigasekitori 11
	Koganemaru	Kogyok 7 × Hinomaru
	Shinseki 1	Shinrik × Shigasekitori
	Norin 27	Kono 35 × Asahi 1
	Asakaze	Takara × Saikai 28 (Norin 27)
Resistant to moderately resistant	Yachikogane	Tosan 38 × Ginbozu-nakate
	Yomohikari	Norin 22 × Ginbozu-nakate
	Koshisakae	Norin 22 × Shin 4
	Hoyoku	Jukkoku × Zensho 26
Susceptible	Takara	Dokai × Shinriki
	Shinzan	Shinriki × Yamakita-bozu
	Imari 1	Saikai 15 × Senbon-asaho
	Tsurugiba	Tokai-asahi × Tozan 36
	Kinmaze	Ryoosaku × Aichinakate-asahi
	Sasahigure	Norin 8 × Tohoku 24
	Koshiji-wase	Norin 22 × Norin 1
	Hokuriku 52	Norin 13 × Norin 21

Resistant Varieties

Field selection, testing and breeding of varieties for resistance began many years ago in Japan. During the past 50 or more years, many resistant varieties have been identified and used for breeding to develop new varieties. Mihashi *et al.* (1953) have listed the following varieties which have been found to be resistant at various periods:

1923–24:	Shigasekitori 11, Sugaippon, Aka-shinriki
1930 :	Shigasekitori, Sengoku 4, Hatsugasumi
1931 :	Magatama, Hatsugasumi, Kameji 3, Hokubu 2
1937–38:	Mii, Zensho 26, Kogyoku 1, Kogyoku-asahi, Hokubu 2
1949–50:	Shinseki 2, 4, 13, Hokubu 2–1014, Hene-i-10, Hen 2-1-24 (Shinseki 1, Shoseki 1, Hatsugasumi)

Other varieties were later selected by various experiment stations in Japan (Table II–6). None of them, however, is highly resistant to the most virulent strains recently identified (Mizukami & Wakimoto, 1969). Kono 35 in Table II–6 is a field selection from Shinriki and has been one of the important sources of resistance to bacterial blight in Japan. The resistant variety Zensho 26,

Fig. II–11. Varietal reactions to bacterial leaf blight at seedling stage; from left, very susceptible (kresek), susceptible, most resistant and at right showing pale yellow symptoms (susceptible).

developed from Shigasekitori, was used for developing other resistant varieties; Hoyoku in 1961, Kokumasari in 1962 and Shiranui in 1964. From Shobei, Kogyoku and other resistant varieties were developed. Kogyoku was further used to develop Sachikaze in 1960 and Nihonbare in 1963. Yoshimura &

Tagami (1967) listed several other varieties resistant to moderately virulent strains: Ohu 244, Fujiminori, Honen-wase, Shin 2, Nakashin 120, Koshihikare, Hogareshirazu, Shirogane etc.

Washio *et al.* (1966) considered varieties Aikoku-soto-sango-kei and Nakashin 120 to be resistant to all three groups of bacterial strains.

More recently, Sakaguchi *et al.* (1968) tested 863 cultivated varieties, domestic and foreign to Japan, and many lines of *Oryza glaberrima* and wild rices. They found Lead rice from Burma and TKM 6 and Nigeria 5 resistant to the most virulent group of strains of the bacterium. One line of *O. latifolia* with genome CCDD and one line of *O. granulata* were also resistant.

TABLE II–7

Reaction of selected resistant varieties to 10 strains of *Xanthomonas oryzae* (disease scales 0 to 9, resistant to susceptible). (IRRI, 1968).

Variety	Bacterial strains									
	B-2	B-6	B-13	B-15	B-38	B-75	B-77	B-57	B-59	B-23
Zenith	3	3	4	3	3	3	3	2	3	2
TKM 6	3	4	1	4	2	3	3	3	4	2
Wase Aikoku 3	0	0	0	0	0	1	1	0	1	0
M. Sungsong	0	0	1	1	1	2	3	1	3	1
Keng Chi-ju	2	3	4	3	2	3	3	2	3	3
Early Prolific	2	3	3	2	1	3	2	2	2	2
Early Prolific	3	3	3	3	3	3	3	3	3	3
Early Prolific	2	3	3	4	3	3	3	1	4	1
Early Prolific	4	3	3	2	4	4	5	4	5	2
Lacrosse × Zenith	2	1	1	2	1	2	2	1	3	1
Lacrosse × Zenith × Nira	2	4	5	2	3	2	3	2	2	2
P.I. 162319	1	1	1	1	1	0	0	0	0	0
3 Baifufugoya	1	2	1	1	1	2	2	1	1	0
P.I. 209938	3	4	3	3	5	3	3	2	3	1
Tainan-iku 512	3	5	4	4	6	3	1	4	3	3
Giza 38	2	4	4	5	1	2	3	3	4	4
Balilla	3	3	5	4	4	3	2	4	3	3
BJ 1	3	3	2	3	2	3	3	3	3	2
Sigadis	2	4	2	5	3	3	3	3	3	2
Semora Mangga	0	0	2	1	2	0	0	0	2	1
Tainan 9	3	5	4	3	3	4	4	5	4	3
221c/BCIII/Br/62/2	3	2	4	4	3	3	3	3	4	1
DZ-78	2	2	2	2	2	3	3	2	3	2
DZ-60	3	2	3	3	3	3	3	3	3	0
DD-96	3	3	4	3	3	3	3	3	3	1
DV-29	3	3	2	3	3	3	3	3	3	2
DV-52	2	1	2	1	1	3	3	2	2	0
DV-2 (ck)	6	8	8	6	9	7	6	9	7	3
JC-70 (ck)	9	9	9	9	9	9	9	9	9	7

During the last few years, several thousand varieties have been tested against many of the virulent strains found in the Philippines (IRRI, 1965, 1966, 1967). Varieties showed wide differences in reaction (Fig. II–11). Table II–7 lists varieties showing high degrees of resistance. These varieties may, however, be susceptible to virulent strains from other tropical countries. For example, some preliminary results (verbal communication) have shown that many of them are susceptible to Indian strains. Varieties BJ1 and Malagkit Sungsong seem to have a broader spectrum of resistance than others. Japanese varieties of the resistant, intermediate and susceptible groups, when tested with Philippine isolates, showed much the same reactions, all being susceptible to moderately so. Testing varieties on an international basis is highly desirable for identifying varieties with a broad basis of resistance.

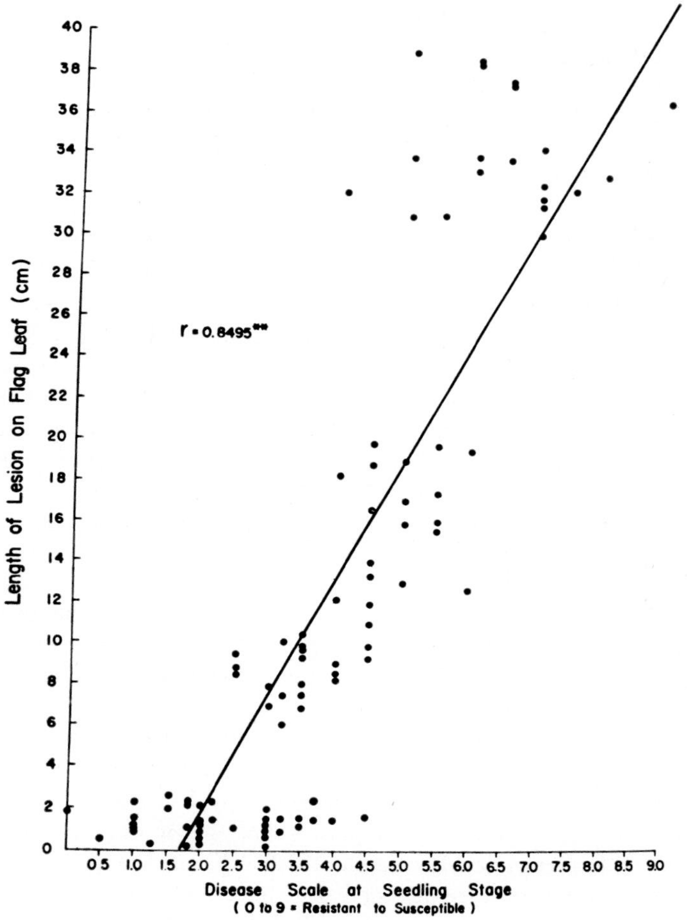

Fig. II–12. Correlation of reaction of rice varieties to bacterial leaf blight between seedling and flowering stages.

Problems in testing for resistance. Testing for resistance by needle inoculation shows the extent of lesion enlargement on different varieties. This is a major characteristic of varietal resistance, but the method will not indicate other important characteristics also related to resistance (field resistance) such as number of water pores or hydathodes, hardness of leaves (which affects ease of wounding), etc.

The age of the culture used, the concentration of inoculum (Kuhara, 1956) and environmental conditions, particularly temperature and moisture, all affect lesion development. Thus a certain degree of variation is found from locality to locality or from season to season in inoculation experiments, although the relative degree of resistance remains.

To save time in testing varietal resistance, seedling leaves are tested instead of the flag leaf. A relatively high correlation (r=0·85) in resistance was found between the two stages (IRRI, 1965) (Fig. II–12). However, in certain varieties, the reaction on seedling leaves may not indicate exactly the reaction on flag leaves (Wakimoto & Yoshii, 1954a). The lower leaves are generally more susceptible than the upper leaves, and the flag leaf and two or three other upper leaves are the most important as far as loss of yield is concerned. Inoculation at the flowering to milky stage, on the middle of the flag leaf, will therefore give the most reliable indication of resistance.

A more important problem is the presence of strains of the pathogen which differ in virulence. In studying varietal resistance, strains should be compared and virulent strains should be used for inoculation so that highly resistant varieties may be identified.

Mechanism of resistance. Not much is known about the mechanism of resistance. Kiryu & Mizuta (1955) analyzed morphological features of resistant, intermediate and susceptible varieties and found that varieties with few, short, narrow leaves, with less horizontal spread of the leaves, are most resistant. Those with luxuriant growth or spreading leaves tend to increase the humidity among the plants while wounds and frequent contact of leaves favour disease dissemination and development. These reactions may be considered as influencing field resistance. The number of water pores and hydathodes may also play an important part.

The nutritional environment inside the host plant seems to be an important factor in determining resistance. When the bacterial cells enter the plant, they soon begin to multiply in both resistant and susceptible varieties, but growth becomes static in resistant varieties and no lesions or only very small ones result. In susceptible varieties, logarithmic multiplication of bacterial cells follows and large lesions develop. When inoculated plants are treated with ether or immersed in water, the size of the lesions is larger than in plants not so treated, or inoculated after treatment. It is assumed that this is due to an increase of amino acids (Mizukami & Murayama, 1960). Fang, Lin & Shu (1963) also reported that susceptible varieties tended to have higher contents of some free amino acids and lower contents of polyphenols and reducing sugars. Higher concentration of sugars and polyphenols have been reported as less favourable to the pathogen. Brief analysis in the Philippines (IRRI, 1967) indicated, however, that resistant varieties have a higher ratio of reducing sugar to total nitrogen than the susceptible varieties. We have also observed that plants grown in the shade develop smaller lesions than those under full sunlight. This further indicates the relation of sugar to lesion development (IRRI, 1968). However, all these correlations may not relate to the real mechanism of resistance.

The production of a phytoalexin which inhibits the multiplication of the bacterium has been detected. It is nonspecific and is decreased by ether treatment. It was distinguishable in resistant varieties but not in susceptible ones (Uehara, 1960).

Inheritance of resistance. Little information on the detailed genetics of resistance is available. In studies on the inheritance of resistance, it was found that genes for resistance may be dominant or recessive, and monogenic or polygenic (Kariya & Washio, 1956). In crosses between Norin 27 (R) × Asahi 1 (S) and Norin 27 × Norin 18 (MR), the gene for resistance in Norin 27 is monogenically dominant over both Asahi 1 and Norin 18. In a cross between Norin 18 and Asahi 1, the moderately resistant Norin 18 showed no dominance over the susceptible Asahi 1. Nishimura & Sakaguchi (1959) found, in crosses between Murasaki-muyozetsu (S) and resistant varieties, phenotypic expression of resistance in the F_2 generation with a ratio of 3 : 1, and linkage of resistant genes at the *lg* locus (no leaf ligule).

Washio *et al.* (1966) more recently reported that resistance to bacterial group A in variety Kidama was controlled by two complementary dominant genes, X_1 and X_2. Resistance in Norin 274 and Katno 60, also to bacterial group A, was conditioned by a dominant gene. Resistance to bacterial group C in Aikoku-sato-sango-kei was controlled by a dominant gene or genes. It was also found that resistance factors for bacterial blight, blast and stripe diseases were independent. Sakaguchi (1967) reported two genes for resistance, Xa_1 and Xa_2. The Kidama group of varieties has Xa_1 controlling resistance to group I of the bacterium. The Rantaj-emas group carries Xa_1 and Xa_2 of which Xa_2 is the principal factor for resistance to group II of the bacterium.

Many crosses between resistant and susceptible varieties made in the Philippines show that resistance may be recessive or dominant depending upon variety combinations and degree of resistance in the resistant parent (IRRI, 1966, 1967).

CONTROL

Various kinds of chemicals have been tested in numerous experiments conducted by both government and private institutions in Japan for controlling bacterial leaf blight.

To determine the proper time of application of bactericides for effective control, a very extensive system of forecasting services has been organized by the central government for insect and disease control. This was begun over 20 years ago and has been elaborated through the years (Ministry of Agriculture and Forestry, 1958).

In spite of all these efforts, no effective chemical control can yet be recommended for general practical use to control the disease, although valuable information has been revealed through these studies.

Forecasting. Several methods of forecasting bacterial leaf blight disease have been studied in Japan. (1) By natural infection: resistant and susceptible varieties are planted in a forecasting field, and periodical observations are made from the nursery until the late stages of growth. Leaves are artificially wounded by needles, and the development of the disease is also observed on *Leersia* spp. (2) By climatic conditions: correlations have been found between the incidence of the disease and the occurrence of floods, typhoons, rain, sunshine, particular temperatures and severe wind, as illustrated by table II–8. (3) By the bacterial

TABLE II–8

CORRELATION BETWEEN INCIDENCE OF BACTERIAL LEAF BLIGHT AND CLIMATIC CONDITIONS. (Saga agric. Exp. Stn, 1960, after Tagami & Mizukami, 1962).

X-value	Forecasting equation	Correlation
Rainfall, end of July	$Y = 1{\cdot}507 + 0{\cdot}043\ X$	0·902
Rainy days, end of July	$Y = 0{\cdot}214 + 0{\cdot}793\ X$	0·605
Rainfall, middle of August	$Y = 2{\cdot}035 + 0{\cdot}0097\ X$	0·915
Rainy days, middle of August	$Y = 1{\cdot}809 + 0{\cdot}596\ X$	0·794
Rainfall, beginning & middle of August	$Y = 1{\cdot}621 + 0{\cdot}009\ X$	0·829
Rainy days, beginning & middle of August	$Y = 0{\cdot}610\ X - 1{\cdot}427$	0·752
Rainy days, end of August	$Y = 0{\cdot}689\ X - 0{\cdot}309$	0·842
Rainfall, middle & end of August	$Y = 1{\cdot}259 + 0{\cdot}0098\ X$	0·924
Rainy days, middle & end of August	$Y = 0{\cdot}435\ X - 0{\cdot}519$	0·958
Rainfall, August	$Y = 0{\cdot}489 + 0{\cdot}011\ X$	0·961
Rainy days, August	$Y = 0{\cdot}481\ X - 2{\cdot}963$	0·976

population: since the bacteria are found in advance of the appearance of disease symptoms, determination of the bacterial population is used in forecasting. Leaves are collected, washed and macerated, and the bacterial cells are concentrated by centrifugation and inoculated to susceptible varieties. The tests are made at the nursery stage and twice later from the maximum tillering stage to the time of ear production. (4) By the bacteriophage population (Tagami, 1958, 1959a, b): the phage is found in water in advance of the bacterial population. Irrigation waters at selected points are sampled. One or two ml of the water is added to 1 or 2 ml of sensitive bacterial suspension, shaken, and 5 to 6 ml of potato semisynthetic agar medium (at 45°C) added. Plaque counts are made after 10–15 hr at 20–25°C. The water may be diluted if the phage population or contamination is high (Tagami & Mizukami, 1962; Wakimoto, 1967).

Forecasting by the phage population is much in advance of other methods, and under normal conditions it is quite accurate. However, all the methods are affected by unusual changes in climate such as flooding, typhoons, extreme heat or drought, all of which may affect the subsequent bacterial population.

Chemical control and other measures. Bordeaux mixture and other copper compounds have been used in Japan since as early as 1909. When the disease is not severe, some benefits are obtained, but copper was after a time found to be toxic to rice plants. Sugar was added and the strength of the Bordeaux mixture was reduced to minimize the toxicity. Mercury compounds were introduced in 1950 and have been used alone or mixed with copper. Since 1955 a large number of antibiotics have been tried, e.g. streptomycin, penicillin, chloramphenicol, cellocidin, aureomycin, blasticidin, etc. They were used sometimes in mixtures with mercury and mercury compounds. They, like copper, cause toxicity and chlorosis at high concentrations, although the addition of iron compounds and other materials may reduce the phytotoxic effect. However, when disease incidence is low, the yield of treated plots is often lower than that of untreated plots because of phytotoxicity (Arata *et al.*, 1961). Chloramphenicol and cellocidin, as well as synthetic compounds such as dithianon, dimethyl-nickel carbamate, fertiazon and phenazine are generally considered less toxic to rice plants, while at the same time being moderately effective in controlling bacterial leaf blight (Yoneyama *et al.*, 1969).

Copper compounds are used mainly for preventing infection; mercury compounds and antibiotics also inhibit lesion development. Since the effects of all these chemicals are short-lived (Mizukami & Seki, 1954), frequent spraying, and in addition a relatively high concentration in the case of antibiotics (400–500μ/ml) are required (Tagami & Mizukami, 1962). Recently, Yoshimura & Tagami (1967) emphasized the importance of the application of chemicals early in the growing season or during the summer lag phase before the bacterial population has had time to build up.

Very little work on chemical control has been done in the tropics. Recently, in tests of antibiotics, streptomycin has been tried in India without any increase in yield despite a very heavy spraying schedule (Indian Council of Agricultural Research, 1966). Treating the field water with bleaching powder has also been tried, adopting the method used for maize stalk rot control, but no conclusive results are yet available (Padmanabhan & Jain, 1966). New systemic compounds such as TF-130 are being tested with promising results, although there are indications that the bacterium soon develops resistance.

No single effective control measure is available for the control of the disease, and integrated measures are therefore suggested in Japan. These include the use of resistant varieties, avoidance of flooding or deep water in the nursery, removal of primary sources of inoculum, spraying with chemicals in both nursery and field, avoidance of the use of nitrogenous fertilizers, etc. The practical usefulness of these measures in the tropics is rather doubtful.

LITERATURE CITED

ALL-INDIA COORDINATED RICE IMPROVEMENT PROJECT. 1968. Progress report, 1968, Vol. 1. New Delhi, Indian Coun. Agric. Res. (Mimeographed).

ALL-INDIA COORDINATED RICE IMPROVEMENT PROJECT. 1969. Progress report, 1969, Vol. 3. New Delhi, Indian Coun. Agric. Res. (Mimeographed).

ANON. 1964. Bacterial blight of rice. *Rice News Teller* **12**: 68.

ARATA, T., HORI, M. & INOUE, Y. 1961. On the phytotoxicity of streptomycin to rice plant. *Ann. phytopath. Soc. Japan* **26**: 78. [Jap.]

BHAPKAR, D. G., KULKARNI, N. B. & CHAVAN, V. M. 1960. Bacterial blight of paddy. *Poona agric. Col. Mag.* **51**: 36–46.

DICKSON, J. G. 1956. Diseases of field crops. 2nd ed. 517 pp., New York, McGraw-Hill Book Co.

CHATTOPADHYAY, S. B. & MUKHERJEE, N. 1968. Occurrence in nature of collateral hosts (*Cyperus rotundus* and *C. defformis*) of *Xanthomonas oryzae*, incitant of bacterial blight of rice. *Curr. Sci.* **37**: 441–442.

DEVADATH, S. 1970. Effect of mixture of inoculations of virulent and less virulent isolates of *Xanthomonas oryzae*. *Curr. Sci.* **39**: 420.

EAMCHIT, S. & OU, S. H. 1969. Effect of temperature on the viability of *Xanthomonas oryzae*. Abs. in *Philipp. Phytopathology* **5**: 2–3.

EAMCHIT, S., OU, S. H. & EXCONDE, O. R. 1969. *Xanthomonas oryzae* strains resistant to antibiotics. Abs. in *Philipp. Phytopathology* **5**: 3.

FANG, C. T., LIN, C. F. & CHU, C. L. 1956. A preliminary study on the disease cycle of the bacterial leaf blight of rice. *Acta phytopath. sin.* **2**: 173–185. [Chin. Engl. summ.]

FANG, C. T., LIN, C. F., CHU, C. L. & SHU, T. K. 1963. Studies on the disease resistance of rice. II. Varietal resistance of rice to bacterial leaf blight and a preliminary analysis of its mechanism. III. Variation of the leaf color of rice in relation to the development of bacterial blight. *Acta phytopath. sin.* **6**: 107–112; 113–118. [Chin. Engl. summ.]

FANG, C. T., REN, H. C., CHEN, T. Y., CHU, Y. K., FAAN, H. C. & WU, S. C. 1957. A comparison of the rice bacterial leaf blight organism with the bacterial leaf streak organism of rice and *Leersia hexandra* Swartz. *Acta phytopath. sin.* **3**: 99–124. [Chin. Engl. summ.]

GOTO, K. FUKTATZU, R. & OKATA, K. 1953. Overwintering of the causal bacteria of rice blight in the rice plant and grasses. (Preliminary report). *Agriculture Hort., Tokyo* **28**: 207–208. [Jap.] Also abs. in *Ann. phytopath. Soc. Japan* **18**: 22.

GOTO, M. 1964. ' Kresek ' and pale yellow leaf, systemic symptoms of bacterial leaf blight of rice caused by *Xanthomonas oryzae* (Uyeda et Ishiyama) Dowson. *Pl. Dis. Reptr* **48**: 858–861.

GOTO, M. 1965a. Phage-typing of the causal bacteria of bacterial leaf blight (*Xanthomonas oryzae*) and bacterial leaf streak (*X. translucens* f. sp. *oryzae*) of rice in the tropics. *Ann. phytopath. Soc. Japan* **30**: 253–257.

GOTO, M. 1965b. Resistance of rice varieties and species of wild rice to bacterial leaf blight and bacterial leaf streak disease. *Philipp. Agric.* **48**: 329–338.

GOTO, M. 1969. Ecology of phage-bacteria interaction of *Xanthomonas oryzae* (Uyeda et Ishiyama) Dowson. *Bull. Fac. Agric. Shizuoka Univ.* **19**: 31–67. [Jap. Engl. summ.]

GOTO, M. & OKABE, N. 1965. Bacteriophages of *Xanthomonas oryzae*, the pathogen of bacterial leaf blight of rice, collected in the Philippines. *Bull. Fac. Agric. Shizuoka Univ.* **15**: 31–37.

GOTO, M. & OKABE, N. 1967a. Colony type mutation associated with the attenuation in *Xanthomonas oryzae* (Uyeda et Ishiyama) Dowson. *Bull. Fac. Agric. Shizuoka Univ.* **17**: 27–30.

GOTO, M. & OKABE, N. 1967b. Host-controlled modification of *Xanthomonas oryzae* phage Bp 1 and Bp 2. *Bull. Fac. Agric. Shizuoka Univ.* **17**: 31–36.

GOTO, M. & OKABE, N. 1967c. Phage-resistant mutants, probable source of lysotypes of *Xanthomonas oryzae* and *X. translucens* f. sp. *oryzae* in nature. *Bull. Fac. Agric. Shizuoka Univ.* **17**: 37–43.

HSU, S. T. 1966. Nutritional requirements *in vitro* of *Xanthomonas oryzae* (Uyeda and Ishiyama) Dowson and its effect on host plant resistance. M.S. Thesis, Univ. Philippines College of Agriculture, 58 pp.

INDIAN COUNCIL OF AGRICULTURAL RESEARCH. 1966. Prog. Rep. All India Coord. Rice Improve. Proj., 1965–66, 60 pp.

INOUE, Y., GOTO, K. & OHATA, K. I. 1957. Overwintering and mode of infection of leaf blight bacteria (*Xanthomonas oryzae*) of rice plant. *Bull. Div. Pl. Breed. Cult. Tokai-Kinki natn. agric. Exp. Stn* **4**: 74–82. [Jap.]

INOUE, Y. & TSUDA, Y. 1959. On the assessment of damage in yield of rice plant attacked by bacterial leaf blight. *Bull. first Agron. Div. Tokai-Kinki natn. agric. Exp. Stn* **6**: 154–167. [Jap. Engl. summ.]

IRRI (INTERNATIONAL RICE RESEARCH INSTITUTE). Annual Reports for 1964–69.

ISAKA, M. 1962. Overwintering of bacterial leaf blight organism in damaged leaf in paddy field. *Proc. Ass. Pl. Prot. Hokuriku* **10**: 90. [Jap.]

ISHIYAMA, S. 1922. Studies on the white leaf disease of rice plants. *Rep. agric. Exp. Stn, Tokyo* 45: 233–251. [Jap. Engl. abs. in *Jap. J. Bot.* **1**: 21, 1922.]

KARIYA, K. & WASHIO, O. 1956. Effect of the selection during early segregating generations for bacterial leaf blight resistance in rice. *Chugoku agric. Res.* **5**: 39–40. [Jap.]

KIRYU, T. & MIZUTA, H. 1955. On the relation between habits of rice plant and varietal resistance of bacterial leaf blight. *Kyushu agric. Res.* **15**: 54–56. [Jap.]

KIRYU, T., NISHIZAWA, T. & KUHARA, S. 1954. On the relation of time passing after injuring of rice plant leaves to the infection of *Bacterium oryzae* (Uyeda et Ishiyama) Nakata. *Bull. Kyushu agric. Exp. Stn* **2**: 125–129. [Jap. Engl. summ.]

KUHARA, S. 1956. Effects of inoculated concentrations of *Xanthomonas oryzae* on the size of diseased spots. *Bull. Kyushu agric. Exp. Stn* **4**: 121–127. [Jap. Engl. summ.]

KUHARA, S., KURITA, T., TAGAMI, Y., FUJII, H. & SEKIYA, N. 1965. Studies on the strain of *Xanthomonas oryzae* (Uyeda et Ishiyama) Dowson, the pathogen of the bacterial leaf blight of rice with special reference to its pathogenicity and phage-sensitivity. *Bull. Kyushu agric. Exp. Stn* **11**: 263–312. [Jap. Engl. summ.]

KUHARA, S., WATANABE, M., TABEI, H. & MUKO, H. 1958. Varietal reaction to bacterial leaf blight disease—pathogenicity of various strains. Abs. in *Ann. phytopath. Soc. Japan* **23**: 9. [Jap.]

KUO, T. T., HUANG, T. C., WU, R. Y. & YANG, C. M. 1967. Characterization of three bacteriophages of *Xanthomonas oryzae* (Uyeda et Ishiyama) Dowson. *Bot. Bull. Acad. sin., Taipei* **8** (Special Issue): 246-254.

KUO, T. T., YANG, C. M., CHOW, T. H. & LIN, Y. T. 1970a. Bacterial leaf blight of rice plant II. The formation and properties of spheroplasts of *X. oryzae*. *Bot. Bull. Acad. sin.*, Taipei **11**: 36–45.

KUO, T. T., LIN, B. C. & LI, C. C. 1970b. Bacterial leaf blight of rice plant III. Phytotoxic polysaccharides produced by *Xanthomonas oryzae*. *Bot. Bull. Acad. sin., Taipei* **11**: 46–54.

KUSABA, T., WATANABE, M. & TABEI, H. 1966. Classification of the strains of *Xanthomonas oryzae* (Uyeda et Ishiyama) Dowson on the basis of their virulence against rice plants. *Bull. natn. Inst. Agric. Sci., Tokyo*, Ser. C, **20**: 67–82. [Jap. Engl. summ.]

LIN, B. C., LI, C. C. & KUO, T. T. 1969. Bacterial leaf blight of rice. I. Serological relationships between virulent strain and weakly virulent strain of *Xanthomonas oryzae*. *Bot. Bull. Acad. sin., Taipei* **10**: 130.

MIHASHI, H., MAKI, Y., NIINO, K. & SEIKE, Y. 1953. *Rep. Ehime agric. Exp. Stn*: 1–75 (From Tagami & Mizukami, 1962.)

MINISTRY OF AGRICULTURE AND FORESTRY. 1958. Guidebook of disease forecasting. Japan, Bureau of Agricultural Development.

MIZUKAMI, T. 1957. On the relationships between *Xanthomonas oryzae* and the roots of rice seedlings. *Agric. Bull. Saga Univ.* **6**: 87–93. [Jap. Engl. summ.]

MIZUKAMI, T. 1961. Studies on the ecological properties of *Xanthomonas oryzae* (Uyeda et Ishiyama) Dowson, the causal organism of bacterial leaf blight of rice plant. *Agric. Bull. Saga Univ.* **13**: 1–85. [Jap. Engl. summ.]

MIZUKAMI, T. 1964. Occurrence of bacterial leaf blight disease in India. *J. Pl. Prot., Tokyo* **18**: 179–181. [Jap.]

MIZUKAMI, T. 1966. Epidemiology of bacterial leaf blight of rice and use of phages for forecasting. Paper presented at the Eleventh Pacific Science Congress, Symposium on Plant Diseases in the Pacific, pp. 15–32.

MIZUKAMI, T. & MURAYAMA, Y. 1960. Studies on the bacterial leaf blight resistance of rice plant. 1. On the relationship between the growth of *X. oryzae* in a rice plant leaf and the free amino acids in it. *Agric. Bull. Saga Univ.* **11**: 75–82. [Jap. Engl. summ.]

MIZUKAMI, T. & SEKI, M. 1954. On the effects of some fungicides upon the infection and the development of lesions of the bacterial leaf blight of rice plant. *Kyushu agric. Res.* **14**: 209–211. [Jap.]

MIZUKAMI, T. & WAKIMOTO, S. 1969. Epidemiology and control of bacterial leaf blight of rice. *A. Rev. Phytopath.* **7**: 51–72.

MUKO, H. & ISAKA, M. 1964. Re-examination of some physiological characteristics of *Xanthomonas oryzae* (Uyeda et Ishiyama) Dowson. *Ann. phytopath. Soc. Japan* **29**: 13–19. [Jap. Engl. summ.]

MUKO, H., KUSABA, T., WATANABE, M., TABEI, H. & TSUCHIYA, Y. 1957. Effect of major environmental factors on the development of bacterial leaf blight of rice. *Ann. phytopath. Soc. Japan* **22**: 10. [Jap.]

MUKO, H., YOSHIDA, K., KUSABA, T., TABEI, H. & TSUCHIYA, Y. 1952. On the varietal resistance of rice plant to bacterial leaf blight. Abs. in *Ann. phytopath. Soc. Japan* **17**: 42. [Jap.]

MUKO, H. & WATANABE, M. 1958. Studies on nutritional physiology of the organism causing bacterial leaf blight of rice. (2) Carbon and nitrogen sources in relation to growth. *Ann. phytopath. Soc. Japan* **23**: 9. [Jap.]

MUKO, H. & YOSHIDA, K. 1951. A needle inoculation method for bacterial leaf blight disease of rice. *Ann. phytopath, Soc. Japan* **15**: 179. [Jap.]

MURATA, N. 1967. Genetic aspects of resistance to bacterial leaf blight in rice and variation of its causal bacterium. *Proceedings of a symposium on rice diseases and their control by growing resistant varieties and other measures:* 39–49. Japan, Agric. For. & Fish. Res. Council, Minist. Agric. For.

NISHIMURA, Y. & SAKAGUCHI, S. 1959. Inheritance of resistance in rice to bacterial leaf blight *Bacterium oryzae* (Uyeda et Ishiyama) Nakata. Abs. in *Jap. J. Breed.* **9**: 58. [Jap.]

PADMANABHAN, S. Y. & JAIN, S. S. 1966. Effect of chlorination of water on control of bacterial leaf blight of rice caused by *Xanthomonas oryzae* (Uyeda et Ishiyama) Dowson. *Curr. Sci.* **35**: 610–611.

PADWICK, G. W. 1950. Manual of rice diseases. 198 pp., Kew, Commonwealth Mycological Institute.

PORDESIMO, A. N. 1958. Bacterial blight of rice. *Philipp. Agric.* **42**: 115–128.

REINKING, O. A. 1918. Philippine economic plant diseases. *Philipp. J. Sci.*, A, **13**: 165–274.

REITSMA, J. & SCHURE, P. S. J. 1950 'Kresek', a bacterial disease of rice. *Contr. gen. agric. Res. Stn Bogor* 117, 17 pp.

SAKAGUCHI, S., SUWA, T. & MURTA, N. 1968. Studies on the resistance to bacterial leaf blight, *Xanthomonas oryzae* (Uyeda et Ishiyama) Dowson, in the cultivated and wild rice. *Bull. natn. Inst. Agric. Sci., Tokyo*, Ser. D, **18**: 1–29. [Jap. Engl. summ.]

SCHURE, P. S. J. 1953. Attempts to control the kresek disease of rice by chemical treatment of the seedlings. *Contr. gen. agric. Res. Stn Bogor* 136, 17 pp.

SHEKHAWAT, G. S. & SRIVASTAVA, D. N. 1968. Variability in Indian isolates of *Xanthomonas oryzae* (Uyeda et Ishiyama) Dowson, the incitant of bacterial leaf blight of rice. *Ann. phytopath, Soc. Japan* **34**: 289–297.

SRINIVASAN, M. C., THIRUMALACHAR, M. J. & PATEL, M. K. 1959. Bacterial blight disease of rice. *Curr. Sci.* **28**: 469–470.

SRIVASTAVA, D. N. 1967. Epidemiology and control of bacterial blight of rice in India. *Proceedings of a symposium on rice diseases and their control by growing resistant varieties and other measures*: 11–18. Japan, Agric. For. and Fish. Res. Council, Minist. Agric. For.

SRIVASTAVA, D. N. & RAO, Y. P. 1964a. Paddy farmers should beware of bacterial blight. *Indian Farming* **14** (6): 32-33.

SRIVASTAVA, D. N. & RAO, Y. P. 1964b. Seed transmission and epidemiology of the bacterial blight disease of rice in North India. *Indian Phytopath.* **17**: 77–78.

SUWA, T. 1960. Single cell culture of *Xanthomonas oryzae* (Uyeda et Ishiyama) Dowson, the causal organism of bacterial leaf blight of rice plant. *Ann. phytopath. Soc. Japan* **25**: 199–201. [Jap. Engl. summ.]

SUWA, T. 1962. Studies on the culture media of *Xanthomonas oryzae* (Uyeda et Ishiyama) Dowson. *Ann. phytopath. Soc. Japan* **27**: 165–171. [Jap. Engl. summ.]

TABEI, H., KUSABA, T., WATANABE, M. & MUKO, H. 1957. Isolation and quantitative determination of streptomycin resistant strains of the bacterial leaf blight organism. *Ann. phytopath. Soc. Japan* **22**: 9. [Jap.]

TABEI, H. & MUKO, H. 1960. Anatomical studies of rice plant leaves affected with bacterial leaf blight (*Xanthomonas oryzae*), in particular reference to the structure of water exudation system. *Bull. natn. Inst. Agric. Sci., Tokyo*, Ser. C, **11**: 37–43. [Jap. Engl. summ.]

TAGAMI, Y. 1958. Quantity of phage of bacterial leaf blight disease organism in paddy field water and occurrence of the disease. *Proc. Ass. Pl. Prot., Kyushu* **4**: 63–64. [Jap.]

TAGAMI, Y. 1959a. Quantity of bacteriophage in the water of the seedbed in relation to the occurrence of the disease in the rice field. Abs. in *Ann. phytopath. Soc. Japan* **24**: 6. [Jap.]

TAGAMI, Y. 1959b. Relation between quantity of bacterial leaf blignt organism and bacteriophage and occurrence of the disease. *J. Pl. Prot., Tokyo* **13**: 389–394. [Jap.]

TAGAMI, Y., KUHARA, S., KURITA, T., FUJII, H., SEKIYA, N., YOSHIMURA, S., SATO, T. & WATANABE, B. 1963. Epidemiological studies on the bacterial leaf blight of rice, *Xanthomonas oryzae* (Uyeda et Ishiyama) Dowson. I. The overwintering of the pathogen. *Bull. Kyushu agric. Exp. Stn* **9**: 89–122. [Jap. Engl. summ.]

TAGAMI, Y., KUHARA, S., KURITA, T., FUJII, H., SEKIYA, N. & SATO, T. 1964. Epidemiological studies on the bacterial leaf blight of rice, *Xanthomonas oryzae* (Uyeda et Ishiyama) Dowson. 2. Successive changes in the population of the pathogen in paddy fields during rice growing period, *Bull. Kyushu agric. Exp. Stn* **10**:23–50. [Jap. Engl. summ.]

TAGAMI, Y. & MIZUKAMI, T. 1962. Historical review of the researches on bacterial leaf blight of rice caused by *Xanthomonas oryzae* (Uyeda et Ishiyama) Dowson. *Special Rep. Pl. Dis. and Insect Pests Forecasting Serv.* 10, 112 pp. [Jap. Engl. transl. by H. Fujii.]

TANAKA, Y. 1963; 1964a. Studies on the nutritional physiology of *Xanthomonas oryzae* (Uyeda et Ishiyama) Dowson. I. On the nitrogen sources. II. On the carbon sources and a new synthetic medium. *Sci. Bull. Fac. Agric. Kyushu Univ.* **20**: 151–155; **21**: 149–153. [Jap. Engl. summ.]

TANAKA, Y. 1964b. Studies on the phage resistant strains of *Xanthomonas oryzae* (Uyeda et Ishiyama) Dowson. *Sci. Bull. Fac. Agric. Kyushu Univ.* **21**: 161–167. [Jap. Engl. summ.]

UEHARA, K. 1960. On phytoalexin produced by the results of the interaction of the rice plant and the leaf blight bacteria (*Xanthomonas oryzae*). *Ann. phytopath. Soc. Japan* **25**: 149–155. [Jap. Engl. summ.]

VZOROFF, V. I. 1938. Summary of the scientific research work of the Institute of Plant Protection for the year 1936. III. Virus and bacterial diseases of plants, the biological, the chemical, and the mechanical methods of plant protection: 40–45. Leningrad, State Publ. Off. Lit. collect. co-op. Farming 'Selkhozgiz.' (Abs in *Rev. appl. Mycol.* **18**: 378–379.)

WAKIMOTO, S. 1954a. Biological and physiological properties of *Xanthomonas oryzae* phage. *Sci. Bull. Fac. Agric. Kyushu Univ.* **14**: 485–493. [Jap. Engl. summ.]

WAKIMOTO, S. 1954b. The determination of the presence of *Xanthomonas oryzae* by the phage technique. *Sci. Bull. Fac. Agric. Kyushu Univ.* **14**: 495–498. [Jap. Engl. summ.]

WAKIMOTO, S. 1956. Hibernation of leaf blight disease bacteria in soil. *Agriculture Hort., Tokyo* **31**: 1413–1414. [Jap.]

WAKIMOTO, S. 1957. A simple method for comparison of bacterial populations in a large number of samples by phage technique. *Ann. phytopath. Soc. Japan* **22**: 159–163. [Jap. Engl. summ.]

WAKIMOTO, S. 1960. Classification of strains of *Xanthomonas oryzae* on the basis of their susceptibility against bacteriophages. *Ann. phytopath. Soc. Japan* **25**: 193–198.

WAKIMOTO, S. 1967. Strains of *Xanthomonas oryzae* in Asia and their virulence against rice varieties. *Proceedings of a symposium on rice diseases and their control by growing resistant varieties and other measures*: 19–24. Japan, Agric. For. and Fish. Res. Council, Minist. Agric. For.

WAKIMOTO, S. & MUKO, H. 1963. Natural occurrence of streptomycin resistant *Xanthomonas oryzae*, the causal bacteria of leaf blight disease of rice. *Ann. phytopath. Soc. Japan* **28**: 153–158. [Jap. Engl. summ.]

WAKIMOTO, S. & YOSHII, H. 1954a. Seasonal change of resistance of rice plant against leaf blight disease. *Sci. Bull. Fac. Agric. Kyushu Univ.* **14**: 475–477. [Jap. Engl. summ.]

WAKIMOTO, S. & YOSHII, H. 1954b. On the variability of virulence of *Xanthomonas oryzae* under successive infection against the resistant or the susceptible variety of rice. *Sci. Bull. Fac. Agric. Kyushu Univ.* **14**: 479–484. [Jap. Engl summ.]

WAKIMOTO, S. & YOSHII, H. 1955. Quantitative determination of the population of a bacterium by the phage technique. *Sci. Bull. Fac. Agric. Kyushu Univ.* **15**: 161–169. [Jap. Engl. summ.]

WASHIO, O., KARIYA, K. & TORIYAMA, K. 1966. Studies on breeding rice varieties for resistance to bacterial leaf blight. *Bull. Chugoku agric. Exp. Stn*, Ser. A, **13**: 55–86. [Jap. Engl. summ.]

WATANABE, M. 1963. Studies on the nutritional physiology of *Xanthomonas oryzae* (Uyeda et Ishiyama) Dowson. II. Effects of carbon and nitrogen sources on multiplication of bacteria. *Ann. phytopath. Soc. Japan* **28**: 201–208. [Jap. Engl. summ.]

WATANABE, M. 1966. Studies on the strains of *Xanthomonas oryzae* (Uyeda et Ishiyama) Dowson, the causal bacterium of the bacterial leaf blight of rice plant. *Bull. Fac. Agric. Tokyo Univ. Agric. & Tech.* **10**: 1–51.

YAMAMOTO, H. & KUSAKA, T. 1965. A correlation between pathogenicities and antibiotic susceptibilities in various strains of *Xanthomonas oryzae*. *Ann. phytopath. Soc. Japan* **30**: 169–173. [Jap. Engl. summ.]

YONEYAMA, K., SHIMENO, K., TAGUCHI, R. & MISATO, T. 1969. Studies on the screening methods of the chemicals for bacterial leaf blight of rice plant. *J. agric. Chem. Soc. Japan* **43**: 851–856. [Jap. Engl. summ.]

YOSHII, H., YOSHIDA, T. & MATSUI, C. 1953. Bacteriophage of the bacterial leaf blight organism of rice. *Ann. phytopath. Soc. Japan* **17**: 177. [Jap.]

YOSHIMURA, S. & IWATA, K. 1965. Studies on examination method of varietal resistance to bacterial leaf blight disease of rice plant. I. Immersion method of inoculation and its applied method. *Proc. Ass. Pl. Prot. Hokuriku* **13**: 25–31. [Jap.]

YOSHIMURA, S., MORIHASHI, T. & SUZUKI, Y. 1959. Major weed-hosts for the overwintering of the bacterial leaf blight organism and its distribution and development in Hokuriku District. *Ann. phytopath, Soc. Japan* **24**: 6. [Jap.]

YOSHIMURA, S. & TAGAMI, Y. 1967. Forecasting and control of bacterial leaf blight of rice in Japan. *Proceedings of the symposium on rice diseases and their control by growing resistant varieties and other measures*: 25–38. Japan, Agric. For. and Fish. Res. Council, Minist. Agric. For.

YOSHIMURA, S. & TAHARA, S. K. 1960. Morphology of bacterial leaf blight organism under (electron) microscope. *Ann. phytopath. Soc. Japan* **26**: 61. [Jap.]

BACTERIAL LEAF STREAK

HISTORY, DISTRIBUTION AND DAMAGE

Reinking (1918) reported a bacterial leaf stripe disease on rice in the Philippines. The symptoms described are: ' In the young stage the stripes are from 0·5 to 1 millimeter wide and from 3 to 5 millimeter long, run lengthwise, and have a watery dark green translucent appearance. In this stage, the disease is usually confined to the portion between the larger veins. . . .' These early symptoms are characteristic of bacterial leaf streak disease. Reinking did not, however, identify the bacterium causing the disease.

Pordesimo (1958) studied the causal organism in the Philippines, and called the organism *X. translucens* (J.J. & R.) Dowson f. sp. *oryzae* (Uyeda & Ishiyama) n. comb. He, however, mis-identified the disease as bacterial leaf blight.

Fang *et al.* (1957) in China were the first to distinguish the disease from bacterial leaf blight and called the disease bacterial leaf streak. They gave the organism a new name, *X. oryzicola*.

Goto (1964) compared the organism with several formae speciales of *X. translucens*, and stated that he was ' inclined to adopt the scientific name *X. translucens* f. sp. *oryzae* Pordesimo instead of *X. oryzicola* Fang *et al.*' However, Bradbury (1971) pointed out that this name was proposed by Pordesimo for *X. oryzae*, and derived from the specific name given to that organism by Uyeda and Ishiyama. It is misleading to use it for a different organism and he therefore proposed the combination *Xanthomonas translucens* (Jones, Johnson & Reddy) Dowson f. sp. *oryzicola* (Fang *et al.*) Bradbury.

Since the name bacterial stripe has been used for a quite separate disease on rice and other cereals caused by *Pseudomonas panici* (Elliott) Goto & Okabe, the name bacterial streak given by Fang *et al.* appears an appropriate one for this disease.

In spite of the fact that there have been few reports concerning it, the disease is widely distributed in Tropical Asia. Besides the Philippines and South China, the disease has been observed in Thailand, Malaysia and India, has been noted in Vietnam (unpublished) and is known to occur in Indonesia and Cambodia. It has not been found in Japan or other temperate countries.

An accurate estimate of the losses caused by the disease is not available. However, under favourable weather conditions, particularly typhoons and rain storms, the disease spreads through entire fields, all the upper leaves become infected and turn brown, and the consequent losses can be compared to those arising from heavy damage caused by bacterial leaf blight (*X. oryzae*).

SYMPTOMS

Beginning with fine translucent streaks as described above, the lesions enlarge lengthwise and may also advance laterally over the larger veins. Bacterial exudates appear on the surface of lesions under humid conditions and dry up to form very small, yellow beads which are often numerous on the linear lesions. Moistened by dew or rain, the bacterial cells in these beads are dispersed and spread by wind to cause new infections on the same leaf or on other leaves. Old lesions become light brown. On susceptible varieties a yellow halo may appear around the lesions. At the later stages of disease development, entire leaves turn brown and die. They become bleached and appear greyish white and are colonised by many saprophytic organisms. At this late stage the disease is indistinguishable from bacterial leaf blight (Plate V).

THE ORGANISM

Fang *et al.* (1957) described the morphology and physiology of the bacterium as follows: ' Rods, $1 \cdot 2 \times 0 \cdot 3$–$0 \cdot 5\mu$, single, occasionally in pairs but not in chains; no spores and no capsules; motile by a single polar flagellum (Fig. II–13). Gram negative. Aerobic, grow favorably at 28°C. Colonies on nutrient agar pale yellow, circular, smooth, margin entire, convex and viscid. Growth on slant filiform. Growth in nutrient broth moderate, surface growth ring form but no definite pellicle. No growth on Cohn's or Fermi's solution. Gelatin liquefied; milk not coagulated but peptonized, reaction of litmus slightly alkaline and litmus mostly reduced. Nitrites not produced from nitrates; hydrogen sulphide and ammonia produced; no indole. Acid but no gas from dextrose, sucrose, xylose and mannose; no acid from lactose, maltose, arabinose, mannitol, glycerol and salicin. Starch not hydrolyzed. Methyl red and V.P. tests negative.'

Fang *et al.* (1957) also described another new species, *X. leersiae* which produced streak symptoms on the common weed *Leersia hexandra* but was only weakly pathogenic to rice. Its morphology and physiology are very similar to those of the rice streak organism.

Virulence. More than 150 isolates of the organism from the Philippines have been compared for virulence on two rice varieties, Peta and Pah Leuad 111. It was shown that some were weakly virulent, producing lesions less than 0·5 cm long, whereas others were very virulent, producing lesions ranging in length from 5 to 10 cm. The majority of the isolates were of intermediate virulence. From the different groups of isolates, 4 of the least virulent, 3 very virulent and 1 intermediate were selected and inoculated separately on 36 selected varieties showing various degrees of disease reaction. The results showed that resistant varieties were resistant to all isolates and susceptible varieties exhibited the longest lesions and were susceptible to all isolates tested. It is believed therefore that strains differ greatly in virulence but that pathogenic races with opposite reaction in a given variety do not exist in the organism (IRRI, 1968; Ou *et al.*, 1971). Another 25 isolates inoculated on 6 varieties showed that the pathogenicity patterns of this organism are very similar to those of *X. oryzae* (IRRI, 1968).

PLATE V

Upper picture—*Bacterial leaf streak* showing narrow linear lesions.

Lower left—*Bacterial leaf streak* in field plots, showing resistant and susceptible varieties.

Lower right—*Bacterial leaf streak* showing young lesions.

PLATE V

TABLE II–9

PROPERTIES OF 6 PHAGES OF *X. translucens* F. SP. *oryzicola*. (Goto & Okabe, 1965).

Property	Phage Sp1	Sp2	Sp3	Sp4	Sp5	Sp6
Serological type	I	I	II	I	I	III
Size (mμ) Head	50–60	50–60	60–70	50–60	50–60	50–55
Tail	10 × 120–160	10 × 120–160	15 × 120–160	10 × 120–160	10 × 120–160	10 × 130–200
Shape	Tadpole	Tadpole	Tadpole	Tadpole	Tadpole	Tadpole
Plaque size (mm diameter)	35 (S7)*1	17 (S2)	210 (S24)	30 (S3)	68 (S24)	63 (S40)
Latent period (min)	45	55	180	45	45	60
Inactivation temperature (°C)	56–58	56–58	62–64	56–58	56–58	56–58
pH stability	5–10*2	5–10	5–10	5–10	5–10	5–10
UV sensitivity	S*3	S	HS	S	S	S
Host range	Intraspecies specific	Intraspecies specific	Intraspecies specific	Intraspecies specific	Intraspecies specific	Intraspecies specific

*1 The isolate in brackets was used for step growth experiment.

*2 pH range in which 50% PFP survived.

*3 S–sensitive, HS–highly sensitive.

Bacteriophage. Six different phages were studied by Goto & Okabe (1965). They are called Sp phages, and are divided into three serological types. Serological type I includes Sp 1, Sp 2, Sp 4 and Sp 5, type II includes Sp 3 and type III includes Sp 6. The properties of the 6 Sp phages are tabulated in Table II–9 (see also Fig. II–13).

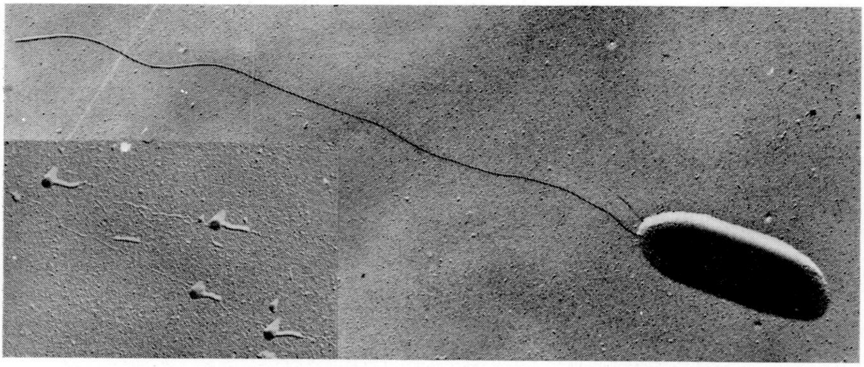

Fig. II–13. *Xanthomonas translucens* f. sp. *oryzicola* (× 22,000 approx.) and phage Sp 1 (left corner, × 28,000 approx.). (Courtesy of Dr. M. Goto.)

Recent studies (IRRI, 1968) indicated that most of these 6 phages are not stable, 9 new phages being produced and detected. They were referred to as Sp-1-1, Sp-1-2, Sp-2-1, Sp-2-2, Sp-2-3, Sp-4-1, Sp-5-1, Sp-6-1 and Sp-6-2. In addition, 18 new phages, based upon host reaction (strains), have been isolated. They were designated as Sp-7 to Sp-24.

Disease Cycle and Effect of Environmental Conditions

The bacterium infects leaves through the stomata. Pordesimo (1958) indicated that a higher percentage of infection was obtained by spraying leaves with a bacterial suspension at midday than in the morning or afternoon, presumably because of the stomatal opening. Detailed anatomical and histological studies on the processes of lesion development are lacking, but it is known that in the early stages the organism invades only the parenchyma cells in between the veins of the leaf.

Soon after lesions develop, a bacterial exudate forms on the surface of the lesions under moist conditions during the night. Under dry conditions these exudates become small, yellow beads which are often numerous on young lesions; they eventually fall into the field water. When the leaves are wet from dew or rain, with the aid of wind the organism spreads from leaf to leaf and often quickly throughout an entire field. In a glasshouse experiment at IRRI, two sets of plants were sprayed with a bacterial suspension; the leaves of one set were disturbed by hand once every morning, the other set was left untouched and more or less still. After a period of three weeks all the leaves in the disturbed set were infected and the number of lesions was 100 times that on the undisturbed set. The leaves of the former set soon began to turn brownish due to heavy infection, but in the undisturbed set there were only relatively few lesions and these were only on the leaves which had been sprayed. Thus the disease in the field often becomes very severe after the rainstorms or typhoons of the rainy season in the tropics. The organism is carried from field to field by irrigation water. Seed harvested from a diseased field may also carry the organism (Faan & Wu, 1965).

The high temperatures prevalent in the tropics permit the growth of the organism throughout the year, although in many areas there is an annual dry, hot period for several months. Central Luzon in the Philippines is one of these areas, but we have found here that wherever there is a little water, either in ponds or streams, a relatively high population of the bacteriophage is present (IRRI, 1967) (Table II–4). This may indicate the presence of the bacterium and indicates, therefore, a source of primary inoculum when the rice crop is planted in the following season.

The bacterial population in irrigation waters is high throughout the rice growing season. In fields of a susceptible rice variety, the population may be very high (estimated to be as many as 50,000 plaques per ml of field water) while with a resistant variety the population is lower (IRRI, 1968).

High temperatures were found to favour the development of leaf streak lesions, but high humidity did not have much effect on lesion development. High humidity did, however, favour the dissemination of the organism, as stated above, and thus affected the number of lesions. This may partly explain the common presence of the disease in the tropics. The nitrogen nutrition of the host plant influences lesion development only slightly (IRRI, 1967).

Host Range

None of the 13 species in 10 genera of gramineous weeds and 10 varieties of sorghum, maize, millet and wheat studied at IRRI (1968) was infected by artificial inoculation. However, all 49 species and strains of wild rice tested can serve as alternative hosts. They include strains of *Oryza spontanea*, *O. perennis*, *O. perennis balunga*, *O. nivara*, *O. breviligulata* and *O. glaberrima*.

Varietal Resistance and Control

Experimental infections may be accomplished by spraying a bacterial suspension on to seedlings. Studies at IRRI (1967) have shown that 3-week-old seedlings may conveniently be used for inoculation tests; that a concentration of bacterial suspension between 0·15 and 0·80 optical density at 640mμ may be used, and within this range the concentration does not influence the results materially; that 15 hours in a high humidity chamber after inoculation is sufficient; and that readings may be made about 2 weeks after inoculation (later readings may confuse results due to lesions developed from secondary infection). It was also found that length of lesion is a more reliable criterion for judging the degree of resistance than the number of lesions because the latter is greatly affected by environmental conditions, such as light which affects the mechanism of stomatal opening. A scale of 10 units is suggested for indicating disease reaction. A scale for disease incidence observed in the field has also been developed. The two scales were found to correlate very closely (r = 0·935) (IRRI, 1967); they are described in Table II–10. Screening for

TABLE II–10

SCALES FOR RATING VARIETAL REACTION TO BACTERIAL LEAF STREAK IN GREENHOUSE INOCULATION AND FIELD VISUAL OBSERVATION. (IRRI, 1968).

Scale	Greenhouse inoculation	Field observation	
0	No lesions	no lesions	Field looks green
1	< 1 mm	sporadically a few lesions	
2	1–2 mm	very few lesions per plant	
3	2–5 mm	few lesions per plant but all over the field	
4	6–10 mm	generally few lesions but some plants heavily infected	
5	11–20 mm	plenty of lesions per plant all over the field	
6	21–30 mm	plenty of lesions and some leaf tips yellow	Field looks yellow
7	31–40 mm	leaf tips yellow	
8	41–60 mm	whole parts of leaves yellow; leaves drying	
9	> 60 mm	leaves drying	

resistant varieties, Goto (1965) tested 102 varieties and found that many of the *japonica* varieties were relatively resistant while most of the *indica* varieties were susceptible. Further tests at IRRI (1968) showed that, among some 700 varieties tested to virulent isolates of the organism, many had a high degree of resistance. They are listed in Table II–11.

TABLE II–11

BACTERIAL STREAK RESISTANT VARIETIES IN THE PHILIPPINES. (IRRI, 1968).

Varieties with reaction Scale 1

1. DZ-60 (IRRI Acc. No. 8558)
2. *Oryza perennis* (100192)
3. Australian *spontanea* (100944)
4. ,, ,, (101146)
5. ,, ,, (101147)
6. ,, ,, (101148)

Varieties with reaction Scale 2

1. Milbuen 5 (3) (2 entries) (49)
2. Hsinchu 56 (77)
3. Ch 242 (87, 7114)
4. Bluerose (551)
5. Norin 37 (2556)
6. Hashikalmi (3397)
7. Charnock (Aus.) (3685)
8. NP–97 (3700)
9. Bmt. 53 R3540 (4132)
10. Huan-sen-goo (4619)
11. Pi 4 (6733)
12. DNJ–142 (8426)
13. DJ–65 (8485)
14. DZ–192 (8518)
15. DZ–179 (8520)
16. DZ–153 (8538)
17. DZ–74 (8556)
18. DZ–39 (8564)
19. DL–10 (8597)
20. DD–89 (8644)
21. DD–96 (2 entries) (8647)
22. DD–100 (8649)
23. DD–113 (8657)
24. Ctg. 1118 (8712)
25. DV–29 (2 entries) (8816)
26. DV–52 (2 entries) (8828)
27. DV–57 (8831)
28. DV–68 (8833)
29. DV–98 (8845)
30. DV–150 (8876)
31. JW–107 (9129)
32. *Oryza glaberrima* (100140)
33. *O. glaberrima* (100143)
34. *O. perennis* (100211)
35. *O. breviligulata* (100927)
36. Australian *spontanea* (101145)
37. *Oryza rufipogon* (101448)

Row *et al.* (1968) screened several hundred varieties in the field in India, and used a scale of 8 units for scoring disease reaction in the field, and a scale of 5 units for artificial inoculation. Eighty-eight varieties including Zenith, Tetep, H4, S67 and CO4 were found resistant in the field. Among the 71 varieties tested by artificial inoculation, Hsinchu 56, Kaohsiung 68 and Nira (Lacrosse × Zenith) were resistant.

Faan & Wu (1965) reported that seed treatment with an organic mercury fungicide controlled the disease in Kwantung, China.

LITERATURE CITED

BRADBURY, J. F. 1971. Nomenclature of the bacterial leaf streak pathogen of rice. *Int. Jnl syst. Bact.* **21**: 72.

FAAN, H. C. & WU, S. C. 1965. Field control of bacterial leaf streak (*Xanthomonas oryzicola*) of rice in Kwantung. *Acta phytophylac. sin.* **4**: 1–6. [Chin. Engl. summ.]

FANG, C. T., REN, H. C., CHEN, T. Y., CHU, Y. K., FAAN, H. C. & WU, S. C. 1957. A comparison of the rice bacterial leaf blight organism with the bacterial leaf streak organism of rice and *Leersia hexandra* Swartz. *Acta phytopath. sin.* **3**: 99–124. [Chin. Engl. summ.]

GOTO, M. 1964. Nomenclature of the bacteria causing bacterial leaf streak and bacterial stripe of rice. *Bull. Fac. Agric. Shizuoka Univ.* **14**: 3–10.

GOTO, M. 1965. Resistance of rice varieties and species of wild rice to bacterial leaf blight and bacterial leaf streak diseases. *Philipp. Agric.* **48**: 329–338.

GOTO, M. & OKABE, N. 1965. Bacteriophages of *Xanthomonas translucens* f. sp. *oryzae*, the pathogen of bacterial leaf streak of rice. *Bull. Fac. Agric. Shizuoka Univ.* **15**: 39–44.

IRRI (INTERNATIONAL RICE RESEARCH INSTITUTE). Annual Report for 1967, 1968.

OU, S. H., FRANCK, G. P. & MERCA, S. D. 1971. Varietal resistance to bacterial leaf streak disease of rice. *Philipp. Agric.* **54**: 8–32.

PORDESIMO, A. N. 1958. Bacterial blight of rice. *Philipp Agric.* **42**: 115–128.

REINKING, O. A. 1918. Philippine economic plant diseases. *Philipp. J. Sci.* Ser. A, **13**: 165–274.

ROW, V. S. R., DEVADATH, S., DATH, A. P. & PADMANABHAN, S. Y. 1968. Studies on the bacterial leaf streak of rice. *Riso* **17**: 327–336.

BACTERIAL STRIPE

This disease was reported on rice by Goto & Ohata (1961) and the organism was referred to as *Pseudomonas setariae*. This was later considered to be identical to the previously known *P. panici*. It was found in Japan, Taiwan and the Philippines. It occurs on seedlings in upland nurseries and causes only little damage.

SYMPTOMS

The disease starts from the lower part of the leaf sheath, where water-soaked, dark green, longitudinal stripes are formed. Under humid conditions, the lesions soon elongate to encompass the whole length of the sheath, turn to reddish or dark brown and sometimes have on the surface glistening scales of bacterial exudate. The lesions are usually a few to about 10 cm long, 0·5 to 1 mm wide, sometimes coalescing to form wider lesions. When the infection is light, plants survive and grow without suffering much damage, but heavy infection may cause stunting and killing of seedlings. Attack may also occur on the young unfolding leaves and cause bud rot; if this happens the plant develops no further and eventually dies. Infection on mature plants has not been observed.

THE ORGANISM

Morphology. The cells are rods, with 1, rarely 2–3 polar flagella, 1·5–2·5 × 0·5–0·8μ; no capsule nor endospore has been observed; Gram negative. On nutrient agar plates, colonies are circular, raised, glistening white by reflected light, slightly iridescent by transmitted light, translucent, the surface smooth, and the margin entire.

Daughter colonies often develop around the mother colony so that the margin later becomes undulate. In nutrient broth turbidity is abundant and uniform, forming a white ring along the wall at the surface and a pellicle.

Physiology. Aerobic; slow liquefaction of gelatin; no hydrogen sulphide produced; nitrate reduced; ammonia produced; soluble starch and cellulose hydrolized; acid produced but no gas in xylose, glucose, fructose, galactose, mannose, glycerol, mannitol and sorbitol; no fermentation in lactose, sucrose, maltose, starch, inulin, dextrin, dulcitol, inositol and salicin; malate, citrate and succinate utilized as sole carbon sources but not tartarate; minimum temperature 2–6°C, optimum 26–30°C, maximum 40–42°C, thermal death point 51–53°C (Goto, 1964).

Nomenclature. The organism is identical with *Pseudomonas panicimiliacei* and *P. setariae* according to Bergey's manual. *Bacterium panici* was the first name given to the organism, by Elliott, who found it on Proso millet. Savulescu transferred it to *Xanthomonas panici*, which was adopted by Bergey's manual, but according to Goto & Okabe (1952) the organism belongs to *Pseudomonas* and should be called *P. panici* (Elliott) Stapp, with other names as synonyms. Goto (1964) further studied the organism by cross-inoculation with 6 isolates from rice, 2 from barley, 2 from Italian millet and 4 from Proso millet and confirmed their earlier conclusion that all isolates are identical and that the organism belongs to *Pseudomonas*. He also stated that the symptoms were almost the same on these hosts except that the disease was rather milder on rice than on the others.

LITERATURE CITED

GOTO, K. & OHATA, K. I. 1961. Bacterial stripe of rice. *Spec. Publs Coll. Agric. natn. Taiwan Univ.* **10**: 49–59.

GOTO, M. 1964. Nomenclature of the bacteria causing bacterial leaf streak and bacterial stripe of rice. *Bull. Fac. Agric. Shizuoka Univ.* **14**: 3–10.

GOTO, M. & OKABE, N. 1952. Studies on the causal organisms of bacterial stripe disease of millet and the brown stripe disease of Italian millet. *Bull. Fac. Agric. Shizuoka Univ.* **2**: 15–24. [Jap.]

BACTERIAL SHEATH ROT

Klement (1955) reported a new bacterial disease of rice in Hungary, attacking the leaves sheathing the panicles, and identified the bacterium as *Pseudomonas oryzicola* sp. nov. It also attacks the stems, producing blurred spots which later turn brown or black, and also the nodes, panicles and seeds which become sterile. The disease frequently develops before the ears emerge, while the panicles are still within the sheaths.

The bacterium forms white colonies and produces a green, fluorescent pigment in culture. It is reported to differ from *Xanthomonas itoana* and *X. oryzae* in liquefaction of gelatin, reduction of nitrate, production of ammonia, formation of hydrogen sulphide, and action in litmus milk and sugar decomposition.

Fang & Ren (1960) also reported *P. oryzicola* in the eastern and northeastern provinces of China. Khu & Bai (1960) reported a brown leaf spot from China, and the causal bacterium is said to be similar to *P. oryzicola* but to differ from it in acid formation in maltose, weak hydrolysis of starch and smaller size, being 1–3 × 0·8–1μ, as compared with 2·5–3·5 × 1·3μ for *P. oryzicola*. It also attacks 30 species of cultivated and wild Gramineae.

Funayama & Hirano (1963) in Japan reported a *Pseudomonas*, identified as *P. oryzicola* Klement, causing brown discoloration of the leaf sheath.

Goto (1965) found a bacterial sheath rot disease in Indonesia. The organism causes water-soaked, dark green, elongated spots on the sheath which enlarge and turn reddish or dark brown and may extend to the culm. In some cases the whole sheath withers and the panicles may develop lesions which produce a white or yellow rot. He compared the bacteriological properties of the isolates from Indonesia with those of two isolates of *P. oryzicola* from Japan and one isolate of *Erwinia carotovora* on *Carica papaya* from the Philippines. He found that both the Indonesian and Japanese isolates were pathogenic to rice while *E. carotovora* was not. However, the Indonesian isolates were closer to *E. carotovora* than to *P. oryzicola*. The two isolates of *P. oryzicola* from Japan showed some relationship with *P. marginalis*.

It is apparent that only scattered information is available on this disease, and it is possible that several organisms are involved which cause similar symptoms.

LITERATURE CITED

FANG, C. T. & REN, H. C. 1960. A bacterial disease of rice new to China. *Acta phytopath. sin.* **6**: 90–92. [Chin. Engl. summ.]

FUNAYAMA, H. & HIRANO, T. 1963. A sheath brown rot of rice caused by *Pseudomonas*. Abs. in *Ann. phytopath. Soc. Japan* **28**: 67–68. [Jap.]

GOTO, M. 1965. A comparative study of the sheath rot bacteria of rice. *Ann. phytopath. Soc. Japan* **30**: 42–45.

KHU, T. C. & BAI, T. K. 1960. Investigations on a new bacterial disease of rice, brown spot. Preliminary communication (distribution, symptoms, and determination of the genus of the causal agent of the disease). *Acta phytopath. sin.* **6**: 93–105. [Chin. Russ. summ.] (*Rev. appl. Mycol.* **43**, 737.)

KLEMENT, Z. 1955. A new bacterial disease of rice caused by *Pseudomonas oryzicola* n. sp. *Acta microbiol. Acad. Sci. Hung.* **2**: 265–274. (*Rev. appl. Mycol.* **36**: 57.)

BLACK ROT AND OTHER BACTERIAL DISEASES OF RICE GRAINS

Black rot was first described by Iwadara (1931) in Japan. It has also been reported from Korea, Manchuria and Taiwan.

The disease is characterized by partial blackening or black spotting on the hulled grain at the apex, sometimes in the middle, or rarely at the base. The organism penetrates the aleurone layer and upper part of the endosperm, kills the infected tissues and turns them black.

The organism was called *Pseudomonas itoana* by Tochinai, and transferred to *Xanthomonas* by Dowson. Tochinai (1932) described the organism as follows: short rod with rounded ends, $1 \cdot 2$–$3 \cdot 5 \times 0 \cdot 5$–$0 \cdot 8\mu$, single or in pairs; flagella one or two, unipolar; no endospore and no capsule; aerobic, yellow-chromagenic, does not liquefy gelatin, diastatic action weak; reduces nitrate; does not produce acid or gas from sugars or glycerin; coagulates milk and reddens litmus slightly on very long standing; produces indol; does not produce ammonia; grows in Uschinsky's solution and feebly in Cohn's solution; optimum temperature 29°C, maximum 38°C, minimum under 10°C, thermal death point 50–51°C; Gram negative.

In Japan Iwadara (1936) found that varieties Chinkobozu No. 1, Bozu No. 6 and Kairyomochi No. 1 were very susceptible and that the disease was severe when the temperature in July and August was high. Experiments showed that the organism attacks the grain at the milky stage, through wounds.

A bacterial grain rot caused by *Pseudomonas glumae* Kurita & Tabei was reported in Japan by Goto, K. & Ohata in 1956 (Hashioka, 1969). Infected grains are scattered on the panicles, but in severe cases more than half of the grains may be attacked. Infected grains are greenish white at first; later they are dirty grey, becoming dirty yellowish brown and drying up.

The bacterium is rod-shaped, $0 \cdot 5$–$0 \cdot 7 \times 1 \cdot 5$–$2 \cdot 5\mu$, 2–4 polar flagella, Gram negative, capsule staining positive but no endospore. Colonies on potato agar plates are faintly yellowish, milky white, chromogenic. Acid produced but no gas from xylose, arabinose, dextrose, fructose, mannose, glycerin etc.; and no gas from sucrose, maltose, lactose, dextrin, starch, inulin etc. Milk coagulated and digested, gelatin liquefied. No production of H_2S and indol, but ammonia produced. Nitrate not reduced. Temperature for growth 11–40°C, optimum 30–35°C (Asuyama, 1939; Hashioka, 1969).

Several other bacteria have been reported to infect stored rice grains, and three may be mentioned.

(1) Cinnamon speck of rice grains—caused by *Xanthomonas cinnamona* (Miyake & Tsunoda) Muko. The grains are cinnamon-coloured, with a distinct odour, are brittle, and the cells disintegrate, often starting at the embryo. The bacterium is fluorescent, rod-shaped, $1 \cdot 1$–$3 \cdot 4 \times 0 \cdot 7$–$1 \cdot 2\mu$, 1 polar flagellum 2–3 times as long as the body, no capsule, stainable with acid stains. Colonies on agar convex, glistening yellow, later uneven zonation of minute granular texture; on gelatin ripply margin, actinomorphic; on broth yellow, slightly bullate, membranous; in liquid culture yellow, membranous on surface of media. It grows well in Uschinsky's solution and Cohn's. It is aerobic, milk not coagulated, no gas, no indol, no ammonia, nitrate not reduced, H_2S produced, reduces pigment. Temperature for growth 0–46°C, optimum 35°C (Hashioka, 1969).

(2) Black-eye spot of rice grains—caused by *Xanthomonas atroviridigenum* (Miyake & Tsunoda) Tagami & Mizukami.

The infected, unhulled grains are smaller than normal, dark yellowish brown to dark grey or even black when severely infected. Hulled rice is also discoloured partly or entirely, yellowish brown, brown or black. Kernels may also be shrunken, embryos black or damaged.

The organism is rod-shaped, $1{\cdot}0$–$1{\cdot}6 \times 1{\cdot}5$–$3{\cdot}0\mu$, with 1 polar flagellum. Single or in pairs, no capsule, Gram negative. Colonies yellow, irregularly round, membranous, radiately bullate, diffuse dark pigment in media. Aerobic, gelatin not liquefied, nitrate reduced, ammonia and H_2S produced but not indol, milk coagulated, litmus reduced, produces green colour in Uschinsky's solution. Temperature for growth about 0–41°C, optimum 30°C (Hashioka, 1969).

(3) Bacillus A, B and C—reported by Suzuki (1935) who also studied the pH and temperature relations. Little information is available.

LITERATURE CITED

ASUYAMA, H. 1939. New diseases and pathogens reported recently on the cultivated plants in Japan. V. *Ann. phytopath. Soc. Japan* **9**: 44–52. [Jap.]

HASHIOKA, Y. 1969. Rice diseases in the world.—III Bacterial diseases. *Riso* **18**: 189–204.

IWADARA, S. 1931. On the black rot of rice grain. *J. Sapporo Soc. Agric. For.* **22**: 458–459.

IWADARA, S. 1936. On the geographical distribution of the black rot of rice grains and the relation of atmospheric temperature to the outbreak of the disease. *Rep. Hokkaido natn. agric. Exp. Stn* 36, 52 pp. [Jap. Engl. summ.]

SUZUKI, H. 1935. Studies on bacteria in the interior of rice seeds. I and II. *Botany Zool., Tokyo* **3**: 749–760. [Jap.]

TOCHINAI, Y. 1932. The black rot of rice grains caused by *Pseudomonas itoana* n. sp. *Ann. phytopath. Soc. Japan* **2**: 453–457.

PART III

FUNGUS DISEASES — FOLIAGE DISEASES

BLAST

History and Distribution

Records of rice blast disease may be traced back to very early times. In China, Soong Ying-Shin, in his book 'Utilization of Natural Resources',

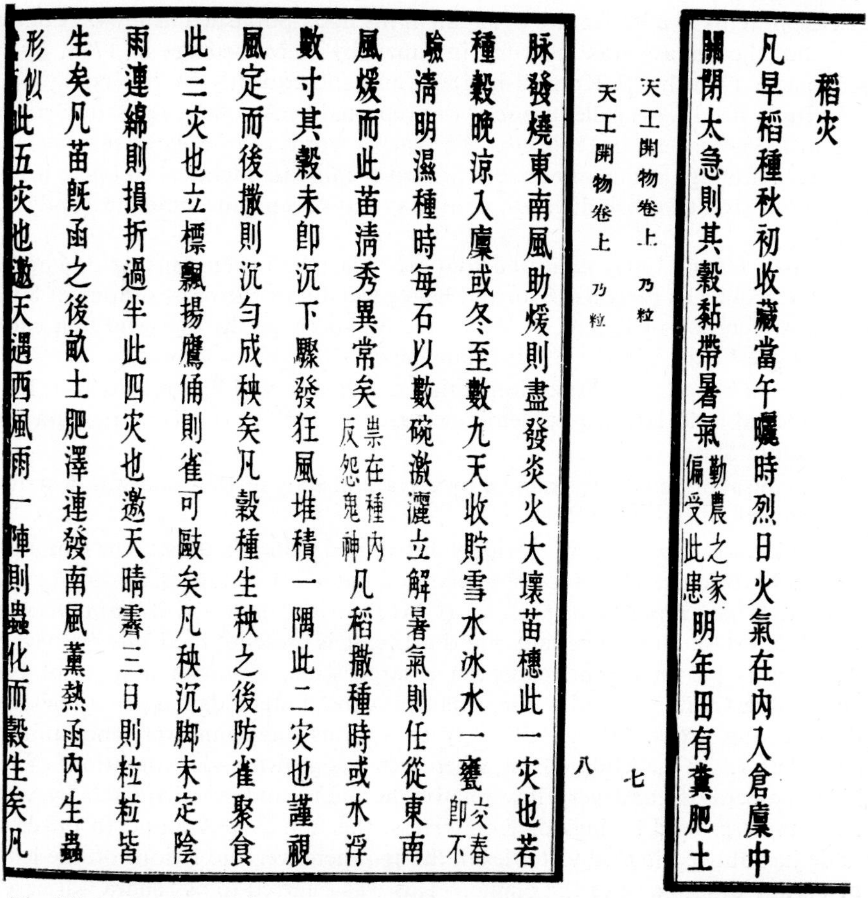
稻災

凡早稻種秋初收藏當午曬時烈日火氣在內入倉廩中關閉太急則其穀黏帶暑氣勤農之家偏受此患明年田有糞肥土

天工開物卷上 乃粒 七

天工開物卷上 乃粒 八

脉發燒東南風助煖則盡發炎火大壞苗穗此一災也若種穀晚涼入廩或冬至數九天收貯雪水冰水一甕交春即不驗清明濕種時每石以數碗激灑立解暑氣則任從東南風煖而此苗清秀異常矣祟在種內反怨鬼神凡稻撒種時或水浮數寸其穀未即沉下驟發狂風堆積一隅此二災也謹視風定而後撒則沉勻成秧矣凡穀種生秧之後防雀聚食此三災也立標飄揚鷹俑則雀可驅矣凡秧沉脚未定陰雨連綿則損折過半此四災也邀天晴霽三日則粒粒皆生矣凡苗既函之後畝土肥澤連發南風薰熱函內生蟲形似此五災也邀天遇西風雨一陣則蟲化而穀生矣凡

Fig. III–1. Portion of the early descriptions of rice maladies by Soong Ying-Shin in his book 'Utilization of Natural Resources' published in 1637 (late Ming Dynasty). The first malady mentioned is perhaps blast disease. It was believed then to be caused by the heat absorbed by the grain during drying in hot sunshine, the grain then being stored before it cooled off. It is perhaps the origin of the strange term 'rice fever disease' which has been used for blast disease extensively in Chinese and Japanese literature. The book was reprinted in Japan in 1771 (cited by 'Systematic Encyclopedia of Agriculture' p. 58, in Japanese, Chuo-Ko-Ron Co., Tokyo, 1965). The China Books Committee reprinted it in 1955 (Chung-Hwa Book Co., Taipei, Taiwan) and the above copy is from this reprint.

published in 1637, described several rice maladies (Fig. III–1). The first disease he described was the blight of rice seedlings, which caused them to appear as if scalded. This trouble was believed at that time to be due to heat which was absorbed into the grain during drying in hot sunshine, the grain thereafter being stored before it had been cooled off. When the seed was planted the next year in a field which had been manured, in the presence of a warm, moist southeast wind, the disease developed. The heat or fever ascribed as the cause of the disease is perhaps the original basis for calling the rice blast disease ' rice fever disease ', a name which has been extensively used in Chinese and Japanese literature.

According to Goto (1955), M. Tsuchiya indicated the occurrence of blast disease in Ichikawa Prefecture of Japan in his book published in 1704. Following him, the disease was recorded in Japan by S. Miayanaga in 1788, by N. Kojima in 1793, by T. Konishi in 1809, and subsequently by others.

In Italy, the disease called *brusone* was reported by Astolfi in 1828, by Brugnatelli in 1838, by Gera in 1846, etc. The early records of *brusone* were somewhat vague. Although some authors considered the disease identical to blast, it may have consisted of other diseases, or it may have been quite different (Padwick, 1950).

Metcalf (1906, 1907) indicated that rice blast had been causing damage in South Carolina as early as 1876, and he regarded it as the most serious of 8 rice diseases recorded in U.S.A. in 1907. He was perhaps the one who first called the disease ' blast ' in the English language.

Padmanabhan (1965b) mentioned that the disease was first recorded in India in 1913 and a devastating epidemic occurred in 1919 in the Tanjore delta area of Madras State.

The causal organism was named by Cavara in Italy in 1891 and was described in Japan by Shirai in 1896.

Rice blast is one of the most widely distributed plant diseases. The Commonwealth Mycological Institute has recorded some 70 countries throughout the world that have reported its presence (CMI Distribution Maps of Plant Diseases, 51, ed. 5, 1968). It may be said that the disease is present practically everywhere that rice is grown commercially on a large scale, although it has not been reported in California and perhaps has not caused much damage in a few other rice-growing areas. Its adaptability to various environmental conditions is remarkable. In the Middle East, where rice was grown under conditions of very high temperature and very low relative humidity and was irrigated by water from underground springs or from rivers, the fungus was found to infect the node just above the field water level, though there were no lesions on the leaves or on any other parts of the plants. This was referred to as ' shara ' disease in Iraq.

Losses

Blast is generally considered as the principal disease of rice because of its wide distribution and its destructiveness under favourable conditions. Rice seedlings or plants at the tillering stage are often completely killed. Heavy infections on the panicles are often detrimental to rice yields. Exact figures for yield losses are, however, few. In 1962 in Japan, leaf blast occurred over an area of 865,000 hectares, of which 847,000 hectares had been sprayed with chemicals. Panicle blast developed on 721,000 hectares of the 909,000 hectares sprayed with chemicals. In 1960, the estimated loss in yield was 273,300 tons, about 24·8% of the total losses due to insects, diseases, cold injury, typhoon and flood. The loss in 1953, an epidemic year, was about 800,000 tons. Annual losses during 1953 to 1960 varied from 1·4% to 7·3% with an average of 2·98% of total yield, even though extensive chemical control was practised (Goto, 1965a). In India, more than 266,000 tons of rice were lost to blast in the 1960–61 season, which was about 0·8% of the total yield (Padmanabhan, 1965a). In epidemic areas of the Philippines, many thousands of hectares have suffered more than 50% losses of yield.

Several studies have been made to estimate the yield losses due to panicle blast and leaf blast (Goto, 1965a; Padmanabhan, 1965a). For estimating losses from panicle blast, Kuribayashi & Ichikawa (1952) developed the formula $y = 0{\cdot}69X + 2{\cdot}8$ where y is the ratio of loss and X is the percentage of panicles affected by blast. Those with one-third or more of the spikelets affected are considered as blasted panicles. Other formulae similar to this have also been calculated in Japan (Goto, 1965a). However, the amount of loss by panicle blast is greatly influenced by the time of infection; the earlier the infection, the larger the loss. More losses also occur when the larger branches of the panicles are infected. Panicle blast is often preceded by varying amounts of leaf blast which also affects the yield.

Estimating yield losses due to leaf blast has also been attempted. Leaf blast causes stunting of plants, and reduces the number of matured panicles, the 1000 grain weight, the weight of brown rice, etc. Correlation studies between stunting and other phenomena and severity of leaf blast, and percentage of loss were made. The time of infection and other factors were found to influence disease severity and eventually yield (Goto, 1965a). The situation is a complicated one and an accurate estimate is difficult, if not impossible, to make.

Padmanabhan (1965a) estimated the yield losses from blast by comparing yields in (1) nonepidemic and epidemic areas; (2) resistant and susceptible varieties which have similar potential yielding ability; and (3) plots protected and unprotected by fungicides. Strict comparisons are not possible because many other factors are involved, but he reported that 1% neck infection reduced the yield by 1·4% in one case, and by 17·4% and 0·4% in other cases.

Symptoms

The fungus produces spots or lesions on leaves, nodes, and different parts of the panicles and the grains, but seldom, if ever, on the leaf sheath in nature (Plates VI, VII).

The leaf spots are typically elliptical with more or less pointed ends. The centre of the spots is usually grey or whitish, and the margin is usually brown or reddish brown. Both the shape and the colour of the spots, however, vary, depending upon environmental conditions, the age of the spots, and the degree

PLATE VI

Upper picture—*Blast* showing three types of leaf lesions—left, resistant; middle, moderately resistant; and right, susceptible; referred to as types 2, 3, and 4, respectively in the standard scales of the International Uniform Blast Nurseries.

Lower picture—*Blast* nursery showing resistant and susceptible lines. Border rows removed.

PLATE VI

of susceptibility of the rice variety. The spots usually begin as small, water-soaked, whitish, greyish or bluish dots. They enlarge quickly under moist conditions on susceptible varieties and remain greyish for some time. Fully developed lesions reach 1–1·5 cm long, 0·3–0·5 cm broad, and usually develop a brown margin. Spots on susceptible varieties growing under moist, shaded conditions show very little brown margin, but instead sometimes have a yellow halo around the spot. On highly resistant varieties, only minute brown specks of pin-head size may be observed. Varieties with intermediate reaction show small round or short-elliptic lesions, a few millimetres long and with a brown

Fig. III–2. Young rice plants completely killed by blast at the tillering stage.

margin. The development of the brown colour usually indicates either a resistant varietal reaction or the existence of some condition unfavourable for lesion development. These small brown lesions have been called chronic lesions while those of a greyish colour, and those which enlarge considerably, have been called acute lesions. Yoshii (1937) differentiated three zones in lesions—the venenate, necrotic and disintegrated zones. The venenate zone is the outer portion of the

PLATE VII

Upper picture—*Blast* disease infection at the base of the panicles (panicle blast or neck rot).

Lower picture—*Blast* disease showing infection at the nodes (node blast).

PLATE VII

lesion, formed by the infiltration of toxic substances secreted by the fungus, and consists of a yellowish zone which blends into healthy tissues. The necrotic zone is usually a brown, narrow streak along the inner side of the venenate zone. The cell inclusions and cell walls are degenerated and discoloured in the necrotic zone while in the disintegrated zone they are completely broken down and destroyed.

In describing various degrees of varietal resistance, many authors have distinguished many kinds of lesions based upon their size, shape and colour (see varietal resistance).

Fig. III–3. Seedlings at the centre of the nursery bed killed by blast.

Numerous spots may occur on a leaf which may soon be killed; this is followed by the drying up of the leaf sheath. Seedlings or plants at the tillering stage are often completely killed in the field (Plate VI).

When the node is infected, the sheath pulvillus rots and turns back and in drying often breaks apart, remaining connected by the nodal septum only (Plate VII). All parts above the infected node die.

Any part of the panicle may be infected, attack producing brown lesions. Areas near the panicle base are often attacked, causing the ‘ rotten neck ’ or ‘ neck rot ’ symptoms, and the panicles often fall over. Panicle branches and glumes are also attacked (Plate VII).

Kuilman (1940) noted that severe attack of blast is responsible for stunting of the plant. The physiology of the process has been studied by Tokunaga *et al.* (1953).

The predominant symptoms of blast disease in any given area depend upon the climatic conditions. In the temperate regions, where long periods of drizzle or light rain occur, leaf blast at the tillering stage is often severe and may kill the plants completely (Fig. III–2). In the tropics, seedlings in nurseries are often vulnerable (Fig. III–3), but after transplanting severe infections are seldom found. ' Neck rot ' occurs wherever environmental conditions are favourable. In Iraq, where the weather is very dry and hot, the disease infects only the node just above the irrigation water (Ou, 1957). Moisture is primarily responsible for these differences in symptom expression.

Leaf spots of blast disease may sometimes be confused with other leaf spot diseases on rice. To assist in identification, a diagrammatic sketch of some of the commoner diseases is given in Fig. III–4.

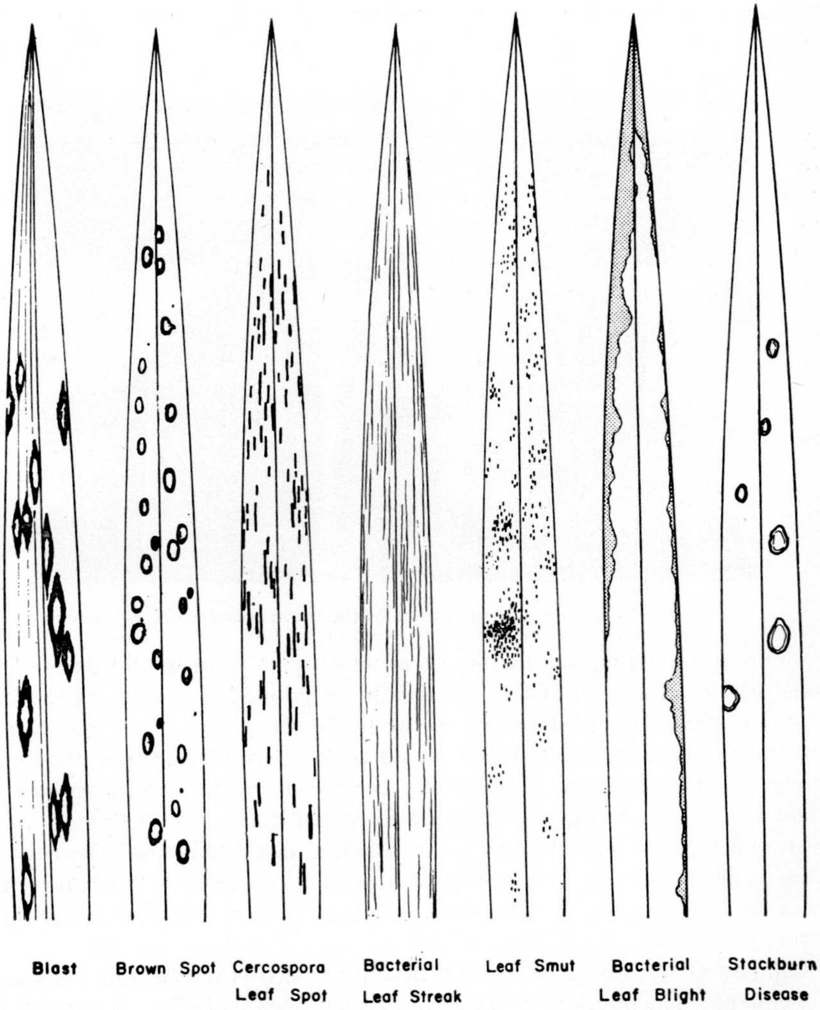

Fig. III–4. Sketches showing typical lesions of some of the common leaf spot diseases of rice.

CAUSAL ORGANISM

Taxonomy and nomenclature. There is a good deal of confusion concerning the taxonomy and nomenclature of the causal organism, and extensive studies are required to clarify the situation.

Two generic names have been used for the fungus, *Pyricularia* and *Dactylaria*. *Pyricularia* was established by Saccardo in 1880 based upon *Trichothecium griseum* Cooke on crab grass, *Digitaria sanguinalis*, from North America. *Pyricularia grisea* (Cooke) Saccardo is the type species of the genus. *Dactylaria* was also established by Saccardo in the same year, and he assigned *Acrothecium purpurellum* Sacc. as the type. *Dactylaria purpurella* was found growing on rotten oak wood. The names *Pyricularia grisea* and *Dactylaria grisea*, *P. oryzae* and *D. oryzae* have all been used at various times by different workers. Asuyama (1965) concluded that the blast fungus on rice and crab grass is reasonably included in the genus *Pyricularia*.

In most of the older literature, the name is spelt as *Piricularia*. Saccardo himself originally used the spelling *Pyricularia* but later changed it to *Piricularia*. Hughes (1958) checked the original publication of the name and found that it was spelt *Pyricularia*. According to the International Rules of Botanical Nomenclature, the spelling *Pyricularia* should therefore be used, and since the publication of Hughes' paper several authors have used the spelling *Pyricularia*.

There is more confusion about the specific name of the fungus. Cavara (1891) first described a species of *Pyricularia* on rice in Italy and named it *P. oryzae*. An identical description of the fungus appeared again in 1892 as *P. oryzae* sp. nov. in Briosi & Cavara, I Funghi parass. No. 188, and again as *P. oryzae* sp. nov. in the same year under the authorship of Cavara. The legend to Briosi & Cavara's exsiccatum of 1892 was reprinted in 1897 under the name *P. oryzae* Briosi & Cavara. The name is often attributed in the literature to Briosi & Cavara, but the name was first published by Cavara alone and he alone should be cited as author (Padwick, 1950). Cavara in 1893 described another new species, *Dactylaria parasitans* on crab grass; this was soon found to be synonymous with *P. grisea*.

In describing *P. oryzae* as a new species, Briosi and Cavara mentioned that it differs from *P. grisea* by its sparse, usually nonseptate hyphae and biseptate and larger conidia. In fact, however, the matter was not as simple as this, and their remarks caused a good deal of controversy later. Shirai in 1896 first described the rice blast fungus in Japan and found that it resembled *Pyricularia oryzae*, although it differed in the number of conidia borne on the conidiophore and the mode of conidial formation. Hori in 1898 compared the Japanese blast fungus with American specimens referred to as *P. grisea* and maintained that *P. grisea* produced 3–5 conidiophores from a stoma, each conidiophore bearing many conidia, while *P. oryzae* produced only one conidiophore with one conidium at its apex. Kawakami (1901, 1902) considered, on the other hand, that *P. grisea* and *P. oryzae* were indistinguishable. In addition to rice and crab grass, he found that the fungus attacked Italian millet (*Setaria italica*), German millet (*S. italica* var. *germanica*) and yellow foxtail (*S. glauca*). Shirai (1905) confirmed Kawakami's view after comparing the fungus on various hosts, though he considered the fungus to be a species of *Dactylaria*. Many other workers (Fulton, 1908; Duggar, 1909; Stevens, 1913; Metcalf, 1913) have shared this view.

Soon after, however, Sawada (1917), as the result of a series of inoculation experiments to investigate host range, considered that there were two species, for which he used the genus *Dactylaria*. *D. oryzae* (Cav.) Sawada was regarded as infecting rice, Italian millet, barley and wheat, and *D. grisea* crab grass and Italian millet. He also described three new species, *D. panici-paludosi* on *Panicum paludosum*, *D. leersiae* on *Leersia hexandra* and *D. costi* on *Costus speciosus*. Nisikado (1917, 1927) also recognized four species, *P. oryzae*, *P. grisea*, *P. setariae* and *P. zingiberi* based upon infectiveness on their respective hosts. Thomas (1940) found that *P. oryzae* from rice could infect wheat, barley, oats, maize and Italian millet, although the fungus from Italian millet failed to infect rice. Many cross-inoculation experiments have been made since by many workers, and the results have often been controversial (see host range). Recent knowledge on the specialization and great variation in pathogenicity of this organism (see below under pathogenic races and variability) may partly explain the controversy.

Asuyama (1965) expressed the opinion that it may be reasonable to include the morphologically related species under one specific name, and to subdivide this species into specialized forms on the basis of pathogenicity. For instance, Rao and Rao described the fungus on *Brachiaria mutica* as *P. oryzae* f. *brachiariae*.

To establish this system of classification, a systematic and comprehensive study is needed. Asuyama suggested the subdivision of the fungus as it occurs on cereals and grasses as follows:

a. rice, barley, barnyard grass (*Echinochloa*), tall fescue (*Festuca*), sugarcane *Leersia oryzoides*, giant reed (*Arundo donax*), ? maize.
b. crab grass (*Digitaria*), St. Augustine grass (*Stenotaphrum*), ? rice.
c. Italian millet, green foxtail (*Setaria*) and ragi (*Eleusine*).
d. bajra (*Pennisetum*).
e. proso millet (*Panicum*).
f. sweet vernal grass (*Anthoxanthum*).
g. *Brachiaria mutica*.
h. *Eriochloa villosa*.
i. *Leersia japonica*.
j. *Panicum repens*.
k. *Zizania latifolia*.

Facing the problem of choosing between *P. grisea* and *P. oryzae* as the specific name, Asuyama (1965) also stated that *P. grisea* is the earliest name given to this group of fungi and, therefore, should be used, according to the rules of nomenclature. But in most of the voluminous literature the fungus is found under the name *P. oryzae*, both on rice and on other plants. *P. oryzae* is thus a familiar and well established name, not easy to discard. In the *Index of plant diseases in the United States* and *Diseases of cereals and grasses in North America* (Sprague, 1950), the name *P. oryzae* is retained for the fungus on rice and *P. grisea* for that on other cereals and grasses. This is a compromise and the problem is still not resolved. The name *Pyricularia oryzae* Cavara is here adopted until further work on this problem is done.

Morphology. The description of conidiophore and conidia (Fig. III–5a) of *P. oryzae* by Briosi and Cavara may be translated as follows: ' sporiferous hyphae usually epiphyllous, scattered, rounded to roundish-awl shaped, a little thickened at the base, septate at that part, not or not so clearly so above, 60–120μ long, 405μ broad, grey; conidia obclavate, tapering at the apex, truncate or extended into a short tooth at the base, 2-septate, slightly darkened, translucent, 20–22 × 10–12μ ' (Padwick, 1950). The fungus on rice in Japan has been described by Shirai (1896), Hori (1899), Sawada (1917) and many others. Nisikado (1917) pointed out that in fact more than one conidium may be borne on a conidiophore (which was supposed to be the generic character) but not in the form of a ' head '. The secondary conidia are formed on the tips of new branches arising just below the joints of the first conidium and conidiophore. One to 20 conidia may be found on a single conidiophore. He further described

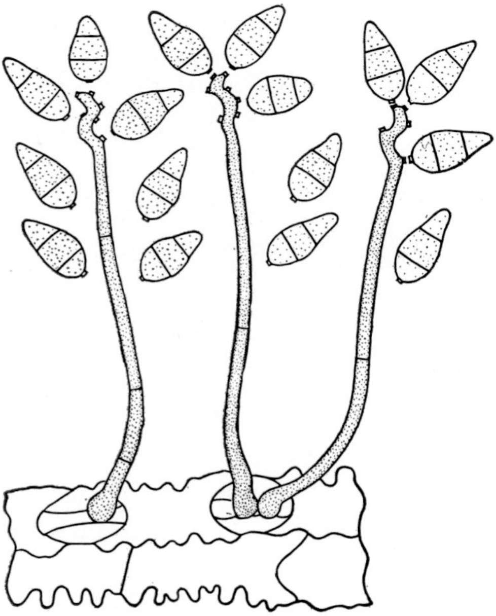

Fig. III–5a.
Pyricularia oryzae, conidia and conidiophores, × 500.
(Courtesy of Dr. M. B. Ellis.)

the conidiophores as being produced in clusters from each stoma, rarely solitary, with 2–4 septa. The basal portion is much swollen, olive to fuliginous, tapering towards the lighter coloured apex. The conidia are variable in size and shape, pyriform to obclavate, base rounded, apex narrowed, 2-septate, rarely 1–3 septate, sometimes slightly constricted at the septa, almost hyaline to pale olive, usually 19–23 × 7–9μ, with a small basal appendage 1·6–2·4μ (av. 2μ). Conidia germinate from the apical or basal cell, and less frequently from the middle cell.

The size of the conidia of the fungus from rice varies considerably among isolates and in different environmental conditions. Aoki (1935) measured 16 isolates on PDA and showed the average length to range from 21·2 to 28·4μ. Tochinai & Shimamura (1932) classified 39 isolates into 9 forms based upon cultural characteristics on steamed rice straw. The conidia of four of the forms were short, the mean value ranging from 19·3 to 22·8μ. The conidia of the other five forms were long, 26·8 to 29·9μ. All isolates from infected spikes or glumes of rice plants were of the long type while most of the isolates from the nodes were of the short type. Some of the measurements of the conidia on infected rice plants are shown in Table III–1. The size of conidia also differs in different hosts (Table III–2).

TABLE III–1

SIZE OF CONIDIA OF *Pyricularia oryzae* PRODUCED ON RICE PLANT LESIONS. (Asuyama, 1965).

Research worker	Range (μ)	Average
Hori (1898)	20–25 × 9–11·5	
Sawada (1917)	16–32 × 7–11	26·0 × 8·8
		25·8 × 8·5
		23·9 × 9·2
		26·1 × 9·8
		27·3 × 8·1
Nisikado (1926)	16·7–23·8 × 6·0–9·5	19·2 × 8·6
	11·5–28·5 × 6·4–11·0	22·8 × 8·7
	17·6–30·6 × 6·6–10·0	23·1 × 8·1
Sueda (1928)	17·0–24·3 × 7·3–9·7	20·9 × 8·8
	17·8–34·0 × 8·1–12·9	25·8 × 10·3
Henry and Andersen (1948)	20–26 × 8–9	
Yamanaka (personal communication)	18·8–25·0 × 7·5–11·3	21·6 × 9·3
	18·0–28·8 × 8·0–11·3	24·1 × 9·7
	23·0–34·5 × 7·5–10·0	27·3 × 9·2

TABLE III–2

SIZE OF CONIDIA OF *Pyricularia* COLLECTED ON VARIOUS PLANTS. (Asuyama, 1965).

Research worker	Host plant	Range (μ)	Average
Sawada (1917)	rice	16–32 × 7·0–11	26·0 × 8·8
	crab grass	17–40 × 6·5–10	23·6 × 8·1
	Leersia japonica	20–35 × 7·0–10	27·0 × 8·6
	Panicum repens	17–28 × 8·5–12	22·5 × 10·2
	Italian millet	21–31 × 7·0–10	25·0 × 8·6
	Costus speciosus	20–30 × 7·5–10	24·0 × 8·6
Nisikado (1926)	rice		19·2 × 8·6
			22·8 × 8·7
			23·1 × 8·1
	crab grass	16·6–27·4 × 6·0–7·6	22·5 × 7·2
	Italian millet	19·9–25·0 × 6·4–8·1	21·0 × 7·5
	green foxtail	17·2–29·0 × 6·0–9·0	23·9 × 8·2
	Zingiber mioga	14·3–20·2 × 7·1–8·3	17·5 × 7·8
Goto, Yamanaka & Kobayashi (1954)	*Zizania latifolia*	21·0–29·4 × 11·2–15·4	24·6 × 14·1
	Zingiber mioga	12·6–20·4 × 6·0–8·4	15·8 × 6·8

Ono & Nakazato (1958) found that conidia on lesions of the acute type were more or less round and small as compared with those on chronic lesions. Narita, Iwata & Yamanuki (1956) inoculated the fungus from rice on various hosts and found that the sizes of conidia produced on these hosts were different (Table III–3). When these conidia were inoculated back to rice, the conidia produced were of the same size.

TABLE III–3

SIZE OF CONIDIA OF RICE BLAST FUNGUS PRODUCED BY INFECTION OF CEREALS AND GRASSES. (Narita, Iwata & Yamanuki, 1965. After Asuyama, 1965.)

Inoculated plant	Size	Inoculated plant	Size
Festuca arundinacea	23·0 × 8·9	*Oryza sativa*	23·5 × 9·6
F. elatior	22·7 × 9·1	*Phalaris arundinacea*	25·7 × 8·7
F. rubra (1)	21·9 × 8·3	*P. canariensis*	25·8 × 9·3
(2)	19·9 × 8·8	*Phleum pratense*	25·4 × 9·2
Holcus lanatus	22·1 × 8·9	*Secale cereale*	24·3 × 9·0
Hordeum vulgare (1)	23·9 × 8·6	*Zea mays*	21·9 × 8·6
(2)	26·4 × 9·7		

According to Nisikado (1926) and Sueda (1928) conidia produced under moist conditions were somewhat longer than those formed under dry conditions. Sueda (1928) and Yamanaka & Kobayashi (1962) observed that conidia formed at higher temperatures (30° and 27°C, respectively), were longer than those formed at lower temperatures (18–20° and 21–22°C). Variation in width was not significant.

Conidia produced on culture media were observed by several authors (Sawada, 1917; Nisikado, 1926; Kulkarni & Patel, 1956; Ono & Nakazato, 1958) to be longer than those on the host plants and to vary according to the kind of media (Table III–4, 5). Ono & Suzuki (1959a) observed that conidia which developed

TABLE III–4

AVERAGE DIMENSIONS OF CONIDIA OF THE RICE BLAST FUNGUS PRODUCED ON HOST LESIONS AND ON PLANT DECOCTION AGAR MEDIA. (Sawada, 1917; after Asuyama, 1965).

On host lesions	On agar media with decoction of				
	Bean	Rice	Crab grass	*Panicum repens*	Tomato
26·0 × 8·8	28·0 × 8·4	28·7 × 8·8	28·9 × 8·7	28·9 × 8·9	21·7 × 8·3

TABLE III–5

AVERAGE SIZE OF CONIDIA PRODUCED ON RICE DECOCTION AGAR OF *Pyricularia* ISOLATES FROM VARIOUS PLANTS. (Nisikado, 1926; after Asuyama, 1965).

Rice	*Setaria*	*Setaria*	Crab grass	*Zingiber mioga*	*Zingiber mioga*
23·2 × 8·7	25·7 × 7·8	24·3 × 7·9	27·5 × 7·7	19·2 × 7·2	18·4 × 6·3

from conidia or conidiophores which had fallen on to the water surface were considerably smaller than normal.

Examining conidia under the electron microscope, Mizusawa (1959) found indications that there were at least three layers in the cell membrane. At the basal appendage, the outer two layers were cut off as an opening. The innermost layer was continuous, closing the opening of the basal appendage. Endoplasmic reticulum and protoplasmic junction, observed in conidia of *Helminthosporium oryzae*, have not been confirmed yet in *Pyricularia*. Horino & Akai (1965) reported the conidial wall to be two-layered. Wu & Tsao (1967) studied the fine structure of *P. oryzae* and reported that the conidial cell wall consisted of 4 layers. They also demonstrated the nucleus with nucleolus, mitochondria, endoplasmic reticulum, vacuoles, and septum pore, as well as two kinds of dense particles, one of which it was suggested might be a lysosome (Fig. III–5b).

The conidia of *P. oryzae* form appressoria at the tips of the germ tubes when they germinate on the host plant, or on a glass slide or cellophane sheet. Hori (1899) first described the appressorium and called it a ' recreating spore '. It was called ' resting spore ' by Kawakami (1901) and ' chlamydospore ' by Suyematsu (1916), Sawada (1917) and Nisikado (1917). Matsumura (1928) explained the mechanism of host entry, and applied the term ' appressorium '. Appressoria vary in size and shape; they are smooth, thick-walled, ranging from 5 to 15μ in diameter, globose, ovoid, or oblong. Suzuki (1951, 1953) separated strains of the fungus by the number and size of their appressoria; he believed appressoria to be related to pathogenicity, strains with large ones being more pathogenic.

Suyematsu (1916), Sueda (1928), Refati (1955), Togashi (1961) and Yamanaka & Kobayashi (1962) observed brown or dark-coloured cells in the conidia under certain cultural conditions but this seems to have no significance in the life cycle of the fungus. Ito & Kuribayashi (1931a, b) and others found intercalary thick-walled cells and regarded them as a kind of chlamydospore.

The number of nuclei in each mycelial cell, conidium and appressorium is not certain. Suzuki (1965, 1967) has long considered the cells to be multinucleate with varying numbers (2–6) of chromosomes and the fungus to be heterokaryotic. Chu & Li (1965) expressed the same view, reporting 4 chromosomes. On the other hand, Yamasaki & Niizeki (1965) observed that most of the cells are uninucleate, although in certain strains a certain percentage of the cells contain 2–6 nuclei. Horino & Akai (1965) and Wu & Tsao (1967) reported that according to their electron microscope observations the cells are uninucleate (Fig. III–5b). Hashioka & Inagaki (1968) further found all the seven species of *Pyricularia* to be uninucleate, rarely binucleate. However, heterokaryon resulted between *P. oryzae* and *P. zizaniae* from hyphal anastomosis and produced various types of conidia, some resembling those of the parents, as well as intermediate and abnormal types.

Giatgong & Frederiksen (1967, 1968, 1969) also found the cells to be uninucleate except for a few barrel-shaped cells which have as many as 4 or more nuclei; they reported that there are 6 chromosomes. In view of recent findings that the fungus is very variable in pathogenicity and other characters, further

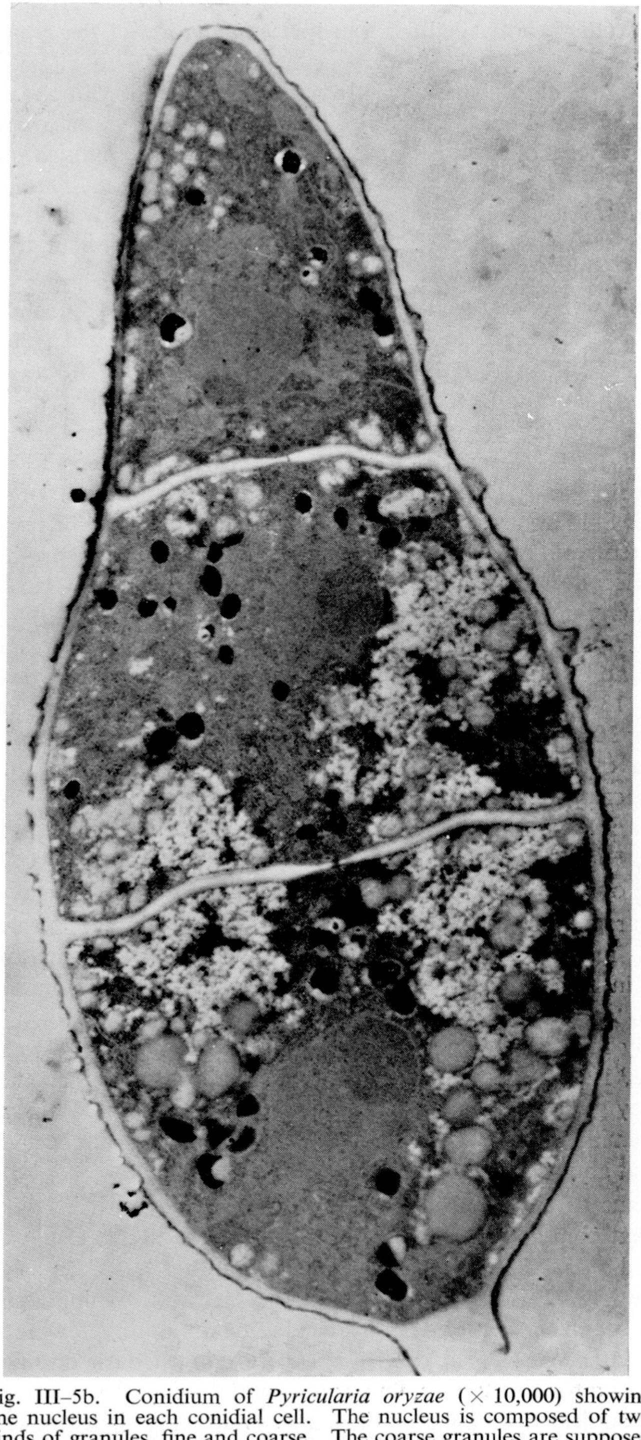

Fig. III–5b. Conidium of *Pyricularia oryzae* (× 10,000) showing one nucleus in each conidial cell. The nucleus is composed of two kinds of granules, fine and coarse. The coarse granules are supposed to be nucleoli. Septum pore and mitochondria and other organelles are also shown. (Courtesy of Mr. H. K. Wu.)

cytological studies are needed for a proper explanation of the great variability of the fungus.

Cultural morphology varies greatly with isolates and with the media used. The amount of aerial mycelium varies from very scant to a thick cottony mass. The colour varies from whitish or cream, through drab, pinkish, grey to dark olivaceous (Fig. III–6).

Physiological responses to the physical environment. Many studies have been made on the response of mycelial growth and sporulation of the fungus to temperature, moisture, light etc.

The optimum temperature for mycelial growth is about 28°C and the growth range from 8–9° to 37°C as shown by various workers (e.g. Suyematsu, 1916; Sawada, 1927; Nisikado, 1927; Abe, 1930; Yoshii, 1936). Some studies have shown that the optimum temperature and growth temperature range vary somewhat with isolates (Konishi, 1933; Tochinai & Shimamura, 1932; Tseng, Yuan & Wu, 1965). Sporulation in culture occurs at temperatures from 10–15 to 35°C (Kuribayashi, 1928; Sueda, 1928) with an optimum of about 28°C (Tochinai & Shimamura, 1932). The mode of sporulation in respect to time varies with temperature. At 28°C the spores were found to be produced rapidly but production decreased after 9 days, whereas sporulation at 16, 20 and 24°C tended to increase even after 15 days (Henry & Anderson, 1948). The release of conidia from conidiophores is not affected by temperatures between 15 and 35°C (Ono & Suzuki, 1959b). The thermal death-point of cultured conidia is 50°C for 13–15 minutes in water, but under dry conditions some conidia survive after 30 hours at 60°C (Sueda, 1928). Hyphae tolerate heat better than cold; about one-fifth survive for 50–60 days at −4° to −6°C (Abe, 1935). By quick freezing, cultures may be kept at −30°C for at least 18 months and, in fact, some workers have kept cultures under liquid nitrogen for a much longer time. Conidia germinate best at 25–30°C, not at 10–15°C (Nisikado, 1927). The optimum is about 25–28°C (Sueda, 1928).

Conidia are produced on rice leaf lesions only when the relative humidity of the air is maintained at 93% or above. The rate of sporulation increases with a rise in relative humidity (Hemmi & Imura, 1939). Conidia germinate freely in water but germination rarely occurs in saturated air. The critical air humidity for spore germination is between 92 and 96% (Hemmi & Abe, 1932b; Abe, 1933b). Eight per cent of the conidia can rise to the surface of water. They germinate there and reproduce conidia after 24 hours of floating (Ono & Suzuki, 1959a). Mycelium grows best at 93% air humidity, growth decreasing at about or below this point (Abe, 1941).

Light seems to affect sporulation greatly. Kuribayashi & Ichikawa (1941) observed that conidia in nature are disseminated at night. Barksdale & Asai (1961) found a diurnal periodicity in spore release. Lesions on attached leaves which were kept at 100% relative humidity released conidia only at night. The release began soon after dark, reached a maximum in a few hours and then slowly decreased, ceasing at dawn. When prolonged dark or light treatment was maintained for 1 or 2 days, few or no conidia were released until the lesions were again subjected to light or darkness. Alternate light and dark periods

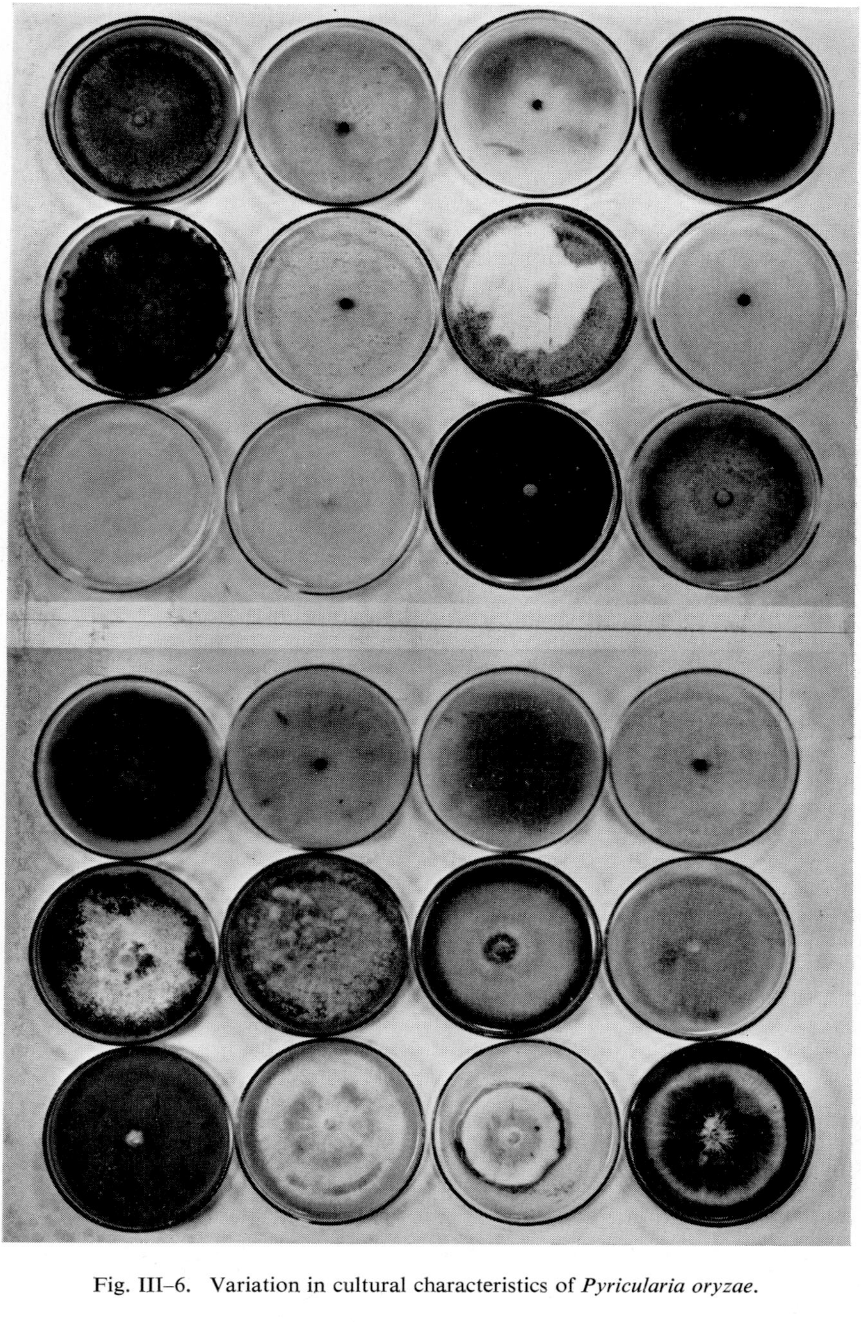

Fig. III–6. Variation in cultural characteristics of *Pyricularia oryzae*.

seem to be necessary. Sporulation in culture decreases with reduction of illumination by either fluorescent light or sunlight. When a culture is started in the dark, sporulation is restrained if the culture is exposed to full light in the later stages of development (Suzuki & Yoshimura, 1963, cited by Hashioka, 1965). Suzuki (Hozumi, 1969) reported that the diurnal periodicity of spore discharge was lost after one day under continuous light and after 3 days in continuous darkness. One hour of light at 1–2 a.m. stopped spore discharge. Mycelial growth increases with decreased light (Imura, 1938, 1940). Direct sunlight suppresses germination of conidia (Sueda, 1928). In diffused light, the percentage of conidial germination is reduced to about half that in darkness. Light also inhibits elongation of germ tubes (Abe, 1931).

Nutritional requirements. The fungus grows well in many media containing plant tissues or plant decoctions. It is, however, difficult to culture on synthetic media; but the addition of a small amount of hot water extract of rice straw stimulates growth and sporulation. This strongly suggests the presence of growth-promoting substances in the straw extract. Leaver, Leal & Brewer (1947) found biotin and thiamine to be necessary for growth, but other vitamins were not required. Tanaka & Katsuki (1951b) and Otani (1952b) obtained the same results. Otsuka, Tamari & Ogasawara (1957d, 1958b), working with 47 isolates of the fungus, found that a few isolates could grow in biotin-deficient media and a few others in thiamine-deficient media.

Straw extract from which biotin was removed by active charcoal stimulated growth even when biotin and thiamine were also separately added. Some constituent in the straw extract which activated biotin was suspected to be present. Tanaka & Katsuki (1956) reported that succinic, malic, and citric acids, known as components of the TCA cycle and contained in rice leaves,

TABLE III–6

SYNTHETIC MEDIA FOR GROWING *Pyricularia oryzae* (Tanaka & Katsuki, 1965).

Medium A		Medium B		Medium C	
Sucrose	15·0 g	Sucrose	30·0 g	Sucrose	15·0 g
NH_4NO_3	1·0	KNO_3	3·0	KNO_3	1·0
KH_2PO_4	0·5	Urea	0·50	KH_2PO_4	0·25
K_2HPO_4	0·5	KH_2PO_4	1·0	K_2HPO_4	0·25
$MgSO_4 \cdot 7H_2O$	0·25	K_2HPO_4	1·0	$MgSO_4 \cdot 7H_2O$	0·25
$CaCl_2 \cdot 6H_2O$	0·05	$MgSO_4 \cdot 7H_2O$	0·50	$CaCl_2 \cdot 6H_2O$	0·05
$FeSO4 \cdot 7H_2O$	0·75 mg	$CaCl_2 \cdot 6H_2O$	0·10	NH_4 citrate $\cdot H_2O$	1·00
$MnSO_4 \cdot 5H_2O$	0·22	$FeSO_4 \cdot 7H_2O$	0·75 mg	Na-glutamate	1·00
$CuSO_4 \cdot 5H_2O$	0·60	$MnSO_4 \cdot 5H_2O$	0·22	Na-acetate	0·25
$ZnCl_2$	7·50	$CuSO_4 \cdot 5H_2O$	0·60	Na-thioglycolate	0·10
$(NH_4)_6Mo_7O_{24} \cdot 4H_2O$	0·06	$ZnCl_2$	7·50	$FeSO_4 \cdot 7H_2O$	0·75 mg
H_3BO_4	0·09	H_3BO_3	0·09	$MnSO_4 \cdot 5H_2O$	0·22
Biotin	5·0 µg	$(NH_4)_6Mo_7O_{24} \cdot 4H_2O$	0·06	$CuSO_4 \cdot 5H_2O$	0·60
Tniamine	2·5 mg	Biotin	5·0 µg	$ZnCl_2$	7·50
H_2O	1000 g	Thiamine	1·0 mg	Biotin	1·0 µg
		H_2O	1000 g	Thiamine	2·0 mg
				H_2O	1000 g

were excellent additive stimulants. Certain free amino acids, such as glutamic and aspartic acids, and trace elements such as manganese, zinc and molybdenum are considered as co-growth factors and stimulants (Tanaka & Katsuki, 1951b, 1952). From their studies they have developed synthetic media (Table III–6), which allow more detailed study of the nutrition of *P. oryzae*.

The utilization of carbon and nitrogen sources has been studied by many investigators (Tochinai & Nakano, 1940; Leaver, Leal & Brewer, 1947; Tanaka & Katsuki, 1951a, b; Otani, 1952b, 1953; Tanaka & Tsuji, 1952; Chinte, 1965). In most of the studies mentioned only one or a few isolates of fungus were used, and the results presented only partial pictures of the situation. Otsuka, Tamari & Ogasawara (1957a, b, c, 1965) made more comprehensive studies, using 47 different isolates and comparing their differences.

Among the carbon compounds, Otani (1953) reported that maltose, sucrose, glucose, inulin and mannitol, as well as organic acids, such as succinic acid, were the best carbon sources, whereas lactose and galactose were not suitable for the isolate which was used as the test organism. Using 32 different carbon compounds and 47 isolates, Otsuka *et al.* (1965) showed that sucrose, glucose, maltose, fructose, lactose, and xylose were the most suitable carbon sources. All isolates utilized starch, dextrin, mannose, arabinose, galactose, rhamnose, raffinose and dulcitol, but not inositol or organic acids. In general, organic acids were not suitable carbon sources. Some strains utilized sorbose, sorbitol, mannitol, inulin and glycerol well, but others not at all. There were especially marked differences among the isolates in the utilization of sorbose, sorbitol, inulin and mannitol. These results demonstrated that utilization of carbon compounds varies considerably with the fungus isolate. Otsuka also tried to correlate differences in utilization of these compounds with differences in pathogenicity (see varietal resistance).

Among nitrogenous compounds, Otani (1952b) reported that KNO_3, $NaNO_3$, glycine, L-alanine, aspartic acid, DL-glutamic acid and asparagine markedly accelerated growth, while $NaNO_2$, cystine, taurine and creatine inhibited growth of the isolate of the fungus he was using. Tanaka, Katsuki & Katsuki (1952, 1953) studied the utilization of ammonium and nitrate nitrogen by the fungus. The inferiority of ammonium-N was found to be due to the fall of pH. At pH 7, the fungus grew well on ammonium-N. Studies by Otsuka, Tamari & Ogasawara (1957a, b, c) involving 23 nitrogen compounds and 47 isolates of the fungus showed that KNO_3, $NaNO_3$, L-aspartic acid, L-asparagine, L-arginine, L-alanine, L-proline, L-serine, glycine, L-histidine and L-glutamic acid were suitable for growth; while oxyproline, L-cystine and L-phenylalanine were not utilized in general. The utilization of L-phenylalanine, L-cystine, L-tryptophan and $NaNO_2$ varied markedly among isolates.

Twenty-four isolates tested at IRRI (1965) also showed great differences in the utilization of various nitrogen and carbon sources, particularly nitrogen.

Based upon nutritional differences and other physiological properties, Otsuka, Tamari & Ogasawara (1957d, 1958a, b, 1965) have classified isolates of the fungus into 13 biochemical types.

METABOLIC PRODUCTS AND SOME PHYSIOLOGICAL PROPERTIES

Two toxins were isolated by Tamari & Kaji (1954) from culture filtrate as well as from severely diseased plant tissues. One was identified as a picolinic acid with $C_6H_5NO_2$ as the molecular formula; the other, an unknown compound, was designated ' piricularin '. Piricularin is stable in water, especially on the acidic side. The physico-chemical properties of piricularin are as follows: the crystal from hot water is a colourless long prism, the melting point is 73·5°C, the molecular formula is $C_{18}H_{14}N_2O_3$ (Tamari & Kaji, 1957). They also established micro-estimation methods for the two toxins (Ogasawara, Tamari & Kaji, 1957b; Ogasawara, Tamari & Suga, 1961). Application of piricularin on mechanically wounded rice leaves produced necrotic spots closely resembling those of blast disease. These artificial spots were also produced by known metabolic inhibitors when similarly applied (Toyoda & Suzuki, 1958). Tamari & Kaji (1959) also found that piricularin exerts a dual deleterious effect on rice plants, causing accumulation of coumarin in addition to its primary toxicity.

The toxins inhibit the growth of rice seedlings as well as germination of the conidia of the fungus. Piricularin may be detoxicated by chlorogenic and ferulic acids. The physiological responses of rice plants to these toxins and their mechanism have been studied in detail by Tamari, Ogasawara & Kaji (1965). Iwasaki *et al.* (1969) isolated another toxic substance which they called pyriculol.

Otsuka, Tamari & Ogasawara (1957d, 1958a, b, 1965) and Otsuka (1959, 1961) reported the production of amino acids and vitamins in culture. In general, *P. oryzae* produced in synthetic media large amounts of fatty amino acids, such as valine, serine and alanine, and acidic amino acids, such as glutamic and aspartic acids. Heterocyclic amino acids, aromatic amino acids and thio-amino acids were produced only in small amounts. The amounts produced varied with isolates. While the organism requires vitamins in variable amounts, it also produces thiamine in different quantities. All strains studied produced riboflavin, pantothenic acid, vitamin B-6 and folic acid in culture.

The fungus shows slight activity in amylase and pectin decomposition and isolates vary greatly in proteolytic and succinic dehydrogenase activity. Cytochrome oxidase was found to play a leading role in the terminal oxidase system in the initial stage of the fungus growth in culture while polyphenol oxidase is prominent in later stages. Different isolates varied in this respect (Otani, 1955; Hirayama, 1933; Takamura, 1957; Tamari, Otsuka, Honda, Kaji & Ogasawara, 1961; Otsuka, Tamari & Ogasawara, 1965).

Starch-hydrolyzing activity was found to vary with different isolates but was generally weak. No H_2S was produced in any isolate, and only a few isolates showed weak gelatin-liquefying activity. All isolates had strong aesculin-decomposing activity and most isolates reduced nitrate (Otsuka, Tamari & Ogasawara, 1957a, b, c, 1965).

Jothianandan & Shanmugasundaram (1969) isolated proteinase from the fungus; Nakajima *et al.* studied the chemical constituents of the cell wall and Wu & Shih studied the mycelial proteins.

Pathogenic races and variability. The existence of strains of *P. oryzae* differing in pathogenicity was first noticed by Sasaki (1922b) who ascertained that rice varieties resistant to his strain A were severely infected by strain B. It was not, however, until 1950, when some varieties such as Futaba, developed by hybridization and known to be resistant for about 10 years, suddenly became very susceptible, that intensive studies on pathogenic races began in Japan. In about 1960, 10 varieties were selected as differentials, two tropical, four Chinese and four Japanese in origin. Thirteen pathogenic races were identified and classified into three groups, called T, C and N (Goto, 1960, 1965b).

In U.S.A. different pathogenic races were first reported by Latterell *et al.* (1954). In further studies with additional isolates from U.S.A. and countries in Asia and Latin America, 15 races were identified (Latterell, Tullis & Collier, 1960). Race 16 was added later (Atkins, 1962).

Pathogenic race studies were initiated in Taiwan in the middle of the 1950s. Sixteen varieties have been used as differentials and 19 races have been identified (Hung & Chien, 1961; Chiu, Chien & Lin, 1965).

Many pathogenic races have also been identified from Korea (Ahn & Chung, 1962; Lee & Matsumoto, 1966), from the Philippines (Bandong & Ou, 1966; IRRI 1967) and from India (Padmanabhan, 1965c), each country using different varieties for race identification.

The number of known races has increased as further work has been done in each country. Japan has now registered 18 races (Hirano, 1967) and Taiwan, 27 (Chien, 1967) and in the Philippines 116 races have been differentiated (IRRI, 1968). Matsumoto *et al.* (1969) tested 327 isolates from 13 countries and reported finding 34 races.

In view of the fact that different sets of varieties have been used for the differentiation of races in various countries, the pathogenicity of races identified in one country cannot be compared with that of races in other countries. In an attempt to simplify the situation, a co-operative effort was made in 1963 between Japan and U.S.A. to develop an international set of differentials. During three years of co-operative studies, many hundreds of isolates collected in Japan and U.S.A. were extensively tested on the 39 differential varieties which have been used in Japan, U.S.A. and Taiwan. Eight varieties were finally selected and 32 race groups characterized, based upon the isolates tested. The races were called international races and were designated as IA, IB etc. to IH to indicate groups, followed by numerals to indicate race numbers (Atkins *et al.*, 1967). Standardization of the numbering of the international races has been suggested to avoid confusion (IRRI, 1967; Ling & Ou, 1969). The eight international differential varieties and the varieties used in various countries are listed in Table III–7.

Goto, I. & Sakai (1963) noticed significant intra-clonal variation in pathogenicity and stated that segregation seems to occur in vegetative isolates from a clone. Kiyosawa (1966c), Niizeki (1967) and Katsuya & Kiyosawa (1969) reported ‘spontaneous mutation’ and change of pathogenicity in high frequency.

TABLE III–7

DIFFERENTIAL VARIETIES FOR *P. oryzae* USED IN DIFFERENT COUNTRIES

International[1]	Japan[2]	U.S.A.[3]	Taiwan[4]	Philippines[5]	India[6]	Korea[7]	Colombia[8]
Raminad Str. 3	Tetep	Zenith	Kung-shan-wu-shan-ken	Kataktara DA–2	AC. 1613	Zenith	Raminad Str. 3
Zenith	Tadukan	Lacrosse	Taichung 65	CI 5309	CR. 906	Ishikari-shiroke	Zenith
NP–125	Usen	Caloro	Pai-kan-tao	Chokoto	Bengawan		NP–125
Usen	Chokoto	Sha-tiao-tsao (P)	Taichung 171	Co 25	S.M. 6	Pi 1	Usen
Dular	Yakeiko	(CI 8970–P)	Chianung 242	Wagwag	Mas	Sensho	Dular
Kanto 51	Kanto 51	Sha-tiao-tsao (s)	Kwangfu 1	Pai-kan-tao	Intan	Kanto 51	Kanto 51
Sha-tiao-tsao (s)	Ishikari-shiroke	CI 5309	Chianung 280	Peta	CR. 907	Ayanishiki	Sha-tiao-tsao (s)
(CI 8970–S)	Homare-nishiki	PI 180061	Taichung line 33	Raminad Str. 3	BJ–1	Norin 17	Caloro
Caloro	Ginga	PI 201902	Kanto 51	Taichung t-c-w-c	S. 67	Norin 22	Aichi-asahi
	Norin 22	Wagwag	Norin 21	Lacrosse		Norin 1	Ishikari-shiroke
	Aichi-asahi	Raminad Str. 3	Sensho	Sha-tiao-tsao (s)		Tonewase	(Napal)*
	Norin 20	(Rexoro)*	Cutsugulcul	Khao-tah-haeng 17			(Bluebonnet 50)*
		(Taichung 65)*	Natala				
			Kao-chio-lin-chou				
			Kaohsiung-ta-li-chen-yu				
			Taichung-ti-chio-wu-chien				

1 Atkins, J. G. *et al.*, *Pythopathology* **57**: 297–301, 1967.
2 Goto, K., The rice blast disease: 237–242. Baltimore, Md, Johns Hopkins Press, 1965.
3 Atkins, J. G., The rice blast disease: 243–244, 1965 and Latterell, F. M. & Marchetti, M. A., *Phytopathology* **58**: 1057, 1968.
4 Chiu *et al.*, The rice blast disease: 245–255. Baltimore, Md, Johns Hopkins Press.
5 Bandong, J. M. & Ou, S. H., *Philipp. Agric.* **49**: 655–667, 1966.
6 Padmanabhan, S. Y., *Curr. Sci.* **34**: 307–308, 1965.
7 Ahn, C. J. & Chung, H. C., *Seoul Univ. J. Biol. Agric.*, Ser. D, **11**: 77–83, 1962.
8 Galvez E., G. E. & Lozano T., J. C., *Phytopathology* **58** : 294–296, 1968.
* Used as supplementary variety.

Recently it was demonstrated that conidia produced by single lesions consisted of many pathogenic races; in addition daughter conidia from a monosporidal culture could also be separated into many pathogenic races (Giatgong & Frederiksen, 1967, 1969; Ou & Ayad, 1968; Chien, 1968; IRRI, 1968,

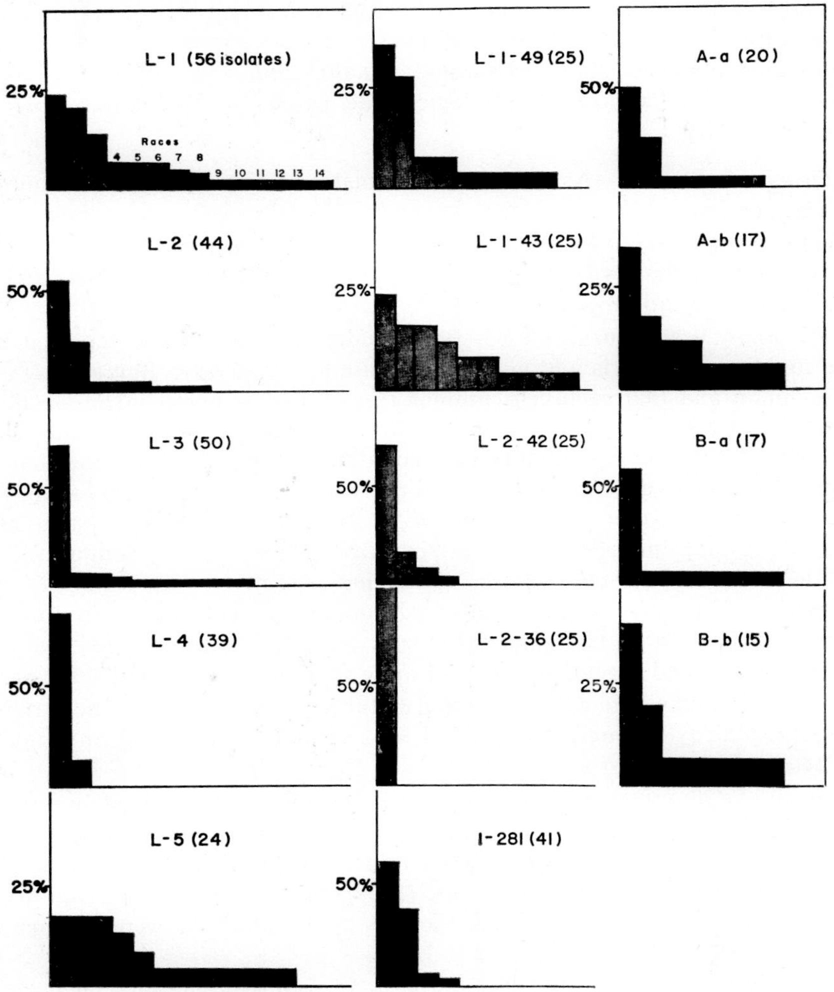

Fig. III–7. Relative population of pathogenic races originating from single lesions (L–1 to L–5), single conidia (L–1–49 to 1–281), and single cells (A–a, apical to B–b, basal) of *P. oryzae*.

1969). Conidia produced in most cultures consist of many races (Fig. III–7). Obviously this is a very significant development and further study and understanding of variation in pathogenicity are needed.

Yamasaki *et al.* (1964) obtained cultures resistant to chemicals by repeated passage through media containing copper sulphate (up to 240 mM), boric acid (up to 647–841 mM) and mercuric chloride (up to 2·2 mM). Some remained resistant for many succeeding transfers, others reverted to the original non-resistant type. Chien (1964) obtained similar results with mercuric chloride, Nakamura & Sakurai (1968) with blasticidin S, and Ursugi *et al.* (1969) with blasticidin S and kasugamycin.

Yamasaki & Niizeki (1966) induced morphological and biochemical mutants by means of chemicals. Kuo *et al.* (1967), using various mutagens and UV light, also obtained many nutritional mutants requiring different kinds of amino acids and noticed morphological and pathogenicity changes from the original cultures.

Origin of variation. As stated above, isolates of *P. oryzae* differ in cultural characteristics and in their nutritional requirements, as well as in pathogenicity. The great variability in pathogenicity among isolates has increased the difficulties of breeding resistant varieties. To assist in future work it is essential to know the cause of variation.

For many years, Suzuki (1965, 1967) studied the fungus and reported that the conidia, appressoria and mycelial cells are heterocaryotic, that anastomosis is common and that each cell contains three to seven nuclei. He concluded that heterocaryosis is the basis of variation. Chu & Li (1965) also found that the cells of both the mycelium and the conidium are multinucleate and that the number of nuclei depends on the conditions under which the fungus has been grown. On the other hand, Yamasaki & Niizeki (1965) reported that most fungal cells are uninucleate. In certain strains, however, they found 13–20% of the cells to be multinucleate, containing 2–6 nuclei. Anastomosis and migration of the nucleus were observed, and nuclei had apparently fused to form heterodiploids. Giatgong & Frederiksen (1967, 1969) observed that only a few conidial and vegetative cells had more than one nucleus and that anastomosis was found within a monoconidial culture. They believe that the variation in pathogenicity among monoconidial lines cannot be explained on the basis of heterocaryosis.

In all the above studies the number of nuclei was determined by nucleus staining. It has been speculated that some of the stained bodies in the cells may have been lipids or other bodies rather than nuclei. Mizusawa (1959), Horino & Akai (1965) and Wu & Tsao (1967) (Fig. III–5) made electron microscope observations and found that the cells of the conidium were uninucleate. It seems reasonable to believe this to be so.

Since the perfect state of the fungus is not known, the other possible mechanisms of genetic change are parasexualism and heterocytosome. Parasexualism has been indicated by Yamasaki and Niizeki, who noted hyphal anastomosis and migration and apparent fusion of nuclei. The haploidization of these heterozygous diploids may bring about further genetical changes. So far, however, no cytological evidence of these genetic changes has been put forward.

Disease Cycle

Infection. Infection follows germination of conidia, formation of appressoria, production of infection tubes from the appressoria and penetration through cuticle and epidermis. The infection hyphae may also enter through stomata (Suada, 1928). Yoshii (1936) observed that the infection tube formed a vesicle after entering the cell and this vesicle gave rise to hyphae. Kawamura & Ono (1948) found that in resistant varieties the invaded cells responded quickly, producing brown granules or resin-like substances, and the hyphae ceased to grow. In susceptible varieties the cells responded slowly and the hyphae grew freely. Hashioka *et al.* (1968) and Locci (1970) studied the processes of spore germination, appressoria formation and penetration into epidermal cells by electron microscopy.

The time required for the conidia to invade the host cells varies with temperature. Hashioka (1965) found that a minimum of 10 hours at 32°C, 8 hours at 28°C, or 6 hours at 24°C were required. Anderson, Henry & Tullis (1947) found maximum infection on plants at temperatures of 24–28°C and with 16–24 hours of continuous wetting. Kahn & Libby (1958) reported that the minimum periods required to initiate infection were 10, 12 and 14 hours at 26·7, 21·1 and 18·3°C respectively. Free water is required for germination and high relative humidity near saturation is necessary for infection. Barksdale & Jones (1965) found that a period of 12·2 hours of dew is required for infection at 60°F, 9·7 hours at 70°F and 7·7 hours at 80°F and derived the equation 1/hours of dew = 0·265 - 12·26/temperature in degrees F, to describe the minimum conditions required for infection. Asai *et al.* (1967) found, however, that the amount of dew, the length of the dew period, and the leaf area at the time of inoculation were the only variables correlated with the average number of lesions per plant. Variation of air temperature between 61 and 95°F had no effect. Suzuki (Hozumi, 1969) found that spores treated with water for 20 minutes to 3 hours and then dried, lost germinability when water was again applied.

The incubation period also varies with temperature: 13–18 days at 9–10°C, 7–9 days at 17–18°C, 5–6 days at 24–25°C, and 4–5 days at 26–28°C (Hemmi *et al.*, 1936). The temperature conditions favourable for infection are similar to those for mycelial growth, sporulation and conidial germination mentioned above.

Light and darkness have been found to affect infection. Infection occurs most easily in darkness and is suppressed under diffused light (Abe, 1931; Hemmi & Abe, 1932b). The acute type of lesion is formed if leaves are kept in the dark previous to inoculation (Ono, 1953). However, disease development decreases under extended darkness. In a study of the effect of sunlight before or after inoculation, Imura (1938) reported that the lengths of incubation periods were LL>LD>DL>DD (LL=light before and after inoculation, LD=light before and darkness after inoculation etc.). The degrees of infection were DL>LD>LL>DD.

Dissemination. Conidia are produced on lesions on the rice plant about 6 days after inoculation. The rate of sporulation increases with increase in relative humidity, and below 93% R.H. no conidia are produced (Hemmi & Imura, 1939). A typical lesion is able to produce 2000–6000 conidia each day for about 14 days under laboratory conditions (IRRI, unpublished data). Kato *et al.* (1970) reported that conidia formation reaches a peak 3–8 days after the appearance of leaf lesions and 10–20 days after the appearance of lesions on rachises. Lesions on the upper five leaves could produce spores for infection at the initial heading stage.

Most of the spores are produced and released during the night, particularly between 2 and 6 a.m. A diurnal periodicity was observed by Barksdale & Asai (1961) and Suzuki (Hozumi, 1969). In the tropics we found a second peak of spore discharge in the afternoon after a monsoon shower. Suzuki (Hozumi, 1969) also observed that water is necessary for spore discharge; the more water droplets retained on the infected leaves, the more spores released. When treated with water, most spores discharged within 2 minutes, and especially during the initial 30 seconds. Ingold (1964) reported that the conidium is probably violently discharged, although to a very short distance, by the bursting of the minute stalk-cell attaching it to the conidiophore. Most spores are trapped near the ground where there is no wind, though they may be collected as high as 24 m above the ground. The stronger the wind, the greater the spore flight (Kuribayashi & Ichikawa, 1952; Ono *et al.*, 1962, 1963). Spores have also been trapped at 7000 feet from an airplane window (Ou, unpublished monthly report to FAO).

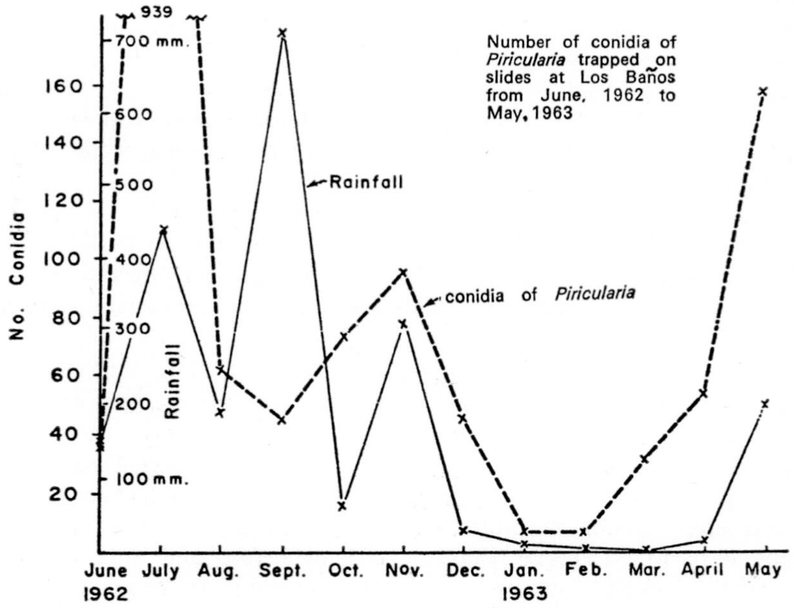

Fig. III–8. Air-borne conidia of *Pyricularia*, and rainfall, at Los Baños, Laguna, Philippines.

The number of spores deposited on a leaf differs greatly according to the position of the leaf and the angle between leaf and stem. More spores are deposited on the third leaf than on the second and still fewer on the top leaf. More spores are deposited on leaves of varieties with nearly horizontal leaves than on leaves of varieties which have leaves at an acute angle with the stem (Suzuki, 1961; Ono, 1965).

During the rice growing season, conidia are disseminated in the field for a relatively short period in temperate regions. For instance, in Japan, the peak of the air-borne population is generally in August (Table III–8). In the tropics, air-borne conidia are present the year round with peaks extending from May to June and November to December (Fig. III–8).

TABLE III–8

RELATION BETWEEN SPORE FLIGHT AND OUTBREAK OF BLAST

(After Ono, 1965)

		Number of spores caught				
		Nakasu village	Toyoshina town	Nakashioda village	Nagano city	Kashiwabara village
July	1*	0	1	2	0	0
	2	0	0	1	0	0
	3	0	0	0	0	0
	4	0	1	0	0	0
	5	0	0	0	1	0
	6	0	3	13	0	1
Aug.	1	0	8	41	0	0
	2	0	30	344	0	0
	3	1	111	229	1	0
	4	0	101	136	2	0
	5	0	48	102	1	0
	6	0	21	40	1	0
Sept.	1	0	73	50	2	0
	2	0	15	16	1	0
	3	0	16	12	1	0
	4	0	53	15	1	0
	5	0	0	0	0	3
	6	0	3	—	0	1
Degree of outbreaks						
leaf blast		0·5	13·0	10·3	0	1·3
neck blast		0·3	47·9	46·6	9·6	1·0
node blast		0	16·7	23·4	7·1	1·0

*1 indicates first 5 days; 2, second 5 days; 3, third 5 days, etc., of the month.

Air-borne conidia are the most important means of dissemination of *P. oryzae* although the fungus may also be disseminated by infected seeds and straw and by conidia which fall into the irrigation water.

Overwintering. In the temperate regions, mycelium and conidia on diseased straw and seed are the principal overwintering organs. Under dry conditions at room temperature, conidia may live for more than a year and mycelium for almost three years. But under moist conditions they will not survive until the next crop (Kuribayashi, 1928; Ito & Kuribayashi, 1931a). The over-wintered conidia infect rice plants readily and overwintered mycelium produces conidia when moistened. In the field the commonest source of primary inoculum is the straw. Ito (1932) found that conidia and mycelium on the surface of straw piles all died before the next spring but those within the piles overwintered readily. The fungus may be found within the embryo, endosperm, and glumes of the seed and sometimes between the glumes and the kernel (Suzuki, 1930). The fungus can be isolated from the seed although it is not a good competitor with other organisms in culture. The fungus may also overwinter on winter cereals and on weed hosts.

In the tropics, overwintering is not important in the disease cycle because air-borne conidia are present throughout the year. The fungus may also live in diseased rice plants and alternative hosts at any time of the year.

EFFECTS OF ENVIRONMENTAL CONDITIONS

Effects of climatic factors. The effect of temperature on infection and development of blast has been studied by many workers. When rice seedlings are grown under different soil temperatures, infection occurs most severely on those grown at 20°C, less at 24°C and 32°C, least at 28°C (Hemmi & Abe, 1932b; Abe, 1933a). Similarly, when both soil and air temperatures are concerned, seedlings become susceptible at low temperatures after several days at 18–20°C prior to inoculation. They become resistant at a moderate temperature (25–28°C) which is the optimum for growth of the host, and less resistant at 30°C (Hashioka, 1943, 1950a). Adult plants at higher soil temperatures, 20–29°C, are more resistant to neck blast than those at 18–24°C (Hemmi, Abe & Inone, 1941). On the other hand, seedling blight of germinating seedlings increases with soil temperature from 20 to 32°C (Hemmi, Abe & Inone, 1941). Thus the severity of seedling blight is correlated with increased growth of the fungus.

Further experiments have shown, however, that the relationship between temperature and host predisposition seems to vary with the period of thermotreatment, the combination of air and soil or water temperatures, the age of the rice leaves, the varieties employed, etc. Plants grown in different water temperatures are less susceptible if grown in cool water during the early period of treatment. If cool treatment is prolonged the reverse is true (Okamoto & Yamamoto, 1955b). Seedlings become susceptible when air and water temperatures are different, particularly under cool air (17°C) and warm water (23°C) (Tasugi & Yoshida, 1959, 1960). The predisposing effect of low temperature varies considerably with the age of leaves at the time of inoculation, even within individual seedlings. This has been explained as due to changes in the amount of carbohydrates and nitrogenous compounds in the leaves (Goto & Yamanaka, 1958; Goto & Ohata, 1960, 1961; Ohata *et al.*, 1966). Blast susceptibility resulting from temperature treatment at 10°C and 30°C also differs with the variety. Certain varieties become susceptible at high temperatures (Shimoyama, 1960). Generally speaking, rice varieties from temperate areas are more liable to be predisposed to blast infection at low temperatures than are tropical varieties (Hashioka, 1944, 1950b). Low night temperature (20°C) has been reported as being essential for infection and development of lesions in India (Sadasivan, Suryanarayanan & Ramakrishnan, 1965).

The humidity of the air and the soil moisture affect greatly the susceptibility of plants to attack, as well as disease development. Rice plants become susceptible when grown in ' dry ' soil, moderately resistant in moist soil and resistant under flooded conditions (Hemmi & Abe, 1932a; Kahn & Libby, 1958). Irrespective of varietal resistance, age, whether the variety is an upland or lowland one, or the part of the plant affected (leaf blast or neck blast), susceptibility to blast is inversely related to soil moisture. Silicification of the epidermis of rice leaves as well as other anatomical characteristics associated with resistance are favoured by low soil moisture which predisposes the plant to infection (Suzuki, 1934a, b, c, 1937). High air humidity favours lesion development. When chronic lesions on leaves are transferred into saturated air, the margin of the lesions becomes like that of the acute type, and progession of the disease is stimulated (Ono, 1953). In the tropics, since variation in temperature is small (Fig. III–9), the air humidity in the macro- and microclimate and dew

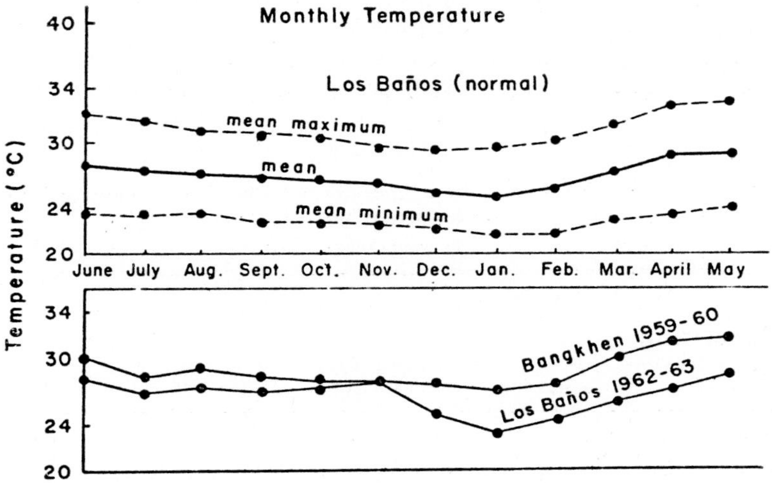

Fig. III–9. Monthly average temperatures at Los Baños, Laguna, Philippines and at Bangkhen, Bangkok, Thailand.

formation are the major controlling factors for disease development. In a seedbed as shown in Fig. III–3, the seedlings at the centre are completely killed by blast while many around the edge remain green. The major difference is between the humidity of the micro-climate in the centre and around the border. Blast is generally not severe in the tropics during the vegetative stages after transplanting. This may be explained by the fact that rains usually come as showers which only last for a few hours in the tropics, whereas long periods of drizzle often occur in the temperate regions. In areas in the tropics where there are frequent and long periods of rains, blast will cause severe damage. The inoculum builds up continuously because of the favourable temperature and also because of the constant availability of host plants, two rice crops often being planted in a year and planting time often being extended over a long period.

Hashioka (1950a) postulated that, in eastern Asia, the prevalence of blast in the temperate regions is conditioned solely by the relation of temperature to the growth of the blast fungus; in the subtemperate regions, it is controlled by temperature and also by the relation of the age of the plant to resistance; in the subtropical and tropical regions, prevalence is governed mostly by factors relating to the resistance of the host.

Imura (1938, 1940) found that slight shading from sunlight at the early stage of lesion development fosters extension of the lesion. Further development is promoted in proportion to the intensity of light and is inhibited by shading. Suryanarayanan (1959) also reported that greater light intensity promoted lesion enlargement when compared with diffused light. Seedlings grown under different amounts of insolation in glasshouses for 9 days showed decreased susceptibility in proportion to the decrease of insolation (Hashioka, 1950a). Asparagine, glutamine and several amino acids in rice leaves increase with decrease of resistance because of insufficient light under artificial shading or cloudy weather (Abumiya & Kobayashi, 1953).

Kikawa (1900) pointed out that wind increased the susceptibility of rice plants to blast and Sakamoto (1940) proved this experimentally.

TABLE III–9

RELATION BETWEEN APPLICATIONS OF NITROGENOUS FERTILIZER AND RICE BLAST DISEASE (POT TEST)

(Kawai, 1962)

Method of application			Number of lesions per leaf	Percentage of plants with neck blast
Amount of $(NH_4)_2SO_4$, (g/pot)		Time of top-dressing (days after transpl.)		
Basal dressing	Top-dressing			
2·0	0	0	10·8	82·3
1·5	0·5	25	6·8	61·2
1·0	1·0	25	3·8	21·5
0·5	1·5	25	4·5	36·6
0·0	2·0	25	6·1	34·2
2·0	0	0	10·0	100·0
1·5	0·5	35	0·5	43·2
1·0	1·0	35	0·6	15·8
0·5	1·5	35	0·0	18·4
0·0	2·0	35	0·0	23·4
2·0	0·0	0	7·1	78·7
1·5	0·5	45	2·3	50·6
1·0	1·0	45	0·9	20·1
0·5	1·5	45	0·7	27·1
0·0	2·0	45	0·4	31·6
2·0	0	0	6·4	81·8
1·5	0·5	55	0·3	63·7
1·0	1·0	55	0·3	42·3
0·5	1·5	55	0·1	56·7
0·0	2·0	55	0·3	51·8

Effects of nutritional factors

Nitrogen. Many experiments over a long period have shown that a high nitrogen supply always induces heavy incidence of blast regardless of the phosphorus or potassium supply (Hori, 1899; Miyazaki, 1928, 1930; Bokura, 1930; Murata, Kuribayashi & Kawi, 1933; Ikata *et al.*, 1938). The intensity of the effect of nitrogen on the disease varies with soil and climatic conditions and also with the method of applying the fertilizer. The effect is great when quick-acting nitrogenous fertilizers, such as ammonium sulphate, are applied in excess and all at once; split applications usually lessen the effect (Table III–9). Applications made too late or at a low air temperature during the early stages of growth also have a greater effect. Marked effects also are found in soils of low fertilizer-holding capacity, such as sandy or shallow soils, while the effect is less in clay soils or deep soils (Nisikado, 1926; Yoshida, 1943; Inoue, 1943; Kawai, 1952). Delayed or large top dressing often causes severe disease (Murata, Kuribayashi & Kawai, 1933; Ito & Sakamoto, 1937–1940; Okamoto, 1950). Heavy applications of green manure also result in severe disease (Nisikado, 1926; Murata, Kuribayashi & Kawai, 1933).

There have been many investigations made in an effort to explain the increased susceptibility induced by a high nitrogen supply. Ito & Sakamoto (1939–1943) and Sakamoto (1948) demonstrated that resistance decreased with increased water permeability of the epidermal cells. This they interpreted as being due to the injurious effects of ammonium accumulation in the cells of plants treated with high nitrogen. Otani (1948, 1950, 1952a) and other workers showed that there was a marked increase of soluble nitrogen, particularly amino acids and amines, in plants receiving much nitrogen. A close correlation was shown to exist between severity of the disease and the soluble nitrogen content of plants under various environmental conditions. Ohata & Kozaka (1962) demonstrated a close correlation between mycelial growth in the epidermal cells of detached leaf sheaths fed with various amino acids and amine solutions and mycelial growth in solutions containing these chemicals. The soluble nitrogen which accumulates in plants may serve as a suitable nutrient for fungus growth. Plants receiving large amounts of nitrogen are found to have fewer silicated epidermal cells and thus lower resistance (Ikeda, 1933; Suzuki, 1934a, b). Kawamura & Ono (1948) reported that dewdrops on leaves of plants receiving large amounts of nitrogen stimulate the germination and appressorium formation of the spores.

Phosphorus. The influence of phosphorus fertilizers on blast disease is usually not great. Many field experiments in Japan have shown that only when the nitrogen supply is high does an increase in phosphorus fertilizer result in an increase in severity of the disease (Kawai, 1933). In phosphate-deficient soil, however, experiments showed that large amounts of phosphate reduced the severity of the disease to some extent (Ikata *et al.*, 1938; Okamoto, 1950). Under conditions in which phosphate is deficient to the point that plant growth is retarded or inhibited, the supply of phosphate reduces the disease to a low level until normal growth occurs. Beyond this point, further application of phosphate causes an increase in the disease (Kozaka, 1965).

Potassium. Early experiments in Japan showed that an ample supply of potassium reduces infection (Miyazaki, 1928, 1930; Ito & Hayashi, 1931), and the supply of large amounts of potassium was recommended as a control measure. Repeated experiments later, however, showed that while an ample supply of potassium sometimes reduces infection it generally causes an increase in the disease. Murata, Kuribayashi & Kawai (1933) and Chiba & Yamashita (1957) reported that a large potassium supply did not reduce the disease in plants receiving high nitrogen. Toyoda & Tsuchiya (1941) reported that in pot tests, large amounts of potassium caused more disease when the nitrogen supply was high than when it was low. Okamoto (1958) in his detailed investigation found that in potassium-deficient soil, a higher potassium supply increased the disease for a time, but reduced it later. In soil rich in potassium, the disease always increased with an increase of potassium level in plants receiving a large amount of nitrogen. The effect of potassium on the disease is therefore complicated because of its interrelationship with nitrogen. In the field, the ratio between potassium and nitrogen changes with the growth of the plants. Sako & Takamori (1958) demonstrated that application of potassium alone increased the disease which was, however, reduced by adding magnesium.

TABLE III–10

Relation between silica content in leaf blades of rice plants and varietal resistance to blast disease

(Yoshii, 1941)

Variety	Silica content		Disease index	
	Percentage	Index	Inoculation	Natural inf.
Kamairazu	2·74	112	100	+++++
Asahi	2·46	100	49	++++
Banshinriki	2·68	109	24	+++
Aikoku	2·74	111	1·1	+
Shinju 1	2·42	98	0·2	+
Sensho	2·27	92	0·2	±
Kameji	2·34	95		+
Ginbozu	2·42	98		++

Not much is known as to why potassium affects disease development. Kawamura & Ono (1948) observed stimulated spore germination and appressorium formation in dewdrops on plants receiving large amounts of potassium.

Silica. Many workers have repeatedly found and emphasized that (1) rice plants with a high silica content and thus a large number of silicated epidermal cells show slight damage from blast disease (Onodera, 1917; Miyake & Ikeda, 1932; Hemmi, 1933; Suzuki, 1934c) and (2) the resistance of rice plants to the disease is increased by the application of silica (Kawashima, 1927; Akai, 1939; Ishizuka & Hayakawa, 1951; Tasugi & Yoshida, 1958). This correlation between silica content and blast resistance exists within a given variety under varying conditions but not necessarily among different varieties (Table III–10) (Ikata, Matsuura & Taguchi, 1931; Yoshii, 1941a, b; Hashioka, 1942).

The physiological function of silica in disease resistance has been investigated by many workers. Hemmi (1933), Suzuki (1934c) and Ito & Sakamoto (1939–1943) believed that the silicated epidermal layer acts to prevent physical penetration of the fungus. Ishibashi (1936), Akimoto (1939), Yoshii (1941a, b) and Ishizuka & Hayakawa (1951) pointed out, however, that the resistance induced by silica may be ascribed only in part to the physical barrier and more significantly to the lower absorption of nitrogen. The degree of resistance was not always correlated with silica content but always with a decrease in nitrogen content. Tasugi & Yoshida (1958) demonstrated the accumulation and deposition of silica around cells injured by fungus penetration at a very early stage of invasion. They assumed that this accumulation served as a barrier against invasion. After penetration, the silicated epidermal layer and silica layer produced during penetration do not inhibit further mycelial growth.

VARIETAL RESISTANCE

Methods of evaluating varietal resistance. Varietal differences in resistance to blast can be and often have been observed in ordinary fields. For instance, during 1900–10, varieties Kameji and Aikoku were found to be highly resistant in Japan while Asahi, Omachi and Shinriki were slightly, moderately and very susceptible respectively (Ito, 1965). However, to obtain a more accurate assessment of the degree of resistance or susceptibility than is possible from field observation, and to handle a large number of varieties in a short period of time, favourable and uniform environmental conditions for maximum disease development must be provided. Many methods have been developed for varietal testing both in the field and by artificial inoculation in the glasshouse. Many scales for measuring the degree of resistance and susceptibility have also been described.

Field tests. Varietal resistance is usually tested at the seedling stage. According to the effects of environmental conditions on disease development discussed above, the most favourable environment for field testing varietal resistance may be summarized as follows. A dry, upland nursery is more favourable than a flooded field. A heavy application of nitrogenous fertilizer is required; at IRRI 120–160 kg nitrogen per hectare have been used, the amount depending upon the original fertility of the soil. A high humidity should be maintained in the microclimate of the nursery. Sprinkling with water, two or three times each day, depending on the weather, can easily maintain the necessary moisture. Infection takes place in the range 16–35°C but the optimum temperature is between 24° and 28°C. In most of the temperate regions tests may be conducted from June to September. In the tropics, the temperature is more or less favourable throughout the year (Fig. III–9) and the test may be made at any time, provided that high moisture is maintained. This has been proved in both the Philippines and Thailand. A high population of air-borne inoculum in temperate regions is limited to a few months, mostly from July to September with the peak in August (Table III–8). In the tropics, the population of air-borne inoculum is high from May–June to November–December (Fig. III–8). Square or rectangular plots of about $0 \cdot 5\ m^2$ for each variety have been used in several places in Japan. Others have used short rows, about 1 m long, using 10 g of seed for each variety (Fernando, Okamoto & Padmanabhan, 1961). Ou (1963) introduced the use of border rows to help retain higher and more uniform moisture in the testing rows and reduced the testing rows to 0·5 m in length (Fig. III–10). A uniform blast nursery testing method was developed (Ou, 1965b) and this has been adopted by more than 25 countries participating in the International Uniform Blast Nursery Programme.

Fig. III–10. A section of a blast nursery showing method of planting.

In breeding rice for blast resistance in Japan, Shigemura (1955) reported that special localities, where blast is epidemic every year, were selected for testing reaction. He also reported the use of a ' late sowing on dry field method ' for testing leaf blast and a ' late sowing and late transplanting method ' for testing neck rot. In Taiwan, such special sites were also selected for testing. However, conditions favourable for blast may be provided artificially, in which case tests may be made in a wide range of localities and seasons in the tropics.

Testing neck blast by means of field experiments is usually difficult because of the great variation in results from test to test (Ou & Nuque, 1963). As is discussed below, such testing seems to be unnecessary.

In the field testing of blast resistance, the plants are exposed to all the races of the fungus present in the locality during the time of the test. However, the races and their prevalence differ from season to season, so that repeated tests are necessary to discover the reaction of varieties to the various races of the fungus present in the same locality. Recent studies (IRRI, 1968;

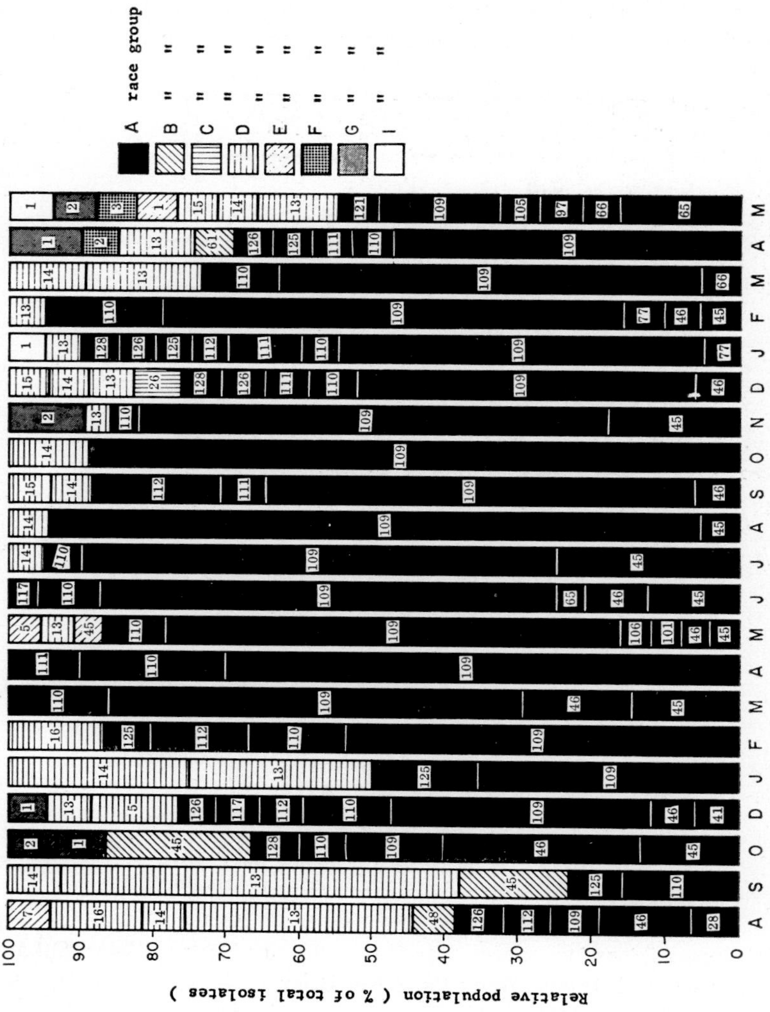

Fig. III–11. Races of *Pyricularia oryzae* Cav. detected in IRRI blast nursery each month from Aug. 1966 to May 1968 and their relative population (based upon international differentials).

Quamaruzzaman & Ou, 1970) revealed that 60 Philippine races or 34 international race groups were detected in the IRRI blast nursery during the period of 20 months when samples were taken. The composition and frequency of the races detected in each month were not the same. Some races were present most of the time, others were found periodically and still others occurred only once (Fig. III–11).

Artificial inoculation. To determine the reaction of rice plants to specific races of *P. oryzae*, and for environmental, nutritional and other specific studies,

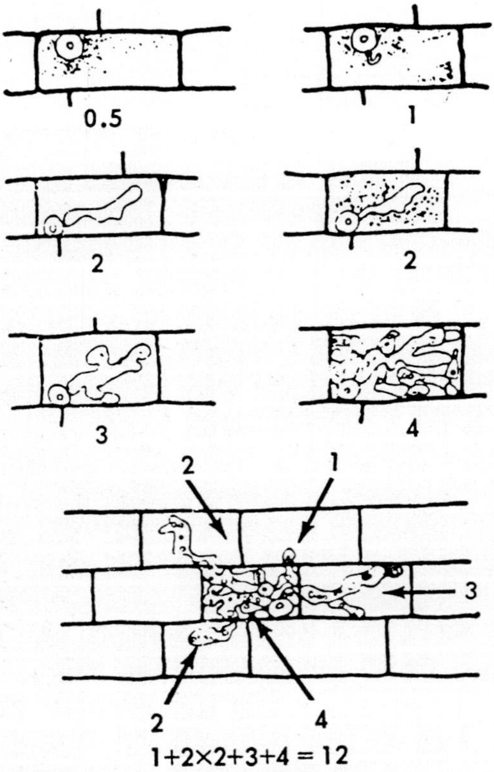

Fig. III–12. Standard for calculating the degree of fungus infection or host resistance. (Takahashi, 1965.)

artificial inoculations are necessary (Anderson & Henry, 1946; Sakamoto, 1949, 1951; Kahn & Libby, 1958; Panzer & Beier, 1958; Goto, 1960; Latterell, Marchetti & Grove, 1965). The most conventional method is to spray a conidial suspension in water on to the leaves, keeping the necessary control of the environmental conditions under which the plants are kept. The use of a dew chamber is most satisfactory. Spraying the suspension in a closed chamber and letting the mist settle on the leaves gives a more uniform deposit of conidia on the leaves than does spraying of the suspension directly on to the plants. Anderson & Henry (1946) suggested the use of adhesive agents which is perhaps not necessary. Anderson & Henry (1947), Kahn & Libby (1958) and Panzer & Beier (1958) used dry spore dust. To prepare the dust, they mixed the conidial suspension with finely ground peat, talc, or perlite, filtered it, dried it for 24 hours at 40°C and powdered it. Pure spore dust may also be prepared by mass culture, the conidia being collected by filtering the air which contains the conidia.

Sakamoto (1949, 1951) introduced the leaf sheath inoculation method of determining varietal resistance, and Takahashi (1958b) used it extensively in studying the mechanism of resistance. Leaf sheaths are cut into 7–10 cm pieces, filled inside with spore suspension and incubated for about 40 hr at 24–28°C. The cells of the inner surface of the sheaths are examined under the microscope, and penetration and the amount of growth inside the invaded cells are used as criteria for determining the degree of resistance, using a standard c ale (Fig. III–12). Many workers inject a spore suspension into the leaf

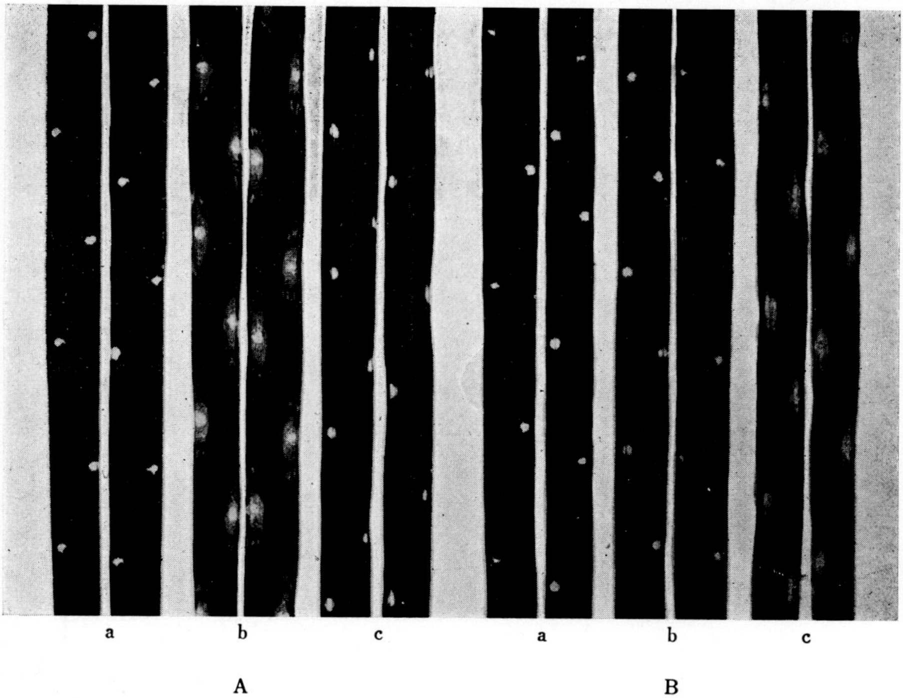

Fig. III–13. Inoculation by leaf punch. A—On variety Aichiasahi, a. not inoculated, b. inoculated with race N–1, c. inoculated with race C–3. B—On Kanto 51, a. not inoculated, b. inoculated with race N–1, c. inoculated with race C–3. (Ohata & Kozaka, 1967, translated.)

sheaths of seedlings and lesions appear on the young leaves which unfold in a few days (Kuribayashi & Terazawa, 1953). When necessary, two or more separate injections, about one week apart, may be made on the same seedlings. Hashioka (1963) tried the ‘ coleoptile test ’ by smearing a conidial suspension on germinating seeds. Ohata & Kozaka (1967) (Fig. III–13) inoculated by punching holes in rice leaves. By this method two or more races of the fungus can be inoculated on single leaves.

Many isolates or strains of the fungus do not produce enough conidia in culture for inoculation, and many kinds of media have been employed by various workers to stimulate spore production. In U.S.A. a special rice polish medium has been frequently used. Barley seeds have been used in Taiwan and Japan. Media have been enriched by highly nitrogenous substances such as peptone to increase spore production. Since isolates differ in their nutritional requirements, no single medium will be suitable for all isolates. At IRRI, it has been found that the leaf petioles of water hyacinth, banana stalks, rice stem nodes and coconut milk agar are favourable media for most isolates. High pH (± 9) also seems to stimulate spore production (IRRI, 1965; Chinte-Sanchez & Ou, 1967). Kato & Dimond (1966) studied the effects on sporulation of light and various chemical compounds which alter nucleic acid metabolism.

The density or concentration of conidia in the water suspension used for artificial inoculation should be regulated in comparative determinations of varietal reaction. Kobayashi & Abumiya (1960) found that the density of conidia affects not only the number but also the type of lesions. Ordinarily, concentrations of 2×10^4–5×10^4 spores/ml were used. Recently, Kiyosawa & Fujimaki (1967) showed that among the concentrations tested (1×10^4–4×10^4), 3×10^4 spores/ml was the optimum concentration for the development of maximum numbers of lesions on the leaves.

Ohata & Kozaka (1967) and Kiyosawa & Fujimaki (1967) found that with certain rice varieties, when conidia of a virulent strain (to which a variety shows a susceptible reaction) and an avirulent strain are mixed in the suspension, the resulting lesions are smaller than when inoculation is with the virulent strain alone. The higher the proportion of the avirulent strain in the mixture, the greater the effect in reducing the disease index.

For neck blast inoculation, Ou & Nuque (1963) injected about 1 ml of spore suspension into the leaf sheaths of emerging panicles (about half way emerged) with an automatic syringe. Using this technique, a relatively large number of panicles could be inoculated and often 100% infection occurred.

Relation between leaf blast and neck blast resistance. It has been reported that certain varieties resistant to blast in the leaf stage were observed to be susceptible to neck rot at a later stage of growth and, conversely, varieties susceptible to leaf blast showed little or no neck blast (Hashioka, 1950a; Podhradszyky & Sudi, 1957; Ito, 1965; Chang, Wang & Yang, 1965). Such observations have led to the belief that resistance to leaf blast and neck rot are not correlated and that different genes are involved. Many experiments have therefore been made in breeding programmes to test for resistance to neck blast as well as leaf blast. Other observations have indicated, however, that reactions to leaf and neck blast are correlated (Ou, 1963; Templeton, Johnston & Henry, 1961). All the observations and experiments mentioned above were made in the field where the nature of the inoculum (race composition) was not known. For a long time, the question of the reaction to blast of different parts of the plant was not clarified (Takahashi, 1965; Ou, 1965a).

Ou & Nuque (1963) injected spore suspensions of numerous isolates of *P. oryzae* on to many rice varieties with known reaction in the leaf stage, and also into the leaf sheaths of the half-emerged panicles. They found that varieties resistant to isolates at the leaf stage showed no neck blast while those susceptible to isolates at the leaf stage had 46–100% neck rot (Table III–11). It is thus evident that resistance to leaf blast is closely correlated with resistance to neck rot. The phenomenon that plants showing early leaf resistance later become susceptible to neck blast is mainly due to the presence of different races of the fungus at the later stage. In the field, many races may be present and they may change in amount at different seasons as mentioned above (IRRI, 1967). When varieties are thoroughly tested for resistance at the leaf stage, further testing at the flowering stage does not seem necessary. However, occasional exceptions may be found (Willis *et al.*, 1968).

TABLE III–11

NUMBER OF NECK INFECTIONS ON PANICLES OF 11 RICE VARIETIES INOCULATED WITH DIFFERENT ISOLATES OF THE BLAST FUNGUS

(Ou & Nuque, 1963)

Variety	Isolates	Seedling reaction	Number Infected/Inoculated	Neck rot (%)
Peta	I–123	S	45/46	98
	I–142	R	0/27	0
	I–150	S	29/29	100
	I–141	R	0/25	0
Tam Vuot	I–31	S	23/27	85
	I–19	R	0/22	0
	I–123	S	31/34	91
Leuang Yai	I–141	S	14/18	78
	I–140	R	0/15	0
FB–86	I–140	S	16/16	100
	I–141	R	0/15	0
Samo Trang	I–141	S	22/48	46
	I–139	R	0/41	0
Leuang Rahaeng 8	I–38	S	11/15	86
	I–42	R	0/28	0
FK–165	I–104	S	26/27	96
	I–101	R	0/18	0
H–5	I–94	S	9/15	60
	I–143	R	0/20	0
BE–3	I–32	S	28/31	90
	I–38	R	0/21	0
CO 25	I–143	S	20/27	74
	I–142	R	0/21	0
Radin Ebos	I–104	S	19/19	100
	I–10	R	0/19	0

Criteria and scales of resistance or susceptibility. Rice leaves infected by the blast fungus show various types of reaction depending upon varietal resistance or susceptibility (Plate VI). Resistant varieties show no symptoms (referred to as type 0 in later discussions) or very minute brown specks (type 1) to larger brown spots about 1 mm diameter (type 2). Intermediate groups show more or less roundish, restricted lesions, 2–3 mm diameter, with a grey centre and brown margin (type 3). Susceptible varieties produce large elliptical lesions 1–2 cm long with a grey centre and brown margin (type 4) and, in the case of very susceptible varieties, large broadly elliptical lesions (type 5). Brown margins are developed in the later stages of lesion development or under less favourable conditions. A yellow halo around the lesion is sometimes observed, often when the plants are kept in darkness or are receiving little light.

These lesion types are used as the main criteria for evaluating varietal reaction. The number of lesions on each leaf or unit leaf area may also be used as a criterion. Some varieties in the blast nursery are completely killed while others have always some green parts remaining, although both groups are very susceptible and have similar lesion types. Kuilman (1940) noted that a virulent attack of blast also caused stunting. This characteristic has also been used occasionally to measure resistance.

Based upon the type, colour and number of lesions, as well as the amount of stunting, workers have designed various scales to classify degrees of resistance or types of reaction. Hashioka (1950a) recognized five classes of lesions, which he designated A, B, C, D and E. These are approximately equivalent to the types 5, 4, 3, 2, 1 and 0, respectively, as described above. In addition he used A_1, A_2 and A_3 etc. to indicate few, several and many lesions respectively. Ono (1953) classified the lesions into four types, which he called the brown spot, white spot, acute and chronic types. Abumiya (1955, 1958) classified blast lesions into several types and described the zones of lesions as: white (w), blackish purple (p), brown (b), necrosis (g) and yellow (y). Lesions are described as: pg, bg, ybg etc. The capital letters G, P, B, Y indicate greater intensity of colours or characteristics. Atkins (personal correspondence) uses six classes (0, 1, 2, 3, 4 and 5), corresponding approximately to the types described above, in artificial inoculation experiments. Podhradszyky (1961) classified lesions into four groups according to the size of the developing spots.

Padmanabhan & Ganguly (1959) classified lesions into five types, A, B, C, D and E, from resistant to susceptible, and added I, II, III to each group to indicate the number of lesions. They further allocated scoring points, 1, 2, 4, 8 and 16 to each type, and 2, 5, and 10 to each lesion number group, respectively. The product of the two scores becomes the final reading of resistance.

An *ad hoc* committee of the FAO-IRC Working Party on Rice Production and Protection for the establishment of Uniform Blast Nurseries (Fernando, Okamoto & Padmanabhan, 1961) recommended a very comprehensive and elaborate system of evaluating resistance. The system considers (a) type of lesions, A, B, C, D, E, F and G (more or less corresponding to 0 (A) to 5 (F & G) above); (b) leaf area damaged, using grades 0, 1, 2, 3, 4, 5 and 6 to represent 0%, 0·5%, 2%, 5%, 11%, 25% and 55% damage respectively; (c) number of lesions, using letters o, f, m and n to represent none, few, many and numerous, respectively; and (d) stunting of plants, using s for slightly stunted and ss for severely stunted. Ou (1961a) used a combined scale of 8 classes for a more complete recording of field reaction. Scales 1, 2, 3 and 4 were based mainly upon lesion types as mentioned above, while scales 5, 6, 7 and 8 were based on the percentage of the leaf killed, the numbers representing up to 25%, 50%, 75% and 100%, respectively.

During the symposium on rice blast held at IRRI in 1963, a scale of seven units, numbered 1 to 7, was adopted for the Uniform Blast Nurseries, in view of the fact that the original scale recommended by Fernando, Okamoto & Padmanabhan (1961) was too complicated. The seven units of the scale were illustrated in colour for easy comparison (Ou, 1965b) and 1000 copies of the illustrations have been distributed to workers all over the world.

In studying pathogenic races, lesion types 1 and 2 described at the beginning of this section are considered to indicate resistance, type 3 an intermediate reaction, and types 4 and 5 susceptibility (Goto, 1960; Latterell, Tullis & Collier, 1960; Hung, Chien & Lin, 1961).

Scales for measuring resistance to neck blast usually take into account the percentage of panicles infected. Partial or complete girdling of the neck, and whether infection is on the main or the smaller branches may be considered separately. As to the level of infection considered to represent resistance or susceptibility, this varies according to the worker and the conditions of the test. Since the percentage of neck infection varies greatly in different trials, results which can be compared with those of other workers can be obtained only when a control variety (or varieties) with known resistance is included in the experiment for comparison.

Selecting varieties for resistance. Numerous field tests and observations have been made in many countries to determine varietal reaction to blast and to select for resistance. As a result, many resistant varieties have been identified and a number of them have been used in breeding programmes (see below under breeding). The reaction of varieties varies from country to country and from locality to locality, as well as from season to season in the same locality. It is not considered useful to review these studies in detail, and some of the reports of the work done have been briefly cited (Ou, 1965a). In general, most of the work in the past was concerned with testing only a relatively small number of varieties, in a few seasons, or in limited geographical areas. The resistant varieties selected were not, therefore, exposed to a large number of pathogenic race or forms, and so they do not have a very broad base for resistance. This may account for the relative lack of success of breeding programmes in the past. To illustrate the changes in varietal reaction which can occur in different seasons and localities, some of the work done in the Philippines is briefly described below.

During 1962–64, 8214 rice varieties were tested in IRRI blast nurseries and 1457 of them showed a highly resistant reaction (IRRI, 1964) (Table III–12). These 1457 varieties were tested seven times in the same nursery during the next two years, at the end of which time only about 450 varieties remained resistant (IRRI, 1966). These 450 varieties were then tested at seven stations

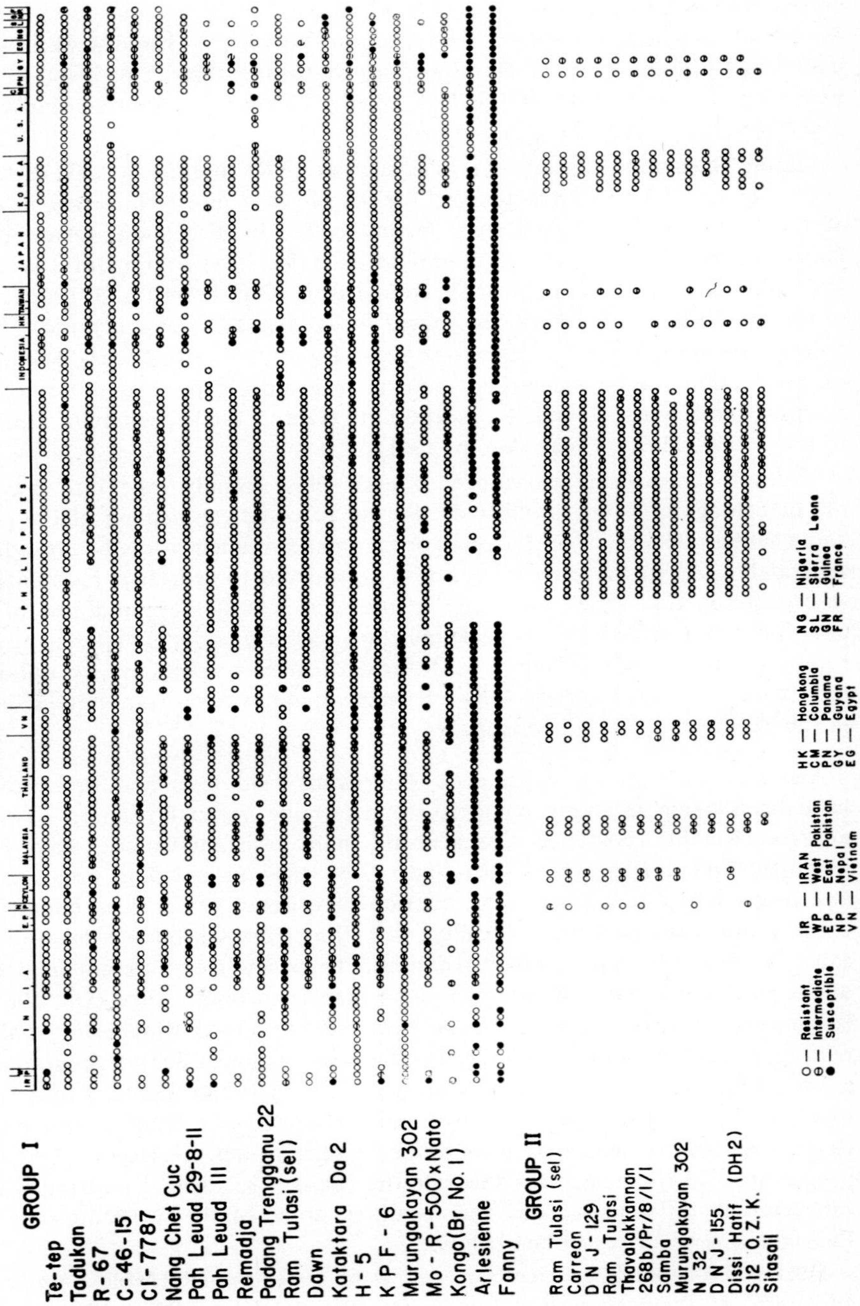

Fig. III-14. **INTERNATIONAL UNIFORM BLAST NURSERIES—SOME RESISTANT & SUSCEPTIBLE VARIETIES 1963—1968 RESULTS**

in different regions of the Philippines. After a few tests in each station, only about 75 varieties were found to exhibit a resistant reaction at all the stations (IRRI, 1966), and some more varieties may be expected to break down in future tests. These results showed very well the change of reaction which can occur in different seasons and localities or, in other words, the great differences in the pathogenic races or forms of the organism to be found at different times and places. However, there is a small number of varieties which so far have shown resistance at all times and in all places in the Philippines.

TABLE III–12

REACTIONS OF WORLD COLLECTION VARIETIES TO BLAST (IRRI, 1964)

Reactions	Number of varieties	Percentage
Highly resistant (Scale 1–2)	1457	17·74
Moderately resistant (Scale 3)	970	11·81
Moderately susceptible (Scale 4)	924	11·25
Susceptible (Scale 5–6)	1869	22·75
Highly susceptible (Scale 7)	2927	35·63
Varied reaction (Scale 1–7)	67	0·82
Total	8214	100·00

To obtain varieties with a broad spectrum of resistance, it is imperative that varieties be tested in a large number of rice-growing countries in a wide range of geographic regions. So that this may be achieved, an international programme is necessary. During the symposium on rice blast disease held at IRRI in 1963, the International Uniform Blast Nurseries scheme initiated by the International Rice Commission of FAO was modified and strengthened (Ou, 1965b). In 1963–64, approximately 258 selected varieties were tested in about 45 stations in 26 countries. During 1964–65, 58 tests were completed at 39 stations in 18 countries, and during 1966–69, 101 results were obtained from 23 countries (Ou, 1966; IRRI, 1968). It is possible that some deviations or errors occurred in this type of co-operative study but a few facts have been definitely established: (1) Some varieties have a broad spectrum of resistance, although none seems to be resistant in all tests (Fig. III–14). Varieties such as Te-tep have been recorded as having a susceptible reaction in only two of the 159 tests which were made in all the stations in several countries during five years. It is one of the varieties used as differentials in Japan and in the U.S.A.–Japan co-operative studies on physiologic races of *P. oryzae* (Atkins *et al.*, 1967) and thousands of isolates must have been inoculated on to the variety. But no susceptible reaction has been recorded. (2) Varietal reactions usually show regional patterns, i.e. any given variety is resistant or susceptible at all the test stations throughout a region or neighbouring regions (Table III–13). (3) Most exotic varieties show a resistant reaction; thus many *japonica* varieties are resistant in the South Asia (India-Pakistan) region while numerous *indica* varieties are resistant in the temperate Asia (Japan-Korea) region. Some varieties from the Southeast Asia region show resistance in South Asia and others in temperate Asia (Table III–14). The race pattern in Southeast Asia seems to be more complicated, and only those varieties with a very broad spectrum of resistance are resistant in that region (Table III–14).

TABLE III–13

GENERAL REACTION TO *P. oryzae* OF SELECTED VARIETIES IN VARIOUS REGIONS INDICATING DIFFERENCES IN PHYSIOLOGIC RACE PATTERNS

(O = resistant, M = intermediate, X = susceptible.) (Ou, 1966)

	Variety	General reaction in South Asia	General reaction in Southeast Asia	General reaction in Temperate Asia
21	Aichi Asahi	O	X	X
111	Aimasari	O	X	X
113	Sasashigure	O	X	X
196	Taichung 170	O	XO	X
20	Norin 22	O	OM X	X
192	Chianan 2	O	OM X	X
193	Tainan 3	O	OM X	X
195	Kaohsiung 64	O	OM X	X
104	221/BCIV/1/45/8	O	XO	O
165	Peta	O	X	O
169	Raminad Str. 3	O	X	O
173	FB 86	O	XM	O
197	Taichung (Native) 1	O	X	O
56	59–334	XO	X	O
74	T 21	X	X	O
151	Patnai 23	XO	X	O
164	Tjere Mas	XO	X	O
175	C 22	X	X	O
11	Te-tep	OX	O	O
12	Tadukan	OX	O	OX
45	C 46–15	O	OM	O
59	Murungakayan 302	O	OX	O
218	Zenith	OX	OM	O
61	Fanny	X	X	X
62	Arlesienne	XO	X	X
83	T.K.M. 6	X	X	X

In general, the results of international blast nurseries (Ou, 1964, 1966; Ling, 1968) show that many varieties are resistant in many testing stations (or to many races) and most varieties are resistant to a varying number of races. No variety seems to be resistant to all races but, on the other hand, no race is able to attack all varieties.

Since 1966, a further 321 varieties selected from the IRRI nursery (mentioned above) have been introduced into the international tests in the hope that more varieties with a broad spectrum of resistance may be identified. So far only 95 test results from 17 countries have been received. About a dozen

TABLE III–14

VARIETAL REACTION TO RICE BLAST DISEASE AT VARIOUS TEST STATIONS IN FOUR MAJOR REGIONS OF ASIA

(O=resistant, M=intermediate, X=susceptible; *=varieties showing broad spectrum of resistance over all regions). (Ou, 1966).

	Variety	India Pakistan	Malaysia Thailand Viet Nam	Philippines Indonesia	Korea Japan
18	Homare Nishiki	OOOO––OOM	MMOOMOMOOMOX	X–XXXMXMOOMM–	XXXXXOXXXXOO
20	Norin 22	OOOO––OOO	MMM–MOOOMOMM	MMXXMMMXO–XMM	XXXXXXXXXXXX
54	H–501	OMOOOO–O–	O–MMXXM––––X	O–MXMMOOOO–––	OXOOOOOOOOOO
112	Kinmaze	OMOO–––OO	–XMMX–MO–MMX	XOOMXXOXOXXMO	X–X–XOXXXX–O
165	Peta	OOOO–––OO	––OOMXMX–––X	XXMXXX–XXXXXX	OOOOOOOOOOOO
186	Taichung 65	OOOOOO–OO	––OOXMMM–MM–	XOOXXXOMOOX–O	X–XXXXXXXXXX
188	Taichung 155	OOOOOO–O–	––OOM–MM––––	X–OXXXOOM–OX–	X–XXXXXXXXXX
191	Chianung Yu 280	OOOOOO–OO	––OOM—MO–OOO	OOOXMOOOO–XXX	OXXOXO?XOOOX
197	Taichung (Native) 1	OOOO––O–M	–––XXXMXXXXX	XXXXXXXXXXXXX	O–OOO?OOOOOO
11	Te-tep*	XXOO––OMO	OOOOOOOOOOOO	OOOOOOOOOOOOO	OMMOOOOOOOOO
12	Tadukan*	OXOO––XOO	OOOMOOOOOOMM	OMOOMMOOOMOOO	OXOMOOOOOOOO
50	H–4*	OXOO––XMM	OOOOOOOOOOOM	OOMMXXOMOOMOO	OOOOOOOOOOOO
59	Murungakayan 302*	OXOOOOOOO	OOOOOOOOMOOO	OOOMXXMMOOMOO	OOOOOOOOOOOO
75	Ram Tulasi (Sel)*	OOOOMXOO–	OOOMOOOOO–OO	M–OOOOOOOOOOM	O–OOOOOOOOOO
145	Hashikalmi	OO––OO–O–	O–OOOMMOMOMO	M–OMXXOOOO–––	OXXOOOO?OOOO
146	Kataktara	OOOOXX–XX	O–OOOOMOMOOO	MOOOOOOOOOOOO	OXOOOOOOOOOO
218	Zenith*	OOOO––OXM	––MOM–MOOOOO	MOMMMMOMOOO–O	OOOOOOOOOOOO
11	Te-tep*	XXOO––OMO	OOOOOOOOOOOO	OOOOOOOOOOOOO	OMMOOOOOOOOO
12	Tadukan*	OXOO––XOO	OOOMOOOOOOMM	OMOOMMOOOMOOO	OXOMOOOOOOOO
45	C 46–15*	OMOOOOOOM	MOOOMOOOXOOO	MOMMOMOOOOOOO	OO?OOOOOOOOO
75	Ram Tulasi (Sel)*	OOOOMXOO–	OOOMOOOOO–OO	M–OOOOOOOOOOM	O–OOOOOOOOOO
106	Kanto 53	OMOO–––MM	O–OXOOOOXOOX	M–OOOOOOO–MMO	O–XOXXOOOOOO
146	Kataktara	OOOOXX–XX	O–OOOOMOMOOO	MOOOOOOOOOOOO	OXOOOOOOOOOO
147	K. P. F. 6	OOOO–––MM	O–OOOOXOXXXX	MOOOOOOOOOOOO	OXXOMOOOOOOO
174	Milbuen 5 (3)	OMOOOO–MM	––OOOOMO–––X	OOOOOOOMOMO–O	O–XOXXOOOOOO
210	Pah Leuad 29–8–11*	OMOO––OO–	––OOOOMO––XX	OOOOOOOOOOO–O	OOOO?OOOO–OO
218	Zenith*	OOOO––OXM	––MOM–MOOOOO	MOMMMMOMOOO–O	OOOOOOOOOOOO
1	CI 7787 (Zenith)*	–OO–––XOO	OOOXOOOOMOOO	MOOMM–OMOOOOO	OOOOOOOOOOOO
10	PI 231129	XX––OX–O–	OOOOMMO––––X	X–XXMMXXOO–––	OOOOOOOOOOOO
45	C 46–15*	OMOOOOOOM	MOOOMOOOXOOO	MOMMOMOOOOOOO	OO?OOOOOOOOO
49	M–302*	OX–O––OOO	OOOMOMMMXXXX	OOOMOOOOMOXXM	O–OOOOO–OOOO
50	H–4*	OXXO––OOO	OOOOOOOOOOOM	OOMMXXOMOOMOO	OOOOOOOOOOOO
51	H–5	OMOOOOOOO	OOOMOOOOXOXX	OOOMMMOOXOXOO	OOOOOOOOOOOO
59	Murungakayan 302*	OXOOOOOOO	OOOOOOOOMOOO	OOOMXXMMOOMOO	OOOOOOOOOOOO
75	Ram Tulasi (Sel)*	OOOOMXOO–	OOOMOOOOO–OO	M–OOOOOOOOOOM	O–OOOOOOOOOO
178	R 67*	OMOO–––OO	–MMOOOOO––OO	OOXOOOOMOXO–O	OOOOOOOOOOOO

varieties have shown promising results with no or very little susceptible reaction in any of the 95 tests (IRRI, 1968; Ou *et al.*, 1970; Fig. III–14).

The use of established races of *P. oryzae* to screen varietal resistance by artificial inoculation is very laborious and incomplete. This is so not only because there are many races and many varieties but also because one or two isolates do not represent the pathogenicity of all isolates belonging to the same race. In other words, isolates belonging to one race do not necessarily have the same pathogenicity to varieties other than the differentials. Furthermore, some isolates may not be stable, as discussed above. Artificial inoculation is, however, very useful in testing new races or specific isolates on specific varieties or hybrid lines.

Several workers have searched for sources of resistance in the wild species of *Oryza* (Kawamura, 1940; Mello-Sampayo & Vianna Silva, 1954; Katsuya, 1959, 1960, 1961; Lin & Lin, 1961). Some species reported to be resistant by one author were found by another to be susceptible, but different strains of wild rice and different races or isolates of the fungus were used. In general, the reactions to blast of wild *Oryza* spp. were much the same as those of cultivated varieties.

Horizontal resistance. Horizontal resistance, field resistance or the non-race-specific type of resistance to blast has not been explored in any detail, but some types of reaction may be considered as horizontal resistance, e.g. (1) the presence only of small lesions (intermediate, type 3 reaction, described above), and (2) the formation of only a small number of lesions (typical lesions present but few in number on each plant). Other types of horizontal resistance may include adult plant resistance (the older leaves are generally more resistant), the angle of leaves to the stem (more conidia are deposited on leaves with a more horizontal position), silica content etc., but much remains unknown on this subject.

Varieties with small lesions (type 3, intermediate reaction) are often found in blast nurseries. However, this reaction is controlled mostly by one or more major genes, as such varieties become susceptible in subsequent tests. Over 400 varieties showed an intermediate reaction in early tests at IRRI, 50% of them breaking down after two more tests. To separate major gene resistance from horizontal resistance in these varieties, they have to be exposed to all races in rice-growing countries in various geographic areas (IRRI, 1968).

Varieties with only a few lesions are also commonly observed in blast nurseries. In most cases, the small number of lesions is due to a low conidial population of the specific races to which the varieties are susceptible. At IRRI, another 400 varieties showed such a reaction in early tests, but when tested again, only about 50 varieties remained resistant. To distinguish whether such a ‘ few-lesion ’ type of resistance is due to a low population of the specific races to which the varieties are susceptible, or is due to field horizontal resistance, the fungus may be isolated from the lesions, cultured and artificially inoculated back on to the original varieties from which it was isolated. In preliminary experiments at IRRI, many varieties, when inoculated as described above, show smaller numbers of lesions than a control variety, Khao-teh-haeng 17 (known to be susceptible to all races known to be present in the Philippines). A more detailed study of Te-tep showed that 35 isolates from that variety produced an average of only about 3 lesions per plant on Te-tep whereas more than 30 lesions were produced on Khao-teh-haeng 17 when inoculated at the same time. Testing the single conidial subcultures of some of the isolates showed that most of the subcultures were not pathogenic to Te-tep, from which they were isolated. This rapid change of pathogenicity in the fungus and the broad spectrum of resistance in Te-tep seem to promise stable resistance to the disease (IRRI, 1969; Ou *et al.*, 1971).

Ezuka *et al.* (1969) also tested many varieties for field resistance and found many resistant varieties which were derived from Norin 22, Shinju and Futaba, and from Ishikari Shiroke.

Sakurai & Toriyama (1967) reported that varieties St 1 and Chugoku 31 developed for resistance to stripe virus showed field resistance to blast. They produced small numbers of lesions when inoculated with several isolates. However, St 1 was reported to be severely infected in other localities in Japan, and in Korea.

Breeding for blast resistance

Breeding for blast resistance in Japan, U.S.A., India, Taiwan and Thailand has been reviewed in detail by Ito (1965), Ito & Hirano (1967), Atkins *et al.* (1965), Padmanabhan (1965b), Chang, Wang & Lin (1965) and Dasananda (1965).

In Japan, systematic rice breeding work began in 1927, and since then a great deal has been done towards developing blast resistant varieties. The variety Norin 6 (Joshu × Senichi) was developed in 1935 and Norin 8 (Ginbozu × Asahi) in 1936. Norin 6 was then found to be resistant to neck blast but susceptible to leaf blast while Norin 8 was the reverse. By crossing Norin 6 with Norin 8, Norin 22 and Norin 23, resistant to both leaf and neck blast, were developed. It was found easy to introduce the resistant gene of Norin 22 to nonresistant varieties. Several resistant varieties have since been developed by various stations to improve further the nonlodging and other characters of both Norin 22 and Norin 23:

Norin 22 × Norin 23	Koganenami
Norin 22 × Norin 1	Hatsunishiki
Norin 22 × Rikuu	Chokai
Norin 23 × Norin 22	Taeho
Norin 22 × Norin 1	Honenwase
Chukyoasahi × Norin 22	Chiyohikari
Chukyoasahi × Norin 22	Yamabiko

An upland variety, Sensho, more resistant to blast, was used in crosses with lowland varieties by Iwatsuki (Ujihara, 1951) and the resistant varieties Shinju and Futaba were developed. From these two resistant varieties, four other varieties, Wakaba, Koganenishiki, Yutakasenbon, and Senbonmasari were developed (Fig. III–15a, b).

Matsuo (1952) found that two Chinese varieties, Toto and Reishiko, of the *japonica* type, were highly resistant. They were used as sources of resistance and to avoid seed sterility which often occurs between *japonica* × *indica* crosses. Koyama (1952) released Kanto No. 51 and Kanto No. 52 from Toto × Ginbozu-chusei and Kanto No. 53, 54 and 55 from Norin 10 × Reishiko. From Kanto

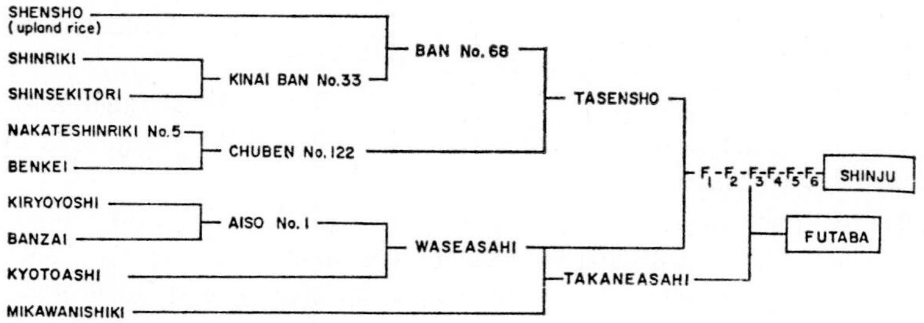

Fig. III–15a. Breeding processes of blast resistant varieties Shinju and Futaba. (Ito, 1965.)

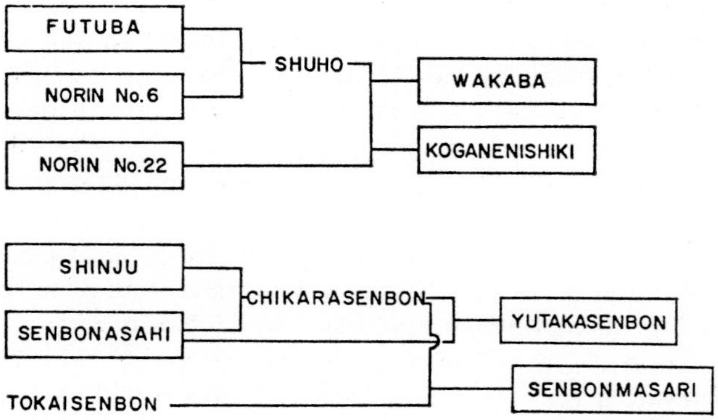

Fig. III–15b. Breeding processes of blast resistant varieties Wakaba, Koganenishiki, and Yutakasenbon. (Ito, 1965.)

No. 53 × Norin 29, Ito *et al.* (1961) selected Kusabue which also has improved quality and yielding ability. Several other varieties were also developed:

Kanto No. 53 × Norin 29 .... Kusabue
Kanto No. 53 × Eiko .... Yakara
Kanto No. 53 × Eiko .... Teine
(Takara × Sensho 26) × Kanto 53 .... Oyodo
$F_3$249 (Norin 25 × Toto × Norin 36) × Norin-mochi 45 .... Mangetau-Mochi
$F_3$249 × Heinoku-Mochi .... Kagura-mochi
Kanto No. 51 × Hatwnishiki.................... Ugonishiki

Oyodo was also resistant to bacterial leaf blight.

Kitamura (1961) found that certain *japonica* varieties showed low sterility when crossed with *indicas*. He developed several resistant varieties by the use of the *indica* variety Tadukan:

Tadukan × Senbonasahi	Pi 1 and Pi 2
Norin 8 × Tadukan	Pi 3 and Pi 4

These Pi varieties are generally poor in yield but are highly resistant to blast, and have been used extensively for breeding purposes in recent years.

The variety Zenith was also used by Kariya as a source of resistance to develop varieties BC–68 and Fukunishiki (Hirano, 1967). Many other resistant varieties have been developed in various experiment stations in Japan.

In U.S.A., the first attempt to control blast by varietal resistance was the introduction of 12 Italian varieties which were supposed to be resistant; but the resistance did not hold in U.S.A. (Metcalf, 1921).

Zenith, Nira and Fortuna were the most resistant varieties known in the early 1940s. These varieties were crossed with Lacrosse (Adair, 1949) and a new variety, Nova, was developed which, after many tests in Arkansas and Texas, was released in 1963. However, it was found to be susceptible to races 7 and 8 of *P. oryzae* in U.S.A.

A programme of testing and breeding for resistance to blast was started in 1959. The resistance of commercial varieties to races present in U.S.A. was determined (Table III-15) and various crosses were made, such as Northrose × Zenith, Mo. R–500 × Nato, etc. Tests for reaction were first made by artificial inoculation with race 6, which is most common in the field, and race 1, to which Zenith is susceptible. Subsequent tests also involved other races and also field tests (Atkins *et al.*, 1965). The variety Dawn was released in 1966 (Bollich *et al.*, 1966).

In Madras State, India, breeding for blast resistance started early in the 1920s and CO 4 and, later, TKM 1, were found resistant. Hybridization projects begun in 1927 and 1928, using CO 4 × ADT 10, resulted in two new resistant varieties, CO 25 and CO 26, which rapidly replaced the popular ADT 10. The crosses CO 4 × CO 13 and CO 4 × GEB–24 produced CO 29 and CO 30. Variety CO 4 has also been crossed with many other varieties in attempts to develop new blast-resistant varieties for Madras.

In Andhra Pradesh State, CO 4 has also been used as a source of resistance and many varieties have been developed:

M 183 × CO 4	Culture	3552
BCP–2 × CO 4	,, AC	5392
BCP–1 × CO 4	,,	1849
CO 19 × CO 4	,,	1807
P 1283 × CO 15	,,	1816
MTU 19 × CO 4	,,	10569
Bam 6 × CO 4	,,	1762
Bam 6 × CO 4	,,	1763
COB × CO 4	,,	6522
COB × CO 4	,,	6517
GEB–24 × CO 4	,,	6578

TABLE III–15

REACTION OF REPRESENTATIVE U.S. RICE VARIETIES TO *Pyricularia oryzae* RACES 1–8, 10, AND 16[1]

(Atkins *et al.*, 1965)

CI No.	Variety	Races of *Pyricularia oryzae* 1	2	3	4	5	6	7	8	10	16
Short grain											
1561–1	Caloro	R	S	S	S	R	S	S	S	R	S
1600	Colusa	R	S	S	S	R	S	S	S	R	
8988	Calrose	R	S	S	S	R	S	S	S	R	
Medium grain											
7787	Zenith	S	R	R	R	R	R	S	R	R	R
8310	Arkrose	R	S	S	S	R	S	S	S	R	S
8318	Magnolia	R	R	S	R	R	S	S	R	R	
8985	Lacrosse	R	S	R	R	R	S	S	S	R	S
8988	Nato	R	R	S	R	R	S	S	R	R	S
9155	Mo. R–500	R	R	R	R	R	R	S	R	R	R
9407	Northrose	R	S	S	S	R	S	S	R	R	S
9416	Gulfrose	S	R	R	R	R	R	S	R	R	R
9459	Nova	R	MR	MR	R	R	MR	S	S	R	R
Long grain											
1344	Fortuna	S	R	S	S	R	S	S	R	R	
2702	Nira	S	R	R	R	R	MR	MR	MR	R	
8990	Bluebonnet 50	S	R	S	R	R	S	S	R	R	S
8993	Century Patna 231	S	R	S	R	R	R	MR	R	R	R
9013	Toro	S	R	S	R	R	S	S	S	R	S
9386	Vegold	MS	R	S	R	R	S	S	S		
9433	Belle Patna	S	R	S	R	R	S	S	S	R	R

[1] Some varietal reactions may be changed by subsequent tests.

[2] *R*, resistant; *MR*, moderately resistant; *MS*, moderately susceptible; *S*, susceptible.

In other states of India (Mysore, Maharashtra, Uttar Pradesh, Kashmir) breeding for blast resistance has also been started, using CO 4 or its progeny and other varieties as the donor of resistance (Padmanabhan, 1965b).

At the Central Rice Research Institute of India at Cuttack, Orissa, numerous varieties have been screened for blast resistance, and the screened materials have been tested in other states (Padmanabhan, 1965b).

In Taiwan, more than 55 foreign introductions, as well as many local selections which showed some degree of resistance, have been used as sources of resistance. Breeding work has been done in the District Agricultural Improvement Stations at Taichung, Chiayi, Kaohsiung, Taipei and Taitung for many years, and numerous hybrid lines have been tested for blast resistance in special localities where blast is epidemic every year. Varieties and lines selected are: from Taichung Station, Taichung 178, 179, 181, 182, 183, 76 and 50; from Chiayi Station, Chianung 242, 280, C113 and C115; from Kaohsiung Station, Kaohsiung 122, 135, 137, 164, 165 and 202; from Taipei Station, Taipei 45; and from Taitung Station, Taitung 83. The majority of these varieties are progenies of crosses among local *ponlai* varieties. Most of the programmes are directed towards breeding for yield, quality, earliness etc., and not specifically for blast resistance, but the hybrid lines produced are tested and selected for resistance to blast.

In Thailand, serious attention to blast disease has been given since 1959. Numerous varieties have been tested for resistance and hybridization and selection have begun. Hundreds of crosses have been made and progenies tested in various regions. Vigorous programmes are in progress but no new varieties have yet been released. However, susceptible old varieties have gradually been eliminated and more resistant ones are now in wider use.

A limited attempt has been made to induce blast resistance by radiation both in Taiwan and Thailand (Huang, 1961; Lin & Lin, 1961; Li *et al.*, 1961; Dasananda, 1965; Yamasaki & Kawai, 1968). X-rays, thermal neutrons and gamma rays have been employed. Results showed some increase in resistance in susceptible varieties but high resistance has not been obtained.

Reviewing the work on breeding for blast resistance in the past, it may be said that considerable success may be recognized. However, in most cases a high degree of resistance has not been achieved.

Some varieties show a breakdown in resistance after a few years, and others become susceptible in other areas. The difficulty arises from the variation in pathogenicity of the fungus, a variability which in the past has been underestimated by workers in this field. The accumulation of information on races and its application to rice breeding began only ten years ago. For a longer lasting resistance in varieties to be developed in the future, varieties with a broad spectrum of resistance must be used, although even with these varieties resistance would be lost after some time, as new races would eventually appear. It is also important, therefore, that two or more sources of resistance complementing each other be used, either in phenotypically similar sister lines to be grown as multilateral varieties as suggested by Borlaug in rust resistant wheat, or with both sources of resistance incorporated into single varieties. Efforts

TABLE III–16

SUMMARIZED DESCRIPTION OF EARLY GENETIC WORK ON BLAST DISEASE RESISTANCE

(Takahashi, 1965)

Author	Varieties used in experiments	Number of crosses	Number of gene pairs	Dominancy	Remarks
Sasaki (1923)	*Japonica*	7	1	Resistance	
Nakatomi (1926)	*Japonica*	6	2	Resistance	
Nakamori (1936a)	*Japonica*	2	2	Resistance	
Nakamori and Kosato (1949)	*Japonica* and *indica*	1	2	Resistance	
Hashioka (1950a)	*Japonica* and *indica*	13	2	Resistance	
Takahashi (1951)	*Japonica*	8	3	Resistance	Gene effects accumulative
Okada and Maeda (1956)	*Japonica* and *indica*	10	3	Resistance	Gene effects accumulative
Oka and Lin (1957)	*Japonica* and *indica*	1	1	Susceptibility	
Abumiya (1959)	*Japonica* and *indica*	42	3	Resistance	Susceptible gene in *Japonica* varieties acts as an inhibitor to resistance

should also be made to look for field or horizontal resistance, as has been done in the case of resistance to late blight in potato, even though we know little about this type of resistance with reference to rice blast disease.

Genetics of Resistance

Sasaki (1923) in Japan began to study the inheritance of resistance to blast in 1917, only a few years after Biffen and Nilsson-Ehle demonstrated that disease resistance in crop plants was controlled by genes. A number of studies have followed and those completed before about 1959 were summarized, as shown in Table III–16, by Takahashi (1965).

From these studies, it appears that the genes controlling resistance vary from one to three pairs and that resistance is dominant in most cases. The lack of agreement among the results of different investigators may have resulted from the use of genetically different materials, the methods used for inoculation, and the criteria used for classifying resistance and susceptibility. All the above experiments were made prior to the establishment of the existence of physiologic races of *P. oryzae* and there was no knowledge as to what race or races were involved. Nakamori (1936a), for instance, observed that some varieties, while resistant at the Gifu Station, Japan, were susceptible at its branch station, and other varieties behaved in the reverse way. Sasaki (1923) classified plants into susceptible and resistant groups based on the presence or absence of lesions; Nakatomi (1926) used the percentage of dead plants; Hashioka (1950a) based his assessments upon grey lesions, brown lesions or no lesions; Okada & Maeda (1956) and Abumiya (1959) used lesion types y, b, g, p, etc. Oka & Lin (1957) employed biometrical analysis and Takahashi (1951, 1965) used sheath inoculation. It is obvious that there was no uniform basis for comparison and that, since the strain of fungus in use was unknown, exact information cannot be derived from these early experiments.

After the recognition of the existence of physiologic races, I. Goto (1959, 1960) employed various races and used the sheath inoculation method. He studied 250 crosses between many varieties and found that in about 50% of the crosses, resistance in the F_1 hybrids was dominant, in 25% it was incompletely dominant and in the other 25% susceptibility was dominant. He concluded that four pairs of genes were responsible for resistance and that each gene was cumulative, although they were somewhat different from each other.

The variety Asahi is susceptible to many races in Japan but is resistant to race Nagano–89. Narita & Iwata (1959) showed that a dominant gene controls resistance in Asahi to the race Nagano–89. Other varieties such as Te-tep and Tadukan which are resistant to most races are only intermediate in resistance to race H–2. This race, however, cannot infect Ishikari Shiroke. These facts seem to suggest the applicability of the 'gene-for-gene' hypothesis of Flor, developed from studies on flax rust. Along this line of thinking Takahashi (1965) proposed a working hypothesis, illustrated by a simplified figure as shown in Fig. III–16. Here the host variety I is most resistant and IV most susceptible, while II and III are intermediate. Each of the strains A, B, C and

D of the fungus has a different door to susceptibility, controlled by resistance genes in the host. Each of the doors is opened by the parasite which has the specific key (virulent gene) to it. A race of the fungus which has only one key specific to door A can invade only variety IV. Another race, having two keys, A and C, may invade varieties II and IV. Interactions among them may result in intermediate reactions.

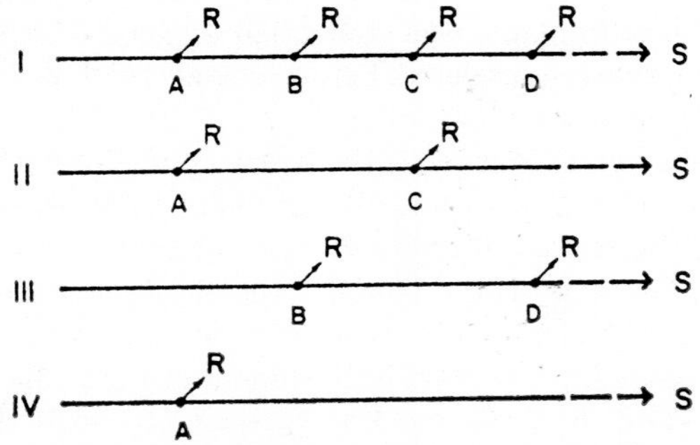

Fig. III–16. Simplified model showing relation between gene action and expression of resistance. (Takahashi, 1965.) (See text for details.)

More recently several experiments with known races have been made to study the inheritance of blast resistance. Atkins & Johnston (1965) studied the crosses Northrose × Zenith, Northrose × Gulfrose, and Northrose–Nato × Gulfrose inoculated with races 1 and 6. Zenith and Gulfrose are susceptible to race 1 and resistant to race 6 while Northrose and Nato are the reverse. When only one race was inoculated, the F_2 population showed a 3 : 1 ratio between resistant and susceptible and the F_3 a 1 : 2 : 1 ratio of resistant, segregating and susceptible. When both races were inoculated, the F_2 population showed a 9 : 3 : 3 : 1 ratio (resistant to races 1 and 6, to 1 only, to 6 only, and to neither, respectively). These and other results of the study suggested the presence of two independent, dominant genes for resistance to races 1 and 6. The resistant gene in Northrose and Nato was designated as *Pi–1* and that in Zenith and Gulfrose, *Pi–6*.

Yamasaki and Kiyosawa carried out a series of experiments on inheritance of resistance to blast (Yamasaki & Kiyosawa, 1966; Kiyosawa, 1966a, b; Kiyosawa, Matsumoto & Lee, 1967; Kiyosawa, 1967a, b, c; 1968a, b; 1969a, b; Kiyosawa & Murty, 1969; Kiyosawa & Yokoo, 1970). They grouped a number of Japanese and Chinese varieties into five types based upon the reactions of these varieties to seven isolates representing seven of the 14 races found in Japan. Gene analyses were made using some varieties belonging to the Aichi Asahi, Kanto 51, To-to and Ishikari Shiroke types, and the applicability of the gene-for-gene hypothesis was proposed. The varieties of the Aichi Asahi type carry one dominant gene *Pi–a* controlling resistance to the fungus strains Ina 72 and Ina 168. Kanto 51 has one gene *Pi–k* controlling medium resistance to the fungus strain P–2b and high resistance to strains Hoku 1, Ken 54–20, Ken 54–04 and Ina 168. The Ishikari Shiroke type carries one gene *Pi–i* controlling medium resistance to strains P–2b, Ina 72, Ken–54–04 and Ina 168. Genes *Pi–k* and *Pi–i* showed variation from complete dominance to incomplete dominance according to environmental conditions. Gene *Pi–a* showed complete dominance. The three genes behaved independently.

Kiyosawa (1966b; 1967a, b) further identified two genes for resistance in the variety Tadukan, the *Pi–a* gene and a new gene, *Pi–ta*; and two genes in Zenith, *Pi–a* and *Pi–z*. *Pi–ta* was also found in variety Pi No. 1 and $Pi\text{–}ta_2$ in variety Pi No. 4; the two genes are closely linked. In Norin 22, a medium-resistant variety, Kiyosawa, Matsumoto & Lee (1967) found one major gene with two or more minor genes and one major and one or more minor genes controlling the resistance to fungus strains Ken 54–04 and Ken Ph–03 respectively. They do not appear to have named the genes. Another gene, *Pi–m*, was identified in the variety Minehikari together with *Pi–a* and *Pi–k* in the Toto group of varieties, and *Pi–ks* in Shinriki. They further found *Pi–k* and $Pi\text{–}k^s$ in Japanese varieties against a fungus strain from the Philippines (Kiyosawa, 1969a), and three genes, *Pi–a*, *Pi–k* and $Pi\text{–}k^p$ in variety Pusur. The inheritance of resistance in variety HR–22 was found to be very complex but a gene, $Pi\text{–}k^h$ was identified (Kiyosawa & Murty, 1969). Two genes, *Pi–a* and $Pi\text{–}z^t$, were found in a cross between Norin 8 and CO25 and the gene $Pi\text{–}z^t$ was also present in TKM1 × Norin 8 (Kiyosawa & Yokoo, 1970). At least 11 resistant genes and alleles have been identified among the varieties that have been tested against the small number of strains of the fungus which they usually use. A resistant variety may have one to three or more of the resistant genes.

Ezuka *et al.* (1969a) tested 373 rice varieties in Japan with seven strains of the fungus and found 11 reaction types. Varieties of the so-called Kanto 51 type of Kiyosawa were further divided into three genotypes: (a) *Pi–k*, (b) *Pi–i* and *Pi–k* and (c) *Pi–i* and *Pi–m*. The Toto type was also divided into three genotypes: (a) *Pi–a* and *Pi–k*, (b) *Pi–a*, *Pi–k* and *Pi–i* and (c) *Pi–a*, *Pi–i* and *Pi–m*. The Pi No. 4 type was divided into: (a) *Pi–ta* and (b) *Pi–a* and $Pi\text{–}ta^2$. They also indicated that several new or unidentified genes seemed to be present in several varieties and strains. The test varieties were classified into 19 presumed genotypes and seven uncertain groups. With the many races of the fungus and many types of resistance in varieties, the genetics of resistance seem to be very complicated.

Hsieh & Chien (1967) and Hsieh *et al.* (1967) also reported four genes for resistance to *P. oryzae* in many varieties and called them *Pi–4*, *Pi–13*, *Pi–22* and *Pi–25* according to the respective fungus races used for tests.

One of the most difficult problems in the study of inheritance of resistance to blast is the variable reaction found in hybrid populations, there often being a full range of reactions from resistance to susceptibility. Where to separate the resistant and susceptible classes depends on the discretion of the investigator, and it may be that analysis of the data should be interpreted on the basis of quantitative rather than qualitative inheritance (Takahashi, 1965). It has also been assumed that the conidia of the fungus isolate used in inoculation tests are uniform in pathogenicity, but in the light of recent experiments (see

TABLE III–17

RECENT STUDIES ON INHERITANCE OF RESISTANCE TO BLAST

Author	Resistant variety	Fungus race	Resistant gene designation	Resistant gene action	Remarks
Atkins & Johnston (1965)	Zenith & Gulfrose	Race 6 (U.S.)	*Pi–6*	Dominant	
	Northrose & Nato	Race 1 (U.S.)	*Pi–1*	Dominant	Independent to *Pi–6*
Yamasaki & Kiyosawa (1966)	Aichi Asahi group	Ina–72 Ina–168	*Pi–a*		
	Kanto 51 group	P–2b, Hoku–1, Ken–54–04, Ina–168	*Pi–k*	Dominant or incomplete	
	Ishikari Shiroke group	P–2b, Ina–72, Ina–168, Ken–54–04	*Pi–i*	Dominant	Independent to *Pi–a* & *Pi–k*
Kiyosawa (1966)	Reishiko	P–2b, Ken–54–20	*Pi–k*	Dominant	
	Shekiyama	P–2b, Ina–72	*Pi–i*		
Kiyosawa, Matsumoto & Lee (1967)	Norin 22	Ken–54–04	One major & 1 or more minor genes		
Kiyosawa (1967a)	Zenith	Ina–72 Ina–168, etc.	*Pi–a*, *Pi–z*		Independent
	Pi No. 1 (Tadukan)	Ina–72 Ina–168	*Pi–a*		
		Hoku–1 Ken–54–20, etc.	*Pi–ta*		
	Pi No. 4 (Tadukan)	Ken–53–33, Ina–72, Hoku–1 Ken–54–04, etc.	*Pi–ta 2*	Dominant	Linked with *Pi–ta*
Kiyosawa (1967c)	Shinriki	Ken–Ph–03	*Pi–ks*		Allelic to *Pi–k*
Kiyosawa (1968)	Minehikari (Toto type)	Hoku–1, Ken–54–20, Ina–168, Ina–72, and mutants	*Pi–a*, *Pi–i*	Dominant	Independent except linkage with *Pi–m* & *Pi–k*
Kiyosawa (1969b)	Pusur	Hoku–1 & 5 other isolates	*Pi–a*, *Pi–k*, *Pi–k*^{p} and others		
Kiyosawa & Murty (1969)	HR–22	Ina–72 and 5 others	*Pi-k*^{h}		
Kiyosawa & Yokoo (1970)	CO25	P–2b & 6 others	*Pi–a*, *Pi–z*^{t}		
Hsieh & Chien (1967)	Pai-kan-tao		*Pi–4*, *Pi–13*, *Pi–22*, *Pi–25*	Dominant	Independent
	Sensho		*Pi–13*, *Pi–22*	Dominant	Independent
	Chianung 280		*Pi–4*, *Pi–22*, *Pi–25*	Dominant	Independent
	IG–65–2		*Pi–4*, *Pi–22*	Dominant	Independent
	IG–65–3		*Pi–4*, *Pi–22*	Dominant	Independent
	H–106		*Pi–4*	Dominant	Independent

above, under races and variability), at least in some isolates the conidia from the same pure culture differ in pathogenicity. The stability of the inoculum must therefore be considered.

In general, genes for resistance to blast are mostly dominant and independent. Recently, data have been accumulated supporting the gene-for-gene theory. A summary of some of the more recent reports is given in Table III–17.

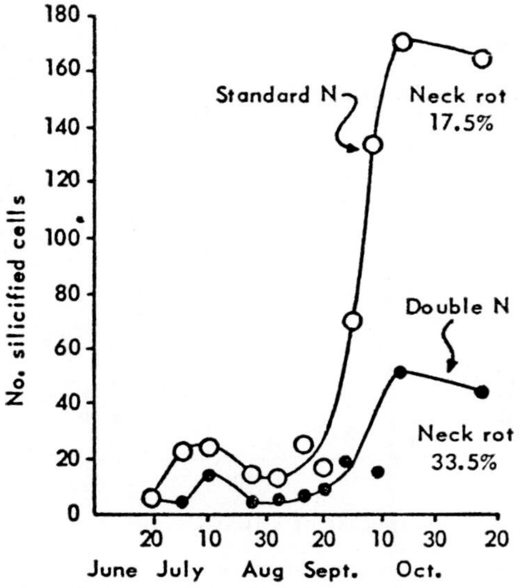

Fig. III–17. Seasonal changes in number of silicified cells in uppermost leaves of rice plants treated with standard and double amounts of nitrogen fertilizer. (After N. Suzuki, 1965.)

Nature of Resistance

A number of studies have been made in attempts to explain the mechanism of resistance (N. Suzuki, 1965). Early works are mostly correlation studies, such as the correlation of silicon content with resistance and the content of nitrogenous compounds, including amino acids and amines or soluble nitrogen, with susceptibility. More recent studies have dealt with host-parasite interactions and more specifically with host variety and fungus race interaction.

Miyake & Adachi (1922) first reported that the variety Bozu, resistant to blast, contains a larger amount of silicon than a susceptible variety. The degree of resistance was found by Miyake & Ikeda (1932) to increase in proportion to the amount of silicates applied and also to the amount of silicon accumulated in the plants. Many investigators confirmed this and further found that (1) heavy nitrogen applications, which induce more blast, decrease silicon accumulation, and (2) the number of silicified cells is larger in older leaves which are also more resistant to blast than younger leaves with less silicified cells (Figs. III–17, III–18). Yoshida, Ohnishi & Kitagishi (1962a, b, c, d) found that 90% or more of the silicon is in the form of silica gel, accumulating in layers alternating with cellulose layers in the epidermis, vascular bundles and sclerenchyma. The function of silica gel in the epidermis may be to control transpiration and to protect the plant from fungus and insect invasion. High silicon content may serve as a part of the mechanism of resistance. But the facts that some varieties with low silicon content are resistant to certain fungus races and those with high silicon are susceptible, and that the fungus is able to penetrate the epidermal cells even in resistant varieties, indicate that the nature of resistance is much more than mechanical protection alone.

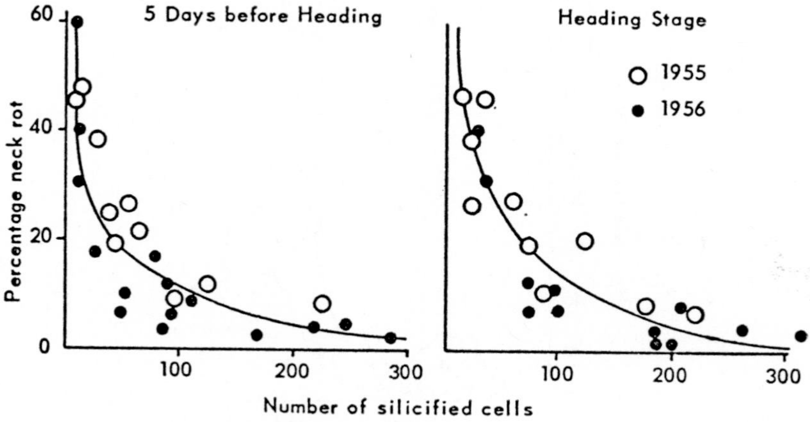

Fig. III–18. Correlation between percentage of neck rot and number of silicified cells in uppermost leaves at two stages of growth. (After N. Suzuki, 1965.)

Ito & Sakamoto (1939) reported that resistance to mechanical puncture of the leaf epidermis is correlated with resistance to blast. They found that puncture resistance was decreased by applications of nitrogen fertilizer and by low soil moisture, but was increased as the plants became older. They (1940) found further that puncture resistance decreases with loss of water or turgidity of the tissues and that it is related to the permeability of the protective membrane. They (1941–43) observed that the application of ammonium sulphate increased permeability, as did the presence of ammonia in the plant.

Otani (1952a) found a close correlation between susceptibility and soluble nitrogen or dibasic amino acids. Tokunaga (1959) and Tokunaga *et al.* (1966) found that the ratio between soluble nitrogen and total nitrogen is closely correlated with susceptibility. Sadasivan, Suryanarayanan & Ramakrishnan (1965) reported more accumulation of free amino acids at low night temperatures, and this was correlated with greater susceptibility. Ohta, Goto & Kozaka (1966) found that total nitrogen changes induced by low temperature treatment closely corresponded to changes in susceptibility while changes in total carbohydrates had the reverse effect. Many other workers also found more soluble nitrogen or amino acids and amines in susceptible varieties or under more favourable conditions for disease development such as low temperature and high nitrogen.

Hori, Arata & Inoue (1960) reported that the distribution of starch in the leaf sheath is related to resistance, longer accumulation indicating more resistance.

Ramakrishnan (1966a, b) reported a general increase in nitrogenous constituents in infected leaves of susceptible varieties but a slight decrease in infected leaves of resistant varieties. Reducing sugars in both resistant and susceptible varieties decreased as infection progressed while the amount of phenols increased.

Otsuka, Tamari & Ogasawara (1965) attempted to relate fungus nutrition with pathogenicity. They indicated that in general, the strains of the fungus which cannot utilize $NaNO_2$, or utilize it poorly, mostly showed strong pathogenicity, whereas those that utilized it well showed weak pathogenicity. Those that utilized tryptophan well were weakly pathogenic, whereas those utilizing cellulose, among carbon sources, showed strong pathogenicity. Kuo *et al.* (1968) reported that in a glutamic acid-requiring mutant pathogenicity was restored by adding the nutrient but mutants requiring arginine and histidine remained avirulent even when these nutrients were added.

Correlation studies yield some information on the apparent relationship between morphological features or chemical constituents of the rice plant and blast resistance but are not direct evidence of cause and effect. To understand the mechanism of resistance, the response of the host cell to the invading fungus should be studied directly.

Kawamura & Ono (1948) observed that when the fungus was inoculated to several resistant and susceptible varieties, many different types of lesions resulted, from minute brown specks in resistant varieties and small spots with brown margin in intermediate types to large grey lesions in susceptible types. In one of the resistant varieties, Kannonsen, the degree of silicification of the epidermis was much lower than in susceptible ones.

Resistance to penetration of the fungus is obviously less important than resistance to its spread within the host plant after penetration. A hypersensitive reaction is common in resistant varieties. Kawamura and Ono were able to isolate *P. oryzae* from hypersensitive lesions two days after inoculation but not after four days. They suggested that the death of the invading hyphae may not be due to starvation but rather to some toxic substances produced during the interaction between parasite and host. Anatomical observations showed that two processes take part in limiting the spread of lesions: (1) the rapid host cell response to the invading hyphae at an early stage of infection; and (2) the appearance of dead cells filled with a resin-like substance, the dead cells showing no sign of contraction. When both processes occur, only minute brown specks are produced; when one of the processes is lacking, medium-size lesions result; and when both are lacking, large susceptible lesions are produced (Figs. III–19, 20, 21).

The inner epidermis of the leaf sheath of the rice plant provides excellent material for studying cell responses because it has large, transparent cells, and it was used extensively by Ito and Sakamoto in their studies. The sheath inoculation method was described by Sakamoto (1951). Takahashi (1958a)

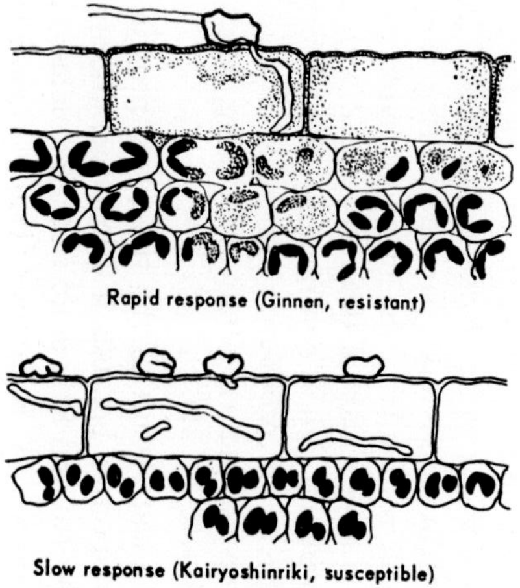

Fig. III–19. Rapid and slow response of rice leaves against infection of *Pyricularia oryzae*. (After N. Suzuki, 1965.)

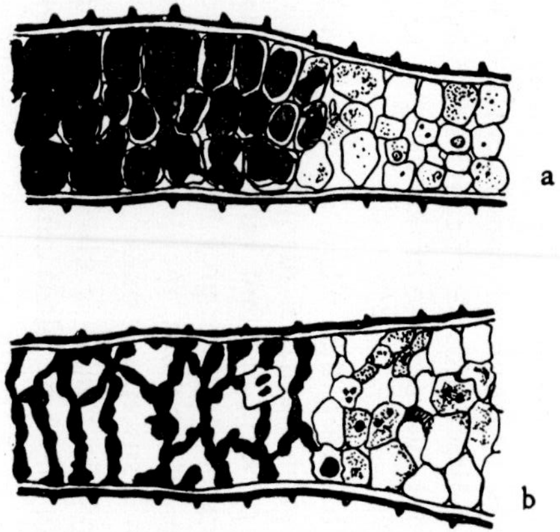

Fig. III–20. Two types of necrotic response of leaf tissue against blast fungus invasion. a. Necrotic tissue in which cells are filled with resin-like substance and do not contract (resistant). b. Necrotic tissue accompanying heavy contraction of dead cells (susceptible). (After N. Suzuki, 1965.)

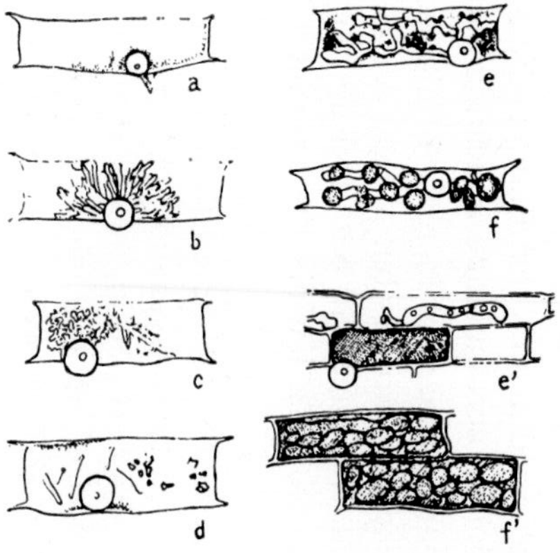

Fig. III–21. Six types of response of inner epidermis leaf cells against *Pyricularia oryzae* infection: a, b, c, d, hyaline to pale yellow response, permitting no fungus growth within the cell (resistant); e, f, e', f', brown to deep brown response, permitting full growth of hyphae within the cell (susceptible). (After N. Suzuki, 1965.)

proposed a standard value to distinguish various degrees of infection or resistance, as mentioned earlier (Fig. III–12). Using the sheath inoculation method, Takahashi (1956) observed the responses of resistant and susceptible varieties as shown and explained in Fig. III–21.

Takahashi (1959) and Ohata, Goto & Kozaka (1963) confirmed the observation of symbiosis between the fungus and susceptible varieties in the early stage of infection as first described by Ito & Sakamoto (1941). In four combinations between two fungus races and two rice varieties, Ohata *et al.* observed that in the resistant combination (variety resistant to fungus), the hyphae had penetrated the cells in 12 hours, when the cells were still alive, but after 18–24 hours they were dead. In susceptible combinations the infected cells remained alive for about 44 hours. Ohata (N. Suzuki, 1965) found this type of reaction also in the case of other race-variety combinations as follows:

Race N–1 × Aichiasahi (susceptible)—symbiosis for 48 hours
N–1 × Reishiko (resistant)—necrotic responses in 18 hours
C–3 × Aichiasahi (resistant)—necrotic responses in 18 hours
C–3 × Reishiko (susceptible)—symbiosis for 48 hours

Doi & Suzuki (1952) and Suzuki, Doi & Toyoda (1953) found that cell membranes in leaf lesions became brown, the brown discoloration being due to the accumulation of phenol, a type of orthodihydroxyphenol as indicated by its reaction to different reagents. In the case of a resistant host, the cell membrane and cell contents were discoloured. In the case of a susceptible host, the brown pigments which appeared around the invading hyphae resembled those produced by the fungus in liquid culture (Fig. III–22).

As mentioned above (Tamari & Kaji, 1954; Tamari, Ogasawara & Kaji, 1965) two toxins, piricularin and *a*-picolinic acid, were isolated from cultures of the fungus and from severely diseased plants. Both are toxic to rice plants, causing stunting of seedlings, leaf spotting and other injurious effects. Tamari & Kaji (1955) found that when combined with chlorogenic acid or ferulic acid, both present in rice plants, they become nontoxic to rice plants. They believed that the biosynthetic ability of chlorogenic acid, which possesses piricularin-detoxifying abilities, is related to rice plant resistance. They also found that in extremely low dilution, 1/1,600,000, piricularin exhibits a stimulative effect on the respiration and growth of rice plants. Resistant varieties generally tend to be more resistant to the inhibitory effect of piricularin and more sensitive to the stimulative effect. This stimulative effect of piricularin gives the rice plant an increase in polyphenol content and this increase is more marked in resistant than in susceptible varieties. The toxin theory has not, however, yet explained why varieties resistant to certain races are susceptible to others.

Host Range

Early in 1898 Hori in Japan successfully infected crab grass (*Digitaria sanguinalis*) with fresh conidia produced on blast lesions on infected rice plants. Kawakami (1901, 1902) infected rice plants with conidia from crab grass. He considered that the fungi on rice, crab grass, yellow foxtail, and Italian millet were identical. Yoshino (1905) and Hara (1916, 1918) also identified the fungus on these hosts and on barley, green foxtail, and common and wild gingers, as the same species, though they made no inoculation tests. Sawada (1917) in Taiwan, however, found that the rice fungus infected barley, wheat and Italian millet, but not crab grass or other grasses. Nisikado (1926) reported that the rice fungus seldom attacked Italian millet and failed to infect

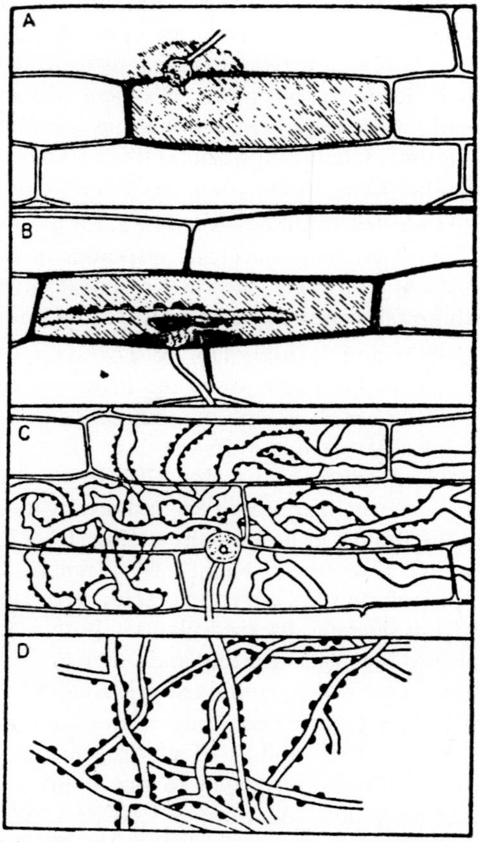

Fig. III–22. Different types of brown discoloration of inner epidermis cells of leaf sheath inoculated with isolate P–2. A, Reishiko (resistant); cell content and membrane colours pale yellow. B. Reishiko at later stage; cell contents gelatinize, colour brown, and arrest hyphal growth. C, Aichi-asahi (susceptible); brown granules appear around hyphae. D, brown granules, resembling those in C, are formed in liquid culture of fungus.

green foxtail, crab grass, proso millet, and the common and wild ginger. McRae (1922) in Pusa, India, showed that the rice fungus would not infect crab grass, *Panicum repens* or Italian millet. Anderson, Henry & Tullis (1947) in U.S.A. reported that the rice fungus also attacked other cereal crops such as barley, wheat and maize. These earlier results illustrate the discrepancies which arose in different studies of the host range of the rice blast fungus. Several other experiments yielded similar contradictory results. Asuyama (1965) reviewed the subject and compiled the accompanying table (Table III–18).

TABLE III–18

SUSCEPTIBILITY OF GRAMINEOUS PLANTS TO *Pyricularia oryzae* (Asuyama, 1965)

Plant	Literature cited[1]	
	Susceptibility[2]	Non-susceptibility
Agropyron repens	1 (C)	
Agrostis palustris, A. tenuis	1 (C)	
Alopecurus pratensis	1 (C)	
Anthoxanthum odoratum	2, 1 (C)	2
Avena byzantina, A. sterilis	1 (C)	
A. sativa	1 (C), 3	8, 2
Bromus catharticus	1 (BC)	
B. inermis	1 (C)	
B. sitchensis	1 (BC)	
Chikushichloa aquatica	4	
Dactylis glomerata	1 (C)	
Digitaria sanguinalis	5, 6	10, 9
Echinochloa crus-galli	7, 6	2
Eleusine coracana		3
E. indica	6	
Festuca altaica, F. rubra	1 (C)	
F. arundinacea	1 (ABC)	
F. elatior	1 (BC)	
Glyceria leptolepis	1 (BC)	
Hierochloe odorata	1 (C)	
Holcus lanatus	1 (BC)	
Hordeum vulgare	8, 2, 7, 1 (BC), 9, 6, 3	10
Leersia japonica		9
L. oryzoides	1 (AC)	
Lolium italicum	1 (C)	
L. multiflorum, L. perenne	1 (BC)	
Oplismenus undulatifolius	6	
Panicum miliaceum	1 (C)	2, 10
P. repens		9
Phalaris arundinacea	1 (BC)	
Phleum pratense	1 (BC)	
P. canariensis	1 (BC)	
Poa annua, P. trivialis	1 (C)	
Secale cereale	8, 2, 1 (BC)	
Setaria italica	2, 7, 1 (C), 10, 9, 6, 11, 3	10
S. viridis	2	2, 10
Sorghum vulgare		2
Triticum aestivum	8, 1 (C), 9, 6, 3	2
Zea mays	8, 12, 7, 2, 1 (BC), 3	
Zizania latifolia	1 (C)	4
Zingiber mioga		10, 9
Z. officinale		10

[1] 1, Narita, Iwata & Yamanuki, 1956; 2, Goto & Yamanaka 1960a; 3, Thomas, 1940; 4, Katsuya, 1961; 5, Hori, 1898; 6, Suzuki & Hashimoto, 1953a; 7, Kuribayashi, Ichikawa & Terazawa, 1953; 8, Andersen, Henry & Tullis, 1947; 9, Sawada, 1917; 10, Nisikado, 1926; 11, Nozu & Yokogi, 1924; 12, Hemmi *et al.*, 1949.

[2] 1(C) shows the result of artificial inoculation, 1(B) of field infection, and 1(A) of natural infection.

At one time, the natural infection of cereals by rice blast was thought to be of minor practical importance because of its rare occurrence, and that of wild grasses as having no bearing on the disease cycle of rice blast. Hemmi *et al.* (1949), however, suspected that some races of *Pyricularia* on crab grass might be capable of infecting rice, as they found one which was pathogenic to rice seedlings. Suzuki & Hashimoto (1953a) observed that the rice fungus occasionally attacked barnyard grass, crab grass, *Eleusine indica* and *Oplismenus undulatifolius* as well as barley, wheat and Italian millet; host range and symptoms varied with the isolates of the fungus present. Kuribayashi, Ichikawa & Terazawa (1953) demonstrated that *Pyricularia* isolates from several grasses infected certain varieties of rice. Iwata & Takakura (1954), Iwata, Yamanuki & Narita (1955) and Narita, Iwata & Yamanuki (1956) found field infection in potted plants of 13 species in 10 genera by exposing them in a rice blast nursery. By artificial inoculation, grasses of 38 species in 23 genera were found susceptible. Katsuya (1961) reported that *Chikushichloa aquatica* was susceptible to one isolate of the rice blast fungus. These results seem to indicate clearly that the host range of the rice blast fungus is considerable.

The reasons for the contradictory results reported in past experiments are mainly variation in the genetic background of the fungus; the races or strains studied; variations in the genetic background of the grass hosts; the use of different clones of the same species; and variation in the environmental conditions, such as soil fertility, under which the tests were made. That the use of different isolates or races of the fungus results in different susceptibility patterns on grasses has been shown by Suzuki (1951, 1953) in barley and crab grass and by Narita, Iwata & Yamanuki (1956) in wheat and creeping red fescue. That variation in the host results in different susceptibility was shown by Narita, Iwata & Yamanuki (1956) in *Festuca arundinacea*, *F. elatior*, *Lolium perenne* and *Anthoxanthum odoratum*. In wheat, variety Norin 67 is susceptible but Honkei 334 is resistant.

Besides those mentioned above, *Pyricularia* was found on a number of cultivated and wild plants in the Gramineae and also on species in the Zingiberaceae, Cannaceae, Musaceae and Cyperaceae. Some of them were reported also to attack rice (Asuyama, 1965). The following have been recorded as hosts—

Gramineae:

Eriochloa villosa (cup grass). Nisikado (1926).

Eremochloa ophiuroides (centipede grass). Johnson (1954).

Leersia japonica. Sawada (1917).

L. hexandra. Veeraraghavan & Padmanabhan (1965).

Panicum repens. Sawada (1917), McRae (1922), Thomas (1941), and Veeraraghavan & Padmanabhan (1965). All found no cross infection with other hosts. The Indian Council of Agricultural Research (1954) reported that it was probably mainly responsible for the carry-over of infection in rice fields.

Phragmites communis (common reed). Kuribayashi, Ichikawa & Terazawa (1953) reported *Pyricularia* from this host pathogenic to a few rice varieties.

Narita, Iwata & Yamanuki (1956), on the other hand, found *P. communis* to be resistant to the rice fungus.

Arundo donax (giant reed). Srinivasan & Vijoyalakshmi (1957) in India found a strain of *P. oryzae* capable of infecting *A. donax* and sugarcane.

Brachiaria mutica (signal grass). McIntosh (1951) and Veeraraghavan & Padmanabhan (1965).

Stenotaphrum secundatum (St. Augustine grass). Johnson (1954) reported *Pyricularia* from this host to be pathogenic to certain rice varieties. It was also studied by Malca & Owen (1957).

Saccharum officinarum (sugarcane). Hastings de Gutierrez (1955) found that cane fields adjacent to heavily diseased rice plantings were infected. Cross inoculations were successful.

Pennisetum typhoides (bajra). Mehta, Singh & Mathur (1952) and Buckley & Allen (1951).

The ‘ Index of Plant Diseases in the United States ’ (1952) listed several other plants as hosts of *P. grisea*: *Agrostis, Andropogon, Digitaria, Eragrostis, Fluminea, Leersia, Muhlenbergia, Panicum, Paspalum, Sorghum* spp. Sprague (1950) added *Cynodon dactylon* and *Hystrix patula*. Campacci (1950) found *Pyricularia* on *Cistichlis*; Teng (1932) on *Eleusine indica*; McRae (1922) on *Panicum ramosum*.

Zingiberaceae:

Zingiber officinale (common ginger). Nisikado (1926).

Z. mioga (wild ginger). Nisikado (1926).

Curcuma aromatica (yellow zedoary). Sawada (1959).

Costus speciosus. Sawada (1917).

Cannaceae:

Canna indica (common garden canna). Roldan (1938), Sawada (1959).

Musaceae:

Musa sapientum (banana). Stehle (1954), Hoette (1936), Wardlaw (1940), Meredith (1962).

Cyperaceae:

Cyperus rotundus. Lutrell (1954).

C. compressus. Thirumalachar, Kulkarni & Patel (1956).

Chemical Control and Disease Forecasting

Chemicals have been used extensively in Japan for controlling blast disease, and virtually all rice fields have been treated with chemicals in recent years. The success of chemical control in Japan is due to a combination of special socio-economic and technical reasons. The government has a very high supporting price for rice (about US$400 per metric ton) and very extensive disease forecasting and agricultural extension services. The average yield of rice per hectare is high. Industry is well developed and capable of producing relatively cheap chemicals. The relatively short-statured rice varieties enable the use of simple equipment to apply the chemicals to individual plants and the well-organized communities enable the use of aircraft for large-scale application even though individual farm holdings are small (less than one hectare on average). The cost of application is *c.* 1·5–4% of the value of the produce. This means that even if blast disease has caused only slight damage to the rice crop, it is economically worthwhile to control it with chemicals.

Copper compounds. The chemicals used in the early years were copper compounds. Bokura (1914, cited by Okamoto, 1965) first found that Bordeaux mixture was effective in controlling blast in the field, and, from 1923 onwards, Bordeaux mixture and other copper compounds were used extensively in the cooler regions of Japan where blast was prevalent. Copper compounds have, however, two defects; they are phytotoxic to rice plants and sometimes reduce yield instead of increasing it, particularly when blast attack is light and the disease causes little damage; and they do not control blast under very severe epidemic conditions.

Mercuric compounds. After World War II, copper fungicides were used in mixtures with phenylmercuric acetate (PMA). The mixture is more effective than is copper alone in controlling blast, and it is less toxic to rice plants. In 1950, the discovery was made by Ogawa (1953) that a mixture of PMA and slaked lime was surprisingly effective, was very low in toxicity and was cheap. It has, therefore, been used extensively. This stimulated the study of various mercury compounds, their activities, formulation and application methods. Okamoto & Yamamoto (1961) and Okamoto & Hamaya (1962) found that of the organomercuric fungicides, which can generally be expressed by the formula R-Hg-X, the compounds most effective against blast have R represented by phenyl (i.e. phenyl-Hg), with X influencing effectiveness only slightly. Ethyl-Hg and methoxyethyl-Hg combinations are less effective than phenyl-Hg regardless of X, although some of them are effective for seed treatment. Phenylmercuric

TABLE III–19

NAME AND USE OF ORGANOMERCURIC COMPOUNDS FOR BLAST CONTROL IN JAPAN
(Okamoto, 1965)

Chemical name	Use
Ethylmercuric chloride	Seed treatment
Methoxyethylenemercuric chloride	Spray, dust, seed treatment
Phenylmercuric chloride	Spray, dust
p-Tolylmercuric chloride	Spray, dust
Phenylmercuric iodide	Spray, dust
Ethylmercuric phosphate	Spray, dust, seed treatment
Phenylmercuric acetate	Spray, dust, seed treatment
Ethylmercuri-*p*-toluene sulfonanilide	Spray, dust, seed treatment
Phenylmercuri-*p*-toluene sulfonanilide	Spray, dust
p-Tolylmercuri-*p*-toluene sulfonanilide	Spray, dust
Phenylmercuric dinaphthylmethanedisulfonate	Spray, dust, seed treatment
Phenylmercuric urea	Spray, dust
Ethylmercuric urea	Spray, dust
Phenylmercuric mercaptobenzothiazole	Spray, dust
2-Oxy-2-phenylmercuric oxyhexachlorodiphenyl-methane	Spray, dust
Phenylmercuric triethanol ammonium acetate	Spray

acetate, phenylmercuric iodide, phenylmercuric *p*-toluene-sulphonanilide, and phenylmercuric fixtan have been the commercial products used most and with equal effectiveness, although some others have also been used (Table III–19).

Phenylmercuric compounds are absorbed by plant tissues and therefore have much stronger preventive action against blast infection and sporulation on lesions, and have a longer effect than copper compounds. Infection is prevented not only on the treated parts of the plants but also on untreated parts, such as new leaves and panicles which emerge after treatment. Though the mechanism is not completely understood, these compounds also have some therapeutic effect which restricts fungus growth in the plant tissues (Okamoto & Yamamoto, 1955a; Okamoto *et al.*, 1958; Okamoto, Yamamoto & Matsumoto, 1959).

Organomercuric compounds are considered to be strong inhibitors of the respiratory enzymes of pathogenic fungi. They react with glutathione and other SH-enzymes, and thus lose their fungicidal activity rather readily. Rice plants are resistant to infection for about two weeks after application (Nakazawa, 1959).

Phenylmercuric compounds have slight or no toxic action on *japonica* rice varieties when applied in normal dosages and it has been found that they may even slightly increase yield when tested in the field even though there are few or no visible blast lesions present. They have, however, a severe toxic action on some *indica* varieties (Okamoto *et al.*, 1960).

These organomercuric compounds are applied as dusts or liquids. In the form of dusts they contain 0·15–0·25% metallic mercury, and are applied at 30–40 kg (45–100 g metallic mercury) per hectare. In liquid form the content is 15–30 ppm, usually 20 ppm, metallic mercury. Materials for spraying are formulated as wettable powders or emulsifiable concentrates, and are applied at 1,000–15,000 l (15–45 g metallic mercury) per hectare.

For liquid seed treatment, they are used as follows:

Organomercuric compound	*Concentration (metallic Hg)*	*Soaking period (dry seeds)*
Ethylmercuric phosphate	12–15 ppm	3– 6 hours
Methoxyethylmercuric chloride	25 ppm	6–12 hours
Phenylmercuric acetate	20 ppm	3– 6 hours
Phenylmercuric fixtan	20 ppm	3– 6 hours

Antibiotics. The use of organomercuric compounds for blast control has reduced losses greatly during the last 15 years. However, as mercuric compounds are highly toxic it is not desirable to use them on food plants, and their use on rice in the field has now been banned in Japan. The poisonous nature of mercuric compounds has led to the development of many antibiotics and non-mercuric compounds in recent years (Fukunaga *et al.*, 1968).

Yoshii (1949) first found an antibiotic which inhibited the growth of *P. oryzae* on rice leaves. It was produced by a species of *Cephalothecium*, and he called it ‘cephalothecin’. Following this, ‘antiblastin’ (Suzuki, 1954), ‘antimycin A’ (Harada, 1955), ‘blastmycin’ (Watanabe *et al.*, 1957) and 'blasticidin-A' (Fukunaga *et al.*, 1955) have been found and tested. But none of these antibiotics could be put to practical use because of chemical instability or high toxicity to fish. In 1955, 'blasticidin-S' was developed by Fukunaga, Misato and their colleagues (Misato, 1961), and was found to have superior qualities for blast control.

Blasticidin-S is produced by *Streptomyces griseo-chromogenes* Fukunaga, the species which also produced blasticidin-A. It forms white, needle-shaped crystals, melts at 235–236°C, and is soluble in water and acetic acid but not in other organic solvents. Its chemical formula is $C_{18}H_{24}O_5N_8H_2O$. Hydrolysis of blasticidin-S produces cytosine, blastidone and levulinic acid.

Blasticidin-S completely inhibits spore germination at 1 ppm, and is thus as effective as PMA. At 1 ppm, blasticidin-S also inhibits mycelial growth, whereas 10 ppm (metallic) of PMA are required. This explains the stronger therapeutic effect of the antibiotic against blast. The number of lesions is smaller when plants are sprayed with 5 ppm of solution of blasticidin-S than with 20 ppm of PMA, when inoculations are made one or two days before spraying. The protective effect of the antibiotic is, however, slightly less than that of PMA, indicating that it decomposes more rapidly. It is able to permeate into rice tissues, which increases its curative or therapeutic action, preventing lesion formation and enlargement, and sporulation.

Blasticidin-S acts on the respiration and metabolism of *P. oryzae*; glutamic acid incorporation in the mycelium is strongly inhibited at 1 ppm, and it seems that inhibition of protein synthesis is a primary action (Misato *et al.*, 1959).

In many compaıative field tests, the control given by using blasticidin-S has been found to be equal to or better than that given by PMA, and it has come into large scale use during the last few years. The concentrations recommended for field use are 20 ppm in sprays and 0·2–0·4% in dusts.

Blasticidin-S is toxic to rice plants only when used in excess. Yellowish chlorotic spots appear within a few days after application and in extreme cases the spots become brownish and necrotic.

To obtain both preventive and curative effects, blasticidin-S and PMA have been used together, each at one half the normal concentration.

The successful development of blasticidin-S has stimulated research, and many other antibiotics, organometallic materials and other compounds have been developed in recent years and are in the field gradually replacing blasticidin-S. Some of them are listed below:

Kasugamycin (Kasumin)
5B (Barium pentachlorophenate)
Kitazin (EBP, O, O-diethyl-S-benzyl thiophosphate)
Hinosan (EDDP, O-ethyl-S-S-diphenyl thiophosphate)
Inezin (ESBS, O-ethyl-S-benzylphenyl phosphonothiolate)
Blastin (PCBA, pentachlorobenzyl alcohol)
Oryzon (CPA, pentachloromandel nitrite)
Rabcon (CPA, pentachlorophenyl acetate)

TABLE III–19a

RELATIVE FUNGICIDAL ACTIVITY OF THE FUNGICIDES IN CURRENT USE IN JAPAN FOR THE CONTROL OF RICE BLAST

(Kozaka, 1969)

Commercial name	Protective activity against		Eradicative activity against			Systemic activity through	
	Penetration	Residual action	Development of lesion	Sporulation	Residual action	Foliage application	Root application
Blasticidin–S	+++	++	++++	+++	++	—	—
Kasumin	+	±	+++++	++++	+++	±	+++
Blastin	++++++	++++++	±	++++	+++	+++	—
Oryzon	++++++	+++++	±	++ ?	++	+ ?	—
Rabcon	++++	+++	+	++ ?	++	± ?	—
Kitazin P	+++	++	+++	+++	++	±	++++
Hinosan	+++	+++	+++	+++	+++	±	+
Inezin	+++	+	+++	+++	++	±	+

Kozaka (1969) reviewed the fungicidal activity of these non-mercuric compounds and summarized the situation as in Table III–19a.

Recently (IRRI, 1969) benlate was found to have strong systemic fungicidal effect against blast.

Many public and private institutions in Japan are engaged in the extensive development of chemicals for controlling blast and other diseases of rice, and new products can be expected to appear from time to time.

Chemical control in the tropics has been tried in several countries but is not in general use, as many technical and socio-economic reasons, as mentioned above, at present militate against the use there of chemicals for blast control.

Disease forecasting. Because of the large quantities of chemicals used in disease control, Japan has a very elaborate forecasting system, the object being to determine as precisely as possible the optimum time at which to apply fungicides, so as to bring in the most profitable return possible.

Many studies of methods of forecasting blast disease have been made, based upon information on the fungus, the host plant, and the environment (Ono, 1965).

To produce a forecasting system based upon environmental (weather) conditions, individual factors such as temperature, moisture, sunshine and wind were studied separately as regards their relation to infection and development of the disease (mentioned above). Then these weather factors were

TABLE III–20

CORRELATIONS OF WEATHER FACTORS WITH NECK BLAST CONTROL

(After Ono, 1965)

Period	Temperature			Humidity	Quantity of precipitation	Days of precipitation	Times of sunshine
	Average	Maximum	Minimum				
May							
Early[1]	+0·315	+0·416			+0·388		
Middle[2]							
Late[3]	+0·474		+0·650				
June							
Early			+0·528				
Middle				+0·373		+0·359	
Late	−0·381	−0·399	+0·432				−0·549
July							
Early				+0·312	+0·529	+0·323	−0·564
Middle				+0·304			
Late			−0·349	+0·637	+0·710		
August							
Early							
Middle		−0·369					
Late	−0·310	−0·374		+0·377			−0·446
September							
Early				+0·406	+0·618	+0·498	
Middle	−0·344				+0·421		
Late				+0·363			

[1] First ten days of the month.
[2] Middle ten days of the month.
[3] Last ten days of the month.

analysed in blast epidemic and normal years in 5-day or 10-day periods from May to September, and correlations calculated, from figures such as those shown in Table III–20. Since weather conditions differ in each prefecture, separate formulae were calculated for forecasting in different areas. For example, at Yamanashi Prefecture, the following formula has been used to forecast neck blast:

$$y = 12703 \cdot 73 - 257 \cdot 34x \qquad r = 0 \cdot 761$$

Here x=average percentage of sunshine in June and July; y=areas of outbreak of the disease; r=correlation coefficient between x and y.

Spore dissemination has been studied in detail, as mentioned above, and it has been found that the numbers of air-borne spores present and outbreaks of blast are closely correlated (Tables III–8, III–21). This fact has been used to evolve one of the most reliable forecasting tools available (Table III–22).

TABLE III–21

CORRELATION OF SPORE FLIGHT TO OUTBREAK OF BLAST DISEASE
(After Ono, 1965)

		Correlation coefficient	
		Neck blast	Node blast
July	1[1]	+0·244	+0·272
	2	+0·263	+0·493
	3	+0·420	+0·542
	4	+0·450	+0·514
	5	+0·595	+0·880
	6	+0·797	+0·814
Aug.	1	+0·800	+0·812
	2	+0·864	+0·811
	3	+0·969	+0·775
	4	+0·961	+0·874
	5	+0·857	+0·827
	6	+0·881	+0·819
Sept.	1	+0·918	+0·823
	2	+0·818	+0·854
	3	+0·507	+0·259
	4	+0·902	+0·774
	5	+0·736	+0·803
	6	+0·203	+0·130

[1] 1 indicates first 5 days; 2, second 5 days; 3, third 5 days, etc., of the month.

Growth conditions, such as the height of the plants and the number of tillers present, and various chemical components of the plant, such as starch accumulation, silicon content or number of silicated cells, etc. have also been studied with regard to their use for forecasting. The amount of leaf blast and the type of lesions, whether of the brown spot or white spot type and whether of the chronic or acute type, have also been used for forecasting neck blast (Ono, 1965).

The forecasting organization in Japan is very extensive. At the national level, the Ministry of Agriculture and Forestry co-operates with the National

Institute of Agricultural Sciences and the National Agricultural Experiment Stations in co-ordinating the forecasting programmes. In each prefecture, two or three officers are responsible for the work, the total number involved being about 130. There are 540 forecasters, about 11 or 12 persons in each prefecture, and 11,000 field officers, about one for every 300 hectares of land. They conduct tests, collect and analyse data, and forecast outbreaks. Information is broadcast regularly twice a month and, in addition, there are special reports and annual conferences. These activities are concerned not only with blast but also with many other diseases, as well as insect pests (Ono, 1965).

TABLE III–22

FORECASTING FORMULAE OF OUTBREAK OF NECK BLAST, BASED ON BLAST FUNGUS SPORES CAUGHT AT NAGANO PREFECTURE

(After Ono, 1965)

	Forecasting formula
July 5[1]	$y = 9{\cdot}22x + 13{\cdot}38 \pm 15{\cdot}09$[2]
6	$y = 1{\cdot}80x + 14{\cdot}96 \pm 13{\cdot}03$
Aug. 1	$y = 0{\cdot}82x + 16{\cdot}66 \pm 13{\cdot}49$
2	$y = 0{\cdot}27x + 15{\cdot}76 \pm 12{\cdot}19$
3	$y = 0{\cdot}40x + 10{\cdot}81 \pm 5{\cdot}47$
4	$y = 0{\cdot}38x + 3{\cdot}73 \pm 8{\cdot}35$
5	$y = 0{\cdot}21x + 14{\cdot}50 \pm 11{\cdot}40$
6	$y = 0{\cdot}43x + 14{\cdot}30 \pm 12{\cdot}08$
Sept. 1	$y = 0{\cdot}71x + 10{\cdot}11 \pm 10{\cdot}72$
2	$y = 3{\cdot}07x + 6{\cdot}48 \pm 13{\cdot}83$
3	$y = 2{\cdot}50x + 11{\cdot}09 \pm 21{\cdot}35$
4	$y = 1{\cdot}09x + 14{\cdot}22 \pm 15{\cdot}81$

[1] 1 indicates first 5 days; 2, second 5 days; 3, third 5 days, etc., of the month.
[2] x=sum of numbers of spores caught in the 5 days; y=percentage of neck blast.

CONTROL BY CULTIVATION PRACTICES

Besides the use of chemicals and resistant varieties, the control of blast disease in Japan also relies on adjusting cultural practices. Often the various methods are employed in a co-ordinated control programme.

Time of planting has been demonstrated to be an important factor in blast development, early plantings in Japan usually having less disease than later plantings. This is explained by the fact that in early plantings the air temperature is too low at the tillering stage and too high at the heading stage for vigorous disease development (Hashioka, 1950, 1950b; Kuribayashi & Ichikawa, 1952). In India, Chandra & Palaniswamy (1963) reported on the relationship between time of planting and incidence of blast and noted that severe blast was correlated with low temperatures, high humidity and heavy dews.

Seedlings raised in upland nurseries are more susceptible to blast even after they are transplanted. This is explained by the lower silicon content of the epidermal cells, as discussed above. Otani (1948, 1950, 1952a) also showed that these seedlings contained more soluble nitrogen, amino acids, and amines, which favoured disease development. Yamada & Ota (1956) showed that

TABLE III–23

EFFECT OF DRAINING ON SUSCEPTIBILITY OF RICE PLANTS TO NECK BLAST*

(Suzuki, 1933)

Plot	Period of draining (++++)			Disease index of neck blast
	Transplanting × … × Booting stage	Booting stage × … × Inoculation	Inoculation × … × Evaluation	
1	++++++++++++++++++++	++++++++++++++++++++	++++++++++++++	606
2	++++++++++++++++++++	++++++++++++++++++++	——————————————	465
3	————————————————————	++++++++++++++++++++	++++++++++++++	323
4	————————————————————	++++++++++++++++++++	——————————————	298
5	++++++++++++++++++++	————————————————————	——————————————	232
6	————————————————————	————————————————————	++++++++++++++	211
7	++++++++++++++++++++	————————————————————	++++++++++++++	195
8	————————————————————	————————————————————	——————————————	100

* Plants used in this experiment were inoculated immediately after heading.

upland seedlings had higher physiological activities, producing more roots and absorbing more nitrogen than seedlings grown in wet nurseries.

As already mentioned, various kinds and amounts of fertilizers affect disease development. In particular nitrogenous fertilizers have a profound influence, and farmers often have to restrict nitrogen fertilizer applications to avoid serious outbreaks of blast. The type of soil should always be considered in relation to the kinds and amounts of fertilizers to be applied (Kozaka, 1965).

Control of irrigation water has also been used as a means of reducing blast damage. In Japan, if the temperature of the irrigation water is 20°C or below there is usually an increase in disease incidence. This also relates, however, to the length of time that the plants, usually as seedlings, are subjected to the cold water and depends also on the amount of nitrogen in the soil (Kozaka, 1965). The time at which the irrigation water is drained off also affects disease development as shown in Table III–23.

In Japan, deep transplanting of seedlings has been found to retard growth, delay maturation and reduce yield, as well as causing more blast infection. Close spacing also often increases the severity of the disease.

LITERATURE CITED

ABE, T. 1930. The relation of temperature and time to the invasion of rice blast fungus. *Ann. phytopath. Soc. Japan* **2:** 277–278. [Jap.]

ABE, T. 1931. The effect of sunlight on infection of rice plants by *Piricularia oryzae*. *Forschn Geb. PflKrankh., Kyoto* **1:** 46–53. [Jap.] (*Rev. appl. Mycol.* **11:** 800.)

ABE, T. 1933a. The influence of soil temperature on the development of blast disease of rice. *Forschn Geb. PflKrankh., Kyoto* **2:** 30–54. [Jap. Engl. summ.] (*Rev. appl. Mycol.* **13:** 264.)

ABE, T. 1933b. The relationship of atmospheric humidity to infection of the rice plant by *Piricularia oryzae* Br. et Cav. *Forschn Geb. PflKrankh., Kyoto* **2:** 98–124. [Jap. Engl. summ.] (*Rev. appl. Mycol.* **13:** 265.)

ABE, T. 1935. On the resistance of conidia of *Piricularia oryzae* to low temperatures. *Ann. phytopath. Soc. Japan* **5:** 206–215. [Jap. Engl. summ.]

ABE, T. 1941. Relation of air humidity to the mycelial development of the rice blast fungus. *Ann. phytopath. Soc. Japan* **10:** 331–332. [Jap.]

ABUMIYA, H. 1955. On the infection types of leaf blast disease of the rice plant. *Jubilee Publ. commem. 60th Birthdays of Profs. Y. Tochinai and T. Fukushi*: 197–201. Sapporo, Japan. [Jap. Engl. summ.]

ABUMIYA, H. 1958. Phytopathological studies on the breeding of rice varieties resistant to blast disease. II. On the process of brown necrosis formation at leaf lesion. *Bull. Tohoku natn. agric. Exp. Stn* **14:** 15–21. [Jap. Engl. summ.]

ABUMIYA, H. 1959. Phytopathological studies on the breeding of rice varieties resistant to blast disease. III. On the resistance for the lesion development in leaf blast with their classifying method and on the inheritance of resistance of the Japanese and foreign rice varieties. *Bull. Tohoku natn. agric. Exp. Stn* **17:** 1–101. [Jap. Engl. summ.]

ABUMIYA, S. & KOBAYASHI, H. 1953. Contents of amino acids and amides in rice leaves in relation to blast disease. *Ann. phytopath. Soc. Japan* **18:** 75. (Also in *Proc. Ass. Pl. Prot., North Japan* **4:** 33–34.) [Jap.]

ADAIR, C. R. 1949. Rep. Arkansas agric. Exp. Stn (unpublished). (Cited by Atkins *et al.*, 1965.)

AHN, C. J. & CHUNG, H. C. 1962. Studies on the physiologic races of rice blast fungus, *Piricularia oryzae*, in Korea. *Seoul Univ. J., Biol. and Agr.*, Series B, **11:** 77–83. [Korean. Engl. summ.]

AKAI, S. 1939. On the ash figures of leaves of rice plants transplanted from the different kinds of nursery beds, and their susceptibilities to blast disease. *Ann. phytopath. Soc. Japan* **9:** 223–235. [Jap. Engl. summ.]

AKIMOTO, S. 1939. Varietal difference in absorption of nitrogen and silica in relation to rice blast. *Agriculture Hort., Tokyo* **14:** 2279–2290. [Jap.]

ANDERSEN, A. L. & HENRY, B. W. 1946. The use of wetting and adhesive agents to increase the effectiveness of conidial suspensions for plant inoculations. *Phytopathology* **36:** 1056–1057.

ANDERSEN, A. L., HENRY, B. W. & TULLIS, E. C. 1947. Factors affecting infectivity, spread, and persistence of *Piricularia oryzae* Cav. *Phytopathology* **37:** 94–110.

AOKI, Y. 1935. On physiologic specialization in the rice blast fungus, *Piricularia oryzae* Br. et Cav. *Ann. phytopath. Soc. Japan* **2:** 107–120. [Jap. Engl. summ.]

ASAI, G. N., JONES, M. W. & RORIE, F. G. 1967. Influence of certain environmental factors in the field on infection of rice by *Piricularia oryzae*. *Phytopathology* **57:** 237–241.

ASTOLFI, G. 1828. Congettura sulla malattia del brusone che infesta del riso. . . . *Ann. Univ. Tecnol., Milano* **6:** 198. (*Spec. Publ. Sect. Mycol. Dis. Surv. U.S. Dep. Agric.* 6: 13, 1954.)

ASUYAMA, H. 1965. Morphology, taxonomy, host range, and life cycle of *Piricularia oryzae*. In *The rice blast disease: Proc. Symp. at IRRI, July, 1963*: 9–22. Baltimore, Maryland, Johns Hopkins Press.

ATKINS, J. G. 1962. Prevalence and distribution of pathogenic races of *Piricularia oryzae* in the U.S. Abs. in *Phytopathology* **52:** 2.

ATKINS, J. G., BOLLICH, C. N., JOHNSTON, T. H., JODON, N. E., BEACHELL, H. M. & TEMPLETON, G. E. 1965. Breeding for blast resistance in the United States. In *The rice blast disease: Proc. Symp. at IRRI, July, 1963*: 333–341. Baltimore, Maryland, Johns Hopkins Press.

ATKINS, J. G. & JOHNSTON, T. H. 1965. Inheritance in rice of reaction to races 1 and 6 of *Piricularia oryzae*. *Phytopathology* **55:** 993–995.

ATKINS, J. G., ROBERT, A. L., ADAIR, C. R., GOTO, K., KOZAKA, T., YANAGIDA, R., YAMADA, M. & MATSUMOTO, S. 1967. An international set of rice varieties for differentiating races of *Piricularia oryzae*. *Phytopathology* **57:** 297–301.

BANDONG, J. M. & OU, S. H. 1966. The physiologic races of *Piricularia oryzae* Cav. in the Philippines. *Philipp. Agric.* **49:** 655–667.

BARKSDALE, T. H. & ASAI, G. N. 1961. Diurnal spore release of *Piricularia oryzae* from rice leaves. *Phytopathology* **51:** 313–317.

BARKSDALE, T. H. & JONES, M. W. 1965. Minimum values of dew period and temperature required for infection by *Piricularia oryzae*. Abs. in *Phytopathology* **55:** 503.

BOKURA, U. 1930. On the relation of the balance of three principal fertilizers to rice blast. *Pl. Prot., Tokyo* **17:** 332–340. [Jap.]

BOLLICH, C. N., ATKINS, J. G., SCOTT, J. E. & WEBB, B. D. 1966. Dawn—a blast resistant, early maturing, long grain rice variety. *Rice J.* **69:** 14–18.

BUCKLEY, T. A. & ALLEN, E. F. 1951. Notes on current investigations, April to June, 1951. *Malay. agric. J.* **34:** 133–141. (*Rev. appl. Mycol.* **31:** 538.)

CAMPACCI, C. A. 1950. Rice blast. *Biologico* **16:** 128–130. (*Rev. appl. Mycol.* **30:** 191.)

CAVARA, F. 1891. Fungi Longobardiae exsiccati sive mycetum specimina in Longobardia collecta, exsiccata et speciebus novis vel criticis, iconibus illustrata. Pugillus I, No. 49. (Cited in Padwick, G. W., 1950.)

CHANDRAMOHAN, J. & PALANISWAMY, S. 1963. Incidence of blast disease of rice in relation to time of planting. *Rice News Teller* **11:** 86–91.

CHANG, T. T., WANG, M. K., LIN, K. M. & CHENG, C. P. 1965. Breeding for blast resistance in Taiwan. In *The rice blast disease: Proc. Symp. at IRRI, July, 1963*: 371–377. Baltimore, Maryland, Johns Hopkins Press.

CHANG, W. L., WANG, M. K. & YANG, S. C. 1965. Reaction of rice varieties and strains to blast in the uniform blast nursery. *Jnl Taiwan agric. Res.* **14** (3): 1–10.

CHIBA, S. & YAMASHITA, K. 1957. Influence of the rate of nitrogen and potassium on the growth of rice plant in potassium-deficient soil. *Rep. Soc. Pl. Prot. North Japan* **8:** 25–27. [Jap.]

CHIEN, C. C. 1964. Studies on the drug resistance of *Piricularia oryzae* against mercuric chloride. *Jnl Taiwan agric. Res.* **13** (3): 44–50. [Chin. Engl. summ.]

CHIEN, C. C. 1967. Studies on the physiologic races of the rice blast fungus, *Piricularia oryzae* Cav. *Bull. Taiwan agric. Res. Inst.* **26:** 1–63.

CHINTE, P. T. 1965. The nutritional requirements and physiological responses of *Piricularia oryzae in vitro*. M.S. thesis, Coll. agric. Univ. Philipp., 76 pp.

Chiu, R. J., Chien, C. C. & Lin, S. Y. Physiologic races of *Piricularia oryzae* in Taiwan. In *The rice blast disease: Proc. Symp. at IRRI, July, 1963*: 245–255. Baltimore, Maryland, Johns Hopkins Press.

Chu, O. M. Y. & Li, H. W. 1965. Cytological studies of *Piricularia oryzae* Cav. *Bot. Bull. Acad. sin., Taipei* **6:** 116–130.

Dasananda, S. 1965. Breeding for blast resistance in Thailand. In *The rice blast disease: Proc. Symp. at IRRI, July, 1963*: 379–396. Baltimore, Maryland, Johns Hopkins Press.

Doi, Y. & Suzuki, N. 1952. Histochemical studies on rice blast lesions. Preliminary report of 1951 work in Section of Plant Pathology, Inst. agric. Sci.: 268–286. [Jap.]

Duggar, B. M. 1909. Fungous diseases of plants. 508 pp., Boston, Ginn & Co.

Ezuka, A., Yunoki, T., Sakurai, Y., Shinoda, H. & Toriyama, K. 1969a. Studies on the varietal resistance to rice blast. 1. Tests for genotype of true resistance. *Bull. Chugoku agric. exp. Stn*, E **4:** 1–32. [Jap. Engl. summ.]

Ezuka, A., Yunoki, T., Sakurai, Y., Shinoda, H. & Toriyama, K. 1969b. Studies on the varietal resistance to rice blast. 2. Tests for field resistance in paddy fields and upland nursery beds. *Bull. Chugoku agric. exp. Stn, E* **4:** 33–54. [Jap. Engl. summ.]

Fernando, L. H., Okamoto, H. & Padmanabhan, S. Y. 1961. Committee on the establishment of uniform blast nurseries. Report presented to the Ninth Meeting of the FAO-IRC Working Party on Rice Production and Protection.

Fukunaga, K., Misato, T., Ishii, I. & Asakawa, M. 1955. Blasticidin, a new antiphytopathogenic fungal substance. *Bull. agric. Chem., Japan* **19:** 181–188.

Fukunaga, K., Misato, T., Ishii, I., Asakawa, M., & Katagiri, M. 1968. Research and development of antibiotics for the rice blast control. *Bull. natn. Inst. agric. Sci., Tokyo*, Ser. C, **22:** 1–94. [Jap. Engl. summ.]

Fulton, H. R. 1908. Diseases affecting rice in Louisiana. *Bull. Louisiana agric. Exp. Stn* 105, 28 pp.

Giatgong, P. & Frederiksen, R. A. 1967. Variation in pathogenicity of *Piricularia oryzae.* Abs. in *Phytopathology* **57:** 460.

Giatgong, P. & Frederiksen, R. A. 1968. Chromosomal number and mitotic division in *Piricularia oryzae*. Abs. in *Phytopathology* **58:** 728.

Giatgong, P. & Frederiksen, R. A. 1969. Pathogenic variability and cytology of monoconidial subcultures of *Piricularia oryzae*. *Phytopathology* **59:** 1152–1157.

Goto, I. 1959; 1960. On the inheritance of resistance to the blast disease. Speeches given at annual conference of the Phytopathological Society of Japan. (Cited by Takahashi, 1965.)

Goto, I. & Sakai, K. I. 1963. Intra-clonal variation of pathogenicity in *Piricularia oryzae. Rep. natn. Inst. Genet. Japan 1962*, **13:** 57–58.

Goto, K. 1955. History of the blast disease and changes in methods of control. *Agric. Improv. Bur., Minist. agric. For., Japan* **5:** 1–2. [Jap.]

Goto, K. 1960. Progress report of the joint research on the races of the rice blast fungus, *P. oryzae. Spec. Bull. Pl. Dis. and Insect Forecast. Serv.*, Fasc. I. **5:** 1–89. [Jap.]

Goto, K. 1965a. Estimating losses from rice blast in Japan. In *The rice blast disease: Proc. Symp. at IRRI, July, 1963*: 195–202. Baltimore, Maryland, Johns Hopkins Press.

Goto, K. 1965b. Physiologic races of *Piricularia oryzae* in Japan. In *The rice blast disease: Proc. Symp. at IRRI, July, 1963*: 237–242. Baltimore, Maryland, Johns Hopkins Press.

Goto, K. & Ohata, K. 1960. Influence of low air temperature on susceptibility of rice plants to blast disease. Abs. in *Ann. phytopath. Soc. Japan* **25:** 26. [Jap.]

Goto, K. & Ohata, K. 1961. Change of environment and outbreak of rice blast disease; fluctuation of resistance after low temperature treatment and physiology of rice plants. *Ann. phytopath. Soc. Japan* **25:** 215. [Jap.]

Goto, K. & Yamanaka, T. 1956. Studies on races of rice blast fungus; results of inoculation experiments. Abs. in *Ann. phytopath. Soc. Japan* **21:** 98–99. [Jap.]

Goto, K. & Yamanaka, T. 1961. Cooperative research on physiological races of *Piricularia oryzae*. *Spec. Rep. 5, Studies on Insect and Dis. Damage*, Spec. Suppl. 1.

Hara, K. 1916. *Piricularia oryzae* on barley. *Byochu-gai zasshi* **3:** 693–694. [Jap.]

Hara, K. 1918. Diseases of the rice plant. 218 pp. [Jap.]

Harada, Y. 1955. Studies on a new antibiotic for rice blast control. Lecture given at the annual meeting of the Agric. Chem. Soc. Japan. (Cited by Okamoto, 1965.)

Hashioka, Y. 1943. Studies on the rice blast disease in the tropics. IV. Influence of temperature of air and soil upon the resistance of the rice plants to the blast disease. *J. Soc. trop. Agric., Taiwan* **15**: 53–65. [Jap. Engl. transl.]

HASHIOKA, Y. 1944. Studies on the rice blast disease in the tropics. VIII. Relation of temperature to leaf blast resistance of the different varieties of rice plants collected from districts in various latitudes. *J. Soc. trop. Agric., Taiwan* **16**: 196–204. [Jap. Engl. summ.]

HASHIOKA, Y. 1950. The microclimate of the paddy field in connection with prevalence of rice blast disease. *J. agric. Met., Tokyo* **6**: 25–29. [Jap. Engl. transl.]

HASHIOKA, Y. 1950a. Studies on the mechanism of prevalence of rice blast disease in the tropics. *Tech. Bull. Taiwan agric. Res. Inst.* 8, 237 pp.

HASHIOKA, Y. 1950b. On the resistance of rice plants to rice blast in relation to air temperature. *Nogaku* **4** (6). [Jap.]

HASHIOKA, Y. 1963. Rice blast varietal resistance in relation to the local environments, with notes on stem nematode and other diseases of rice in Thailand. Final report to F.A.O., 66 pp. (Mimeographed.)

HASHIOKA, Y. 1965. Effects of environmental factors on development of causal fungus, infection, disease development, and epidemiology in rice blast disease. In *The rice blast disease: Proc. Symp. at IRRI, July, 1963*: 153–161. Baltimore, Maryland, Johns Hopkins Press.

HASHIOKA, Y. & INAGAKI, K. 1968. Karyological characters of the different species and an interspecific heterokaryon of *Pyricularia*. Abs. in *Ann. phytopath. Soc. Japan* **34**: 391. [Jap.]

HASHIOKA, Y., IKEGAMI, H. & MURASE, T. 1968. Fine structure of the rice blast. III. The mode of invasion of *Pyricularia oryzae* into rice epidermal cells. *Res. Bull. Fac. Agric. Gifu Univ.* **26**: 23–30.

HASTINGS DE GUTIERREZ, L. 1955. Presence of *Piricularia oryzae* on sugar cane in Costa Rica. *Pl. Dis. Reptr* **38**: 695.

HEMMI, T. 1933. Experimental studies on the relation of environmental factors to the occurrence and severity of blast disease in rice plants. *Phytopath. Z.* **6**: 305–324.

HEMMI, T. *et al.* 1936. Studies on the rice blast disease. IV. Relation of the environment to the development of blast disease and physiologic specialization in the rice blast fungus. *Mater. Rural Improv. Dep. Agric. For., Japan* 105, 145 pp. (Cited by Hashioka, 1965.)

HEMMI, T. *et al.* 1949. Studies on the blast fungus of Indian corn. *Ann. phytopath. Soc. Japan* **13** (3–4): 23–25. [Jap.]

HEMMI, T. & ABE, T. 1932. Studies on the rice blast disease. II. Relation of the environment to the development of blast disease. *Bull. Dep. agric. For. Japan* 47, 204 pp. [Jap.]

HEMMI, T., ABE, T. & INOUE, Y. 1941. Studies on the rice blast disease. VI. Relation of the environment to the development of blast disease and races of the blast fungus. *Bull. agric. Develop. Min. agric. For., Japan* 157: 1–232. [Jap.]

HEMMI, T. & IMURA, J. 1939. On the relation of air humidity to conidial formation in the rice blast fungus, *Piricularia oryzae*, and the characteristics in the germination of conidia produced by the strains showing different pathogenicity. *Ann. phytopath. Soc. Japan* **9**: 147–156. [Jap. Engl. summ.]

HENRY, B. W. & ANDERSEN, A. L. 1948. Sporulation by *Piricularia oryzae*. *Phytopathology* **38**: 265–278.

HIRANO, T. 1967. Recent problems in rice breeding for blast resistance in Japan. *Symp. on rice diseases and their control by growing resistant varieties and other measures*: 103–111. Tokyo, Agric. For. and Fish. Res. Council.

HIRAYAMA, S. 1933. On the oxidase and dehydrase in phytopathogenic fungi. *Proc. Acad. Sci. Japan* **2**: 639.

HOETTE, S. 1936. Pitting diseases of bananas in Australia. *Proc. R. Soc. Vict.*, N.S., **48**: 90–95. (*Rev. appl. Mycol.* **16**: 195.)

HORI, M., ARATA, T. & INOUE, Y. 1960. Studies on the forecasting method of blast disease. VI. Forecasting by the degree of accumulated starch in the sheath of rice plant. *Ann. phytopath. Soc. Japan* **25**: 2. [Abs. Jap.]

HORI, S. 1899. Blast disease of rice plants. *Spec. Rep. imp. agric. Exp. Stn Tokyo* 1: 1–36. [Jap.]

HORINO, O. & AKAI, S. 1965. Comparison of the electron micrographs of conidia of *Helminthosporium oryzae* Ito and Kurib. and *Piricularia oryzae* Cavara. *Trans. mycol. Soc. Japan* **6**: 41–46. [Jap. Engl. summ.]

HSIEH, S. C. & CHIEN, C. C. 1967. Recent status of rice breeding for blast resistance in Taiwan with special regard to races of the blast fungus. *Proc. Symp. on rice diseases and their control by growing resistant varieties and other measures*: 68–81. Tokyo, Agric. For. and Fish. Res. Council.

HSIEH, S. C., LIN, M. H. & LIANG, H. L. 1967. Genetic analysis in rice. VIII. Inheritance of resistance to races 4, 22 and 25 of *Pyricularia oryzae*. *Bot. Bull. Acad. sin., Taipei* **8** (Spec. No.): 255–260.

HUANG, C. H. 1961. Induction of mutations for rice improvement in Taiwan. *JCRR, Taiwan, Pl. Ind. Ser.* 22: 59–76.

HUGHES, S. J. 1958. Revisiones hyphomycetum aliquot cum appendice de nominibus rejiciendis. *Can. J. Bot.* **36**: 727–836.

HUNG, C. S. & CHIEN, C. C. 1961. Investigation of physiological races of the blast fungus (*Piricularia oryzae*). *Spec. Publ. Taiwan agric. Res. Inst.* 3: 31–37.

IKATA, S., TAGUCHI, S., YOSHIDA, M. & IGUCHI, H. 1938. Results of experiments on the control of rice blast disease. Report 5. *Farm Bull. Dep. Agric. For., Japan* 130. [Jap.]

IKEDA, M. 1933. Influence of application of various nitrogenous fertilizers on silica content of rice plants. *Tottori Soc. agric. Sci.* **4**: 265–270. [Jap. Engl. summ.]

IMURA, J. 1938. On the effect of sunlight upon the enlargement of lesions of the blast disease. *Ann. phytopath. Soc. Japan* **8**: 23–33. [Jap. Engl. summ.]

IMURA, J. 1940. On the influence of sunlight upon the incubation period and the development of the blast disease and the *Helminthosporium* disease of the rice plant. *Ann. phytopath. Soc. Japan* **10**: 16–26. [Jap. Engl. summ.]

INGOLD, C. T. 1964. Possible spore discharge mechanism in *Pyricularia*. *Trans. Br. mycol. Soc.* **47**: 573–575.

INOUE, Y. 1943. On the epidemic outbreak of rice blast in 1941 (Oita Prefecture). *Ann. phytopath. Soc. Japan* **12**: 181–190. [Jap.]

INTERNATIONAL RICE RESEARCH INSTITUTE (IRRI). Annual reports for 1964–69.

ISHIBASHI, H. 1936. Effects of silica on rice plant growth. *J. Sci. Soil Manure, Tokyo* **10**: 245–256. [Jap.]

ISHIZUKA, Y. & HAYAKAWA, Y. 1951. Resistance of rice plant to the imochi disease (rice blast disease) in relation to its silica and magnesia contents. *J. Sci. Soil Manure, Tokyo* **21**: 253–260. [Jap. Engl. summ.]

ITO, R. *et al.* 1961. On the new rice variety ‘Kusabue’. *J. Kanto-Tosan agric. Exp. Stn* **18**: 23–33. [Jap. Engl. summ.]

ITO, R. 1965. Breeding for blast resistance in Japan. In *The rice blast disease: Proc. Symp. at IRRI, July, 1963*: 361–370. Baltimore, Maryland, Johns Hopkins Press.

ITO, S. 1932. Primary outbreak of the important diseases of rice plants and common treatment for their control. *Rep. Hokkaido agric. Exp. Stn* 28, 211 pp. [Jap. Engl. summ.] (*Rev. appl. Mycol.* **12**: 532.)

ITO, S. & HAYASHI, H. 1931. On the relation of silica supply to rice blast. *J. Sapporo Soc. agric. Sci.* **22**: 460–461. [Jap. Engl. summ.]

ITO, S. & KURIBAYASHI, K. 1931a. Studies on the rice blast disease. *Farm Bull. Dep. Agric. For., Japan* 30, 81 pp. [Jap.]

ITO, S. & KURIBAYASHI, K. 1931b. Studies on the rice blast disease. I. *Noji Kairyo Shiryo, Min. agric. For., Japan* 30, 44 pp. [Jap.]

ITO, S. & SAKAMOTO, M. 1939–43. Studies on rice blast. *Rep. Hokkaido Univ. Bot. Lab. Fac. Agric.*

IWASAKI, S., NOZOE, S., OKUDA, S., SATO, S. & KOZAKA, T. 1969. Isolation and structural elucidation of a phytotoxic substance produced by *Pyricularia oryzae* Cav. (Pyriculol). *Tetrahedron Letters* 45: 3977–3980.

IWATA, T. & TAKAKURA, K. 1954. New host plants of rice blast fungus. Abs. in *Ann. phytopath. Soc. Japan* **18**: 150. [Jap.]

IWATA, T., YAMANUKI, S. & NARITA, T. 1955. Pathogenicity of rice blast fungus to gramineous plants. Abs. in *Ann. phytopath. Soc. Japan.* **19**: 175. [Jap.]

JOHNSON, T. W. 1954. *Piricularia oryzae* on *Eremochloa ophiuroides* in Florida. *Pl. Dis. Reptr* **38**: 796.

JOTHIANANDAN, D. & SHANMUGASUNDARAM, E. R. B. 1969. Studies on a proteinase of *Piricularia oryzae*. *Phytopath. Z.* **65**: 318–324.

KAHN, R. P. & LIBBY, J. L. 1958. The effect of environmental factors and plant age on the infection of rice by the blast fungus, *Piricularia oryzae*. *Phytopathology* **48**: 25–30.

KATO, H. & DIMOND, A. E. 1966. Factors affecting sporulation of the rice blast fungus, *Piricularia oryzae*. *Phytopathology* **56**: 864–865.

KATO, H., SASAKI, T. & KOSHIMIZU, Y. 1970. Potential for conidium formation of *Pyricularia oryzae* in lesions on leaves and panicles of rice. *Phytopathology* **60**: 608–612.

KATSUYA, K. 1959. Susceptibility of wild and foreign cultivated rice to the blast fungus, *Piricularia oryzae. Ann. Rep. 1958, Natn. Inst. Genet., Japan* **9**: 48–49.

KATSUYA, K. 1960. Susceptibility of wild rice to the blast fungus, *Piricularia oryzae. Ann. Rep. 1959, Natn. Inst. Genet., Japan* **10**: 76–77.

KATSUYA, K. 1961. Studies on the susceptibility of Oryzeae to the blast fungus, *Piricularia oryzae. Ann. phytopath. Soc. Japan* **26**: 153–159. [Jap. Engl. summ.]

KATSUYA, K. & KIYOSAWA, S. 1969. Studies on mixture inoculation of *Pyricularia oryzae* on rice. 2. Inter-strain difference of mixture inoculation effect. *Ann. phytopath. Soc. Japan* **35**: 299–307.

KAWAI, I. 1952. Ecological and therapeutic studies on rice blast. *Noji Kairyo Gijutsu Shiryo* **28**: 1–145. [Jap. Engl. summ.]

KAWAKAMI, T. 1901. On the rice blast disease. *J. Sapporo Soc. Agric. For.* **2**: 1–47. [Jap.]

KAWAKAMI, T. 1902. On the blast disease of rice. *J. Sapporo Soc. Agric. For.* **3**: 1–3. [Jap.]

KAWAMURA, E. 1940. Reaction of certain species of the genus *Oryza* to the infection of *Piricularia oryzae. Kyushu Univ. Sci. Fak. Terkult Bull.* **9**: 157–166. [Jap. Engl. summ.]

KAWAMURA, E. & ONO, K. 1948. Studies on the relation between the pre-infection behavior of rice blast fungus, *Piricularia oryzae*, and water droplets on rice plant leaves. *Bull. natn. agric. Exp. Stn* **4**: 1–12. [Jap.]

KAWASHIMA, R. 1927. Influence of silica on rice blast disease. *J. Sci. Soil Manure, Tokyo* **1**: 86–91. [Jap.]

KIKKAWA, S. 1900. The rice blast disease in San-in region in 1899. *Agric. Exp. Stn Min. Agric. Comm. Spec. Rep.* [Jap.]

KITAMURA, A. 1961. Genetic study on sterility caused by hybridization between *japonica* and *indica. Recent Developments in Thremmatology Bull.* **3**: 23–25. [Jap.]

KIYOSAWA, S. 1966a. Studies on inheritance of resistance of rice varieties to blast. II. Genetic relationship between the blast resistance and other characters in the rice varieties Reishiko and Sekiyama 2. *Jap. J. Breed.* **16**: 87–95.

KIYOSAWA, S. 1966b. Studies on inheritance of resistance of rice varieties to blast. III. Inheritance of resistance of a rice variety Pi No. 1. *Jap. J. Breed.* **16**: 243–250.

KIYOSAWA, S. 1966c. On spontaneous mutation of pathogenicity in *Piricularia oryzae. Pl. Prot., Tokyo* **20**: 159–162. [Jap.]

KIYOSAWA, S. 1967a. The inheritance of resistance of the Zenith type varieties of rice to the blast fungus. *Jap. J. Breed.* **17**: 99–107.

KIYOSAWA, S. 1967b. Inheritance of resistance of the rice variety Pi No. 4 to blast. *Jap. J. Breed.* **17**: 165–172.

KIYOSAWA, S. 1967c. Genetic studies on host-pathogen relationship in the rice blast disease. *Proceedings of a Symposium on rice diseases and their control by growing resistant varieties and other measures*: 137–153. Tokyo, Agric. For. Fish. Res. Council.

KIYOSAWA, S. 1968a. Genetic relationship among blast resistance and other characters in hybrid of Korean rice variety, Diazi Chall (Butamochi), with Aichi Asahi. *Jap. J. Breed.* **18**: 88–93.

KIYOSAWA, S. 1968b. Inheritance of blast-resistance in some Chinese rice varieties and their derivatives. *Jap. J. Breed.* **18**: 193–204.

KIYOSAWA, S. 1969a. Inheritance of resistance of rice varieties to a Philippine fungus strain of *Pyricularia oryzae. Jap. J. Breed.* **19**: 61–73.

KIYOSAWA, S. 1969b. Inheritance of blast-resistance in West Pakistani rice variety, Pusur. *Jap. J. Breed.* **19**: 121–128.

KIYOSAWA, S. & FUJIMAKI, H. 1967. Studies on mixture inoculation of *Pyricularia oryzae* on rice. I. Effects of mixture inoculation and concentration on the formation of susceptible lesions in the injection inoculation. *Bull. natn. Inst. agric. Sci., Tokyo*, Ser. D, **17**: 1–20.

KIYOSAWA, S., MATSUMOTO, S. & LEE, S. C. 1967. Inheritance of resistance of rice variety Norin 22 to two blast fungus strains. *Jap. J. Breed.* **17**: 1–6.

KIYOSAWA, S. & MURTY, V. V. S. 1969. The inheritance of blast-resistance in Indian rice variety, HR-22. *Jap. J. Breed.* **19**: 269–276.

KIYOSAWA, S. & YOKOO, M. 1970. Inheritance of blast resistance of the rice variety, Toride 2, bred by transferring resistance of the Indian variety, CO25. *Jap. J. Breed.* **20**:181–186.

KOBAYASHI, T. & ABUMIYA, H. 1960. Studies on the resistance of rice plant against infection of blast fungus. I. Influence of the inoculum density on the number of lesions and on the destruction of the leaves under artificial inoculations. *Bull. Tohoku natn. agric. Exp. Stn* **19**: 21–27.

KONISHI, S. 1933. On the physiologic specialization in the rice blast fungus, *Piricularia oryzae* Br. et Cav. *Forschn Geb. PflKrankh., Kyoto* **2**: 55–77. [Jap. Engl. summ.]

KOYAMA, T. 1952. On the breeding of highly resistant varieties to rice blast by the hybridization between Japanese varieties and foreign varieties in Japanese type of rice. *Jap. J. Breed.* **2**: 25–30. [Jap.]

KOZAKA, T. 1965. Control of rice blast by cultivation practices in Japan. In *The rice blast disease: Proc. Symp. at IRRI, July, 1963*: 421–438. Baltimore, Maryland, Johns Hopkins Press.

KOZAKA, T. 1969. Chemical control of rice blast in Japan. *Rev. Pl. Prot. Res.* **2**: 53–63.

KUILMAN, L. W. 1940. Uitstoeling en bloei van de Rijstplant. III. Vatbaarheid van de (op voedingsoplossing gekweekte) varieteit Oentoeng voor *Piricularia oryzae* Cav. *Meded. alg. Proefstn Landb., Batavia* **45**: 27–32. [Engl. summ.] (*Rev. appl. Mycol.* **22**: 77.)

KULKARNI, N. B. & PATEL, M. K. 1956. Study of the effect of nutrition and temperature on the size of spores in *Piricularia setariae* Nishikado. *Indian Phytopath.* **9**: 31–38. (*Rev. appl. Mycol.* **36**: 642.)

KUO, T. T., LEE, Y. S., YUAN, H. F. & LI, H. W. 1968. Nutritional aspect of host-parasite relationship in the rice blast fungus. *Bot. Bull. Acad. sin., Taipei* **9**: 36–45.

KUO, T. T., YANG, L. H., YEH, P. Z. & LI, H. W. 1967. Studies on induced mutants of *Piricularia oryzae* and on their pathogenicity. *Bot. Bull. Acad. sin., Taipei* **8**: 20–29.

KURIBAYASHI, K. 1928. Studies on overwintering, primary infection, and control of rice blast fungus, *Piricularia oryzae*. *Ann. phytopath. Soc. Japan* **2**: 99–117. [Jap. Engl. summ.]

KURIBAYASHI, K. & ICHIKAWA, H. 1941. Relation of the floating conidia of rice blast fungus to prediction of the disease. *J. Pl. Prot., Tokyo* **28**: 309–314; 404–416. [Jap.]

KURIBAYASHI, K. & ICHIKAWA, H. 1952. Studies on forecasting of the rice blast disease. *Spec. Rep. Nagano agric. Exp. Stn* **13**: 1–229. [Jap.]

KURIBAYASHI, K., ICHIKAWA, H. & TERAZAWA, M. 1953. Study of physiological specialization in rice blast fungus. *Rep. Nagano agric. Exp. Stn for 1951 and 1952*, 31 pp. (Mimeographed.)

KURIBAYASHI, K. & TERAZAWA, H. 1953. Injection as an artificial inoculation method in rice blast disease. *Proc. Ass. Pl. Prot., Hokuriku* **3**: 9–10. [Jap.]

LATTERELL, F. M., TULLIS, E. C. & COLLIER, J. W. 1960. Physiologic races of *Piricularia oryzae* Cav. *Pl. Dis. Reptr* **44**: 679–683.

LATTERELL, F. M., TULLIS, E. C., OTTEN, R. J. & GUBERNICK, A. 1954. Physiologic races of *Piricularia oryzae*. Abs. in *Phytopathology* **44**: 495.

LATTERELL, F. M., MARCHETTI, M. A. & GROVE, B. R. 1965. Coordination of effort to establish an international system for race identification in *Piricularia oryzae*. In *The rice blast disease: Proc. Symp. at IRRI, July, 1963*: 257–274. Baltimore, Maryland, Johns Hopkins Press.

LEAVER, F. W., LEAL, J. & BREWER, C. R. 1947. Nutritional studies on *Piricularia oryzae*. Abs. in *J. Bact.* **54**: 401–408.

LEE, S. C. & MATSUMOTO, S. 1966. Studies on the physiologic races of rice blast fungus in Korea during the period of 1962–1963. *Ann. phytopath. Soc. Japan* **32**: 40–45. [Jap. Engl. summ.]

LI, H. W., HU, C. H., CHANG, W. T. & WENG, T. S. 1961. The utilization of X-radiation for rice improvement. In *Symposium on the effects of ionizing radiations on seeds and their significance for crop improvement, Karlsuhe, 1960*: 483–493.

LIN, K. M. & LIN, P. C. 1960. Radiation-induced variation in blast disease (*Piricularia oryzae*) resistance in rice. *Jap. J. Breed.* **10**: 19–22. [Jap. Engl. summ.]

LIN, K. M. & LIN, P. C. 1961. Survey of intervarietal variation in resistance to blast disease. *Spec. Pub. Taiwan agric. Res. Inst.* 3: 38–44.

LING, K. C. 1968. Results of 1966 and 1967 international uniform blast nursery tests. *Int. Rice Commn Newsl.* **17** (3): 1–23.

LING, K. C. & OU, S. H. 1969. Standardization of the international race numbers of *Pyricularia oryzae*. *Phytopathology* **59**: 339–342.

LOCCI, R. 1970. Fungal diseases of rice. 1. Investigation on the rice blast fungus by scanning electron microscopy. *Riso* **19**: 99–110.

LUTTRELL, E. S. 1954. An undescribed species of *Piricularia* on sedges. *Mycologia* **46**: 810–814.

MALCA, M. I. & Owen, J. H. 1957. The gray leaf spot disease of St. Augustine grass. *Pl. Dis. Reptr* **41**: 871–875.

MATSUMOTO, S., KOZAKA, T. & YAMADA, M. 1969. Pathogenic races of *Piricularia oryzae* Cav. in Asian and some other countries. *Bull. natn. Inst. agric. Sci., Tokyo*, Ser. C, **23**: 1–36.

MATSUO, T. 1952. Genecological studies on cultivated rice. *Bull. natn. Inst. agric. Sci. Tokyo*, Ser. D, **3**: 1–111. [Jap. Engl. summ.]

MATSUURA, G. 1928. On the mode of host entry of *Piricularia oryzae* (Preliminary report). *J. Pl. Prot., Tokyo* **15**: 571–574. [Jap.]

MCINTOSH, A. E. S. 1951. Annual report of the Department of Agriculture, Malaya, for the year 1949. 87 pp. (*Rev. appl. Mycol.* **30**: 508–509.)

MCRAE, W. 1922. Report of the imperial mycologist. *Sci. Rep. Pusa agric. Res. Inst. 1921–22:* 44–50. (*Rev. appl. Mycol.* **2**: 258–260.)

MEHTA, P. R., SINGH, B. & MATHUR, S. C. 1952. A new leaf spot disease of bajra (*Pennisetum typhoides* Stapf and Hubbard) caused by a species of *Piricularia. Indian Phytopath.* **5**: 140–143. (*Rev. appl. Mycol.* **33**: 600.)

MELLO-SAMPAYO, T. & VIANNA E. SILVA, M. 1954. Ensaios preliminares sôbre a determinaçao de resistência de algumas formas cultivadas de arroz à *Piricularia oryzae* Br. et Cav. 38 pp., Ministerio de Economia, Comissao Reguladora do Comércio de Arroz, Lisbon. [Engl. summ.] (*Rev. appl. Mycol.* **34**: 481–482.)

MEREDITH, D. S. 1962. *Piricularia musae* Hughes in Jamaica. *Trans. Br. mycol. Soc.* **45**: 137–142.

METCALF, H. 1906. A preliminary report on the blast of rice, with notes on other rice diseases. *Bull. S. Carolina agric. Exp. Stn* 121, 43 pp.

METCALF, H. 1907. The pathology of the rice plant. *Science, N.Y.*, N.S., **25**: 264–265.

METCALF, H. 1909. The present status of rice blast. *Science, N.Y.*, N.S., **29**: 911.

METCALF, H. 1913. Diseases of rice. In Cook, M. T., *Diseases of tropical plants*: 99–166.

METCALF, H. 1921. The story of a plant introduction. *J. Wash. Acad. Sci.* **11**: 474.

MISATO, C. 1961. Blasticidin-S. Japan Pl. Prot. Ass., Tokyo, pp. 1–55.

MISATO, T. *et al.* 1959. Antibiotics as protectant fungicides against rice blast. 2. The therapeutic action of blasticidin-S. *Ann. phytopath. Soc. Japan* **24**: 302–306. [Jap. Engl. summ.]

MIYAKE, Y. & IKEDA, M. 1932. Influence of silica application on rice blast. *J. Sci. Soil Manure, Tokyo* **6**: 53–76. [Jap.]

MIYAZAKI, K. 1928; 1930. Morphological and physiological studies on rice blast occurring on plants treated with excessive nitrogen. Parts 1; 2. *Agriculture Hort., Tokyo* **3**: 754–764; **5**: 439–445. [Jap.]

MIZUSAWA, Y. 1959. Electron microscopy of ultrathin sections of conidia of *Piricularia oryzae. J. Pl. Prot., Tokyo* **13**: 159–160. [Jap.]

MURATA, J. KURIBAYASHI, K. & KAWAI, I. 1933. Studies on the control of rice blast. III. Influence of homemade manure on blast disease. *Nogi Kairyo Shiryo* **64**: 1–138. [Jap. Engl. summ.]

NAKAJIMA, T., TAMARI, K., MATSUDA, K., TANAKA, H. & OGASAWARA, N. 1970. Studies on the cell wall of *Piricularia oryzae*. Part II. The chemical constituents of the cell wall. *Agric. Biol. Chem.* **34**: 553–574.

NAKAMORI, E. 1936. On the location variability of resistance to rice blast disease. *Agriculture Hort., Tokyo* **11**: 103–114. [Jap.]

NAKAMURA, H. & SAKURAI, H. 1968. Tolerance of *Piricularia oryzae* Cav. to blasticidin-S. *Bull. Agric. Chem. Insp. Stn, Japan* **8**: 21–25. [Jap. Engl. summ.]

NAKATOMI, S. 1926. On the variability and inheritance of the resistance of rice plants to rice blast disease. *Jap. J. Genet.* **4**: 31–38. [Jap.]

NAKAZAWA, M. 1959. Studies on the mechanism of the protective action of organomercuric fungicides, with special references on their therapeutic action. *Bull. Aichi agric. Exp. Stn* 15, 124 pp. [Jap. Engl. summ.]

NARITA, S. & IWATA, T. 1959. Annual report on disease and pest forecasting in crops. Hokkaido Pref. Exp. Stn. (Mimeographed.)

NARITA, S., IWATA, T. & YAMANUKI, S. 1956. Studies on the host range of *Piricularia oryzae* Cav.; Report I. *Rep. Hokkaido Pref. agric. Exp. Stn* 7: 1–33. [Jap. Engl. summ.]

NIIZEKI, H. 1967. On some problems in rice breeding for blast resistance with special reference to variation of blast fungus. *Recent Adv. Breed.* **8**: 69–76. [Jap.]

NISIKADO, Y. 1917. Studies on the rice blast fungus. *Ohara Inst. Landw. Forschn Ber.* **1**: 171–218.

NISIKADO, Y. 1926. Studies on rice blast disease. *Bull. Bur. Agric., Min. agric. For. Japan* **15**: 1–211. [Jap.]

NISIKADO, Y. 1927. Studies on rice blast disease. *Jap. J. Bot.* **3**: 239–244.

OGASAWARA, N., KAJI, J. & TAMARI, K. 1957. Biochemical studies on the blast disease of rice plants. Part 6. Determination of piricularin. *J. agric. Chem. Soc. Japan* **31**: 460–463. [Jap.]

OGASAWARA, N., TAMARI, K. & SUGA, M. 1961. Biochemical studies on the blast disease of rice plants. Part 20. Mechanism of spore germination of *Piricularia oryzae* (2). Presence in spores of picolinic acid and the changes occurring during spore germination. *J. agric. Chem. Soc. Japan* **35**: 1412–1416. [Jap.]

OGAWA, M. *et al.* 1953. Studies on blast control by Ceresan lime. *Chugoku-Shikoku agric. Res.* **3**: 1–5. [Jap.]

OHATA, K., GOTO, K. & KOZAKA, T. 1963. Observations on the reaction of rice cells to the infection of different races of *Piricularia oryzae*. *Ann.phytopath. Soc. Japan* **28**: 24–30. [Jap. Engl. summ.]

OHATA, K. & KOZAKA, T. 1962. Susceptibility to rice blast of leaf sheath of rice plants fed with several amino acids. Abs. in *Ann. phytopath. Soc. Japan* **27**: 73. [Jap.]

OHATA, K. & KOZAKA, T. 1967. Interaction between two races of *Piricularia oryzae* in lesion-formation in rice plants and accumulation of fluorescent compounds associated with infection. *Bull. natn. Inst. agric. Sci., Tokyo*, Ser. C, **21**: 111–132. [Jap. Engl. summ.]

OKA, H. I. & LIN, K. M. 1967. Genic analysis of resistance to blast disease in rice (by biometrical genetic method). *Jap. J. Genet.* **32**: 20–27.

OKADA, M. & MAEDA, H. 1956. Inheritance of resistance to leaf blast in crosses between foreign and Japanese varieties of rice. *Bull. Tohoku agric. Exp. Stn* **10**: 59–68. [Jap. Engl. summ.]

OKAMOTO, H. 1950. The influence of phosphate application on the susceptibility to blast disease of rice plants. *Ann. phytopath. Soc. Japan* **14**: 6–8. [Jap. Engl. summ.]

OKAMOTO, H. 1958. Potassium in relation to rice blast and *Helminthosporium* leaf spot. *Symposium on potassium fertilizer*: 54–62. [Jap.]

OKAMOTO, H. 1965. Chemical control of rice blast in Japan. In *The rice blast disease: Proc. Symp. at IRRI, July, 1963*: 399–407. Baltimore, Maryland, Johns Hopkins Press.

OKAMOTO, H. & HAMAYA, E. 1962. Studies on the volatilization of organo-mercuric compounds and their sorption by rice leaves; with special reference to their phytotoxicity and effectiveness against rice blast. *Bull. Chugoku agric. Exp. Stn*, A (7): 273–298. [Jap. Engl. summ.]

OKAMOTO, H. & YAMAMOTO, T. 1955a. Annual report of plant disease test in 1954: 63–72 *Chugoku agric. Exp. Stn.* (Mimeographed.)

OKAMOTO, H. & YAMAMOTO, T. 1955b. Outbreaks of rice blast disease under different water temperatures during different intervals. Abs. in *Ann. phytopath. Soc. Japan* **20**: 35. [Jap.]

OKAMOTO, H. & YAMAMOTO, T. 1961. Annual report of plant disease control tests in 1960: 29–51. *Chugoku agric. Exp. Stn.* (Mimeographed.)

OKAMOTO, H., YAMAMOTO, T., HAMAYA, E. & MARKS, G. C. 1960. Studies on the phytotoxicity of various organo-mercuric compounds to Japanese and exotic varieties of rice plants and the efficacy of these compounds against rice blast when applied in the field. *Bull. Chugoku agric. Exp. Stn* **4**: 225–282. [Jap. Engl. summ.]

OKAMOTO, H., MATSUMOTO, K. & YAMAMOTO, T. 1959. Studies on the influences of environmental conditions and application methods on the efficacy of fungicides applied against rice blast in the field. VII. The influences of the application intervals and concentration on efficacy of mercury and copper dusts and some antibiotics for the control of rice blast. *Chugoku agric. Res.* **15**: 31–40. [Jap. Engl. summ.]

OKAMOTO, H., MATSUMOTO, K., YAMAMOTO, T. & SEKIGUCHI, Y. 1958. Studies on the influences of environmental conditions and application methods on the efficacy of fungicides applied against rice blast in the field. I–V. *Chugoku agric. Res.* **12**: 1–174.

ONO, K. 1953. Morphological studies on blast diseases and sesame leaf spot of rice plants. *J. Hokuriku Agric.* **2** (1), 75 pp. [Jap. Engl. summ.]

ONO, K. 1965. Principles, methods, and organization of blast disease forecasting. In *The rice blast disease: Proc. Symp. at IRRI, July, 1963*: 173–194. Baltimore, Maryland, Johns Hopkins Press.

ONO, K. & Nakazato, K. 1958. Morphology of the conidia of *Piricularia* from different host plants, produced under different conditions. Abs. in *Ann. phytopath. Soc. Japan* **23**: 1–2. [Jap.]

ONO, K. & SUZUKI, H. 1959a. Spore formation of blast fungus on water surface. I. *Ann. phytopath. Soc. Japan* **24**: 3–4. [Jap.]

ONO, K. & SUZUKI, H. 1959b. Studies on dissemination of conidia of rice blast fungus. *Proc. Ass. Pl. Prot., Hokuriku* **7**: 6–19. [Jap.]

ONO, K., SUZUKI, H., ONUMA, M. & INOUE, E. 1962; 1963. Data of investigations concerning spore flight of blast fungus. *Ann. Rep. Pl. Path. 2nd Lab., natn. Hokuriku agric. Exp. Stn.*

ONODERA, I. 1917. Chemical studies on rice blast (*Dactylaria parasitans* Cavara). *J. Sci. agric. Soc.* **180**: 606–617. [Jap.]

OTANI, Y. 1949. Studies on the relation between the principal components of the rice plant and its susceptibility to blast disease. I. Experiments with the rice seedlings raised in different kinds of nursery beds. *Agric. Sci. North Temperate Region* **2**: 269–280. [Jap. Engl. summ.]

OTANI, Y. 1950. Studies on the relation between the principal components of the rice plant and its susceptibility to the blast disease. II. Abs. in *Ann. phytopath. Soc. Japan* **14**: 37. [Jap.]

OTANI, Y. 1952a. Studies on the relation between the principal components of the rice plant and its susceptibility to the blast disease. III. *Ann. phytopath. Soc. Japan* **16**: 97–102. [Jap. Engl. summ.]

OTANI, Y. 1952b. Growth factors and nitrogen sources of *Piricularia oryzae* Cavara. *Ann. phytopath. Soc. Japan* **17**: 9–15. [Jap. Engl. summ.]

OTANI, Y. 1953. Carbon sources of *Piricularia oryzae* Cavara. *Ann. phytopath. Soc. Japan* **17**: 119–120. [Jap. Engl. summ.]

OTANI, Y. 1955. On the proteolytic enzyme of *Piricularia oryzae* Cavara. *Jubilee Publ. in Commem. 60th birthdays of Profs. Y. Tochinai and T. Fukushi*: 316–322. Sapporo, Japan.

OTSUKA, H. 1957. Studies on amylase of Aspergilli. *Sci. Educ. Niigata Univ.* **6**: 53–59.

OTSUKA, H. 1959. Biochemical studies on rice blast disease. X. The biochemical classification of *Piricularia oryzae* Cavara (7). On the production of vitamin B group by *Piricularia oryzae* Cav. *J. agric. Chem. Soc. Japan* **33**: 1013–1018. [Jap.]

OTSUKA, H. 1961. The biochemical classification of *Piricularia oryzae* Cav. *Mem. Niigata Univ. Fac. Educ.* **3** (2), 105 pp. [Jap. Engl. summ. (4 pp.)]

OTSUKA, H., TAMARI, K. & OGASAWARA, N. 1957a. Biochemical studies on the rice blast disease. X. Biochemical classification of *Piricularia oryzae* Cav. (1). *J. agric. Chem. Soc. Japan* **31**: 791–794. [Jap.]

OTSUKA, H., TAMARI, K. & OGASAWARA, N. 1957b. Biochemical studies on the rice blast disease. X. Biochemical classification of *Piricularia oryzae* Cav. (2). *J. agric. Chem. Soc. Japan* **31**: 794–798. [Jap.]

OTSUKA, H., TAMARI, K. & OGASAWARA, N. 1957c. Biochemical studies on the rice blast disease. X. Biochemical classification of *Piricularia oryzae* Cav. (3). *J. agric. Chem. Soc. Japan* **31**: 886–890. [Jap.]

OTSUKA, H., TAMARI, K. & OGASAWARA, N. 1957d. Biochemical studies on the rice blast disease. X. Biochemical classification of *Piricularia oryzae* Cav. (4). *J. agric. Chem. Soc. Japan* **31**: 890–893. [Jap.]

OTSUKA, H., TAMARI, K. & OGASAWARA, N. 1958a. Biochemical studies on rice blast disease. X. Biochemical classification of *Piricularia oryzae* Cav. (5). *J. agric. Chem. Soc. Japan* **32**: 890–893. [Jap.]

OTSUKA, H., TAMARI, K., OGASAWARA, N. & HONDA, R. 1958b. Biochemical studies on rice blast disease. X. Biochemical classification of *Piricularia oryzae* Cav. (6). *J. agric. Chem. Soc. Japan* **32**: 893–897. [Jap.]

OTSUKA, H., TAMARI, K. & OGASAWARA, N. 1965. Variability of *Piricularia oryzae* in culture. In *The rice blast disease: Proc. Symp. at IRRI, July, 1963*: 69–109. Baltimore, Maryland, Johns Hopkins Press.

OU, S. H. 1961. Some aspects of rice blast and varietal resistance in Thailand. *Int. Rice Commn Newsl.* **10** (3): 11–17.

OU, S. H. 1963. Rice blast disease and breeding for its resistance in Thailand, with notes on other rice diseases. *F. A. O. Expand. tech. Assist. Prog. Rep.* 1673, 26 pp.

OU, S. H. 1964. Results of the FAO-IRC 1962–63 uniform blast nursery tests. *Int. Rice Commn Newsl.* **13** (3): 22–30.

OU, S. H. 1965a. Varietal reactions of rice to blast. In *The rice blast disease: Proc. Symp. at IRRI, July, 1963*: 223–234. Baltimore, Maryland, Johns Hopkins Press.

OU, S. H. 1965b. A proposal for an international program of research on the rice blast disease. In *The rice blast disease: Proc. Symp. at IRRI, July, 1963*: 441–446. Baltimore, Maryland, Johns Hopkins Press.

OU, S. H. 1966. International uniform blast nurseries, 1964–1965 results. *Int. Rice Commn Newsl.* **15** (3): 1–13.

OU, S. H. 1957. Identification of the so-called ' Shara ' disease of rice in Iraq. (Mimeographed.)

OU, S. H. & AYAD, M. R. 1968. Pathogenic races of *Pyricularia oryzae* originating from single lesions and monoconidial cultures. *Phytopathology* **58**: 179–182.

OU, S. H. & NUQUE, F. L. 1963. The relation between leaf and neck resistance to the rice blast disease. *Int. Rice Commn Newsl.* **12** (4): 30–35

OU, S. H., NUQUE, F. L. & EBRON, T. T. 1970. The international uniform blast nurseries, 1968–69 results. *Int. Rice Commn Newsl.* **19** (4): 1–13.

PADMANABHAN, S. Y. 1965a. Estimating losses from rice blast in India. In *The rice blast disease: Proc. Symp. at IRRI, July, 1963*: 203–221. Baltimore, Maryland, Johns Hopkins Press.

PADMANABHAN, S. Y. 1965b. Breeding for blast resistance in India. In *The rice blast disease: Proc. Symp. at IRRI, July, 1963*: 343–359. Baltimore, Maryland, Johns Hopkins Press.

PADMANABHAN, S. Y. 1965c. Physiologic specialization of *Piricularia oryzae* Cav., the causal organism of blast disease of rice. *Curr. Sci.* **34**: 307–308.

PADMANABHAN, S. Y. & GANGULY, D. 1959. Breeding rice varieties resistant to blast disease caused by *Piricularia oryzae* Cav. I. Selection of resistant varieties from genetic stock. *Proc. Indian Acad. Sci.*, Sect. B, **50**: 289–304.

PADWICK, G. W. 1950. Manual of rice diseases. 198 pp., Kew, Commonwealth Mycological Institute.

PANZER, J. D. & BEIER, R. D. 1958. A simple field inoculation technique. *Pl. Dis. Reptr* **42**: 172–173.

PODHRADSZKY, J. 1961. Provokativ vizsgalati modszerek Rizsfajtak *Piricularia*-reziezt-enciajanak elbiralasara. *Novenytermeles* **10**: 67–76 [Engl. summ.] (*Rev. appl. Mycol.* **40**: 606–607.)

QUAMARUZZAMAN, MD. & OU, S. H. 1970. Monthly changes of the pathogenic races of *Pyricularia oryzae* Cav. in a blast nursery. *Phytopathology* **60**: 1266–1269.

RAMAKRISHNAN, L. 1966a. Studies in the host-parasite relations of blast disease of rice. I. Changes in N-metabolism. *Phytopath. Z.* **55**: 297–308.

RAMAKRISHNAN, L. 1966b. Studies in the host-parasite relations of blast disease of rice. II. Changes in the carbohydrates and phenolics. *Phytopath. Z.* **56**: 279–287.

REFATTI, E. 1955. Osservazioni su la morfologia, la germinabilita e su una parziale pigmentazione delle spore di *Piricularia oryzae* Br. et Cav. in coltura. *Ann. Sper. agr.*, N.S., **9** (4) Suppl.: vii–xxi. [Engl. summ.] (*Rev. appl. Mycol.* **35**: 924.)

ROLDAN, E. F. 1938. New or noteworthy lower fungi of the Philippine Islands. II. *Philipp. J. Sci.* **66**: 7–13.

SADASIVAN, T. S., SURYANARAYANAN, S. & RAMAKRISHNAN, L. 1965. Influence of temperature on rice blast disease. In *The rice blast disease: Proc. Symp. at IRRI, July, 1963*: 163–171. Baltimore, Maryland, Johns Hopkins Press.

SAKAMOTO, M. 1940. On the facilitated infection of the rice blast fungus, *Piricularia oryzae* Cav. due to the wind. I. *Ann. phytopath. Soc. Japan* **10**: 119–126. [Jap. Engl. summ.]

SAKAMOTO, M. 1948. On the relation between nitrogenous fertilizer and resistance to rice blast. Abs. in *Ann. phytopath. Soc. Japan* **13**: 53. [Jap.]

SAKAMOTO, M. 1949. On inoculation of the leaf sheath of rice with the blast fungus. *Bull. Tohoku Univ. Inst. agric. Res.* **1**: 120–129. [Jap. Engl. summ.]

SAKAMOTO, M. 1951. On the new method of inoculation of rice plants with the blast fungus *Piricularia oryzae* Cav. *Rep. Tohoku Univ. Inst. agric. Res.* 1–2: 15–23.

SAKURAI, Y. & TORIYAMA, K. 1967. Field resistance of the rice plant to *Piricularia oryzae* and its testing method. *Proceedings of a symposium on rice diseases and their control by growing resistant varieties and other measures*: 123–135. Agric. For. Res. Council, Minist. Agric. For., Japan.

SASAKI, R. 1922b. Existence of strains in rice blast fungus. I. *J. Pl. Prot., Tokyo* **9**: 631–644. [Jap.]

SASAKI, R. 1923. Existence of strains in rice blast fungus. II. *J. Pl. Prot., Tokyo* **10**: 1–10. [Jap.]

SAWADA, K. 1917. Blast of rice plants and its relation to the infective crops and weeds, with the description of five species of *Dactylaria*. *Spec. Bull. Taiwan agric. Exp. Stn* 16, 78 pp. [Jap.]

SAWADA, K. 1927. Lecture on the rice blast disease. *Bull. Govt Res. Inst. Dep. agric. Formosa* 45, 86 pp. [Jap.]

SAWADA, K. 1959. Descriptive catalogue of Taiwan (Formosa) fungi. XI. *Spec. Publ. natn. Taiwan Univ. Col. Agric.* 8, 268 pp.

SHIGEMURA, C. 1955. Some aspects of rice breeding for blast resistance in Japan. Paper presented at the Sixth Meeting of the FAO-IRC Working Party on Rice Breeding.

SHIMOYAMA, M. 1960. Effect of temperature and shading on lesion types of rice blast disease in rice seedlings. Abs. in *Ann. phytopath. Soc. Japan* **25**: 1. [Jap.]

SHIRAI, M. 1896. Notes on plants collected in Suruga, Totomi Yamato and Kii. *Bot. Mag., Tokyo* **10**: 111–114.

SHIRAI, M. 1905. Supplemental notes on the fungus which causes the disease so-called ' Imochibyo ' of *Oryza sativa* L. *Bot. Mag., Tokyo* **19**: 19–28. [Jap.]

SPRAGUE, R. 1950. Diseases of cereals and grasses in North America. 538 pp., New York, Ronald Press.

SRINIVASAN, K. V. & VIJAYALAKSHMI, U. 1957. *Piricularia oryzae* Cav. on *Arundo donax* L. *Sci. Cult.* **23**: 490–491. (*Rev. appl. Mycol.* **37**: 584.)

STEHLE, H. 1954. Quelques notes sur la botanique et l'écologie végétale de l'Archipel des Caraibes. *J. Agric. trop. Bot. appl.* **1**: 71–110. (*Rev. appl. Mycol.* **34**: 517.)

STEVENS, F. L. 1913. The fungi which cause plant disease. 754 pp., New York, MacMillan.

SUEDA, H. 1928. Studies on the rice blast disease. *Rep. Dep. Agric. Govt Res. Inst., Formosa* 36: 1–130. [Jap.]

SURYANARAYANAN, S. 1959. Mechanism of resistance of paddy (*Oryza sativa L.*) to *Piricularia oryzae* Cav. I. General considerations. *Proc. natn. Inst. Sci. India*, Part B, **24**: 285–292.

SUEMATSU, N. 1916. On the artificial culture of rice blast fungus (*Dactylaria parasitans* Cav.). *Bot. Mag., Tokyo* **30**: 97–99. [Jap.]

SUZUKI, H. 1930. Experimental studies on the possibility of primary infection of *Piricularia oryzae* and *Ophiobolus miyabeanus* internal of rice seeds. *Ann. phytopath. Soc. Japan* **2**: 245–275. [Jap. Engl. summ.] (*Rev. appl. Mycol.* **9**: 556.)

SUZUKI, H. 1934a. Studies on the influence of some environmental factors on the susceptibility of the rice plant to blast and *Helminthosporium* diseases and on the anatomical characters of the plant. I. Influence of differences in soil moisture. *J. Coll. Agric. Tokyo imp. Univ.* **13**: 45–108.

SUZUKI, H. 1934b. Studies on the influence of some environmental factors on the susceptibility of the rice plant to blast and *Helminthosporium* diseases and on the anatomical characters of the plant. II. Influence of differences in soil moisture and amounts of nitrogenous fertilizer given. *J. Coll. Agric. Tokyo imp. Univ.* **13**: 235–275.

SUZUKI, H. 1934c. Studies on the influence of some environmental factors on the susceptibility of the rice plant to blast and *Helminthosporium* diseases and on the anatomical characters of the plant. III. Influence of differences in soil moisture and in the amounts of fertilizer and silica given. *J. Coll. Agric. Tokyo imp. Univ.* **13**: 277–331.

SUZUKI, H. 1937. Studies on the relation between the anatomical characters of the rice plant and its susceptibility to blast disease. *J. Coll. Agric. Tokyo imp. Univ.* **14**: 181–264.

SUZUKI, H. 1951; 1953. Studies on the classification of strains in *Piricularia oryzae*. Reports for 1951; 1953. (Mimeographed.)

SUZUKI, H. 1954. Studies on antiblastin (I–IV). Abs. in *Ann. phytopath. Soc. Japan*: **18** 138. [Jap.]

SUZUKI, H. 1961. Deposition of spores of blast fungus on rice leaves. *Proc. Ass. Pl. Prot., Hokuriku* **9**: 41–45. [Jap.]

SUZUKI, H. 1965. Origin of variation in *Piricularia oryzae*. In *The rice blast disease: Proc. Symp. at IRRI, July, 1963*: 111–149. Baltimore, Maryland, Johns Hopkins Press.

SUZUKI, H. 1967. Studies on biologic specialization in *Pyricularia oryzae* Cav. 235 pp., Tokyo Univ. agric. tech. Inst. Pl. Path.

SUZUKI, H. 1969. Studies on the behaviour of the rice blast fungus spore and the application for forecasting method of the rice blast disease. *Bull. Hokuriku agric. Exp. Stn* **10**: 114–118.

SUZUKI, H. & HASHIMOTO, Y. 1953. Pathogenicity of the rice blast fungus to plants other than rice. I, II. Abs. in *Ann. phytopath. Soc. Japan* **17**: 94, 168. [Jap.]

SUZUKI, N. 1965. Nature of resistance to blast. In *The rice blast disease: Proc. Symp. at IRRI, July, 1963*: 277–301. Baltimore, Maryland, Johns Hopkins Press.

SUZUKI, N., DOI, Y. & TOYODA, S. 1953. Histochemical studies on the lesions of rice blast caused by *Piricularia oryzae* Cav. II. On the substance in the cell membrane of rice reacting red in colour with diazo reagent. *Ann. phytopath. Soc. Japan* **17**: 97–101. [Jap. Engl. summ.]

TAKAHASHI, Y. 1951. Phytopathological and plant breeding investigations on determining the degree of blast resistance in rice plants. *Rep. Hokkaido pref. agric. Exp. Stn* 3: 1–61. [Jap. Engl. summ.]

TAKAHASHI, Y. 1956. Studies on the mechanism of the resistance of rice plants to *Piricularia oryzae*. II. Pathological changes microscopically observed in host cells in which fungus hyphae do not grow well. *Bull. Yamagata Univ. agric. Sci.* **2**: 83–100. [Jap. Engl. summ.]

TAKAHASHI, Y. 1958a. A method for forecasting blast disease using the leaf sheath inoculation technique. *Pl. Prot., Tokyo* **12**: 339–345. [Jap.]

TAKAHASHI, Y. 1958b. Studies on the mechanism of resistance to blast disease (*Piricularia oryzae*) in paddy. *Bull. Yamagata Univ. Fac. Agric.* 13. [Jap.]

TAKAHASHI, Y. 1959. Studies on the mechanism of the resistance of rice plant to *Piricularia oryzae*. *J. Yamagata agric. For. Soc.* **13**: 17–28.

TAKAHASHI, Y. 1965. Genetics of resistance to rice blast disease. In *The rice blast disease: Proc. Symp. at IRRI, July, 1963*: 303–329. Baltimore, Maryland, Johns Hopkins Press.

TAMARI, K. & KAJI, J. 1954. Biochemical studies of the blast fungus (*Piricularia oryzae* Cav.), the causative fungus of the blast disease of rice plants. I. Studies on the toxins produced by blast fungus. *J. agric. Chem. Soc. Japan* **29**: 185–190. [Jap. Engl. summ.]

TAMARI, K. & KAJI, J. 1957. Biochemical studies of the blast disease of rice plant. IV. The functions of oxygen and nitrogen in the piricularin molecule. *J. agric. Chem. Soc. Japan* **31**: 387–390. [Jap.]

TAMARI, K. & KAJI, J. 1959. Biochemical studies of the blast disease of rice plant. XIII. The accumulation of coumarin in stunted rice plants caused by the ill-effects of piricularin. *J. agric. Chem.Soc. Japan* **33**: 181–183. [Jap.]

TAMARI, K., OTSUKA, H., HONDA, R., KAJI, J. & OGASAWARA, N. 1961. Biochemical classification of *Piricularia oryzae* Cav. VIII. On the terminal oxidases of the respiration system of *Piricularia oryzae*. *Bull. Niigata Univ. Fac. Agric.* **13**: 66–75. [Jap.]

TANAKA, S. & KATSUKI, F. 1952. Biochemical studies on susceptibility of the rice plants to the blast disease. III. Determination and isolation of glutamic acid and of aspartic acid from the leaves of rice plants. *J. Chem. Soc. (Pure Chem. Sect.) Japan* **73**: 868–870. [Jap.]

TANAKA, S. & KATSUKI, H. 1951a. Growth factors of the fungus causing rice blast disease. I. Biotin as the principal factor. *J. Chem. Soc. (Pure Chem. Sect.) Japan* **72**: 231–233.

TANAKA, S. & KATSUKI, H. 1951b. Growth factors of the fungus causing rice blast disease. II. Thiamine and pteroylglutamic acid as supplementary growth factors. *J. Chem. Soc. (Pure Chem. Sect.) Japan* **72**: 285–287. [Jap.]

TANAKA, S. & KATSUKI, H. 1952. On the chemical constituents of plants susceptible to the blast disease. I. *Bull. Chem Soc. Japan* **73**: 259.

TANAKA, S., KATSUKI, H. & KATSUKI, F. 1952. Biochemical studies on the blast diseaes of rice plants. IV. Effect of different nitrogen sources in the culture media upon the nutritional absorption of the blast disease fungus. *Ann. phytopath. Soc. Japan* **16**: 103–106. [Jap.]

TANAKA, S. & TSUJI, J. 1952. Studies on pentose metabolism of *Piricularia oryzae*. Abs. in *Ann. phytopath. Soc. Japan* **16**: 31. [Jap.]

TANAKA, S., KATSUKI, H. & KATSUKI, F. 1953. Biochemical studies on the blast disease. V. Effect of the different nitrogen sources in the culture media upon the mycelial constituents and amylase activity of the blast disease fungus. *Ann. phytopath. Soc. Japan* **17**: 54–56. [Jap. Engl. summ.]

TASUGI, H. & YOSHIDA, K. 1958. On the relation between silica and resistance of rice plants to rice blast. *Monbu-Sho Shiken Kenkyu Hokoku* **48**: 31–36. [Jap.]

TASUGI, H. & YOSHIDA, L. 1959; 1960. Relation between rice blast resistance and temperature environment. Abs. in *Ann. phytopath. Soc. Japan* **24**: 1; **25**: 1. [Jap.]

TEMPLETON, G. E., JOHNSTON, T. H. & HENRY, S. E. 1961. Rice blast. *Arkans. Fm Res.* **10**: 12.

TENG, S. C. 1932. Fungi of Nanking. II. *Contrib. Biol. Lab. Sci. Soc. China*, Bot. Ser., **8**: 5–48. (*Rev. appl. Mycol.* **12**: 395–396.)

THIRUMALACHAR, M. J., KULKARNI, N. B. & PATEL, M. K. 1956. Two new records of *Piricularia* species from India. *Indian Phytopath.* **9**: 48–51.

THOMAS, K. M. 1940. Detailed administration report of the government mycologist, Madras, for the year 1939–40, 18 pp. (*Rev. appl. Mycol.* **20**: 148–150.)

THOMAS, K. M. 1941. Detailed administration report of the government mycologist, Madras, for the year 1940–41: 53–74. (*Rev. appl. Mycol.* **21**: 362–364.)

TOCHINAI, Y. & NAKANO, T. 1940. Studies on the nutritional physiology of *Piricularia oryzae* Cavara. *J. Fac. agric. Hokkaido Univ.* **44**: 183–229.

TOCHINAI, Y. & SHIMAMURA, M. 1932. Studies on the physiologic specialization in *Piricularia oryzae* Br. et Cav. *Ann. phytopath. Soc. Japan* **2**: 414–441.

TOGASHI, K. 1961. Chlamydospore and appressorium formation in *Piricularia oryzae.* Abs. in *Ann. phytopath. Soc. Japan* **26**: 60. [Jap.]

TOKUNAGA, Y. 1959. Studies on the relationships between metabolism of the rice plant and its resistance to blast disease. I. Correlation of nitrogen and sugar contents of rice plant to blast disease. *Bull. Tohoku agric. Exp. Stn* **16**: 1–5. [Jap. Engl. summ.]

TOKUNAGA, Y., FURUTA, T. & SHIMOYAMA, T. 1953. Process and occurrence of stunting in rice plants caused by blast disease. Abs. in *Ann. phytopath. Soc. Japan* **18**: 76–77. [Jap.]

TOKUNAGA, Y., KATSUBE, T. & KOSHIMIZU, Y. 1966. Studies on the the relationship between metabolism of rice plant and its resistance to blast disease. 3. Correlation of nitrogen metabolism and blast disease in rice plant. *Bull. Tohoku natn. agric. Exp. Stn* **34**: 37–79. [Jap. Engl. summ.]

TOYODA, S. & SUZUKI, N. 1958a. Lesion-like browning of rice leaves induced by several metabolic inhibitors. *Pl. Prot., Tokyo* **12**: 346–348. [Jap.]

TOYODA, T. & TSUCHIYA, S. 1941. Studies on the control of rice blast. I. *Rep. Yamagata agric. Exp. Stn*: 1–146. [Jap.]

TSENG, T. C., YUAN, C. S. & WU, L. C. 1965. Temperature response of *Piricularia oryzae* Cav. isolated in different seasons in Taiwan. *Bot. Bull. Acad. sin., Taipei* **6**: 93–100.

UJIHARA, M. 1951. On the breeding process of blast resistant varieties in the Inshashi Experiment. Farm. *Bull. Aichi Pref. agric. Exp. Stn* **5**: 11–24. [Jap.]

URSUGI, Y., KATAGIRI, M. & FUKUNAGA, K. 1969. Resistance in *Pyricularia oryzae* to antibiotics and organophosphorus fungicides. *Bull. natn. Inst. agric. Sci., Tokyo*, Ser. C, **23**: 93–112. [Jap. Engl. summ.]

VEERARAGHAVAN, J. & PADMANABHAN, S. Y. 1965. Studies on the host range of *Piricularia oryzae* Cav. causing blast disease of rice. *Proc. Indian Acad. Sci.*, Sect. B, **61**: 109–120.

WARDLAW, C. W. 1940. Banana diseases. XIII. Further observations on the condition of banana plantations in the Republic of Haiti. *Trop. agric. Trin.* **17**: 124–127. (*Rev. appl. Mycol* **19**: 661–663.)

WATANABE, K. *et al.* 1957. Blastymycin, a new antibiotic from *Streptomyces* sp. *J. Antibiotics*, Ser. A, **10**: 39–45.

WILLIS, G. M., ALLOWITZ, R. D. & MENVIELLE, E. S. 1968. Differential susceptibility of rice leaves and panicles to *Piricularia oryzae.* Abs. in *Phytopathology* **58**: 1072.

WU, H. K. & TSAO, T. H. 1967. The ultrastructure of *Piricularia oryzae* Cav. *Bot. Bull Acad. sin., Taipei* **8** (Spec. No.): 353–363.

WU, L. C. & SHIH, C. L. 1967. Studies on the fungal proteins in *Piricularia oryzae* Cav. *Bot. Bull. Acad. sin., Taipei* **8**: 54–65.

YAMADA, N. & OTA, H. 1956. Difference in the growth of rice plants between early and late seasonal culture. *Agriculture Hort., Tokyo* **31**: 769–774. [Jap.]

YAMANAKA, S. & KOBAYASHI, T. 1962. Temperature in relation to the size of conidia in *Piricularia oryzae.* Abs. in *Ann. phytopath. Soc. Japan* **27**: 57–58. [Jap.]

YAMASAKI, Y. & KAWAI, T. 1968. Artificial induction of blast-resistant mutations in rice. In *Rice breeding with induced mutations. Tech. Rep. int. atom. Energy Ag.* 86: 65–73.

YAMASAKI, Y. & KIYOSAWA, S. 1966. Studies on inheritance of resistance of rice varieties to blast. I. Inheritance of resistance of Japanese varieties to several strains of the fungus. *Bull. natn. Inst. agric. Sci., Tokyo*, Ser. D, **14**: 39–69.

YAMASAKI, Y. & NIIZEKI, H. 1965. Studies on variation of the rice blast fungus *Piricularia oryzae* Cav. I. Karyological and genetical studies on variation. *Bull. natn. Inst. agric. Sci., Tokyo*, Ser. D, **13**: 231–274. [Jap. Engl. summ.]

YAMASAKI, Y. & NIIZEKI, H. 1966. Studies on variation of the rice blast fungus, *Piricularia oryzae* Cav. II. Induction of morphological and biochemical mutations by chemicals, with special reference to mutants relating to aromatic and hydrogen sulphide metabolism. *Bull. natn. Inst. agric. Sci., Tokyo*, Ser. D, **14**: 1–24. [Jap. Engl. summ.]

YAMASAKI, Y., TSUCHIYA, S., NIIZEKI, H. & SUWA, T. 1964. Studies on drug-resistance of the rice blast fungus, *Piricularia oryzae* Cav. I. Acquisition of the resistance in successive passages through media containing drugs and the nature of such acquired resistance. II. Cross-resistance of drug-resistant strains. III. Search for $CuSO_4$-resistant cells in the population of parental fungus P–2. IV. Reversing effects of various chemicals on permanent resistance to $CuSO_4$-resistant strains. V. Physiological studies on the resistance to $CuSO_4$. VI. Physiological resistance to H_3BO_3. VII. Pathogenicity of drug-resistant strains. *Bull. natn. Inst. agric. Sci., Tokyo*, Ser. D, **11**: 1–110. [Jap. Engl. summ.]

YOSHIDA, M. 1943. On the epidemic outbreak of rice blast in 1941 (Hokuriku District.) *Ann. phytopath. Soc. Japan* **12**: 167–180. [Jap. Engl. summ.]

YOSHIDA, S., OHNISHI, Y. & KITAGISHI, K. 1962a. Histochemistry of silica in rice plants. I. A new method for determining the localization of silicon within plant tissues. *Soil Sci. Pl. Nutr., Japan* **8**: 30–35.

YOSHIDA, S., OHNISHI, Y. & KITAGISHI, K. 1962b. Chemical forms, mobility and deposition of silicon in rice plants. *Soil Sci. Pl. Nutr., Japan* **8**: 107–113.

YOSHIDA, S., OHNISHI, Y. & KITAGISHI, K. 1962c. Histochemistry of silicon in rice plants. II. Localization of silicon within rice tissues. *Soil Sci. Pl. Nutr., Japan* **8**: 36–41.

YOSHIDA, S., OHNISHI, Y. & KITAGISHI, K. 1962d. Histochemistry of silicon in rice plants. III. The presence of cuticle-silica double layer in the epidermal tissue. *Soil Sci. Pl. Nutr., Japan* **8**: 1–5.

YOSHII, H. 1936. Pathological studies of rice blast caused by *Piricularia oryzae*. I. Some studies on the physiology of the pathogen. II. The mode of infection of the pathogen. *Ann. phytopath. Soc. Japan* **6**: 199–218. [Jap. Engl. summ.]

YOSHII, H. 1937. Pathological studies of rice blast caused by *Piricularia oryzae*. III. Pathohistological observations of diseased plants. *Ann. phytopath. Soc. Japan* **6**: 289–304. [Jap. Engl. summ.] (*Rev. appl. Mycol.* **16**: 557.)

YOSHII, H. 1941a. Studies on the nature of rice blast resistance. 2. The effect of combined use of silicic acid and nitrogenous manure on toughness of the leaf blade of rice and its resistance to rice blast. *Bull. Sci. Fak. Teck. Kyushu Imp. Univ., Japan* **9**: 292–296. [Jap. Engl. summ.]

YOSHII, H. 1941b. Studies on the nature of rice blast resistance. 4. Relation between varietal resistance to rice blast and some physical and chemical properties of the leaf blade of rice. *Ann. phytopath. Soc. Japan* **11**: 81–88. [Jap. Engl. summ.]

YOSHII, K. 1949. Studies on *Cephalothecium* as a means of artificial immunization of agricultural crops. *Ann. phytopath. Soc. Japan* **13**: 37–40. [Jap.]

YOSHINO, K. 1905 A list of parasitic fungi collected in Higo Province. *Bot. Mag., Tokyo* **19**: 87–103; 199–222. [Jap.]

BROWN SPOT

HISTORY AND DISTRIBUTION

Breda de Haan (1900) first described this leaf-spotting fungus and named it *Helminthosporium oryzae*. Hori (1901) later described it in Japan and also used the name *H. oryzae* with Miyabe and Hori as the authors. The original description in Hori (1901) and also Hori's later emendation (1918) were in Japanese but were translated into English by T. Tanaka (1922). Kurosawa (1900, 1911) made inoculations and claimed that he first used the names *H. oryzae* and ' nai-yake ' (seedling blight) in 1900. Hara (1916) called the disease ' sesame leaf blight ', a name in use in Japanese literature up to the present day.

Subramanian & Jain (1966) transferred *H. oryzae* to *Drechslera* as *D. oryzae*. Ito & Kuribayashi (1927) found the perfect state of the fungus in culture and gave it the name *Ophiobolus miyabeanus*. Drechsler (1934) considered the fungus to belong to his new genus *Cochliobolus* and Dastur (1942) formally transferred it to that genus. The name *Cochliobolus miyabeanus* (Ito & Kuribayashi) Drechsler ex Dastur is used at present.

During the 1920s and early 1930s, Nisikado and Hemmi in Japan and Ocfemia in the Philippines studied the disease. Not much detailed work was done for some time after that period, although the disease was reported frequently in various countries. Since 1950, however, many workers in Japan have carried out extensive research on the disease. Biochemical aspects were studied by Akai and his colleagues, and by Oku, and physiological aspects were studied by Asada and Baba and others in connection with a physiological disease called ‘akiochi’.

The disease has a world-wide distribution, and it has been reported in all rice-growing countries in Asia, America and Africa (C.M.I. Distribution Map 92).

Damage

The disease causes blight of seedlings grown from heavily infected seeds. Ocfemia (1924a) reported 10–58% seedling mortality in the Philippines in 1918, and Tucker (1927) recorded 15% killed seedlings in Puerto Rico. Padmanabhan *et al.* (1948) also reported seedling blight leading to severe losses as a result of epiphytotics in India. Muller (1953) reported from Central America that the disease sometimes was severe, accompanied by 100% seed infection resulting in seedling blight the next season. The disease may weaken the seedlings and older plants. When the grains are infected, it lowers grain quality and weight. Bedi & Gill (1960) determined the loss of grain weight to be 4·58–29·1%. Sundararaman (1922) reported in Madras in 1918–19 that rice was seriously affected by the disease. The most dramatic aspect of the disease so far recorded was that it was considered to be a major factor contributing to the Bengal famine of 1942 (Report of the Famine Inquiry headed by Sir John Woodhead, 1945), the losses then amounting to 50–90% (Ghose *et al.*, 1960).

In the light of recent studies (Baba, 1958; Sato, 1964–65; Takahashi, 1967) in connection with the physiological disease ‘akiochi’ and the fact that the fungus can rarely be found on rice plants growing in normal soils, it seems that brown spot is associated only with abnormal or poor soil, of which it serves as an index. Baba (1958) stated that yield losses in the disease complex were due to the root rot associated with ‘akiochi’ rather than to brown spot. In his study of the disease, Goto (1958) stated that it is almost impossible to distinguish between the losses due to brown spot and those caused by abnormal soil factors. He noted that when plants cultured in pots are artificially infected after heading, significant reduction in yield seldom happens, although the plants may be affected by leaf spots to a degree rarely observed in nature. Actually, the symptoms of potassium deficiency are somewhat similar to those of brown spot, so that it is often very difficult to ascribe symptoms to fungus attack or soil conditions. Under conditions favourable to infection (abnormal soil conditions) and given suitable environmental factors, however, the disease may cause additive damage, lowering grain quality and causing seedling blight if infected grains are sown. In Southeast Asia considerable areas suffer damage from disease complexes involving *C. miyabeanus*.

SYMPTOMS

The most conspicuous symptoms of the disease are on the leaves and the glumes. Symptoms may also appear on the coleoptile, the leaf sheaths, panicle branches, and more rarely on roots of young seedlings, and stems.

Typical spots on the leaves are oval, about the size and shape of sesame seeds. They are relatively uniform and fairly evenly distributed over the leaf surface. The spots are brown, with grey or whitish centres when fully developed. Young

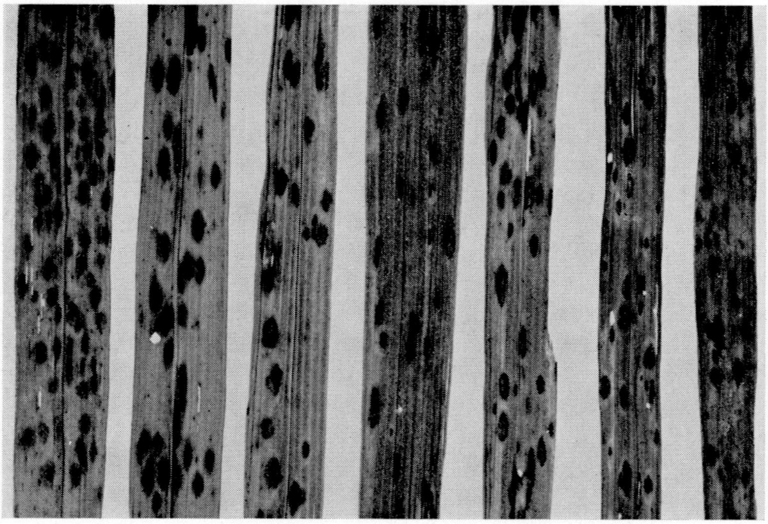

Fig. III–23. Brown spot on rice leaves (× 1).

or undeveloped spots are small and circular, and may appear as dark brown or purplish brown dots. On susceptible varieties the spots are much larger and may reach 1 cm or more in length. Sometimes numerous spots occur and as a result the leaf withers. Concentric lines or zones on the spot have been observed occasionally (Goto, 1954; Sato, 1965).

Black or dark brown spots appear on the glumes and in severe cases the greater portion or the entire surface of some of the glumes may be covered. Under favourable climatic conditions, dark brown conidiophores and conidia develop on the spots to give a velvety appearance. In such cases, the fungus may penetrate the glumes and leave blackish spots on the endosperm. (Fig. III–23, 24.)

Coleoptiles may become infected from diseased seeds. The spots are small, brown, circular or oval. Young roots may also be infected and show blackish lesions. Nodes and internodes are rarely infected.

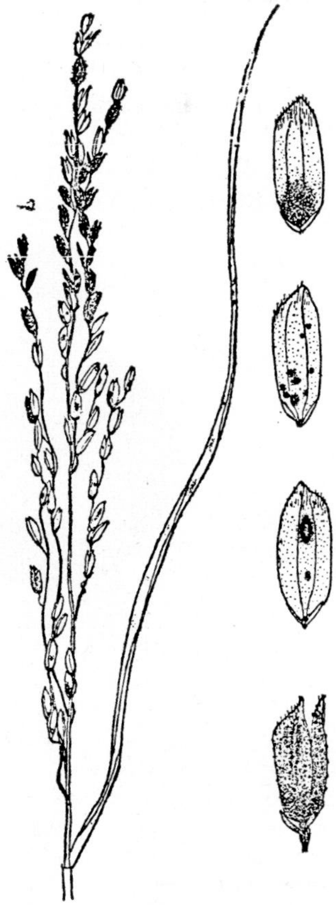

Fig. III–24. Brown spot on panicle and grains. (After Padwick, 1950.)

THE ORGANISM

Morphology. The original description by Breda de Haan is incomplete. The following is extracted from the description of the fungus from Louisiana by Drechsler (1934): Black velvety mycelial mats on the glumes of affected spikelets, which are distributed irregularly and usually sparsely through an otherwise healthy panicle, are composed of prostrate hyphae and more or less erect sporophores. The hyphae are abundant, branching and anastomosing, dark brown or olivaceous, 8–15μ or more in diameter. The sporophores arise as lateral branches from the hyphae, and change from olivaceous at the base to light fuliginous and at the tip even to subhyaline. They are 150–600 $\times$4–8μ; geniculations are not always well defined or conspicuous. Conidia measure 35–170$\times$11–17μ, and there may be as many as 13 septa in large ones; those of moderate length have been regarded as characteristic of the fungus. Typically conidia are slightly curved, widest at the middle or somewhat below the middle, the distal portion tapering towards the hemispherical apex where its width approximates half the median width. The proximal portion tapers towards the rounded off base but the diminution in diameter is usually perceptibly less. Fully matured conidia are fuliginous or brownish, with a moderately thin peripheral wall which is further attenuated at the apex as well as immediately around the rather inconspicuous hilum, visible within the contour of the base. Matured conidia germinate with two polar germ tubes, one from each of the thin-walled regions, while less mature, subhyaline spores may produce germ tubes from intermediate segments (Fig. III–25).

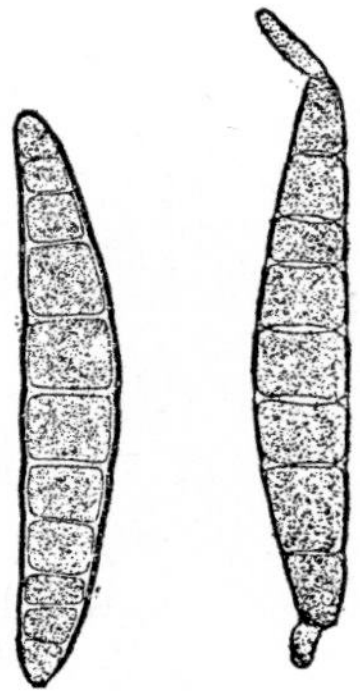

Fig. III–25. *Cochliobolus miyabeanus*. Conidia from grain, the spore on the right germinating after 16 hours. (Padwick, 1950.)

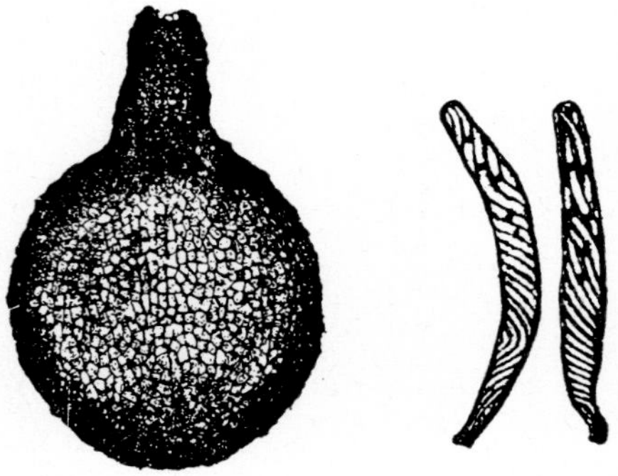

Fig. III–26. *Cochliobolus miyabeanus*. Perithecia and asci. (After Wei, 1957 reproduced from Ito & Kuribayashi, 1927.)

The perfect state as described by Ito & Kuribayashi (1927) is as follows: The perithecia are globose to depressed globose, with the outer wall dark yellowish brown and pseudoparenchymatous, 560–950 × 368–377μ. The asci are cylindrical to long fusiform, 142–235 × 21–36μ. The ascospores are filamentous or long cylindrical, hyaline or pale olive-green, coiled together, 6–15 septate, 250–469 × 6–9μ (Fig. III–26).

TABLE III–24

MEASUREMENTS OF CONIDIOPHORE AND CONIDIA OF *H. oryzae* IN DIFFERENT COUNTRIES (Wei, 1957)

Locality	Author	Conidiophore	Conidia
Java	Breda de Haan		90 × 16μ
Japan	Nisikado	68–688 × 7·6–20μ	15–132 × 10–26μ
India	Sundararaman	70–175 × 5·6– 7μ	45–106 × 14–17μ
China	Wei	99–345 × 7–11μ	24–122 × 7–23μ
U.S.A.	Drechsler	150–600 × 4– 8μ	35–170 × 11–17μ

Conidial measurements vary in different countries (see Table III–24, Wei, 1957). Nisikado & Miyake (1922) considered the Japanese and American strains to be morphologically and physiologically different. However, the differences are perhaps only variations among strains. Moreover, the shape of the conidia is affected by environmental conditions (Emoto, 1925).

A fungus on dead parts of rice plants identified as *Helminthosporium macrocarpum* Grev. by von Thümen in 1889 was considered distinct from *H. oryzae* by Drechsler (1923) and Tanaka, T. (1922), but Padwick (1950) thought they were in all probability the same. The present fungus differs from *H. sativum* Pamm., King & Bakke, the conidia of which have blunter, less tapering apices, and from *H. monoceras* Drechsler and *H. turcicum* Pass., the conidia of which have protruding hila (Padwick, 1950).

The fungus further varies on artificial media because of differing strains and varying cultural conditions. Sakamoto (1934) found catenulate conidia, 1- or 2-celled, rarely 3- or more-celled, oval to elliptical, sometimes oblong to cylindrical, light coloured or hyaline. Wei (1934) also observed, in culture, 0–3 septate, hyaline spores arranged as in *Hormodendrum*, $9{\cdot}5$–32×4–$5{\cdot}5\mu$. The fungus, like other Helminthosporia, saltates frequently depending upon temperature and the nature of the medium (Matsuura, 1930, 1932; Hiroe, 1934; Raychaudhuri, 1947; Misra & Mukerjee, 1962b; Misra & Chatterjee, 1964). Akai (1933) observed autolysis and Hiroe (1935) pseudo-myceliolysis in culture.

Yagi & Hirata (1958) found that the so-called septa of the conidia vanish soon after germination. The contents of each cell were rounded into a spherical body and these bodies were linked together by slender connectors, forming a necklace-like chain. Before germination these spherical bodies completely fill the conidia and the area of contact of two of these bodies gives an optical appearance of a cell wall or septum. As germination advances, the spherical bodies lose their contents, apparently sending nutrients into the germ tube. Under phase contrast and electron microscopes, Matsui & Nozu (1962) found that the conidium wall has three layers, the inner layer enclosing the protoplast and forming a spherical sac. There is cytoplasmic junction between adjacent protoplasts. The septum may be considered a sort of deep invagination of the inner wall.

Goto (1954) found each cell of the mycelium and conidia to have 1–14 nuclei, 2 and 4 nuclei appearing in highest frequency in the material studied. He also suggested the chromosome number of the fungus to be two. Vorraurai & Giatgong (1970) observed 1–9 nuclei in each cell, mostly two. They also found the haploid chromosome number to be 7 and the diploid 14 in the binucleate cells.

Physiology. The fungus grows over a wide temperature range. Nisikado (1923) found the optimum temperature for mycelial growth to be 27–30°C, and for conidial germination, 25–30°C. Conidia were formed between 5° and 35–38°C. Nisikado (1926b) also found that conidia germinated at pH $2{\cdot}6$–$10{\cdot}9$, mycelium grew best at pH $6{\cdot}6$–$7{\cdot}4$, and conidia were produced at pH 4–10. Ocfemia (1924a) found the optimum temperature to be 28°C, minimum 16°C and maximum 40°C while mycelial growth was best at pH $8{\cdot}6$–$8{\cdot}8$, the minimum being $2{\cdot}4$–$2{\cdot}6$.

Nutritional studies have been made by several workers. Das & Baruah (1947) found peptone to be the best of the five nitrogen sources used; glucose, sucrose and starch gave maximum growth at 20%, 10% and 5%, respectively. Stimulated growth was noticed by adding unhusked paddy or rice extract to the medium. Tanaka, H. (1956, see under Akai *et al.*, 1954–56, XVIII) noticed in Czapek's medium that xylose, glucose, fructose, mannose, galactose, sucrose and maltose gave good growth. Maltose produced the best and glycerin the poorest growth. There was no growth on soluble starch or lactose. Misra & Mukerjee (1962a) found maximum growth on mannitol, good growth on sucrose and glucose, and minimum growth on lactose. Among nitrogen sources, peptone was the best, while calcium nitrate, potassium nitrate and urea also supported growth. They found sporulation was best on sucrose, next best on glucose; nitrogen sources other than peptone inhibited sporulation. Mycelial growth increased with increase of sucrose up to saturation. Sporulation occurred up to 0·5%, with none at 2%. In the case of nitrogen sources, mycelium increased up to 0·04–0·045% and sporulation was inhibited at a level higher than 0·1%. Chen *et al.* (1964) found that among carbon sources, the amount of growth obtained was in the descending order of maltose, lactose, soluble starch, sucrose, xylose, glucose, fructose, galactose and arabinose. Among inorganic nitrogen sources, ammonium salts were better than nitrates. Of 22 amino acids and amides tested, 19 were favourable. One of the reasons for the differences in the results of these studies may be that different strains of the fungus were used.

Tanaka, S. & Yoda (1952) in their study of the metabolism of the brown spot organism found that the accumulation of mannitol gradually replaced the soluble sugar present in the mycelium at an earlier stage of its development. The mannitol may serve as a reserve nutrient. The growth of the fungus in culture was accompanied by utilization of the nitrate in the medium. It inverted cane sugar and assimilated glucose more rapidly than fructose.

Asada (1956, I) found biotin and thiamine and some minor elements to be essential for the growth of the fungus and suggested a synthetic medium as follows: KNO_3, 10g; KH_2PO_4, 5g; $MgSO_4 \cdot 7H_2O$, 2·5g; sucrose, 50g; thiamine, 1·25mg; biotin, 1·25μg, $FeSO_4 \cdot 7H_2O$, 1·5mg; $MnSO_4 \cdot 7H_2O$, 0·11mg; $ZnCl_2$, 15mg; $CaCl_2 \cdot 6H_2O$, 5mg; $CuSO_4 \cdot 5H_2O$, 0·15mg; H_2O, 1 litre; pH, 5·7.

The activities of the enzymes of *C. miyabeanus* and their relation to pathogenesis have been studied by Akai (1951), Akai & Ueyama (1955, studies X; 1956, studies XXIV), Asada (1956, II; 1958, VII; 1959, V, VIII), Oku (1958–60, I–VIII), Tanaka, H. (1965), Nozu (1959) and others.

The effect of copper sulphate, azo-pigments, sugars, amino acids etc. on conidial germination and growth have been studied by Akai *et al.* (1954–56, V, IX, XII, XIII, XIV, XXI, XXIII).

The toxic effect of the culture filtrate of this organism was first noticed by Hemmi & Matsuura (1928) on *Vicia faba.* Goto (1953, I; 1958) found that it had a toxic effect on onion cells, on rice plants, and on the older mycelium itself. Asada (1956, II), however, stated that the toxicity of the filtrate was not observed by him and that the injurious action of the fungus on the rice plant was not due to a toxin but to pectic enzymes, xylanase and cellulase in the filtrate. Orsenigo (1957) extracted from the filtrate and purified cochliobolin, which was toxic to seedlings and inhibited the growth of roots and coleoptiles at 30 ppm. Nakamura & Oku (1960, see under Oku, 1958–60, IX) detected in the diseased leaves the substance ophiobolin, which was toxic to roots, coleoptiles and leaves and caused wilting of plants at 2–5 ppm. Terashima *et al.* (1962) isolated ergosterol from the mycelium.

The existence of strains of the organism differing in morphology, cultural characters, physiology, sporulation etc. was noticed many years ago by Nisikado (1926a). Tochinai & Sakamoto (1937) studied 132 monosporic strains on four kinds of media and tested them on 15 rice varieties. They separated ten growth types, noticed morphological variations, saltations, and a wide range of pathogenicity from extremely virulent, through moderate to weak. Nawaz & Kausar (1962) reported similar findings. Padmanabhan (1953), however, considered that there was no specialization in pathogenicity. Vorrauri & Giatgong (1970) noticed that single conidial isolates from the same culture and hyphal tip cell isolates from the same cells showed variation in pathogenicity.

Misra & Chatterjee (1963) found great differences in sporulating ability and pathogenicity between two isolates. Matuo (1948) indicated that conidia produced on rice plants growing in solutions with little or no potassium have increased pathogenicity. Honda & Sakamoto (1968) and Honda (1969) found that sporulation of *H. oryzae* requires both light and dark periods. However, the requirement varies with different isolates. Isolate H_1 sporulated even under continuous darkness while isolate HA_2 sporulated only in darkness after exposure to light. They also found that near-ultraviolet is responsible for stimulating sporulation, and the duration of exposure rather than the dosage is important. Blue light inhibits sporulation.

Disease Cycle

Overwintering and longevity. The fungus overwinters commonly in infected plant parts. In Hokkaido, Northern Japan, Kuribayashi (1929) found conidia on infected grain to be viable after 396–859 days (average period 2 years). Mycelium in infected tissue was viable after 1044–1076 days (average 3 years); under certain conditions it may survive in soil. Nisikado *et al.* (1938) found that the fungus remained viable for 28–29 months at 30°C but was unable to survive at 35°C for more than 5 months. Suzuki (1930) found the fungus not only within discoloured seeds but also in apparently healthy ones. He also found that the fungus can survive in the seed for at least 4 years, and considered that infected seed causes the first occurrence of the disease the following spring. In India, Padmanabhan *et al.* (1953, II) tested various infected plant parts and showed that the fungus was viable from one year to the next growing season only on the seed. Page *et al.* (1947) tested the longevity of conidia under controlled conditions and showed that at low temperature (2°C) 81% of the conidia were viable after 100 days, whereas at 31°C only 6% survived. At 31°C, conidia kept at 20% R.H. remained alive for 6 months, but for only 1 month at 96% R.H. They believed that under warm, humid conditions the conidia would not live very long.

Primary infection through diseased seed is probably most common, although diseased seeds do not necessarily always give rise to infected seedlings. The coleoptile and sometimes roots are often infected from diseased seeds, but lesions may not show on subsequently developed leaves on account of rapid growth of the leaves under normal conditions. Leaf spots mostly arise from secondary infection. Infection may also take place from the soil (Thomas, 1940; Ganguly, 1946).

Penetration, development and conidial formation. The conidia usually germinate by germ tubes from the apical and basal cells, sometimes from other cells. The germ tube is covered with a mucilaginous sheath which adheres to a solid surface and an appressorium is formed at the tip. Infection pegs form under the appressoria and penetrate directly into the epidermis. The germ tubes may also enter the leaf through the stomata without forming appressoria (Nisikado, 1922; Tullis, 1935). Horino & Akai (1966) observed 98% penetration through the motor cells and 2% through stomata. Nisikado (1943) observed that infection of the grains could take place through the base of small hairs, the fungus then penetrating into the surrounding epidermal cells.

Ocfemia (1924a) observed the first visible symptoms within 24 hr after inoculation and Sato (1964–65) saw the colour of infected cells change within 17–20 hr.

The infection processes and the formation of lesions were observed closely by Goto (1958) and Sato (1964–65). The processes include: (1) formation of appressoria; (2) hyphae attack the middle lamella; (3) penetration into the cells; and (4) mycelial development in the cells. At the time of appressorium formation, protoplasmic streaming in the host cells increased, and the nuclei moved to a position near the appressorium. At the time of the separation of the middle lamella, yellowish granules appeared. Then 2 or 3 cells died and penetration followed. Ultimately a minute spot, visible to the naked eye, appeared and on susceptible varieties and under favourable conditions, medium or large spots were formed. Horino & Akai (1968) made detailed observations of the penetration process in susceptible and resistant varieties under the electron microscope. Locci (1969) observed spore germination, appressorium formation and spore formation on the leaf surface by scanning electron microscopy.

Fukatsu & Kakizaki (1955) found conidiophores to be initially produced on the 'disintegrated zone,' usually through stomata. The zone enlarges greatly when the leaf dies. In a moist atmosphere, they appear in 5–14 hr. Sato (1964–65) found the formation of conidia on various types of spots to be none or few on the minute type (0·5mm), small numbers developing slowly on the medium type (0·6–1mm), and large numbers developing quickly on the large type spot (2×1mm). Under favourable conditions aerial mycelium grew out from the lesions and formed secondary conidia on large lesions, fewer on the medium type and none on the minute type.

Sreeramulu & Seshavataram (1962) and Padmanabhan *et al.* (1948–66, V) showed the numbers of air-borne conidia over rice fields to be highest during

October to January in India. These air-borne spores are responsible for secondary infection.

Host-Parasite Relationship

A great many studies have been made on the biochemistry and pathophysiology of brown spot (Akai *et al.*, 1955–58, I–V; Akai *et al.*, 1958–66, I–VI; Asada, 1956–59; Oku, 1958–60, 1962, 1965).

Akai *et al.* (1958–66, I–VI) found accumulation of starch surrounding the diseased spots. The accumulation was due to a decreased activity of β-amylase and increased activity of invertase near the spot. Substances which inhibit starch decomposition were found in the filtrate of the culture which inhibits the β-amylase activity. These substances were believed to diffuse from the diseased lesion and inhibited β-amylase activity in the tissues surrounding the spots. Artificial spots were produced by placing drops of solutions of chemical respiratory inhibitors, metal salts and phenols on the leaves. Some of these induced spots and inhibited starch translocation, others induced spots but did not affect starch movement, and still others had no effect. The first group seemed to inhibit β-amylase activity. They also noticed an increase in catalase within 24 hours after inoculation.

Asada (1957, III, VI, etc.) noticed in plants from both normal and ‘ akiochi ’ soil that when they were artificially inoculated, the soluble and insoluble nitrogen decreased, reducing sugar and other soluble carbohydrates were also reduced, but insoluble carbohydrates increased. Thiamine in leaves from normal soil was destroyed at an early stage of infection. High thiamine content was found in akiochi plants. The spots did not enlarge in an oxygen-free atmosphere. Asada (1956–59, III, VI, VII, VIII, 1962, 1967) found pectic enzyme, cellulase and xylanase in the culture filtrate. In normal leaves oxidative phosphorylation takes place efficiently while in akiochi leaves substances for respiration are consumed without producing energy-rich phosphate compounds. Dehydrogenase activity is higher in the infected leaves. In general, the low metabolic activity in akiochi plants was correlated with susceptibility. This lower activity may be due to uncoupling of oxidative phosphorylation caused by respiration. The uncoupling may be caused by the decay of the roots of the akiochi plants.

Shishiyama *et al.* (1965) proposed that the process of metabolic changes in the fungus and the plant during infection was as follows: ‘ The spore germination using reserved substrates through the dicarboxylic acid cycle ⇀ change of respiration system in the germinating spore from the DCA to TCA cycle ⇀ formation of appressorium ⇀ invasion into the host cell ⇀ disturbance of host metabolism by the toxic substance like ophiobolin and acceleration of the respiration pathway in the host plant ⇀ occurrence of abnormal metabolism of carbon and nitrogen in the host tissue ⇀ temporary accumulation of amino acid and organic acid brought by the disturbance ⇀ use of the amino acid and organic acid by the fungus, deamination, decarboxylation and transamination of the amino acids in the host cells, or the synthesis of abnormal protein such as γ-globulin-like substance ⇀ decrease of glutamate and aspartate → inhibition of energy metabolism of the fungus due to synthesis of abnormal protein or formation of necrotic barrier → interruption of fungal extension, i.e., stoppage of disease spot development.'

Shishiyama *et al.* (1969) reported profound changes in the content of aspartic and glutamic acids in leaves after inoculation. The content of glutamic acid increased 2 days after and decreased 4 days after inoculation of susceptible plants. They also studied related enzymes.

Oku (1965) studied in detail the mechanism of disease development and resistance and interpreted the mechanism briefly as follows: The pathogen invades the host cell, produces a toxin, ophiobolin, which kills the host cells. In the host cells which suffered a sublethal dose of the toxin, an increase of phenolic substances takes place. The accumulated phenolics are oxidized into quinone by the polyphenoloxidase which is secreted by the fungus. The quinones thus produced are polymerized into brown pigment very rapidly by some enzymatic action of the fungus. Then the deposited brown polymer spreads into necrotic lesions to form the typical spot symptom of the disease. Because of the toxicity of the accumulated oxidation products of polyphenols, the fungus in the necrotic lesions cannot grow further and the spots remain limited. He believes, therefore, that the behaviour of phenolic substances after infection seems to be related to disease resistance. The reducing agents, such as ascorbic acid or glutathione, may also play an important role in resistance to the disease.

Oku & Nakanishi (1962) found a phytoalexin-like substance in infected tissues.

Effect of Environmental Conditions

Soils and fertilizers. The brown spot disease has been known for some time to be associated with abnormal soils deficient in nutritional elements, or with soils in a much reduced condition in which toxic substances accumulate.

Through the extensive studies of akiochi disease by Baba and his co-workers (Baba, 1958; Baba & Harada, 1954; Baba *et al.*, 1951–57, I–XII), it was found that akiochi and brown spot occurred together in parallel in conditions in which silica, potassium, manganese or magnesium was deficient or where hydrogen sulphide was evolved to cause root rot. Brown spot easily attacked rice plants when nitrogen deficiency was induced after the middle of the growth period, and the reverse also applied. Phosphorus had, contrary to other elements, a positive correlation with susceptibility, i.e. plants were less susceptible with low phosphorus. When hydrogen sulphide was introduced, it inhibited the absorption of nutrients and water. The magnitude of the effect is in the order K_2O, SiO_2, NH_4–N, MnO_2, H_2O, MgO and CaO at the middle of the growth period of rice. The occurrence of akiochi and brown spot due to H_2S was attributed to the selective inhibition of nutrients and to the disturbance of nutrient balance (K_2O/N, SiO_2/N etc.). Addition of harmful organic acids such as butyric and acetic acids had similar effects.

In their tests of varietal resistance, it was found that resistance to brown spot is positively correlated with resistance to H_2S, causing root rot. Resistance to brown spot can, therefore, be measured by testing varieties in dilute H_2S solution.

It was also found that the oxidation-reduction potential (Eh) of the cell sap of the rice plant was lowered by a lack of some nutrients such as K, Mn, Si and Mg, and by an excess of P or N, and also by the addition of H_2S or other reducing agents. The occurrence of both brown spot and akiochi seems to be closely related to the lowering of the Eh value of the plant, because resistance to both diseases was lowered in all the cases mentioned above. Varieties resistant to the two diseases generally showed higher Eh in the leaf sap.

Misawa (1955) reported that nitrogen deficiency in culture solution was more conducive to brown spot than was deficiency of P or K. In a deficient solution, N was also more effective than P or K in reducing the number and size of leaf spots.

Sato and his group (1957–60, I–III; Sato, 1964–65) also studied brown spot in conjunction with akiochi in muck and loam soils and in pot cultures. They classified spots into minute, medium and large types. Similar results were obtained with respect to the relationship of nutrients to the development of the disease. Potassium deficiency in pot cultures was found to have the most profound effect, large lesions similar to those occurring in muck soil being produced. Only minute and medium-sized lesions were produced under other nutritional conditions.

Chattopadhyay & Dickson (1960) reported that deficiency and excess of nitrogen both increased brown spot development.

Tanaka & Akai (1962) noted that in nutrient solutions an excess of N and K or the presence of Mn, I or Zn decreased susceptibility while an excess of P, deficiency in N, K and Mg, and presence of Cd and Co increased susceptibility. Horino & Akai (1966) noticed that granular substances were deposited in the cells during the infection process; in plants with an excess of K and N, they were abundant, but only a few such granular substances were found when K and N were deficient. An abundance of granular substances indicates higher resistance, while a deficiency of these substances indicates increased susceptibility.

These studies seem to show that brown spot disease is brought about by deficiency of one or more nutrient elements, directly or indirectly, and in practice brown spot and deficiency symptoms are often inseparable. Baba used brown spot as an index of akiochi disease. This also led Goto, I. (1958) to think that it is almost impossible to distinguish between the losses due to brown spot and to soil conditions. The fungus may even be considered a saprophyte on plants weakened physiologically. The leaf spot symptoms on potassium-deficient plants are somewhat similar to brown spot lesions but are without the fungus.

Temperature, moisture and light. Ocfemia (1924b) reported that seedlings were infected at 16–36°C. At the highest temperature (36°C) the rice outgrew the fungus and was less injured. Hemmi & Nojima (1931) found the optimum temperature for infection to be 25–30°C, while Sherf *et al.* (1947) reported it as 20–25°C.

Katsura (1937) showed that at 25°C, a relative humidity of over 89% is required for successful inoculation by conidia, and Sherf *et al.* (1947) found that free water on the leaf surface favoured infection. Hemmi & Suzuki (1931) reported that seedlings were more susceptible in dry than in wet soil. Su (1938) also observed increased infection with reduced water supply.

In glasshouse tests, Ishaque & Talukdur (1962) demonstrated that both the number of spots and the disease index (which also considers the size of the spots) were highest on dry soil, least on wet soil, and intermediate on moist soil.

Naito (1937) reported that partial shading before, during and after inoculation increased infection. Imara (1938) showed that the lesions of brown spot reached initial maximum in darkness, but at a later stage, lesion enlargement is accelerated in medium shade. Imara (1940) also found that the incubation period was shortened by shading the plants before or after inoculation.

Baba & Takahashi (1957) found that a higher or lower water temperature than the optimum (30–32°C) caused selective inhibition of nutrient absorption similar to the effect of adding H_2S, and thus increased susceptibility to brown spot. High air temperature at night and high light intensity in midsummer have similar effects. High air humidity and low soil moisture not only inhibit the absorption of silica and potassium but also reduce the SiO_2 and K_2O content of the leaves, increasing their susceptibility to brown spot.

Host Range

Nisikado & Miyake (1922) reported that wild grasses, including *Cynodon dactylon* and *Digitaria sanguinalis*, were infected by the fungus upon artificial inoculation. Ocfemia (1923) also obtained infection on many species of grasses in 23 genera. Tochinai & Sakamoto (1937) found that some strains of the fungus infected maize and naked barley, while wheat, common barley and oats were comparatively resistant. Thomas (1940) reported that the fungus readily attacked wheat and *Setaria italica*, slight infection was obtained on *Eleusine coracana*, oats and maize but none occurred on sorghum. Chattopadhyay & Chakrabarti (1953) found that *Leersia hexandra* was an alternative host in nature. Su (1936) noticed a *Helminthosporium* sp. similar to *H. oryzae* on *Panicum colonum* in close proximity to rice fields. The various species of grasses infected by artificial inoculation have seldom or never been found to be attacked under natural conditions. When suitable conditions are provided artificially, many grasses and cereal crops can be infected to different degrees. They may not, however, be of great significance in the life cycle of the fungus. On the other hand, Shaw (1921) found that a *Helminthosporium* sp. from wheat attacked rice, and one from rice attacked maize, sorghum, oats, barley and sugarcane.

Varietal Resistance

Varietal reaction and method of testing and scoring resistance. Selection of varieties for resistance to brown spot started many years ago (Suematsu, 1921; Chiu, 1936). Ganguly (1946) found Dakar Nagra 273–32, Patnai 549–

33 and 23, Kalma 219 and Nagra 41–14 to be resistant in Bengal. Yoshii & Matsumoto (1951) tested 20 varieties introduced into Japan. Asada *et al.* (1954) also tested many varieties and found Hainan No. 217 and Chiu Tiu Chiu to be the most resistant. Akai *et al.* (1954–56, I) used Kameji as a resistant and Magatausa as a susceptible variety in their studies. Kawai & Kakisaki (1955) found that Hukubozu, Diakokuwase, Yamahaku, Hinomaru and Norin 17 were resistant. Padmanabhan *et al.* (1948–66, III, VIII) made a systematic effort in India to search for resistant varieties. Among 490 varieties tested Ch 13, Ch 45, T-141, T-498-2A, Co 20, BAM10, T-998m, T-2112, T-2118, and T-960 were considered resistant. Kanjanasoon & Sitthichai (1967) tested 344 varieties in Thailand and found five of them to be resistant: Taichung (N) 1, Muey Nawng 62M, Leuang Yai 34, Leuang Thong 82 and a cross, PJP 54×IRH8-BKN 56–7–106. In U.S.A., ‘HO nurseries’ (HO= *Helminthosporium oryzae*) were established in Texas for several years. Variety CI 9515 was found to be highly resistant. It was crossed with Texas Patna and TP4–9 and selections were further crossed with Century Patna 231 and Bluebonnet. Several HO lines selected were resistant to brown spot, and also blast, but showed low yield potential. In 1966, a new variety, Dawn, was released having high blast resistance and moderate resistance to brown spot (Atkins, personal communication).

Artificial inoculations as well as field tests and a combination of the two methods have been used for determining varietal reaction. Artificial inoculation usually requires less time and space and better control of environmental conditions can be achieved. Conditions favourable for infection and disease development have been discussed above (Hemmi & Suzuki, 1931; Hemmi & Nojima, 1931). Sherf *et al.* (1947) found that infection occurred in the temperature range 20–30°C, optimum 20–25°C. Ten hours of high humidity were required for infection at 22°C. Predisposing the plant in high humidity prior to inoculation did not increase infection but the presence of free water on the leaf surface did. In the field, evening inoculation was more favourable than inoculation in the morning or at noon.

Sato (1964–65) observed that the susceptibility of the rice plant to the disease increases with age. His work confirmed the results of Padmanabhan & Ganguly (1954), Goto (1958) and others. In the early stage of the development of the rice plant, only minute spots were formed and this was not affected by soil or nutrition. After ear formation, large spots developed on the lower leaves, but such spots were not observed in good soil. In a given leaf, only minute spots appeared before unfolding but these spots suddenly became enlarged 7 days after unfolding. Fazli & Schroeder (1966) found that the rice kernel was more susceptible at the flowering and milky stages than at the soft dough or mature stages.

Many strains of the fungus do not produce sufficient conidia in culture media for artificial inoculation. Sherf *et al.* (1947) found that mycelial fragments could be used as inoculum though they are not as infectious as are conidia. The inoculum was prepared by growing the fungus in liquid media; the mycelium was collected by filtering, dried at 40°C for 24 hr, and powdered into dust form. The mycelial dust was mixed with limestone (500 mesh) just before inoculation.

Akai, Shishiyama & Nishimura (1965) reported that the number and size of spots developed were positively correlated with the concentration of spores in the suspension used for inoculation. For uniform results in varietal reaction tests, a standard density of spores is therefore necessary.

As mentioned earlier, Baba (1958) suggested the use of dilute H_2S solution for testing varietal resistance to root rot in connection with akiochi and brown spot. This may prove to be the most convenient method if the correlation between root rot and varietal resistance to brown leaf spot can be firmly established.

Other factors which affect the infection and development of this disease have been discussed above under environmental conditions.

In scoring degrees of resistance, the number of spots per leaf or per unit leaf area and the size of spots have been used. Various scales may be established. Asada *et al.* (1954) distinguished large (A), medium (B) and small (C) spots and used the formula $\frac{(A+1/3B+1/4C)}{0\cdot 5\ ab} \times 100$ to calculate an index (a and b being length and width, respectively, of the leaf). Goto (1958) also distinguished small and large spots and suggested a large: small spot ratio. Padmanabhan *et al.* (1948–66, III) distinguished between underdeveloped and developed spots. Sato (1964–65) further divided plants into minute spot (less than 0·5 mm long), medium spot (0·6–1 mm) and large spot (1 × 2 mm or more) categories. Four, five or six units have been used by these workers. The size of spots is usually considered the more important criterion.

Aluko (1970) established a scale of six units for assessing varietal resistance as follows:

1. HR (highly resistant)—only small dark specks of pinhead size are produced on leaves, few or many, sometimes unrecognizable, no necrotic (collapsed cells) spots.
2. R (resistant)—large brown specks, 0·5–1 mm in diameter, no necrotic spots.
3. MR (moderately resistant)—small, roundish, necrotic spots, about 1mm in diameter, surrounded by brown margin.
4. MS (moderately susceptible)—typical brown spot lesions, circular or oval spots 1–4 mm long with a large necrotic centre and brown or purplish margin, usually relatively few (<50 per leaf) on leaves.
5. S (susceptible)—numerous (50–100) developed lesions as in group 4 or often larger, a small portion (<25%) of leaf area may be killed by coalescence of lesions.
6. VS (very susceptible)—large, quickly expanding lesions, about 5 mm long and more in number (>100 per leaf) than grade 5; leaf killed by coalescence reaching over 25% leaf area.

Inheritance of resistance. Very little is known concerning the inheritance of resistance to brown spot. Nagai & Hara (1930), working with a Korean strain of the causal fungus, found that resistance was a dominant character, whereas Adair (1941) found indications that resistance was recessive, several genes possibly being involved.

Mechanism of resistance. Some anatomical features of rice leaves have been observed to be related to resistance. Suzuki (1934), Chattopadhyay & Chakrabarti (1957) and Misra & Prasad (1964) have shown that thicker epidermal cells and more silicated cells are positively correlated with resistance. Chattopadhyay & Chakrabarti found, however, that the total SiO_2 content is not so correlated. Misra & Prasad (1964) further noticed that in resistant varieties, the stomata opened for a shorter period. These factors may account for part of the mechanism of resistance but they are possibly not very significant, as resistance is shown after the penetration of the plant by the fungus. The exact mechanism after penetration is not known, but the rate of accumulation of polyphenols and their oxidation products as suggested by Oku (1965) may be connected with the phenomenon of resistance as discussed above under host-parasite relationships. Oku (1962) found a phytoalexin-like substance in infected tissue. Sinha & Trivedi (1969) reported that inoculation of two susceptible rice varieties with an avirulent strain reduced the disease index by 83–85%. The same effect was obtained by pretreating plants with liquid in which germinating spores had been incubated for 24 hours.

CONTROL

Nisikado in 1918 and Tisdale in 1922 started the treatment of seeds by the hot water method, and many others experimented with the same method. In most rice-growing countries it would not, however, be practical at the ordinary farmer's level. Nisikado also tried seed treatment with $CuSO_4$ in 1921. Since then many workers have investigated seed treatment and numerous other chemicals, including other copper compounds, formalin, organic mercurics etc., have been tried. In areas where seed infection is common, chemical seed treatment is likely to be useful in reducing the damage on seedlings, and many fungicides developed for seed treatment should be effective.

Field spraying with fungicides to prevent secondary air-borne infection has also been tried (Chattopadhyay, 1951; Chattopadhyay & Chakrabarti, 1961; Mukerji & Bagchi, 1964). An antibiotic, ' funicularin ', has also been tried on seedlings (Yoshii *et al.*, 1958). The practical usefulness of such spraying is doubtful.

Akai *et al.* (1954–56, XV, XVII, XIX) soaked or grew seed in 2-methyl 1, 4-naphthaquinone (vitamin K3) (10^{-2}–2×10^{-2}%), Na-pentachlorophenate (0·01%), boric acid (2×10^{-4}%), β-indole acetic acid and other substances and found that the size of spots was reduced. Srivastava (1966) obtained reduction in lesion length by treating seedlings with sulphanilamide (100 μg/ml) or griseofulvin (25 μg/ml). Naha (1963) studied the movement of cycloheximide in rice plants and suggested the possibility of systemic control. Baldacci (1938) tried to ' vaccinate ' rice plants as a means of controlling the disease but without positive results.

Grummer & Roy (1966) reported that intervarietal mixture, i.e. planting alternate rows of varieties of varying reaction to different strains of the fungus, reduced the spread of the pathogen.

Field sanitation, crop rotation, adjustment of planting date, proper fertilization, good water management and soil amendment have all been suggested and tried as control measures. From recent knowledge of the disease, it appears that to provide proper nutrition in the soil is the most important factor in control, brown spot being more of a nutritional or physiological disorder than a pathological disease. It is not a problem in normal rice soils in Southeast Asia. In areas where the disease occurs and the associated abnormality of the soil is not easily corrected, resistant varieties should be sought for.

LITERATURE CITED

ADAIR, C. R. 1941. Inheritance in rice of reaction to *Helminthosporium oryzae* and *Cercospora oryzae*. *Tech. Bull. U.S. Dep. Agric.* 772, 18 pp.

AKAI, S. 1933. Ueber autolyse bei *Ophiobolus miyabeanus* Ito & Kuribayashi. *Forschn Geb. PflKrankh., Kyoto* **2**: 257–278. [Jap. Engl. summ.]

AKAI, S. 1951. On the cellulase activity in the mycelium of *Cochliobolus* (*Ophiobolus*) *miyabeanus*. *Forschn Geb. PflKrankh.* **4**: 64–70. [Jap.]

AKAI, S. 1965. *Helminthosporium* blight of rice plant with special reference to pathological physiology of the affected plants. *Ann. phytopath. Soc. Japan* **31** (Commem. issue): 193–199. [Jap.]

AKAI, S. *et al.* 1954–56. Studies on *Helminthosporium* blight of rice plants. I–XXIV.

I. On the resistance of varieties of rice plants to *Helminthosporium* blight, by S. AKAI & Y. ASADA. *Forschn Geb. PflKrankh.* **5**: 1–14, 1954.

II. Relation of the combination-ratio of nitrogenous and potassium fertilizers to the susceptibility of rice plants to *Helminthosporium* blight, by S. AKAI & S. MORI. *Mem. Coll. Agric. Kyoto Univ.* **69**: 1–11, 1954.

V. Effect of copper sulphate on the germination of the causal fungus, *Cochliobolus* (*Ophiobolus*) *miyabeanus*, by S. AKAI & S. ITOI. *Bot. Mag., Tokyo* **67** (787–788): 1–5, 1954.

VI. Observations on some isolates of *Cochliobolus* (*Ophiobolus*) *miyabeanus*, by S. AKAI & M. TANAKA. *Mem. Coll. Agric. Kyoto Univ.* **69**: 13–29, 1954.

VII. On the relation of silicic acid supply to the outbreak of *Helminthosporium* blight or blast disease in rice plants, by S. AKAI. *Ann. phytopath. Soc. Japan* **17**: 109–112, 1953. [Jap.]

VIII. Effect of copper sulphate on the growth of the mycelium of *Cochliobolus* (*Ophiobolus*) *miyabeanus*, by S. AKAI & S. ITONI. *Jubilee Publ. Commem. 60th Birthdays of Profs. Y. Tochinai and T. Fukushi*: 24–27, 1955, Sapporo, Japan.

X. On the catalase activities of diseased leaves of rice plants infected by *Cochliobolus miyabeanus*, by S. AKAI & A. UEYAMA. *Forschn Geb. PflKrankh* **5**: 87–94, 1955.

XII. Influence of azo-pigments upon the metabolism of the causal fungus, by S. AKAI & H. YASUMORI. *Ann. phytopath. Soc. Japan* **19**: 11–14, 1954. [Jap.]

XIII. Effect of expressed cell sap of leaves of rice plants upon the conidia germination of *Cochliobolus miyabeanus*, by Y. ASADA & S. AKAI. *Forschn Geb. PflKrankh.* **5**: 63–66, 1955. [Jap.]

XIV. Behaviour of conidia of *Cochliobolus miyabeanus* in dew drops formed on leaves of rice and some other plants, by K. TATSUYAMA. *Forschn Geb. PflKrankh.* **5**: 67–70, 1955. [Jap. Engl. summ.]

XV. On the influence of 2-methyl-1, 4-naphthaquinone (Vitamin K 3) upon the outbreak of *Helminthosporium* blight of rice plants, by S. AKAI, H. YASUMORI, H. OKA & T. TABUCHI. *Forschn Geb. PflKrankh.* **5**: 105–112, 1955. [Jap. Engl. summ.]

XVII. Effect of pentachlorophenol compounds upon the susceptibility of rice plants to *Helminthosporium* leaf spot, by S. AKAI & H. OKU. *Forschn. Geb. PflKrankh.* **5**: 159–163, 1956. [Jap. Engl. summ.]

XVIII. On the influence of carbon sources upon the growth of *Cochliobolus miyabeanus*, by H. TANAKA. *Forschn Geb. PflKrankh.* **5**: 165–170, 1956. [Jap. Engl. summ.]

XIX. Chemotherapeutic application of some compounds to rice plants and the outbreak of *Helminthosporium* leaf spot, by S. AKAI. *Forschn Geb. PflKrankh.* **5**: 45–46, 1956.
XXI. The effects of sugars on the conidial germination of *Cochliobolus miyabeanus*, by A. TOYAMA. *Forschn Geb. PflKrankh.* **6**: 25–32, 1956. [Jap. Engl. summ.]
XXII. On the antifungal effect of 2-methyl-1, 4-naphthaquinone (Vitamin K 3) to *Cochliobolus miyabeanus*, the causal fungus of Gomahagare-disease of rice plants, by S. AKAI & H. OKU. *Forschn Geb. PflKrankh.* **6**: 33–36, 1956. [Jap. Engl. summ.]
XXIII. The effect of some amino acids upon the conidial germination of *Cochliobolus miyabeanus*, by A. TOYAMA. *Forschn Geb. PflKrankh.* **6**: 47, 1956. [Jap.]
XXIV. On the β-amylase activity in mycelium of *Cochliobolus miyabeanus*, by A. NAKAITA. *Forschn Geb. PflKrankh.* **6**: 48, 1956. [Jap.]

AKAI, S. *et al.* 1955–58. Studies on the pathological physiology of plants. I–V.
I. Change of respiration and carbon assimilation in leaves of rice plant infected by *Cochliobolus miyabeanus*, by S. AKAI & H. TANAKA. *Forschn Geb. PflKrankh.* **5**: 95–104, 1955.
II. The free amino acids of rice plants, Kameji and Magatama, and their change in diseased ones due to the attack of the *Helminthosporium* blight fungus, *Cochliobolus miyabeanus*, by S. AKAI, J. SHISHIYAMA & H. EGAWA. *Forschn Geb. PflKrankh.* **6**: 7–10, 1956. [Jap. Engl. summ.]
IV. On the susceptibility of rice plants cultured with nutrient solution containing amino acids to the *Helminthosporium* blight caused by *Cochliobolus miyabeanus*, by S. AKAI, J. SHISHIYAMA & H. EGAWA. *Ann. phytopath. Soc. Japan* **23**: 165–168, 1958. [Jap. Engl. summ.]
V. Characteristic components in the leaves of *Helminthosporium* blight-diseased rice, by S. AKAI, J. SHISHIYAMA, H. OGURA, H. EGAWA & E. YOSHINAGA. Abs. in *Ann. phytopath. Soc. Japan* **22**: 5, 1957. [Jap.]

AKAI, S. *et al.* 1958–66. On the mechanism of starch accumulation in tissues surrounding spots in leaves of rice plants due to the attack of *Cochliobolus miyabeanus*. I–VI. [Jap. Engl. summ.]
I. Observation on starch accumulation in tissues surrounding spots, by S. AKAI, H. TANAKA & K. NOGUCHI. *Ann. phytopath. Soc. Japan* **23**: 111–116, 1958.
II. On the activities of β-amylase and invertase in tissues surrounding spots, by H. TANAKA & S. AKAI. *Ibid.* **25**: 80–84, 1960.
III. The formation of artificial spots and the inhibition of starch decomposition in rice leaves by various chemical compounds, by H. TANAKA & S. AKAI. *Ibid.* **25**: 156–164, 1960.
IV. Inhibitory effect of culture filtrate of the causal fungus on starch decomposition, by H. TANAKA & S. AKAI. *Ibid.* **30**: 136–138, 1965.
V. Some characteristics of starch-decomposition-inhibiting substances found in culture filtrate of the fungus by H. TANAKA & S. AKAI. *Ibid.* **30**: 139–144, 1965.
VI. The degree of staining with various dyes and its relationship to the starch accumulation in the surrounding region of the spots, by H. TANAKA, S. AKAI & T. HISAKAWA. *Ibid.* **32**: 251–259, 1966.

AKAI, S., SHISHIYAMA, J. & NISHIMURA, R. 1965. Effect of spore density on the pathogenicity of *Helminthosporium oryzae* to rice leaves. *Ann. phytopath. Soc. Japan* **30**: 166–168. [Jap.]

AKAI, S. & UEYAMA, A. 1961. Studies on some isolates of *Cochliobolus miyabeanus*, II. *Shokubutsu Byogai Kenkyu, Kyoto* **7** (3): 23–33.

ASADA, Y. 1956–59. Studies on the susceptibility of akiochi (autumn-declined) rice plant to *Helminthosporium* blight. I–X.
I. Essential vitamins, minor elements and suitable synthetic media of *Cochliobolus miyabeanus*. *Ann. phytopath. Soc. Japan* **21**: 68–70, 1956. [Jap. Engl. summ.]
II. Toxicity of culture filtrate to rice leaves and activities of pectic enzyme, cellulase and xylanase of *Cochliobolus miyabeanus*. *Ann. phytopath. Soc. Japan* **21**: 191–193, 1956. [Jap. Engl. summ.]
III. Changes of nitrogen compounds, carbohydrates, reducing ascorbic acid and respiration accompanied by the infection of *Cochliobolus miyabeanus* and existence of hyphae in diseased spots. *Ann. phytopath. Soc. Japan* **22**: 103–106, 1957. [Jap. Engl. summ.]
V. On the influence of the carbon sources, the amount of amino acids and C–N ratio upon the secretion of the perthophytic enzymes of *Cochliobolus miyabeanus*. *Forschn Geb. PflKrankh.* **6**: 109–113, 1959. [Jap. Engl. summ.]

VI. Some physiological observations on the diseased rice plant affected by *Cochliobolus miyabeanus*. *Ann. phytopath. Soc. Japan* **22**: 178–182, 1957. [Jap. Engl. summ.]
VII. Changes of the activities of oxidative enzymes in rice leaves accompanied with the infection of *Cochliobolus miyabeanus* and the presence of these enzymes in the culture filtrate of the fungus. *Mem. Ehime Univ.*, Sect. VI, **3**: 157–163, 1958.
VIII. Production of oxalic acid by the pathogenic fungus and decomposition of pectin. *Forschn Geb. PflKrankh.* **6**: 114–116, 1959. [Jap. Engl. summ.]
X. Some physiological observations on the diseased rice plants affected with *Cochliobolus miyabeanus*, by Y. Asada & T. Tachibana. *Ann. phytopath. Soc. Japan* **24**: 213–218, 1959. [Jap. Engl. summ.]

Asada, Y. 1962. Studies on the susceptibility of ' akiochi ' (autumn-declined) rice plants to *Helminthosporium* blight. *Mem. Ehime Univ.*, Sect. VI, **8**: 1–103. [Jap. Engl. summ.]

Asada, Y. 1967. Pathological research on *Helminthosporium* leaf spot. In *Proc. symp. rice diseases and their control by growing resistant varieties and other measures*: 181–189. Agric. For. Fish. Res. Council, Tokyo, 1967.

Asada, Y., Akai, S. & Fukutomi, M. 1954. Varietal differences in susceptibility of rice plants to *Helminthosporium* blight (Preliminary report). *Jap. J. Breed.* **4**: 51–53. [Jap. Engl. summ.]

Baba, I. 1958. Nutritional studies on the occurrence of *Helminthosporium* leaf spot and ' akiochi ' of the rice plant. *Bull. natn. Inst. agric. Sci., Tokyo*, Ser. D, **7**: 1–157. [Jap. Engl. summ.]

Baba, I. *et al.* 1951–57. Studies on the nutrition of the rice plant with reference to *Helminthosporium* leaf spot. I–XII. [Jap. Engl. summ.]
I. The susceptibility of rice as influenced by the aeration and the kinds of nutrient elements supplied, by I. Baba, Y. Takahashi & I. Iwata. *Proc. Crop Sci. Soc. Japan* **20**: 163–166, 1951.
II. The nutrient absorption of rice as affected by the hydrogen sulphide added to culture solution, by I. Baba, Y. Takahashi & I. Iwata. *Ibid.* **21**: 98–99, 1952.
III. The nutrient absorption as affected by the water temperature.
IV. Nitrogen metabolism as affected by hydrogen sulphide.
V(a). Growth and nutrient absorption as affected by night temperature, by I. Baba, Y. Takahashi & I. Iwata. *Ibid.* **21**: 233–238, 1953.
V(b). Plant growth and susceptibility to the disease as affected by the addition of iron to the culture solution, by I. Baba, Y. Takahashi & I. Iwata. *Ibid.* **22**: 41–42, 1953.
VI. Nutrients absorption as affected by the light intensity, by I. Baba, Y. Takahashi & I. Iwata. *Ibid.* **22**: 49–50, 1954.
VIII. Varietal differences of the rice plant in the growth retardation and in the increased disease susceptibility caused by hydrogen sulphide, by I. Baba, Y. Takahashi & I. Iwata. *Ibid.* **23**: 10–15, 1954.
IX. Absorption of nutrients by root and metabolism of nitrogen sulphide, by I. Baba, Y. Takahashi, I. Iwata & K. Tajima. *Ibid.* **23**: 272, 1955.
XI. Absorption and translocation of nutrients as influenced by soil moisture and air humidity, by I. Baba, I. Iwata, Y. Takahashi & A. Kitaka. *Ibid.* **24**: 169–172, 1956.
XII. Absorption and translocation of nutrients as influenced by hydrogen sulphide and their relationships with the growth of the plant, by I. Baba, I. Iwata & Y. Takahashi. *Ibid.* **25**: 222–224, 1957.

Baba, I. & Harada, T. 1954. Physiological diseases of rice plant in Japan. *Jap. J. Breed.* **4** (Studies, Rice Breeding): 101–151.

Baba, I. & Takahashi, Y. 1957. Growth and susceptibility to *Helminthosporium* leaf spot disease of rice plants grown in different soils under different water temperature. *J. Agric. Met., Tokyo* **13**: 41–44.

Baldacci, E. 1938. Nuove ricerche sulla ' vaccinazione ' delle piante. *Atti Ist. bot. Univ. Pavia*, Ser. IV, **10**: 189–205.

Bedi, K. S. & Gill, H. S. 1960. Losses caused by the brown leaf-spot disease of rice in the Punjab. *Indian Phytopath.* **13**: 161–164.

Breda de Haan, J. 1900. Vorläufige Beschreibung von Pilzen, bei tropischen Kulturpflanzen beobachtet. *Bull. Inst. bot. Buitenzorg* **6**: 11–13.

Chattopadhyay, S. B. 1951. Effect of spraying with different fungicides on the control of secondary air-borne infection of paddy plants by *Helminthosporium oryzae* Breda de Haan. *Sci. Cult.* **16**: 533–534.

CHATTOPADHYAY, S. B. & CHAKRABARTI, N. K. 1953. Occurrence in nature of an alternate host (*Leersia hexandra* Sw.) of *Helminthosporium oryzae* Breda de Haan. *Nature, Lond.* **172**: 550.

CHATTOPADHYAY, S. B. & CHAKRABARTI, N. K. 1957. Relationship between anatomical characters of leaf and resistance to infection of *Helminthosporium oryzae* in paddy. *Indian Phytopath.* **10**: 130–132.

CHATTOPADHYAY, S. B. & CHAKRABARTI, N. K. 1961. Application of fungicides in the control of secondary air-borne infection of *Helminthosporium oryzae* Breda de Haan. *Indian Phytopath.* **14**: 88–92.

CHATTOPADHYAY, S. B. & DICKSON, J. G. 1960. Relation of nitrogen to disease development in rice seedlings infected with *Helminthosporium oryzae*. *Phytopathology* **50**: 434–438.

CHEN, Y. S., REN, H. C. & FANG, C. T. 1964. Studies on the carbon and nitrogen nutrition of *Piricularia oryzae* and *Helminthosporium oryzae*. *Acta phytopath. sin.* **7**: 165–174. [Chin.]

CHIU, W. F. 1936. Studies on helminthosporiose of rice III. Further studies on the varietal resistance and susceptibility of rice to *Helminthosporium oryzae* Breda de Haan. *Bull. Coll. Agric. For. Nanking*, N.S., **48**, 10 pp.

DAS, C. R. & BARUAH, H. K. 1947. Experimental studies on the parasitism of rice by *Helminthosporium oryzae* Breda de Haan and its control in field and storage. *Trans. Bose Res. Inst.* **16**: 31–46. (*Rev. appl. Mycol.* **27**: 294.)

DASTUR, J. F. 1942. Notes on some fungi isolated from ' black point ' affected wheat kernels in the Central Provinces. *Indian J. agric. Sci.* **12**: 731–742.

DRECHSLER, C. 1923. Some graminicolous species of *Helminthosporium*. I. *J. agric. Res.* **24**: 641–740.

DRECHSLER, C. 1934. Phytopathological and taxonomic aspects of *Ophiobolus*, *Pyrenophora*, *Helminthosporium*, and a new genus, *Cochliobolus*. *Phytopathology* **24**: 953–985.

EMOTO, S. 1925. On the variation of the spore shape of *Helminthosporium* affected by the environment. *J. scient. agric. Soc., Tokyo* **276**: 384–386. [Jap.]

FUKATSU, R. & KAKIZAKI, M. 1955. Studies on the brown spot of rice plant. I. Sporulation on the diseased spot. *Ann. phytopath. Soc. Japan* **19**: 117–119. [Jap. Engl. summ.]

GANGULY, D. 1946. *Helminthosporium* disease of paddy in Bengal. *Sci. Cult.* **12**: 220–223.

GHOSE, R. L. M., GHATGE, M. B. & SUBRAHMANYAN, V. 1960. Rice in India (Revised edition). 474 pp., New Delhi, Indian Council of Agricultural Research.

GOTO, I. 1953–54. Studies on the leaf spot of the rice plant caused by *Ophiobolus miyabeanus* Ito et Kurib. I–VI.

I. Effects of the culture-filtrate of the causal fungi upon the plant. Observations by the sheath-inoculation method. *Bull. Yamagata Univ.*, Agric. Sci., **1**: 241–249, 1953. [Jap. Engl. summ.]

II. Histological observations of the invading process, formation of lesions and comparative observations with the blast disease. *Sci. Rep. Inst. agric. Res. Tohoku Univ.* 4: 67–76, 1952.

III. Comparative histological observation with the blast disease. *Bull Inst. agric. Res. Tohoku Univ.* **4**: 271–278, 1952. [Jap. Engl. summ.]

IV. Respiratory changes in affected leaf and the observation of the lesion in paddy field. *Bull. Yamagata Univ.*, Agric. Sci., **1**: 382–391, 1954.

V. On the concentric line of lesion. *J. Yamagata agric. For. Soc.* **6**: 7–8, 1954. [Jap.]

VI. Observations on the nuclear phenomena in the causal fungi. *Bull. Yamagata Univ.*, Agric. Sci., **2**: 23–33, 1954.

GOTO, I. 1958. Studies on the *Helminthosporium* leaf blight of rice plants. *Bull. Yamagata Univ.*, Agric. Sci., **2**: 237–388. [Jap. Engl. summ.]

GRUMMER, G. & ROY, S. K. 1966. Intervarietal mixture of rice and incidence of brown spot disease (*H. oryzae* Breda de Haan). *Nature, Lond.* **209**: 1265–1267.

HARA, K. 1916. Ine no goma-hagare-byo. [The sesame-like leaf blight of rice plant.] *Nogyo Sekai* **11** (9). [Jap.]

HEMMI, T. & MATSUURA, I. 1928. Experiments relating to toxic action by the causal fungus of helminthosporiose of rice. (Preliminary report.) *Proc. imp. Acad. Japan* **4**: 185–187.

HEMMI, T. & NOJIMA, T. 1931. On the relation of temperature and period of continuous wetting to the infection of the rice plant by *Ophiobolus miyabeanus*. *Forschn Geb. PflKrankh.* **1**: 84–89. [Jap. Engl. summ.]

HEMMI, T. & SUZUKI, H. 1931. On the relation of soil moisture to the development of *Helminthosporium* disease of rice seedlings. *Forschn Geb. PflKrankh.* **1**: 90–98. [Jap. Engl. summ.]

HIROE, I. 1934. Experimental studies on the saltation in fungi (Preliminary report). VIII. On the mechanism of the occurrence of ' island type of saltation.' *Trans. Tottori Soc. agric. Sci.* **5**: 134–143. [Jap. Engl. summ.]

HIROE, I. 1935. Experimental studies on the saltation in fungi (Preliminary report). IX. On the biological characters of pseudo-myceliolyse. *Ann. phytopath. Soc. Japan* **4**: 178–190. [Jap. Engl. summ.]

HONDA, Y. 1969. Studies on the effects of light on the sporulation of *Helminthosporium oryzae*. *Bull. Inst. agric. Res. Tohoku Univ.* **21**: 63–132. [Jap. Engl. summ.]

HONDA, Y. & SAKAMOTO, M. 1968. Effects of light on the sporulation of *Helminthosporium oryzae*. *Bull. Inst. agric. Res. Tohoku Univ.* **19**: 201–214. [Jap. Engl. summ.]

HONDA, Y. & SAKAMOTO, M. 1968. On the blue light inhibition of sporulation in *Helminthosporium oryzae*. *Ann. phytopath. Soc. Japan* **34**: 328–335. [Jap. Engl. summ.]

HORI, S. 1901. Ine no hagare-byo. [Leaf blight of rice plant.] *Bull. cent. agric. Exp. Stn Nishigahara, Tokyo* **18**: 67–84. (Engl. transl. and abs. by Tanaka, T. in *Mycologia* **14**: 81–89, 1922.)

HORINO, O. & AKAI, S. 1966. Influence of amount of nitrogen and potassium on host entry and infection of *Helminthosporium oryzae*. *Ann. phytopath. Soc. Japan* **32**: 10–13. [Jap. Engl. summ.]

HORINO, I. & AKAI, S. 1968. Studies on the pathological anatomy of rice plants infected by *Helminthosporium oryzae* Breda de Haan.
I. Behavior of the causal fungus on the coleoptile of rice seedlings and its ultrafine structure.
II. Fine structure of hyphae in tissues and coleoptile of rice plants. *Ann. phytopath. Soc. Japan* **34**: 51–55; 231–234.

IMAM FAZLI, S. F. & SCHROEDER, H. W. 1966. Kernel infection of Bluebonnet 50 rice by *Helminthosporium oryzae*. *Phytopathology* **56**: 507–509.

IMURA, J. 1938. On the influence of sunlight upon the lesion enlargement of the *Helminthosporium* disease of rice seedlings. *Ann. phytopath. Soc. Japan* **8**: 203–211. [Jap. Engl. summ.]

IMURA, J. 1940. On the influence of sunlight upon the incubation period and the development of the blast disease and the *Helminthosporium* disease of the rice plant. *Ann. phytopath. Soc. Japan* **10**: 16–26. [Jap. Engl. summ.]

ISHAQUE, M. & TALUKDAR, M. J. 1962. Studies on the effect of soil moisture on the development of leaf spot disease of rice caused by *Helminthosporium oryzae*. *Pakistan J. biol. agric. Sci.* **5**: 10–17.

ITO, S. & KURIBAYASHI, K. 1927. Production of the ascigerous stage in culture of *Helminthosporium oryzae*. *Ann. phytopath. Soc. Japan* **2**: 1–8.

KANJANASOON, P. & SITTHICHAI, T. 1967. Varietal difference in *Helminthosporium* leaf spot and some problems of control measure in Thailand. In *Proc. symp. rice diseases and their control by growing resistant varieties and other measures*: 191–195. Agric. For. Fish. Res. Council, Tokyo, 1967.

KATSURA, K. 1937. On the relation of atmospheric humidity to the infection of the rice plant by *Ophiobolus miyabeanus* Ito and Kuribayashi and to the germination of its conidia. *Ann. phytopath. Soc. Japan* **7**: 105–124. [Jap. Engl. summ.]

KAWAI, I. & KAKISAKI, T. 1955. Studies on the brown spot of the rice plant. I. Environment to outbreak and its control. *Nogyo Kairyo Gijutso* **70**: 32. [Jap. Engl. summ.]

KURIBAYASHI, K. 1929. Overwintering and primary infection of *Ophiobolus miyabeanus* (*Helminthosporium oryzae*) with special reference to the controlling method. *J. Pl. Prot., Tokyo* **16**: 25–36; 77–85; 143–153. [Jap.]

KUROSAWA, E. 1900. Ine Nai-yake-byo ni tukite. [On seedling rot of rice plant.] *Bull. agric. Soc. Shizuoka Prefecture* 39. [Jap.] (Diehl, W. W. 1954. *U.S. Dep. Agric. Pl. Dis. Courier.*)

KUROSAWA, E. 1911. Notes on some diseases of rice and camphor tree. I. On the ' Naiyake' disease of rice plant. *Miyabe Festschrift*: 47–51. (Abs. in *Just's Jber.* **41**: 174, 1915.)

LOCCI, R. 1969. Scanning electron microscopy of *Helminthosporium oryzae* on *Oryza sativa*. *Riv. Patol. veg., Pavia*, Ser. IV, **5**: 179–183.

MATSUI, C., NOZU, M., KIKUMOTO, T. & MATSUURA, M. 1962. Electron microscopy of conidial cell wall of *Cochliobolus miyabeanus*. *Phytopathology* **52**: 717–718.

MATSUURA, I. 1930. Experimental studies on the saltation in fungi. (Preliminary report.) I. On the saltation of *Ophiobolus miyabeanus* Ito et Kuribayashi parasitic on rice plant. I. *Trans. Tottori Soc. agric. Sci.* **2**: 64–82. [Jap. Engl. abs. in *Jap. J. Bot.* **3**: 68.]

MATSUURA, I. 1932. Experimental studies on the saltation in fungi. (Preliminary report.) VII. On the mechanism of the occurrence of the ' island ' type of saltation. *J. Pl. Prot., Tokyo* **19**: 409–428. [Jap.]

MATSUO, T. 1948. On the effect of potassium deficiency in the soil upon the occurrence of ' helminthosporiose ' of rice plant. *Ann. phytopath. Soc. Japan* **13**: 10–13. [Jap. Engl. summ.]

MISAWA, T. 1955. Studies on the *Helminthosporium* leaf spot of rice plant. *Jubilee Publ. commem. 60th birthdays of Profs. Y. Tochinai and T. Fukushi*: 65–73. Sapporo, Japan. [Jap. Engl. summ.]

MISRA, A. P. & CHATTERJEE, A. K. 1963. Comparative study of two isolates of *Helminthosporium oryzae* Breda de Haan. *Indian Phytopath.* **16**: 275–281.

MISRA, A. P. & CHATTERJEE, A. K. 1964. Induced saltations in *Helminthosporium oryzae* Breda de Haan. *Indian Phytopath.* **17**: 218–221.

MISRA, A. P. & MUKHERJEE, A. K. 1962a. Effect of carbon and nitrogen nutrition on growth and sporulation of *Helminthosporium oryzae* Breda de Haan. *Indian Phytopath.* **15**: 211–215.

MISRA, A. P. & MUKHERJEE, A. K. 1962b. Saltation in *Helminthosporium oryzae* Breda de Haan. *Curr. Sci.* **31**: 27–28.

MISRA, A. P. & PRASAD, Y. 1964. The nature of resistance of paddy to *Helminthosporium oryzae* Breda de Haan. *Indian Phytopath.* **17**: 287–295.

MUKERJEE, S. K. & BAGCHI, B. N. 1964. Control of secondary air-borne infection of *Helminthosporium* disease of paddy. *Rice News Teller* **12**: 103–105.

MULLER, A. S. 1953. Plant disease problems in Central America. *Pl. Prot. Bull. F.A.O.* **1**: 136–138.

NAGAI, I. & HARA, S. 1930. On the inheritance of variegation disease in a strain of rice plant. *Jap. J. Genet.* **5**: 140–144. [Jap. Engl. abs. in *Jap. J. Bot.* **5**: 41, 1930.]

NAHA, P. M. 1963. Movement of cycloheximide in rice plants. *Proc. natn. Inst. Sci. India*, Part B, **29**: 191–196.

NAITO, N. 1937. On the effect of sunlight upon the development of the *Helminthosporium* disease of rice. *Ann. phytopath. Soc. Japan* **7**: 1–13. [Jap. Engl. summ.]

NAWAZ, M. & KAUSAR, A. G. 1962. Cultural and pathogenic variation in *Helminthosporium oryzae*. *Biologia, Lahore* **8**: 35–48.

NISIKADO, Y. 1926. Determination of hydrogen-ion concentration and its application to the studies of plant disease. *Agric. Studies* **9**: 50–112. [Jap.]

NISIKADO, Y. & MIYAKE, C. 1922. Studies on the helminthosporiose of the rice plant. *Ber. Ohara Inst. landw. Forsch.* **2**: 133–194.

NISIKADO, Y., HIRATA, K. & HIGUTI, T. 1938. Studies on the temperature relations to the longevity of pure culture of various fungi pathogenic to plants. *Ber. Ohara Inst. landw. Forsch.* **8**: 107–124.

NISIKADO, Y. & NAKAYAMA, T. 1943. Notes on the pathological anatomy of rice grains affected by *Helminthosporium oryzae*. *Ber. Ohara Inst. landw. Forsch.* **9**: 208–213.

NOZU, M. 1959. Histochemical demonstrations of dehydrogenases in the rice plant tissues affected by *Cochliobolus miyabeanus* and in the causal fungus. *Ann. phytopath. Soc. Japan* **24**: 114–118. [Jap. Engl. summ.]

OCFEMIA, G. O. 1923. The *Helminthosporium* disease of rice. Abs. in *Phytopathology* **13**: 53.

OCFEMIA, G. O. 1924a. The *Helminthosporium* disease of rice occurring in the Southern United States and in the Philippines. *Am. J. Bot.* **11**: 385–408.

OCFEMIA, G. O. 1924b. The relation of soil temperature to germination of certain Philippine upland and lowland varieties of rice and infection by the *Helminthosporium* disease. *Am. J. Bot.* **11**: 437–460.

OKU, H. 1958–60. Biochemical studies on *Cochliobolus miyabeanus*. I–IX.

I. Utilization of organic acids and amino acids as the carbon source. *Ann. Takamine Lab.* **10**: 234–240, 1958.

II. Enzymes concerning amino acid utilization. 1. On the amino acid oxidase. *Forschn Geb. PflKrankh., Tokyo* **6**: 104–108, 1959. 2. On the transaminase and the amino acid decarboxylase. *Ibid.* **6**: 126–131, 1959.

III. Some oxidizing enzymes of the rice plant and its parasites and their contribution to the formation of the lesions. *Ann. phytopath. Soc. Japan* **23**: 169–175, 1958. [Jap. Engl. summ.]

IV. Fungicidal action of polyphenols and the role of polyphenoloxidase of the fungus. *Phytopath. Z.* **38**: 342–354, 1960.

V. β-Glucosidase activity of the fungus and the effect of phenolglucoside on the mycelial growth. *Ann. Takamine Lab.* **11**: 190–192, 1959.

VI. Breakdown of disease resistance in rice by reducing agents. *Ann. phytopath. Soc. Japan* **25**: 92–98, 1960. [Jap. Engl. summ.]

VIII. Properties of the polyphenyl oxidase produced by the fungus. *Ann. Takamine Lab.* **12**: 261–265, 1960.

IX. Detection of ophiobolin in the diseased rice leaves and its toxicity against higher plants by M. Nakamura & H. Oku. *Ibid.* **12**: 266–271, 1960.

OKU, H. 1962. Histochemical studies on the infection process of *Helminthosporium* leaf spot disease of rice plant with special reference to disease resistance. *Phytopath. Z.* **44**: 39–56.

OKU, H. 1965. Host-parasite relationship in *Helminthosporium* leaf spot disease of rice plant. *Ann. Rep. Sankyo Res. Lab.* 17: 35–56.

OKU, H. & NAKANISHI, T. 1962. Relation of phytoalexin-like antifungal substance to resistance of the rice plant against *Helminthosporium* leaf disease. *Ann. Rep. Takamine Lab.* 14: 120–128. [Jap. Engl. summ.]

ORSENIGO, M. 1957. Estrazione e purificazione della cochliobolina, una tossino prodotta de *Helminthosporium oryzae*. *Phytopath. Z.* **29**: 189–196.

PADMANABHAN, S. Y. 1953. Specialization in pathogenicity of *Helminthosporium oryzae*. *Proc. 40th Indian Sci. Congress*, Part 4, Abs. 18.

PADMANABHAN, S. Y. & GANGULY, D. 1954. Relation between the age of rice plant and its susceptibility to *Helminthosporium* and blast diseases. *Proc. Indian Acad. Sci.*, Sect. B, **29**: 44–50.

PADMANABHAN, S. Y. *et al.* 1948–66. *Helminthosporium* disease of rice. I–VIII.

I. Nature and extent of damage caused by the disease, by S. Y. Padmanabhan, K. R. R. Chowdhry & D. Ganguly. *Indian Phytopath.* **1**: 34–47, 1948.

II. Source and development of seedling infection, by S. Y. Padmanabhan, D. Ganguly & M. S. Balakrishnan. *Ibid.* **6**: 95–105, 1953.

III. Breeding resistant varieties—selection of resistant varieties from genetic stock, by D. Ganguly & S. Y. Padmanabhan. *Ibid.* **12**: 99–100, 1959.

IV. Effect of cultural extract of the pathogen on the reaction of rice varieties to the disease, by D. Ganguly & S. Y. Padmanabhan. *Ibid.* **15**: 133–140, 1962.

V. A study of the spore population of *Helminthosporium oryzae* over rice fields, by G. H. Chandwani, M. S. Balakrishnan & S. Y. Padmanabhan. *J. Indian bot. Soc.* **42**: 1–14, 1963.

VII. A study of the meteorological factors associated with the epiphytotic of 1942 in Bengal, by S. Y. Padmanabhan. *Oryza* **1**: 101–111, 1963.

VIII. Breeding resistant varieties: selection of resistant varieties of early duration from genetic stock, by S. Y. Padmanabhan, D. Ganguly & G. H. Chandwani. *Indian Phytopath.* **19**: 72–75, 1966.

PADWICK, G. W. 1950. Manual of rice diseases. 198 pp., Kew, Commonwealth Mycological Institute.

PAGE, R. M., SHERF, A. F. & MORGAN, T. L. 1947. The effect of temperature and relative humidity on the longevity of the conidia of *Helminthosporium oryzae*. *Mycologia* **39**: 158–164.

RAYCHAUDHURI, S. P. 1947. Saltation in *Helminthosporium oryzae* Breda de Haan. *Sci. Cult.* **13**: 77–78

SAKAMOTO, M. 1934. Catenulate conidia formation in *Ophiobolus miyabeanus* Ito and Kuribayashi. *Trans. Sapporo nat. Hist. Soc.* **13**: 237–240.

SATO, K. 1964; 1965; 1965. Studies on the blight diseases of rice plant. *Bull. Inst. agric. Res. Tohoku Univ.* **15**: 199–237; 239–342; **16**: 1–54. [Jap. Engl. summ.]

SHAW, F. J. F. 1921. Report of the Imperial Mycologist. *Sci. Rep. agric. Res. Inst. Pusa, 1920–21*: 34–40.

SHERF, A. F., PAGE, R. M., TULLIS, E. C. & MORGAN, T. L. 1947. Studies on factors affecting the infectivity of *Helminthosporium oryzae*. *Phytopathology* **37**: 281–290.

SHISHIYAMA, J., AKAI, S., EGAWA, H. & OUCHI, S. 1965. Some aspects of the metabolic system of *Helminthosporium* blight fungus in relation to the host parasite interaction. *Mycopath. Mycol. appl.* **26**: 31–48.

SHISHIYAMA, J., EGAWA, H., MAYAMA, S. & AKAI, S. 1969. Role of amino acids in the development of *Helminthosporium* blight disease of rice plant and some enzyme activities relating to amino acid metabolism in the host-parasite interaction. *Mem. Coll. Agric. Kyoto Univ.* 95: 7–34.

SINHA, A. K. & TRIVEDI, N. 1969. Immunization of rice plants against *Helminthosporium* infection. *Nature, Lond.* **223**: 963–967.

SREERAMULU, T. & SESHAVATARAM, V. 1962. Spore content of air over paddy fields. I. Changes in a field near Pentapadu from 21 September to 31 December 1957. *Indian Phytopath.* **15**: 61–74.

SRIVASTAVA, S. N. S. 1966. The systemic control of brown spot disease of rice with sulphanilamide and griseofulvin. *Indian Phytopath.* **19**: 272–275.

SU, M. T. 1936. Report of the Mycologist, Burma, Mandalay, for the year ending 31st March, 1936. *Rep. Oper. Dep. Agric. Burma, 1935–36*: 35–39.

SU, M. T. 1938. Report of the Mycologist, Burma, Mandalay for the year ending 31st March, 1938. *Rep. Oper. Dep. Agric. Burma, 1937–38*: 45–54.

SUBRAMANIAN, C. V. & JAIN, B. L. 1966. A revision of some graminicolous Helminthosporia. *Curr. Sci.* **35**: 352–355.

SUEMATSU, N. 1921. On the resistant varieties. *Ann. phytophath. Soc. Japan* **1**: 53–56. [Jap.]

SUNDARARAMAN, S. 1922. *Helminthosporium* disease of rice. *Bull. agric. Res. Inst. Pusa* 128: 1–7.

SUZUKI, H. 1930. Experimental studies on the possibility of primary infection of *Piricularia oryzae* and *Ophiobolus miyabeanus* internal of rice seeds. *Ann. phytopath. Soc. Japan* **2**: 245–275. [Jap. Engl. summ.]

SUZUKI, H. 1934. Studies on the influence of some environmental factors on the susceptibility of the rice plant to blast and *Helminthosporium* disease and on the anatomical character of the plant. I. Influence of differences in soil moisture. *J. Coll. Agric. Tokyo imp. Univ.* **13**: 45–108.

TAKAHASHI, Y. 1967. Nutritional studies on development of *Helminthosporium* leaf spot. In *Proc. symp. rice diseases and their control by growing resistant varieties and other measures:* 157–170. Agric. For. Fish. Res. Council, Tokyo, 1967.

TANAKA, H. 1965. The activity of some carbohydrates produced by *Cochliobolus miyabeanus*, the causal fungus of *Helminthosporium* leaf spot of rice plant. *Ann. phytopath. Soc. Japan* **30**: 192–196. [Jap. Engl. summ.]

TANAKA, H. & AKAI, S. 1963. On the influence of some nutritional elements on the susceptibility to *Helminthosporium* leaf spot of rice plants. *Ann. phytopath. Soc. Japan* **28**: 144–152. [Jap. Engl. summ.]

TANAKA, S. & YODA, A. 1952. The metabolism of the organism of *Helminthosporium* disease of the rice plant. *J. Chem. Soc. Japan* (*Pure Chem. Sect.*) **73**: 8–10. [Jap.]

TANAKA, T. 1922. New Japanese fungi. Notes and translations XI. *Mycologia* **14**: 81–89.

TERASHIMA, N., HAMASAKI, T. & HATUDA, Y. 1962. Studies on the metabolic products of *Cochliobolus miyabeanus*. I. Isolation of ergosterol of the mycelia. *Trans. Tottori Soc. agric. Sci.* **14**: 69–70. [Jap. Engl. summ.]

THOMAS, K. M. 1940. Detailed administration report of the government mycologist, Madras, for the year 1939–40, 18 pp.

TOCHINAI, Y. & SAKAMOTO, M. 1937. Studies on the physiological specialization of *Ophiobolus miyabeanus* Ito and Kuribayashi. *J. Fac. Agric. Hokkaido Univ.* **41**: 1–96.

TUCKER, C. M. 1927. Report of the plant pathologist. *Rep. Puerto Rico agric. Exp. Stn, 1925*: 24–40.

TULLIS, E. C. 1935. Histological studies of rice leaves infected with *Helminthosporium oryzae*. *J. agric. Res.* **50**: 81–90.

VORRAURAI, S. & GIATGONG, P. 1970. Pathogenic variability and cytological studies on *Helminthosporium oryzae*. *Ninth natn. conf. agric. Sci. Feb. 1970.* Bangkok, Thailand. (Mimeographed.)

WEI C. T. 1934. Rice diseases. *Bull. Nanking Coll. Agric. For.*, N.S., **16**: 1–41. [Chin. Engl. summ.]

WEI, C. T. 1957. Manual of rice pathogens. 267 pp., Peking, Science Press. [Chin.]

YAGI, H. & HIRATA, K. 1958. On the structure of conidia of *Cochliobolus miyabeanus*. *Ann. phytopath. Soc. Japan* **23**: 135–138. [Jap. Engl. summ.]

YOSHII, H., ASADA, Y., KISO, A. & AKITA, T. 1958. Antifungal antibiotic 'Funicularin' produced by *Bacillus funicularis* and its effect upon the susceptibility of rice seedlings to *Helminthosporium* blight. *Ann. phytopath. Soc. Japan* **23**: 150–154. [Jap. Engl. summ.]
YOSHII, H. & MATSUMOTO, M. 1951. Studies on the resistance to helminthosporiose of the rice varieties introduced to Japan. I. *Sci. Rep. Matsuyama agric. Coll.* 6: 23–60. [Jap. Engl. summ.]

DOWNY MILDEW

Saccardo described the fungus *Sclerospora macrospora* in 1890 from a specimen on *Alopecurus* sp. from Australia. The fungus was first found to attack rice plants by Yamada (1911) in Japan and by Brizi (1912, 1919) in Italy. Morphological studies have resulted in several changes in the nomenclature of the fungus, as discussed below.

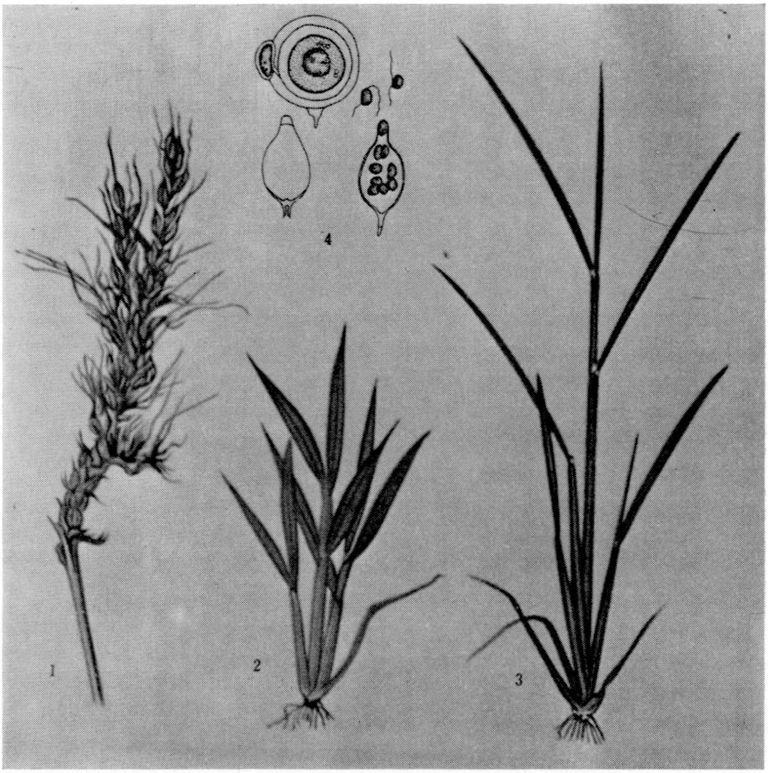

Fig. III–27. Downy mildew. 1. Infected panicle. 2. Infected seedling. 3. Healthy seedling. 4. Oospores, sporangia and zoospores. (Kawada & Goto, 1958.)

Besides Japan and Italy, the fungus has been found on rice in Australia, northeastern and eastern China, Taiwan, and Arkansas, U.S.A. The fungus has a wide host range and attacks numerous cereal crops and grasses in many other parts of the world.

Downy mildew does not cause serious damage to the rice crop, and has received little attention until recently, when it has been studied in detail by Akai and his associates in Japan.

SYMPTOMS

In the earlier stages of the development of the disease, new leaves of infected plants produce chlorotic, yellow or whitish spots or patches. In severe cases, the leaves may be distorted or twisted. The symptoms are usually more pronounced at flowering time. The panicles appear irregular, being unable to emerge from the leaf sheath freely, and becoming contorted and sometimes spiral (Fig. III–27). The leaf sheath may also be twisted. The panicle is reduced in size and usually remains green longer than normal. The floral parts may be abortive or reduced to a few tufts of hairs.

Akai & Fukutomi (1964a) reported that symptoms disappeared in winter when the temperature drops below 4·5°C and reappear at 8–10·5°C or higher. On the other hand, the disease can be controlled by high temperatures (30–35°C) (Yamanaka & Kawai, 1961). The active range of temperature for the disease appears to be 15–25°C with an optimum at about 20°C or slightly below. Akai & Fukutomi (1964b, c) found that chloroplasts gradually fade in the chlorotic spots, and that infection reduces silicification of leaf cells.

CAUSAL ORGANISM

Nomenclature. Yamada (1911) called the fungus on rice *Sclerospora macrospora* Sacc. while Brizi (1912, 1919) named it *S. oryzae* Brizi. Because the sporangial stage is more like that of *Phytophthora* than that of *Sclerospora*, it was called *Phytophthora oryzae* (Brizi) Hara by Hara (1930) and *P. macrospora* (Sacc.) Ito & Tanaka by Tanaka & Ito (1940). The interesting study of Thirumalachar, Shaw & Narasimhan (1953) led to the creation of a new genus, *Sclerophthora*, which has the sporangial stage like that of *Phytophthora* and the oogonial stage of *Sclerospora*. The fungus is now called *Sclerophthora macrospora* (Sacc.) Thirum., Shaw and Naras. by many workers. The taxonomy of the fungus was reviewed in detail by Akai & Fukutomi (1964e).

Morphology. The following description of the genus *Sclerophthora* is by Thirumalachar *et al.*: Mycelium parasitic in higher plants, hyaline, coenocytic; sporangial stage *Phytophthora*-like; sporangiophores hyphoid, very little differentiated from the hyphae within the host, simple or sympodially and successively branched; sporangia large, lemon-shaped, apically poroid, borne singly at the apices of the sporangiophores, germinating in water by the division of the cell contents into biciliate zoospores. Oogonial stage like that of *Sclerospora*; oogonial wall thickened and confluent with the wall of the oospore. Oospore germination indirect, by sporangial formation.

The measurements of the various structures of the fungus on rice are as follows (Tasugi, 1953); sporangia, 60–114$\times$28–57 μ (av. 83·1$\times$42·2μ) (largest at 21°C, smaller below or above); zoospores, 9–16 μ; cystospores, 9–15 μ; oogonia, 46–80$\times$42–69 μ (av. 58·3–64·7$\times$52·8–60·2μ) (larger before fertilization and smaller after spores mature according to Akai & Fukutomi (1964e));

antheridia, 20–48 × 10–29 μ (av. 32·2 × 20·5μ); oospores 36–67 × 32–64 μ (av. 46·8–54·2 × 42·7–50·3μ).

Physiology. Mature sporangia are produced on infected tissue in a few hours at optimum temperature (18–23°C) and at pH 5–6. The temperature range for sporangia production is between 7 and 32°C. Oospores germinate in water between 10 and 26°C, optimum 19–20°C, at pH 7 (Tasugi, 1953). In the field, oospores germinate where there is sufficient moisture. The maximum temperature for germination in the field is 25–26°C. Soil extract has not been found to promote germination. Zoospores are strongly attracted to germinating rice, barley, sorghum and beet seed (Takatsu & Toyama, 1957; Toyama & Takatsu, 1964b). Streptomycin and gibberellin enhance oospore germination which takes place best in water (Akai & Takatsu, 1964).

Akai & Tokura (1964a, b, c) cultured the organism on a synthetic medium consisting of asparagine, 2g; KNO_3, 3g; KH_2PO_4, 1g; $MgSO_4 \cdot 7H_2O$, 0·5g; $FeCl_3 \cdot 6H_2O$, 0·1g; Zn, 0·2ppm; Mn and Cu, 0·1ppm; glucose, 10g; water, 1 litre. They found that nitrate and amino acids are well utilized for mycelial growth at pH 6·5 while ammonium-nitrogen is not. Sucrose, maltose and starch were found to be best for mycelial growth. Media containing 0·3–0·6% KNO_3 and 1–2% glucose or sucrose are favourable for sporangium formation, which is increased by the addition of phosphate buffer (pH 6·24) or calcium carbonate. Some microelements also affect growth and reproduction. Ferric chloride (2×10^{-5}M), lithium chloride, sodium chloride, potassium chloride, magnesium chloride and zinc chloride (10^{-3}–10^{-5}M) promote mycelial growth. Magnesium, zinc and molybdenum enhance sporangium formation.

Ono (1963) noted variation in cultural characteristics among various isolates of the fungus.

Disease Cycle

The fungus overwinters as oospores in infected leaves. These oospores germinate in early spring, and infect wild grasses and rice seedlings in the nursery. Sporangia and oospores produced on these diseased plants soon germinate and cause secondary infections. Takatsu & Toyama (1957) found oospores germinating in fields flooded in the autumn but not in dry soil. Katsura, Tokura & Furuya (1954) considered wild grasses as an important source of primary infection of rice seedlings in the seedbed.

Akai & Fukutomi (1964d) reported infection through the plumule at an early stage of seed germination and later through the first leaf. They traced the infection process on germinating seed and noticed that infection took place through the epiblast or ventral scale, and that the hyphae were attracted to the growing point of the seedling (Fig. III–28). The symptoms appear on the second leaf. Yamanaka & Kawai (1964a) also observed that hyphae migrated into the parenchyma of the growing point but finally degenerated. The hyphae invade the vascular strands, continue to grow, and form oospores.

Katsura (1965) reported that the optimum temperature for infection was 18–20°C, at which the infection processes were completed in 15–20 hr by

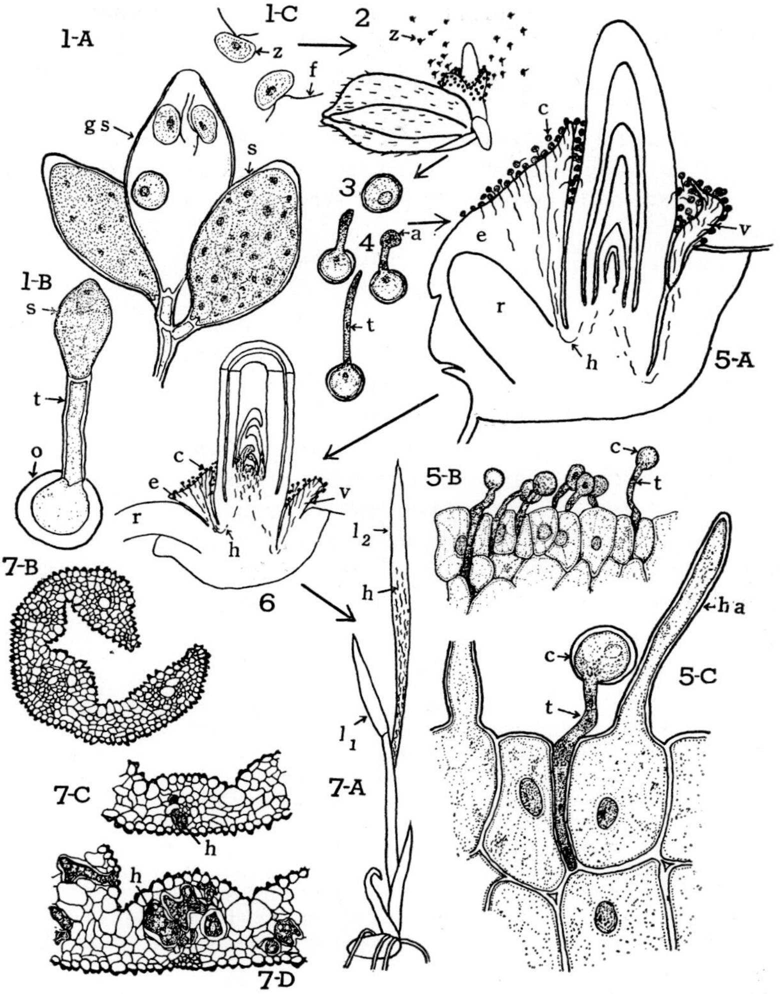

Fig. III–28. Schematic illustration of the infection process of plumules of rice plants by *Sclerophthora* (*Sclerospora*) *macrospora*. 1–A. Zoosporangia (s), germinating zoosporangium (gs). 1–B. Zoosporangium formed on the tip of germ tube (t) of germinated oospore (o). 1–C. Biciliate zoospores (z) and cilium (f). 2. Chemotaxis of zoospores towards the epiblast and ventral scale of germinating rice seed. 3. Encysted zoospore (cystospore). 4. Germination of encysted zoospores with germ tube (t), appressorium (a). 5–A. Germination of cystospores (c) on the epiblast (e) and ventral scale (v) and the intrusion by germ tubes into the tissues. h: hypha, r: primary root (radicle). 5–B, C. Direct penetration of germ tubes (t) of cystospores (c) without appressorium into epiblast through the suture of epidermal cells. ha: hair. 6. Spreading of the intruded hyphae (h) passing through the tissues of epiblast (e) or ventral scale (v) to the growing point of plumule. 7. Distribution of hyphae in the tissues of leaves of young seedling at the primary stage of disease occurrence. 7–A. Distribution of hyphae in the second leaf (1_2). 7–B. Cross-section of the first leaf (1_1) (no hyphae and no symptoms are found). 7–C. Cross-section of the middle part of the second leaf with no distinct symptoms (a few slender hyphae are found in the tissue). 7–D. Cross-section of the second leaf with severe symptoms (a number of hyphae and hyphal vesicles are found). (Akai & Fukutomi, 1964d.)

oospores. Infection by oospores may be completed in three hours (Yamanaka & Kawai, 1964b).

Host Range and Varietal Resistance

The fungus has a wide host range, and more than 43 genera in the Gramineae are affected, including other cereal crops such as oats, barley, wheat, rye, millet and maize (Akai & Fukutomi, 1964e). Very little is known about varietal resistance.

Control

Usually the disease does not warrant special control measures. Yamanaka & Kawai (1964d) reported that infected seedlings became healthy after 20 days at 35°C or after 30 days if the temperature did not fall below 30°C at night.

Covering seeds with rice hulls treated with fungicides prevents infection from taking place on the young shoots of germinating seeds (Adachi & Yamada, 1964). Covering seeds with ordinary soil to a depth of about 10 mm also protects young seedlings from infection (Toyama & Takatsu, 1964a). Chemical treatment of rice seed and the use of chemicals in the autumn have been tried (Yamanaka & Kawai, 1964c; Ichikawa, 1964; Fukuda & Isaka, 1964; Ogusu, Yoshida & Yokoyama, 1964; Teramoto & Suzuki, 1964).

LITERATURE CITED

Adachi, M. & Yamada, K. 1964. Coating of seeds with fungicides for controlling the plumular infection by the downy mildew of rice plants and covering seeds in seedbed with charred rice hulls treated with fungicides. In Studies on the downy mildew of rice plants, II, editor S. Akai. *Spec. Res. Rep. Dis. Insect Forecasting* 17: 132–147. [Jap.]

Akai, S. & Fukutomi, M. 1964a. Masking of symptoms, change of chlorophyll content, and the formation of reproductive organs of *Sclerophthora macrospora* in leaves of affected biennial gramineous plants. *Ibid.*: 1–17. [Jap. Engl. summ.]

Akai, S. & Fukutomi, M. 1964b. Symptoms and chloroplasts in leaves of rice plants affected by the downy mildew. *Ibid.*: 18–32. [Jap. Engl. summ.]

Akai, S. & Fukutomi, M. 1964c. Changes in the content and localization of silicate deposition in leaves of rice plants affected by *Sclerophthora macrospora*. *Ibid.*: 33–42. [Jap. Engl. summ.]

Akai, S. & Fukutomi, M. 1964d. Mechanism of the infection of plumules of rice plants by *Sclerophthora macrospora*. *Ibid.*: 47–54. [Jap. Engl. summ.]

Akai, S. & Fukutomi, M. 1964e. Taxonomic position of the pathogen of the downy mildew in rice plants. *Ibid.*: 112–125. [Jap.]

Akai, S. & Takatsu, S. 1964. Some experiments on the oospore germination of *Sclerophthora macrospora* II. *Ibid.*: 84–88. [Jap.]

Akai, S. & Tokura, R. 1964a. Relation of nitrogen and carbon sources to the mycelial growth and sporulation of *Sclerophthora macrospora*. *Ibid.*: 89–96. [Jap. Engl. summ.]

Akai, S. & Tokura, R. 1964b. Effects of phosphate and calcium carbonate on the sporulation of *Sclerophthora macrospora*. *Ibid.*: 97–106. [Jap. Engl. summ.]

Akai, S. & Tokura, R. 1964c. Effect of microelements on mycelial growth and sporulation of *Sclerophthora macrospora* in synthetic media. *Ibid.*: 107–111. [Jap. Engl. summ.]

Brizi, U. 1912. Sopra una nuova malattia crittogamica del riso. *Atti. Soc. ital. Progr. Sci.* **5**: 859.

Brizi, U. 1919. La peronospora del riso. *Natura, Milano* **10**: 168.

Fukuda, T. & Isaka, M. 1964. Results of demonstrative investigation of treatment of downy mildew of rice plant with streptomycin. *Proc. Ass. Pl. Prot. Hokuriku* **12**: 33–34. [Jap.]

Hara, K. 1930. Rice diseases. Tokyo. [Jap.]

ICHIKAWA, H. 1964. Damage by downy mildew of rice and its control in autumn. *Agriculture Hort., Tokyo* **39**: 764–768. [Jap.]
KATSURA, K. 1965. Downy mildew of rice plant: infection and disease development. *Ann. phytopath. Soc. Japan* **31** (Commem. Issue): 186–192. [Jap.]
KATSURA, K., TOKURA, R. & FURUYA, T. 1954. Studies on the downy mildew of rice plant. Wild grasses as the source of primary infection to rice plant in nursery bed. *Sci. Rep. Fac. Agric. Saikyo Univ.* 6: 49–66. [Jap. Engl. summ.]
OGUSU, I., YOSHIDA, K. & YOKOYAMA, S. 1964. A trial use of streptomycin emulsion on downy mildew of rice plant in paddy field. *Proc. Ass. Pl. Prot., Kyushu* **10**: 97–98. [Jap.]
ONO, K. & YAMAMOTO, T. 1963. Separation of downy mildew fungus and difference of cultural character of isolate. *Proc. Ass. Pl. Prot. Hokuriku* **11**: 9–10. [Jap.]
SACCARDO, P. A. 1890. Fungi aliquot Australienses. *Hedwigia* **29**:154–156; *Sylloge Fungorum* **9**: 342, 1891.
TAKATSU, S. & TOYAMA, A. 1957. Studies on the downy mildew (*Sclerophthora macrospora*) of rice plant. I. On the oospore germination and the infection to rice seedlings by the oospore. *Ann. phytopath. Soc. Japan* **22**: 123–128. [Jap. Engl. summ.]
TANAKA, I. & ITO, S. 1940. *Phytophthora macrospora* (Sacc.) S. Ito and I. Tanaka on wheat plant. *Ann. phytopath. Soc. Japan.* **10**: 126–138. [Jap. Engl. summ.]
TASAUGI, H. 1953. Studies on the downy mildew of rice plant caused by *Phytophthora macrospora* Ito and Tanaka. *Bull. natn. Inst. Agric. Sci., Tokyo*, Ser. C, **2**: 1–48. [Jap. Engl. summ.]
TERAMOTO, M. & SUZUKI, K. 1964. Some experiments on the control of the downy mildew of rice plants. *Spec. Res. Rep. Dis. Insect Forecasting* 17: 168–173. [Jap.]
THIRUMALACHAR, M. J., SHAW, C. G. & NARASIMHAN, M. J. 1953. The sporangial phase of the downy mildew on *Eleusine coracana* with a discussion of the identity of *Sclerospora macrospora* Sacc. *Bull. Torrey bot. Club* **80**: 299–307.
TOYAMA, A. & TAKATSU, S. 1964a. The effect of covering seeds with soil on the infection of plumules of rice plants by the downy mildew fungus. *Spec. Res. Rep. Dis. Insect Forecasting* 17: 59–64. [Jap. Engl. summ.]
TOYAMA, A. & TAKATSU, S. 1964b. Observations on the chemotaxis of zoospores of the rice downy mildew fungus, *Sclerophthora macrospora* and its relationship to the infectivity. *Ibid.*: 74–80. [Jap.]
YAMADA, G. 1911. *Sclerospora* Krankheit der Reispflanzen. *Spec. Bull. imp. Coll. Agric. For. Morioka*, 10 pp. [Jap.]
YAMANAKA, I. & KAWAI, T. 1961. Control of disease by temperature and chemical treatment after infection with downy mildew of rice plant. *Proc. Kansai Pl. Prot. Soc.* **3**: 41–43. [Jap.]
YAMANAKA, I. & KAWAI, T. 1964a. Behaviour of hyphae of *Sclerophthora macrospora* in tissues of affected rice plants. *Spec. Res. Rep. Dis. Insect Forecasting* 17: 43–46. [Jap. Engl. summ.]
YAMANAKA, I. & KAWAI, T. 1964b. Effect of temperature of the zoosporangium formation, its germination in *Sclerophthora macrospora* and the infection of host plants. *Ibid.*: 65–73. [Jap.]
YAMANAKA, I. & KAWAI, T. 1964c. Chemical treatment of seeds of rice plants for the control of plumular infection by the downy mildew. *Ibid.*: 148–157. [Jap.]
YAMANAKA, I. & KAWAI, T. 1964d. On the thermal therapy of the downy mildew of rice plants, II. *Ibid.*: 158–163. [Jap. Engl. summ.]

NARROW BROWN LEAF SPOT

HISTORY, DISTRIBUTION AND DAMAGE

Miyake (1910) first described this disease in Japan and named the fungus *Cercospora oryzae* Miyake. Earlier, Metcalf (1906) had mentioned the presence of a *Cercospora*, probably the same fungus, on leaves of rice plants in North America.

The disease has been reported from Burma, China, India, Indonesia, Malaysia,

Philippines, Thailand and other countries in Asia; from U.S.A., Brazil, Colombia, Dominican Republic, Guatemala, Nicaragua, Puerto Rico, Surinam, Venezuela etc. in America; and from Africa, Australia (N.T.) and Papua-New Guinea. Its world-wide distribution except in Europe is shown in CMI Distribution Map 71.

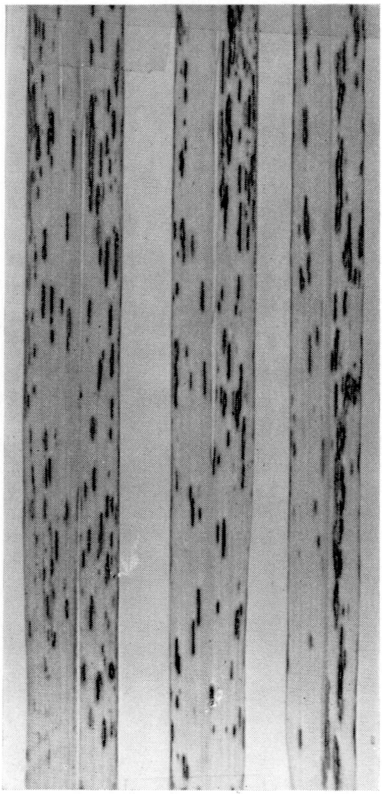

Fig. III–29. Narrow brown leaf spot on rice leaves.

A great deal of attention was paid to the disease in U.S.A. in the 1930s and 1940s. It was considered important because some commercial varieties, such as Blue Rose, at that time were very susceptible, and efforts were made to develop resistant varieties. Overwater (1960) reported 40% loss in Surinam during 1953–54. Tullis (1937) believed that damage is confined to the reduction of the effective leaf surface of the plant, while Ryker & Jodon (1940) stated that the disease causes premature killing of leaves and sheaths, and predisposes the plant to lodging. Generally heavy damage only occurs on very susceptible varieties.

Yoshida in 1948 found *Sphaerulina oryzina* Hara to be the perfect state of *C. oryzae* Miyake (Deighton, 1967). *S. oryzina* was described in 1918, but its relationship to *C. oryzae* was not then suspected.

SYMPTOMS

The disease produces short, linear, brown lesions, most commonly on leaves. It also occurs on leaf sheaths, pedicels and glumes. The lesions are *c.* 2–10 mm long, *c.* 1 mm wide. They tend to be narrower, shorter and darker brown on resistant varieties and wider and lighter brown on susceptible ones. The lesions usually appear in large numbers during the later stages of growth (Fig. III–29).

The symptoms are somewhat similar to those of white leaf streak (see p. 218) described recently and caused by *Ramularia oryzae*.

CAUSAL ORGANISM

Miyake's (1910) description of *Cercospora oryzae* may be translated as follows: Conidiophores emerging from stomata, solitary or in groups of two

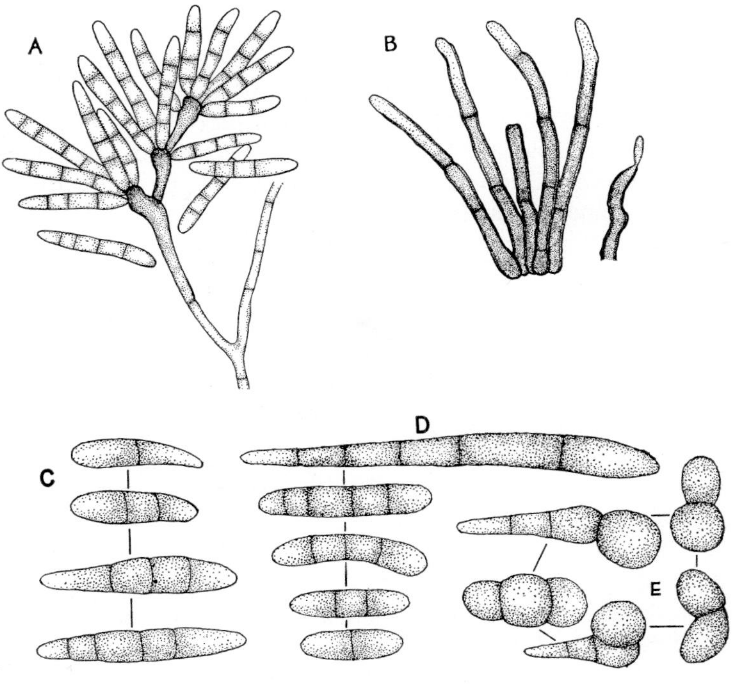

Fig. III–30. *Cercospora oryzae*. A. Conidiophores (secondary) and conidia on culture medium. B. Conidiophores on host. C. Conidia from host. D. Conidia from culture. E. Abnormal conidia produced in culture. (From Tasugi & Ikeno, 1956.)

or three, dark, paler at apex, 3 or more septate, 88–140×4·5μ. Conidia cylindrical to clavate, 3–10 septate, 20–60×5μ. Tasugi & Ikeno (1956) found the conidiophores to be 21·5–103×4–6·3μ (mostly 34·3–55·8×4·3–4·8μ) on the host plant in nature and 8·6–85·6×4·3–8·6μ (30·6–42·9×4·8–6·4μ) in culture; and conidia 12·9–47·2×3·9–6·3μ (25·7–43·3×4·3–5·2μ) on the host and 10·6–72·9×3·3–6·4μ (27·7–38·6×4·3–5·3μ) in culture, hyaline or light olive. Secondary conidiophores are formed in culture (Fig. III–30). Ganguly (1946) noted catenulate conidia in India.

The perfect state, *Sphaerulina oryzina*, was described in Japanese by Hara (1918); his description may be translated as follows: Perithecia scattered or gregarious, globose or subglobose, black, cellulo-membranaceous, with a papilliform mouth, immersed in the epidermal tissue of the host plant, 60–100μ in diameter. Asci cylindrical or club-shaped, round at the top, stipitate, 50–60×10–13μ. Ascospores biseriate, spindle-shaped, straight or slightly curved, 3-septate, hyaline, 20–23×4–5μ (Fig. VI–10A).

The fungus grows on bean agar and produces conidia in culture (Ryker, 1943). Tasugi & Ikeno (1956) found that growth was best on potato and soybean agar, while production of conidia was most luxuriant on rice straw agar. It grows in the temperature range 6–33°C, the optimum being 25–28°C. The optimum pH lies between 5·7 and 7·1.

Based upon reactions on 8 differential varieties, several physiologic races of the fungus have been reported. Ryker (1943) separated isolates of the fungus into 5 definite races and a number of sub-races. Chilton & Tullis (1946) found a new race, race 6. Ryker (1947) later reported races 7 and 8 and Ryker & Cowart (1948) races 9 and 10. The 8 differential varieties used are Blue Rose, Blue Rose 41, Fortuna, Caloro, Colusa, Zenith, Delitus and Southern Red Rice.

Sridhar (1970) reported the fungus on *Panicum repens*.

Disease Cycle

The fungus enters the host tissues through the stomata, becomes established in the parenchyma immediately beneath the stomata and spreads longitudinally in the epidermal cells. The mycelium is mostly intracellular, except that adjacent to the conidiophores which arise from substomatal hyphae and emerge through the stomata (Tullis, 1937).

A brief experiment by Chakrabarti (1964) indicates that rice plants grown in a culture solution with a high potash content are more susceptible to infection by the disease.

Varietal Resistance

Varietal resistance has been tested under conditions of natural infection in the field and by artificial inoculation, plants kept in a moist chamber or in the open after sundown being sprayed with a conidial suspension (Ryker, 1943).

Tullis (1937) first systematically tested varieties and hybrid lines in 1935 and 1936. Among the more than 100 varieties tested, 58 appeared to be resistant, including Nira, Tokalon and C.I. Nos. 461, 2711, 2738, 4603 and 4966. Blue Rose, Edith, Lady Wright and Early Prolific were susceptible. Ryker & Jodon (1940) classified rice varieties into highly susceptible, intermediate and highly resistant groups. Highly susceptible were Blue Rose, Early Prolific, Lady Wright, Edith, Honduras and Carolina Gold. The highly resistant varieties were Rexoro, Fortuna, Nira, Iola and C.I. 461 and 4440. The intermediate group includes Caloro, Colusa, Vintula, Delitus and Blue Rose selection 2854–3. The resistance or susceptibility of these varieties varies somewhat because of their reaction to different pathogenic races of the fungus. Other resistant varieties are also known, Asahi and Kamrose being resistant to all strains in U.S.A. (Adair & Cralley, 1950).

Several varieties developed since then are resistant to the disease — Delrex (Rexoro × Delitus) (Ryker & Cowart, 1948); Sunbonnet, a reselection from Bluebonnet (Rexoro × Fortuna) (Jodon, 1953); selection 44C507 (Rexoro × Purple Leaf × Magnolia) (Jodon, 1954); Toro (Jodon, 1955) etc.

The inheritance of resistance to the disease has been studied in U.S.A. Ryker & Jodon (1940) found, in 5 out of 6 crosses, that the F_2 population showed a 3 : 1 ratio between resistant and susceptible. In the sixth cross a single recessive gene controlled resistance. Further work, however, showed that a single dominant factor for resistance does not satisfactorily explain all cases. Adair (1941) found the factor for resistance to be dominant over susceptibility in several resistant varieties; while in variety Supreme Blue Rose at least one factor dominant for susceptibility was carried. Jodon, Ryker & Chilton (1944) reported that in 35 out of 48 crosses a single dominant factor for resistance was involved. In others, two, three or even more factors seemed to be concerned.

Ryker (1941) found two linked factors for resistance to races 1 and 2, respectively, and noted that plants susceptible to race 1 were resistant to race 2 and *vice versa*. There were few plants showing cross-over. Ryker & Chilton (1942) confirmed that the resistance of Blue Rose 41 to race 1 depends on a single dominant factor; that the moderate resistance of the parent Blue Rose to race 2 is likewise dependent on a single dominant factor; and that the two factors are closely linked and the linkage is rarely broken. Jodon & Chilton (1964) found no linkage between resistance and morphological characteristics of the rice plant.

LITERATURE CITED

ADAIR, C. R. 1941. Inheritance in rice of reaction to *Helminthosporium oryzae* and *Cercospora oryzae*. *Tech. Bull. U.S. Dep. Agric.* 722, 18 pp.

ADAIR, C. R. & CRALLEY, E. M. 1950. 1949 rice yield and disease control tests. *Rep. Ser. Ark. agric. Exp. Stn* 15, 20 pp.

CHAKRABARTI, N. K. 1964. The relation between host nutrition and *Cercospora* leaf spot of rice. *Sci. Cult.* **30**: 458–459.

DEIGHTON, F. C. 1967. *Sphaerulina oryzina*, the perfect state of *Cercospora oryzae*. *Trans. Br. mycol. Soc.* **50**: 499.

GANGULY, D. 1946. A note on the occurrence of *Cercospora oryzae* Miyake on paddy in Bengal. *Sci. Cult.* **11**: 573–574.

HARA, K. 1918. Diseases of the rice plant. Tokyo. [Jap.]

HARA, K. 1959. A monograph of rice diseases. 3rd ed. Tokyo. [Jap.]

JODON, N. E. 1953. Louisiana releases Sunbonnet through Seed Growers' Assn. *Rice J.* **56**: 40.

JODON, N. E. 1954. Breeding for improved varieties of rice and other cereal grains. *Rice J.* **57**: 32–36.

JODON, N. E. 1955. Toro and Sunbonnet, two new midseason, long grain rice varieties compared. *Rice J.* **58**: 8–11.

JODON, N. E., RYKER, T. C. & CHILTON, S. J. P. 1944. Inheritance of reaction to physiologic races of *Cercospora oryzae* in rice. *J. Amer. Soc. Agron.* **36**: 497–507.

JODON, N. E. & CHILTON, S. J. P. 1946. Some characters inherited independently of reaction of physiologic races of *Cercospora oryzae* in rice. *J. Amer. Soc. Agron.* **38**: 864–872.

METCALF, H. 1906. A preliminary report on the blast of rice, with notes on other rice diseases. *Bull. S. Carol. agric. Exp. Stn* 121, 43 pp.

MIYAKE, I. 1910. Studien über die Pilze der Reispflanze in Japan. *J. Coll. Agric. imp. Univ. Tokyo* **2**: 237–276.
OVERWATER, C. 1960. Tien jaren Prins Bernhard polder, 1950–1960. *Surin. Landb.* **8**: 159–218. (*Rev. appl. Mycol.* **40**: 465, 1961.)
RYKER, T. C. 1941. Linkage in rice of two resistant factors to *Cercospora oryzae*. Abs. in *Phytopathology* **31**: 19.
RYKER, T. C. 1943. Physiologic specialization in *Cercospora oryzae*. *Phytopathology* **33**: 70–74.
RYKER, T. C. 1947. New pathogenic races of *Cercospora oryzae* affecting rice. Abs. in *Phytopathology* **37**: 19–20.
RYKER, T. C. & CHILTON, S. J. P. 1942. Inheritance and linkage of factors for resistance to two physiological races of *Cercospora oryzae* in rice. *J. Amer. Soc. Agron.* **34**: 836–840.
RYKER, T. C. & COWART, L. E. 1948. Development of *Cercospora*-resistant strains of rice. Abs. in *Phytopathology* **38**: 23.
RYKER, T. C. & JODON, N. E. 1940. Inheritance of resistance to *Cercospora oryzae* in rice. *Phytopathology* **30**: 1041–1047.
SRIDHAR, R. 1970. A new collateral host of *Cercospora oryzae*. *Pl. Dis. Reptr* **54**: 272–273.
TASUGI, H. & IKENO, S. 1956. Studies on the morphology, physiology and pathogenicity of *Cercospora oryzae* Miyake, the causal fungus of narrow brown spot of rice plant. *Bull. natn. Inst. agric. Sci., Tokyo*, Ser. C, **6**: 167–178. [Jap. Engl. summ.]
TULLIS, E. C. 1937. *Cercospora oryzae* on rice in the United States. *Phytopathology* **27**: 1007–1008.

WHITE LEAF STREAK

Deighton & Shaw (1960) described this new disease from Papua and New Guinea in the South Pacific, and named the organism *Ramularia oryzae*. It was recorded also from the Solomon Islands, North Borneo (now Sabah, part of Malaysia), Nigeria and Sierra Leone, where the organism was found on *Oryza glaberrima* and *O. barthii*. The disease may be present in other countries in the South Pacific but may have been mistaken for the narrow brown leaf spot disease which has somewhat similar symptoms.

Leaf spots are visible on both surfaces of rice leaves, oblong-linear, 1–2·5 mm (rarely 3 mm) long, 0·5 mm wide, white or greyish white, surrounded by a brown, very narrow, margin. Younger lesions may show a white streak only on the upper surface while many of the streaks on the lower surface are brown all over. The lesions are usually numerous and on heavily infected leaves may be contiguous, though each remains clearly delimited by its own narrow brown margin.

Internal mycelium consists of hyaline, branched, septate hyphae, 1–2μ wide, ramifying between and penetrating and killing the cells of the mesophyll layer. There are no stromata and the external mycelium, emerging from the stomata, is hyaline, septate, branched, creeping over the leaf surface and bearing the rather distant, erect, short conidiophores, which are usually at some distance from the stomata.

Conidiophores are amphigenous, hyaline, continuous, erect, straight, 4–20μ long, 1·75–2·54μ wide, with conspicuous, flat, often slightly prominent conidial scars 0·75–1·25μ across. Several conidia may be produced in succession. Conidia are hyaline, cylindric, straight, slightly tapered at each end, the wall minutely and irregularly roughened, catenate, sometimes forming branched chains, 0–3-septate (mostly 1-septate) with a conspicuous, thickened hilum, 15–40 × 2·5–4·75μ, most frequently 20–25μ long (Fig. III–31).

The organism differs from *Cercospora oryzae* in its smaller and less septate conidia.

Raciborski (1900) reported a similar fungus from Java and named it *Napicladium janseanum* Racib.

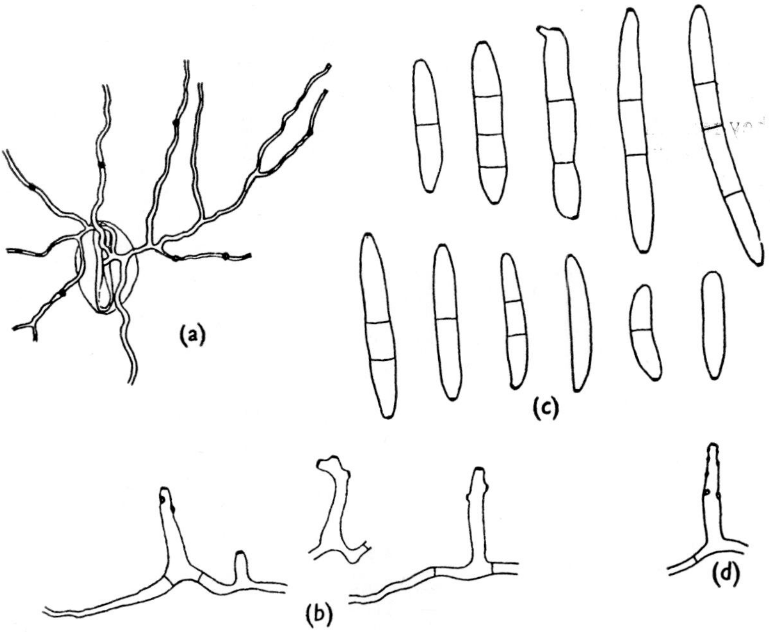

Fig. III–31. *Ramularia oryzae*. a, Diagrammatic drawing showing branching of the repent superficial hyphae (septa not shown) after emergence from a stoma: the position of the conidiophores is shown by black circles. × 580. b, Conidiophores; c, Conidia; d, A conidiophore showing 8 lateral scars. b–d × 1000. (Deighton & Shaw, 1960.)

LITERATURE CITED

DEIGHTON, F. C. & SHAW, D. 1960. White leaf streak of rice caused by *Ramularia oryzae* sp. nov. *Trans. Br. mycol. Soc.* **43**: 516–518.

RACIBORSKI, M. 1900. Parasitische Algen und Pilze Javas. II: 41, 46. (Padwick, G. W. 1950. Manual of rice diseases, p. 179.)

LEAF SMUT

This disease was first described by Butler (1913) and the causal fungus was identified by H. & P. Sydow (1914) as a new species of *Entyloma*, *E. oryzae* H. & P. Sydow. The fungus has been erroneously identified as *Ectostroma oryzae* Sawada (1912) and *Sclerotium phyllachoroides* Hara (1918) (Tullis, 1934; Asuyama, 1935).

It does not usually cause much damage but is widely distributed in rice-growing areas of the world. It has been reported from Afghanistan, Burma, China, Hong Kong, India, Indonesia, Japan, Malaysia, Philippines and Thailand in Asia; from Argentina, Colombia, Cuba, Dominican Republic, Surinam, U.S.A. and Venezuela in America; and from Egypt, France, Ghana, Northern Australia and Papua-New Guinea (see CMI Distribution Map 451).

Symptoms

The fungus causes small, leaden-black spots on both sides of the leaves. The spots are short-linear, rectangular or angular-elliptical, *c.* 0·5–5 mm long, 0·5–1·5 mm broad. Numerous spots (sori) may be found on the same leaf but they remain distinct from each other. Such heavily infected leaves may turn yellow and split. The spots are covered by the epidermis but after soaking in water for a few minutes, the epidermis ruptures, revealing the black mass of spores beneath (Fig. III–32).

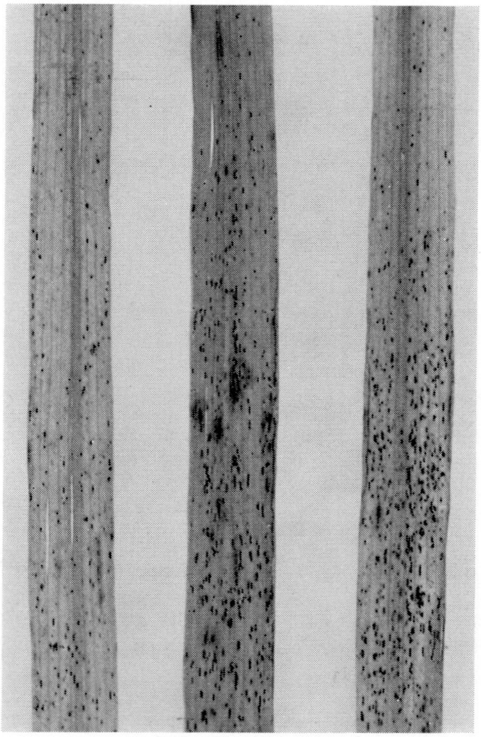

Fig. III–32. Leaf smut (Approx. × 1).

The Organism

The sorus under the epidermis appears as a carbonaceous mass as the spores are held tightly together unless soaked in water for some time. For this reason the sorus has been misinterpreted as a sclerotium. With proper preparation the closely agglutinated spores are seen as angular-globose to angular-ovate, smooth, pale brown, measuring about $6–15 \times 5–9\mu$, with epispores about $1 \cdot 5\mu$ thick.

The spores germinate by a short promycelium, $6–20 \times 5–10\mu$ and bearing 3–7 sporidia. These sporidia are elliptical or obclavate, light brown, $10–15 \times 2–2 \cdot 5\mu$ and may produce Y-shaped secondary sporidia (Kuribayashi, 1934; Worawisitthumrong, 1963) (Fig. III–33).

Zundel (1939) considered *E. oryzae* to be a synonym of *Entyloma lineatum* (Cke.) Davis, and Fischer (1953) referred to it as *E. dactylidis* (Pass.) Cif.

The best temperature for germination is 28–30°C. The fungus overwinters on diseased leaves and the next summer produces sporidia which are capable of causing infection.

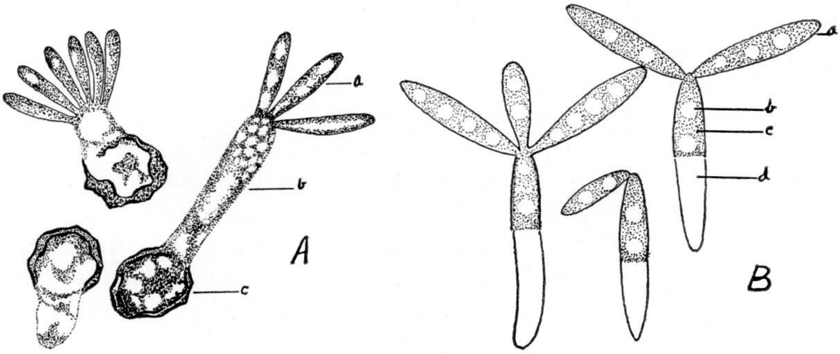

Fig. III–33. *Entyloma oryzae*. A. Germinating teliospores. a, primary sporidium. b, promycelium. c, teliospore. B. Secondary sporidia developing from tip of deciduous primary sporidia. a, secondary sporidium, b, vacuole. c, protoplasm. d, empty portion of sporidium resulting from migration of protoplasm to secondary sporidia. (Worawisitthumrong, 1963. Courtesy of Dr. G. E. Templeton.)

LITERATURE CITED

ASUYAMA, H. 1935. New diseases and pathogens reported in the year 1934 on our cultivated plants in Japan. *Ann. phytopath. Soc. Japan* **4**: 191–197. [Jap.]

BUTLER, E. J. 1913. Diseases of rice. *Bull. agric. Res. Inst. Pusa* 34, 37 pp.

FISCHER, G. W. 1953. Manual of the North American smut fungi. New York.

HARA, K. 1918. Diseases of rice. 2nd ed. Tokyo. [Jap.]

HASHIOKA, Y. 1970. Rice diseases in the world—V. Smuts (Fungal diseases 2). *Riso* **19**: 11–15.

KURIBAYASHI, K. 1934. On the germination of causal fungus of the rice leaf smut. *Ann. phytopath. Soc. Japan* **4**: 68–69. [Jap.]

SAWADA, K. 1912. Investigation of the paddy seedling decay in Formosa. *Spec. Rep. agric. Exp. Stn Formosa* 3, 84 pp. [Jap.]

SYDOW, H. & SYDOW, P. 1914. Novae fungorum species. *Ann. Mycol.* **12**: 195–204.

TULLIS, E. C. 1934. Leaf smut of rice in the United States. *Phytopathology* **24**: 1386.

WORAWISITTHUMRONG, A. 1963. Studies on the life history of *Entyloma oryzae* Sydow and Sydow. M.S. thesis, Univ. Arkansas.

ZUNDEL, G. L. 1939. Studies on the Ustilaginales of the world. IV. Smuts collected in Minas Geraes, Brazil by A. S. Muller. *Mycologia* **31**: 572–589.

STACKBURN DISEASE

This disease was first described by Godfrey (1916) from Louisiana and Texas. The fungus on infected rice leaves was present in great abundance and appeared like black rust of wheat. However, only sclerotia and mycelium were found. Later Godfrey (1920) reported the same fungus also in and on seeds, causing seedling blight in test tube experiments and also to some extent in glasshouse plantings. Tullis (1936) found the fungus on discoloured seeds, observed conidia, and referred it tentatively to *Trichoconis caudata* (App. & Str.) Clem. Ganguly (1947) studied the disease in India and called the associated fungus *T. padwickii*, differentiating it from *T. caudata* because of its longer and more rigid conidial appendage and the different host range. *T. padwickii* has for some time been the generally accepted name for this fungus, but Ellis (1971) has recently transferred the species to *Alternaria* as *A. padwickii* (Ganguly) M. B. Ellis, because of its coloured conidia and the method by which they are formed. Pavgi *et al.* (1966) have reported another species, *T. indica* Pavgi & R. A. Singh, from leaves and glumes of rice in India.

The disease was known only in U.S.A. until Padwick & Ganguly (1945) reported it from India. It is now known to occur in many other countries (see CMI Distribution Map 314) including China and most of the countries of Southeast Asia, and also Egypt, Nigeria, Madagascar and Surinam.

The damage caused by the disease usually occurs when the seeds are attacked, the seeds being spotted or discoloured. Under favourable conditions the percentage of infected seed may be high as reported by Suryanarayana *et al.* (1963) in India, and we have observed it in Southeast Asia. Heavily infected seeds result in seedling blight or weakening of the seedlings. Leaf spots usually do not cause much damage.

SYMPTOMS

Typical spots on the leaves are large, oval or circular, with dark brown, relatively narrow and distinct margins which circle the spots like a ring. The centre of the spot is at first pale brown, gradually becoming almost white and bearing minute black dots, the sclerotia. The spots vary in size, from 0·3 to 1 cm long, and sometimes are surrounded by a second ring. Usually only a few spots on a few leaves are observed in the field.

Grains infected by the fungus show pale brown to whitish spots with a dark brown border of relatively large size, on the glumes. The spots bear black dots in the centre. Several other organisms may cause similar spots on the grains and they are often difficult to separate from each other. The fungus may penetrate the glumes and invade the kernels causing discoloration, or the kernels may even become shrivelled and brittle.

The fungus also attacks both the roots and coleoptile of germinating seeds or young seedlings. The spots are dark brown to black and often coalesce to reach several mm in length. As the decay proceeds, small, discrete, black bodies are formed on the surface of the darkened area. Heavily infected seedlings eventually wither and die; those less severely affected may outgrow and recover from the disease.

CAUSAL ORGANISM

Ganguly (1947) described *T. padwickii* as follows: Leaf spots amphigenous, round to oval, 3–9 mm in diameter, with a dark brown margin and a dull grey centre; mycelium well developed, profusely branched, hyaline at young stage; mature hyphae creamy-yellow, 3·4–5·7μ thick, septate at regular intervals of 20–25μ, branches arising at right-angles to the main axis and constricted at the point of origin, the first septum being placed just near the point of origin.

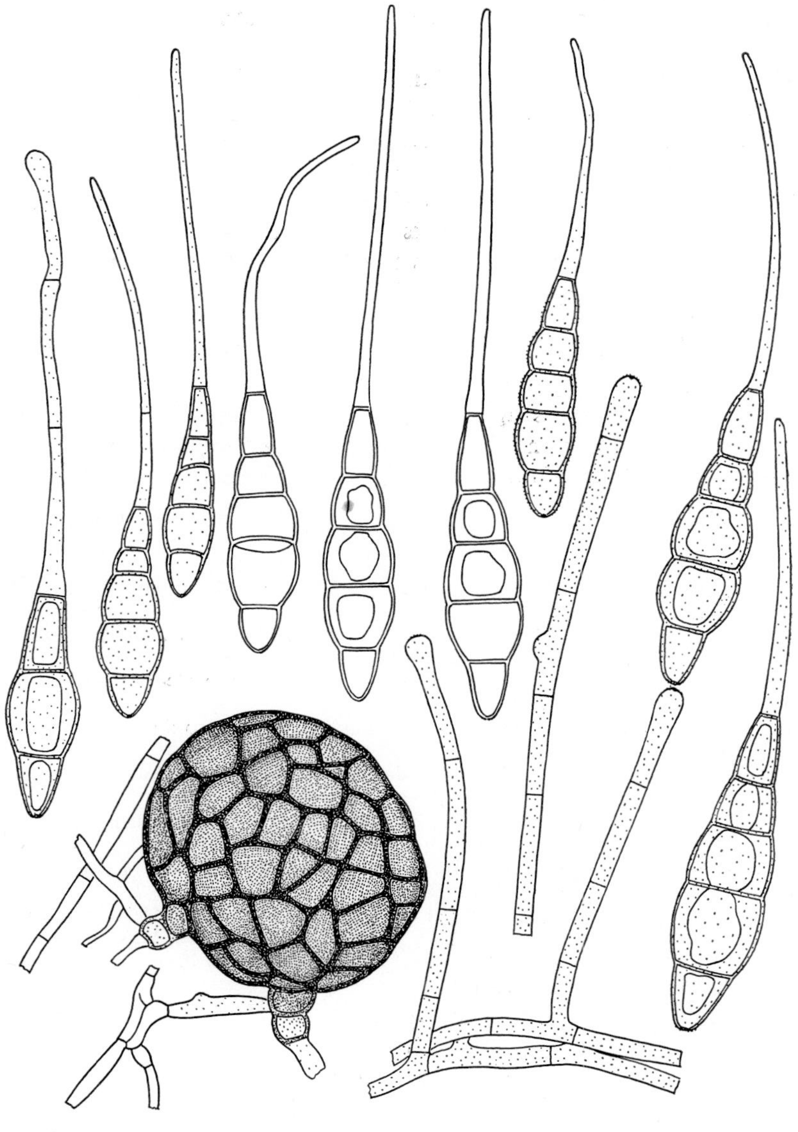

Fig. III–34. *Alternaria* (*Trichoconis*) *padwickii*. × 650. (Courtesy of Dr. M. B. Ellis.)

Sclerotia black, almost spherical, partly embedded within the host tissue, with reticulated walls and connected by fibrils, measuring 124 (52–195)μ. Conidiophores not sharply distinguishable from mature hyphae, partly erect, 100–175μ long and 3·4–5·7μ broad, apex monosporous. Conidia elongately fusoid, with a long appendage at the tip, non-deciduous, 3 to 5-septate, creamy-yellow, constricted at septa, thick-walled, straight, with second or third cell from the base larger than the rest, 103·2–172·7μ long including the appendage and 8·5–19·2μ broad; appendage at the tip of conidium is almost equally as long as the conidium proper, rigid, septate, 2–5μ thick, straight or slightly curved (Fig. III–34).

Little is known of the life cycle of the fungus. Tisdale (1922) stated that the fungus doubtless lives through the winter in the soil and on old rice straw, and infects the rice plants the following season. We have isolated the fungus from 60% of discoloured grains from Thailand and this perhaps is another important source of primary infection. Tullis (1936) indicated that the fungus gains entry through the glumes and attacks the kernel before the rice is mature. Ganguly (1947) inoculated the fungus to 7 rice varieties and found infection to be low in unwounded leaves, but high in wounded leaves, indicating that the fungus is a weak pathogen.

Padwick (1950) has seen the same fungus attacking leaves of a wild grass in paddy fields.

Special control measures against the disease are not known and are not necessary under normal conditions. Hot-water seed treatment was suggested by earlier workers, and fungicide treatments were tried by Suryanarayana *et al.* (1963).

LITERATURE CITED

Ellis, M. B. 1971. Dematiaceous Hyphomycetes: 495. Kew, Commonwealth Mycological Institute.

Ganguly, D. 1947. Studies on the stackburn disease of rice and identity of the causal organism. *J. Indian bot. Soc.* **26**: 233–239.

Godfrey, G. H. 1916. Preliminary notes on a heretofore unreported leaf disease of rice. Abs. in *Phytopathology* **6**: 97.

Godfrey, G. H. 1920. A seed-borne sclerotium and its relation to a rice seedling disease. Abs. in *Phytopathology* **10**: 342–343.

Padwick, G. W. 1950. Manual of rice diseases. Kew, Commonwealth Mycological Institute.

Padwick, G. W. & Ganguly, D. 1945. Stackburn disease of rice in Bengal. *Curr. Sci.* **14**: 328–329.

Pavgi, M. S., Singh, R. A. & Dular, R. 1966. Some parasitic fungi of rice from India. II. *Mycopath. Mycol. appl.* **30**: 314–322.

Suryanarayana, D., Nath, R. & Prabha Lal, S. 1963. Seed-borne infection of stackburn disease of rice—its extent and control. *Indian Phytopath.* **16**: 232–233.

Tisdale, W. H. 1922. Seedling blight and stackburn of rice and the hot water seed treatment. *Bull. U.S. Dep. Agric.* 1116, 11 pp.

Tullis, E. C. 1936. Fungi isolated from discolored rice kernels. *Tech. Bull. U.S. Dep. Agric.* 540, 12 pp.

LEAF SCALD

Leaf scald was first described from Japan by Hashioka and Ikegami in 1955, though it may have been found earlier. It has also been reported from Costa Rica (De Gutierrez, 1960) and Guatemala (Schieber, 1962), and is apparently common in Latin America. It was also found in Thailand by Hashioka and we noticed it in Vietnam, but it is not common in Southeast Asia.

The disease is relatively unimportant but has recently been increasing in Japan and may at times have caused some damage in Latin America.

SYMPTOMS

The disease usually occurs on mature leaves, mostly near the tips, but sometimes starts at the margin of other parts of the blade. The lesions are oblong or diamond-shaped, water-soaked blotches, more or less restricted by the veins, which develop into large ellipsoid or oblong olive areas encircled by dark brown narrow bands accompanied by light brown halos. Individual lesions are 1–5 cm long, 0·5–1 cm broad. Later, the scalded areas turn to light greyish olive. The successive bands of dark brown margins and lighter inner areas exhibit a characteristic zonation (Fig. III–35). The continuous

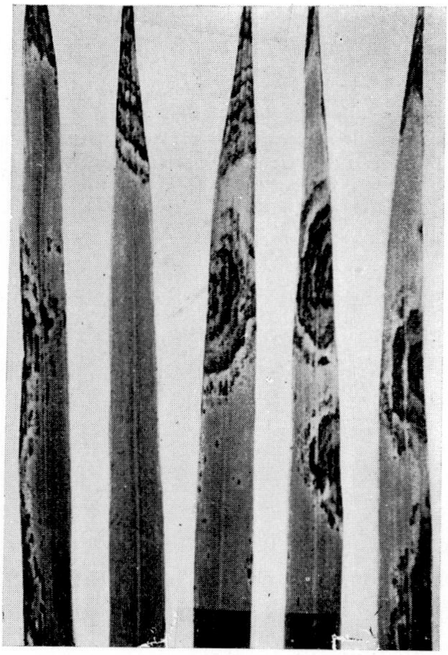

Fig. III–35. Leaf scald symptoms. (Hashioka & Ikegami, 1955.)

enlargement and coalescing of lesions may result in blight of a large part of the leaf blade, and finally badly affected leaves dry and turn to a bleached straw colour with a brown margin and faint zonation. On the leaf sheath and the young inflorescence, only browning occurs.

De Gutierrez (1960) in Costa Rica described the fungus as causing decay of the coleoptile, with red brown infection and root rot. She further reported a head blight which caused considerable sterility and flower deformation and discoloration of the glumes.

THE CAUSAL ORGANISM

The fungus is called *Rhynchosporium oryzae* Hashioka & Yokogi (1955). The conidia are borne on superficial stromata arising on lesions. They are fusiform or oblong-fusoid, tapering toward both ends, curved, 10·8–13·2× 3·7–4μ, septate near the middle, but often the two cells are somewhat unequal, rarely 2-septate, not constricted at the septum, epispore very thin (Fig. III–36). The non-beaked and slightly curved conidia are easily distinguished from those of *R. secalis* (Oud.) Davis and *R. orthosporum* Cald.

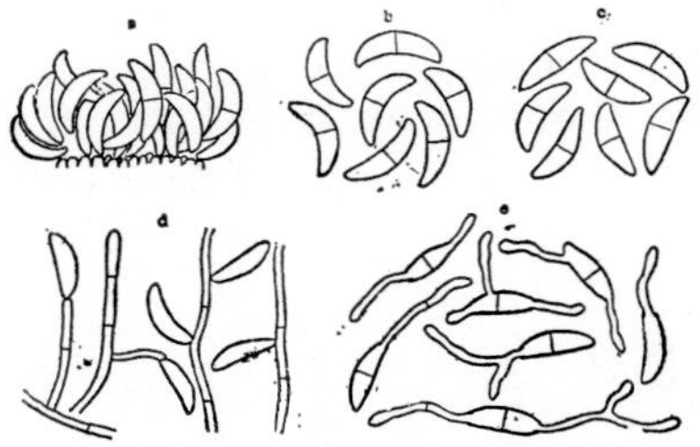

Fig. III–36. *Rhynchosporium oryzae*. a. A superficial stroma arising on a lesion; b. Conidia on a lesion; c. Conidia in culture; d. Conidial formation on hyphae in culture; e. Germination of conidia in distilled water. (Hashioka & Ikegami, 1955.)

Inoue & Takeda (1957) reported the perfect state of the fungus and suggested that it belongs to the genus *Phragmosperma*, but did not publish a name. They described the perithecia as scattered under the epidermis, globose or slightly flattened, brown, 110–140×80–120μ with papillate ostiole; asci clavate to cylindrical, 8-spored, 25–45×7–11μ; ascospores fusoid, slightly curved, 3-(rarely 4-) septate, slightly constricted, hyaline, 10–16×2·5–3·5μ.

The fungus grows poorly on synthetic media but vigorously on potato agar and special media containing vitamin B_1 (Hashioka & Makino, 1956). It grows well under a rather wide range of temperatures, 20–27°C, the optimum being 20°C. It is presumed that this disease can occur even in the cooler mountainous regions of Japan.

Varieties Blue Bonnet, Nato, Magnolia and Zenith were more severely attacked than Pandhori No. 4 and Asahi in El Salvador (Ancalmo & Davis, 1962). Increasing the nitrogen content of the soil favours the disease (Hashioka & Ikegami, 1955; Hashioka & Makino, 1956; De Gutierrez, 1960). Makino & Hashioka (1959) tested the effects of some fungicides on the germination of the conidia of the fungus.

LITERATURE CITED

ANCALMO, O. & DAVIS, W. C. 1962. Diseases of rice new to El Salvador. *Pl. Dis. Reptr* **46**: 293.

DE GUTIERREZ, L. C. 1960. Leaf scald of rice, *Rhynchosporium oryzae*, in Costa Rica. *Pl. Dis. Reptr* **44**: 294–295.

HASHIOKA, Y. & IKEGAMI, H. 1955. The leaf scald of rice. *Jubilee Publ. commem. 60th Birthdays of Profs. Y. Tochinai and T. Fukushi*: 46–51. Sapporo, Japan. [Jap. Engl. summ.]

HASHIOKA, Y. & MAKINO, M. 1956. Relation of nitrogen nutrition of rice plants to the susceptibility to four foliage diseases. *Res. Bull. Fac. Agric. Gifu Univ.* **6**: 58–66.

INOUE, Y. & TAKEDA, S. 1957. Perfect stage of the so-called rice leaf scald fungus. Abs. in *Ann. phytopath. Soc. Japan* **22**: 26. [Jap.]

MAKINO, M. & HASHIOKA, Y. 1959. Germinability and drug tolerance of conidia of the rice leaf-scald fungus (*Rhynchosporium oryzae* Hashioka et Yokogi). *Sci. Rep. Fac. Agric. Meijo Univ.* **3**: 17–21. [Jap. Engl. summ.]

SCHIEBER, E. 1962. *Rhynchosporium* leaf scald of rice in Guatemala. *Pl. Dis. Reptr* **46**: 202.

COLLAR ROT

Four species of *Ascochyta* have been reported to occur on the rice plant: *A. oryzae* Catt., *A. oryzina* Hara, *A. graminis* Sacc. and *A. miurai* Hara (Hara, 1959). In the past, all have been considered as weak parasites or as saprophytes growing on dead plant parts. Recently, however, a species of *Ascochyta* was

Fig. III–37. Collar rot symptoms.

found to cause considerable damage under suitable conditions in a small area near Bangkok, Thailand. The disease was called collar rot and the fungus identified as *A. oryzae* Catt. (Kanjanasoon, 1962).

Symptoms

The disease first appears as small brown lesions at the collar joining the leaf blade and leaf sheath, usually on the uppermost expanded leaf. The lesions expand and turn dark brown, and subsequent rotting causes the leaf blade to separate from the sheath and drop off. In severe cases, several leaves on the same plant may dry up and fall in this way (Fig. III–37).

The Causal Organism

The translation of Cattaneo's description of *A. oryzae* is as follows: *Pycnidia* always covered by the epidermis, cellular-membranaceous, black, with protruding apex, the interior filled with a mass of spores; spores linear-oblong, rounded at both ends, divided into two locules by a transverse median septum seen with difficulty, 4-nucleate, pale yellow, $15 \times 4\mu$ (translation from Padwick, 1950).

The fungus produces abundant pycnidia on sterilized rice stems (Kanjanasoon, 1963).

Hara (1959) devised the following key in an attempt to separate the 4 species of *Aschochyta* found on rice:

I. Conidia oblong or cylindrical
- A. Septation not distinct, oil drops 4, $15 \times 4\mu$ — *A. oryzae*
- B. Septation distinct, oil drops 2, $16–21 \times 4–5\mu$ — *A. oryzina*

II. Conidia fusoid
- A. Conidia $10–12 \times 3–5\mu$ — *A. graminis*
- B. Conidia $6–9 \times 2{\cdot}5–3\mu$ — *A. miurai*

LITERATURE CITED

Hara, K. 1959. A monograph of rice diseases. 3rd ed. Tokyo, Japan. [Jap.]

Kanjanasoon, P. 1962. A note on a new leaf disease of rice in Thailand. *Int. Rice Commn Newsl.* **11**(3): 22.

Kanjanasoon, P. 1963. Pycnidia formation of the collar rot fungus of rice on different media. *Int. Rice Commn Newsl.* **12**(2): 8–10.

Padwick, G. W. 1950. Manual of rice diseases. 198 pp., Kew, Commonwealth Mycological Institute.

RUSTS

Two rust species have been recorded on rice: *Puccinia graminis* Pers. f. sp. *oryzae* Fragoso and *Uromyces coronatus* Yosh.

P. graminis f. sp. *oryzae* was first named *P. oryzae* by Briosi, but without a technical description, and as *P. graminis* by Sydow. Diehl (1944) reviewed the literature and considered *P. graminis* Pers. f. sp. *oryzae* Fragoso to be a valid name. The fungus has been reported on rice from Italy, Spain and

Rumania. Its morphology is apparently similar to that of *P. graminis* on other cereals.

Uromyces coronatus was reported by Liou & Wang (1935) from Chekiang, China, on rice and *Zizania aquatica* L. The description is: Uredia amphigenous, minute, scattered, surrounded by epidermis, powdery; urediospores ovate, oblong or elliptic, finely echinulate, 25–35 × 12–24μ; paraphyses clavate, intermixed. Telia amphigenous, also on sheath, short-oblong, rounded, sometimes linear, surrounded by epidermis, black; teliospores obovate, oblong or ellipsoidal, rounded and cornate at the apex, attenuated to the pedicel, brownish, 25–35 × 16–22μ; wall smooth, 1·5–2μ, pedicel 38μ long.

Neither of the above rusts is of economic importance on rice.

LITERATURE CITED

DIEHL, W. W. 1944. Bibliography and nomenclature of *Puccinia oryzae*. *Phytopathology* **34**: 441–442.

LIOU, T. N. & WANG, Y. C. 1935. Materials for study of rusts of China. V. *Contr. Inst. Bot. natn. Acad. Peiping* **3**: 433–451.

PART IV

FUNGUS DISEASES—DISEASES OF STEM, LEAF SHEATH AND ROOT

STEM ROT

History

The rice stem rot fungus was first described from Italy by Cattaneo (1876) in the sclerotial state and was named *Sclerotium oryzae* Catt. In the same paper, under Pyrenomycetes, he also described *Leptosphaeria salvinii* Catt. from rice stubble. Cavara (1889) later reported *Helminthosporium sigmoideum* Cav. on rice, also in Italy. The genetical relations of the three were not then known.

The disease was reported from several countries in Asia and from U.S.A. during the 1910s and 1920s, the causal organism being known as *S. oryzae*.

Tullis (1932) found conidia and conidiophores of *Helminthosporium sigmoideum* Cav. on agar cultures of *S. oryzae*. They were also found on rice seedlings grown aseptically in test tubes and inoculated with a pure culture of *S. oryzae*. These conidia similarly infected rice seedlings and characteristic sclerotia of *S. oryzae* were produced. Tullis (1933a, b) further demonstrated the genetic connections among *L. salvinii*, *H. sigmoideum* and *S. oryzae* by cultural and inoculation experiments, and showed that they are the three states of the same fungus. The disease was found to cause considerable losses and was studied in detail in U.S.A., particularly in respect to the causal organism and varietal resistance.

Cralley & Tullis (1934) described a new organism, *Helminthosporium sigmoideum* Cav. var. *irregulare* Cralley & Tullis, very similar to *S. oryzae* and differing mainly in its irregular sclerotia. The A-strain of *S. oryzae* reported by Park & Bertus (1934) and Sakurai's (1917) *Sclerotium* No. 3 were probably the same organism.

Nakata & Kawamura (1939) studied the sclerotial diseases of rice in detail, including both forms of the stem rot fungus, and during the 1950s many studies were made in Japan concerning the ecology of stem rot.

The two organisms have similar general morphology, cause similar diseases and often occur together in the same field. They are treated as one disease by many workers though there are differences between them as will be noted below. It may be noted that, although the perfect state (*Leptosphaeria salvinii* Catt.) of *H. sigmoideum* has been found, no such state of *H. sigmoideum* var. *irregulare* is known.

Distribution

Stem rot was reported in Italy in 1876, in Japan in 1910 (Miyake), in India in 1913 (Shaw) and 1918 (Butler), in Ceylon in 1920 (Bryce), in Vietnam in

1921 (Vincens), in U.S.A. in 1921 (Tisdale), and in the Philippines in 1924 (Anon.). Since then the disease has also been reported from most of the other rice-growing countries including Bulgaria in Europe, Kenya, Madagascar and Mozambique in Africa, and Brazil, Colombia, Guiana etc. in Latin America. It has been found in all the countries of Southeast Asia and occurs in almost every field where rice has been grown for many years.

Ono & Suzuki (1960) reported that *H. sigmoideum* was present in poorly drained land while *H. sigmoideum* var. *irregulare* occurred in well drained soil in Japan. Lo & Hsieh (1964) found the former abundant in the second rice crop and the latter in the first crop in Taiwan.

The two organisms often occur together in the same field. A recent survey in the Philippines (Hsieh, 1966) showed that among the 299 samples collected from different parts of the country, about 64% of the samples contained both organisms, 17% *H. sigmoideum* alone and 19% *H. sigmoideum* var. *irregulare* alone.

Damage

The disease was reported to cause heavy damage to the rice crop in the early literature. Miyake (1910) stated that in severe cases the attacked plants became weak, fell to the ground, the formation of rice grains was incomplete and the loss was very great. Butler (1918) stated that stem rot is probably responsible in the aggregate for considerable losses, but is easily overlooked owing to the obscurity of the symptoms. Vincens (1921) reported that *S. oryzae* was one of the commonest fungi on rice in Vietnam. In one instance, the loss in yield was estimated at 50%, and in another at even more. Rhind (1924) reported the disease to be widely distributed in Burma, where losses were great. In the Philippines, losses have been estimated at from 30 to 80% in Tarlac Province (Hernandez, 1923) and in Arkansas, U.S.A. (Anon., 1930), at as high as 75% in certain fields, with an annual average of *c.* 10% in severely infested areas. Chauhan *et al.* (1968) reported 18·24–56·43% loss in yield in variety CO.13 and 0·01–39·74% in Chakia 59 in India. In recent years in Japan, the area showing considerable damage was reported to be 51,000–122,000 hectares and the annual loss in yield was estimated to be 16,000–35,000 tons.

The losses due to the disease are difficult to assess. It causes decay of the leaf sheath and culm, which contributes to lodging, besides causing loss in yield and lowering of milling quality due to the light and chalky character of the grain. On the other hand, Corbetta (1953) reported *H. sigmoideum* to be endemic and virtually ubiquitous in Italy but seldom causing appreciable damage there unless the infected plants were damaged by adverse weather conditions. Observations in the Philippines and other countries of Southeast Asia also reveal the enormous quantities of the causal organisms in rice fields. In the later stages of growth, most of the stems of the rice plants are covered with numerous appressoria or other structures of the fungi. These stems are, however, seldom infected before harvest, although the fungi are present in most of the stubble after the rice harvest. In a survey in which 120,000 tillers from stubble from around 24 townships in the Philippines were examined,

67% of the tillers were found to bear at least one of the stem rot organisms and all the 299 samples were infected to some extent. This type of phenomenon has also been observed in other countries. Recently, Hsieh (1966) reported the importance of the presence of wounds on plants in the development of the disease. Ike & Watanabe (1953) found that artificially induced lodging accelerated the damage caused by the disease and varieties naturally subject to lodging showed low resistance to stem rot. These facts seem to indicate that the severity of the disease may depend largely upon contributing factors, such as lodging or other types of wounds to the plant, rather than upon the presence of the causal organism alone.

SYMPTOMS

The disease usually begins to appear in the field during the later stages of growth of the rice plant. It starts with a small, blackish, irregular lesion on the outer leaf sheath near the water line (Fig. IV–1). Sometimes the sclerotium

Fig. IV–1. Stem rot—young lesions on leaf sheath. (Ono & Suzuki, 1960.)

which initiated the lesion may at this stage still be attached to the affected part. The lesion enlarges as the disease progresses, the fungus penetrates into the inner leaf sheath, and finally the leaf sheath is partially or entirely rotted, when many sclerotia are formed. Soon after the fungus reaches the inner leaf sheath, it comes into contact with the culm (or stem). Numerous appressoria or patches of infection cushions (in the case of *H. sigmoideum*, see below) are formed on the culm. Brownish-black lesions appear and finally one or two internodes of the stem rot and collapse, only the epidermis remaining intact. As might be expected, such infected stems lodge. On splitting infected internodes, dark greyish mycelium may be found within the hollow stem and small black sclerotia can be seen to be dotted all over the inner surface (Fig. IV–2). The next lower internode may, however, be completely free from any sign of the organisms. The presence of the characteristic sclerotia is usually a positive and easy way of diagnosing the disease.

Attack on the stems increases in intensity as the plants approach maturity,

Fig. IV–2. Numerous small sclerotia formed inside the stubble of a rice plant infected with stem rot disease (× 3).

and reaches its peak at harvest time. Early infected plants yield poorly. In the tropics, where sufficient moisture is often present after harvest, the fungi grow further and more sclerotia are formed.

Shaw (1913) in India and Park & Bertus (1932) in Ceylon reported that infected plants produce excessive numbers of tillers from the base of the stems. This has not been observed elsewhere. Such tillering was probably due to the prevailing wet, tropical conditions and to the presence of a rice variety able to produce tillers from the lower uninfected nodes.

Occasionally the disease appears on young transplanted plants in the tropics. The outer leaf sheaths are rotted at the water line, killing the leaves above, while the lower parts of the sheaths remain firm and healthy for some time.

The Organism

Morphology. Tullis (1933b) and Cralley & Tullis (1935) studied some of the authentic and other materials and described the various states of the two organisms in detail. The measurements of the various structures given by different authors, using different materials, vary somewhat. No measurements

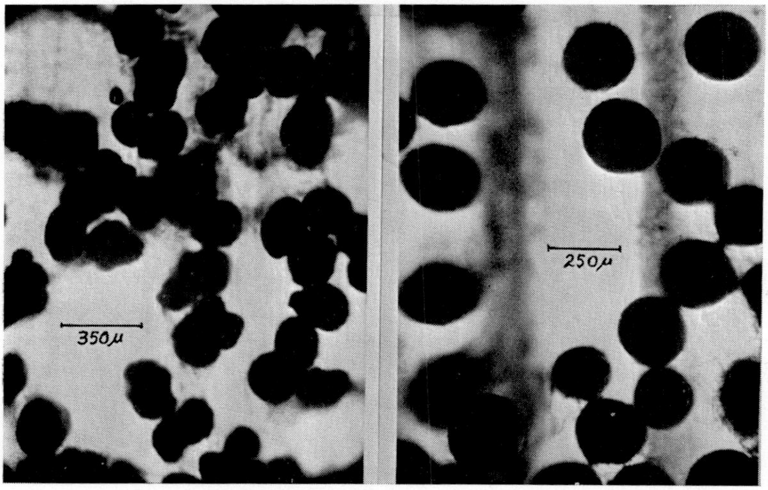

Fig. IV–3. Relative size and shape of the sclerotia of *Leptosphaeria salvinii* (right) and *Helminthosporium sigmoideum* var. *irregulare* (left).

were given by Cattaneo (1876) in the original description of *S. oryzae*. For *L. salvinii*, Cattaneo (1876) gave measurements of 350–400μ for the perithecia; asci, 120μ long; spores, 60μ long, 9μ broad. The original description by Cavara (1889) of *H. sigmoideum* gives the measurements: conidiophores, 100–150 × 5μ; conidia 55–65 × 11–14μ. Park & Bertus (1932) found the sclerotia of *H. sigmoideum* to be 175–580μ in diameter. The descriptions of Tullis (1933b) and Cralley & Tullis (1935) are as follows:

For *Leptosphaeria salvinii* Catt.: *Mycelium*—Hyphae white to olivaceous, septate, profusely branched, 2–5μ in diameter. In culture, mycelium white at first, later smoky to black at surface of medium. On host, mycelium white inside culm, olivaceous outside. Numerous irregular olivaceous appressoria form on culm; range 14–30 × 8–24μ.

Sclerotial state (*Sclerotium oryzae*)—Sclerotia spherical or nearly so, black at maturity, surface nearly smooth, at times covered with cottony weft of white mycelium; 180–280μ, mostly 230–270μ (Fig. IV–3).

Conidial state (*Helminthosporium sigmoideum*)—Conidiophores dark-coloured, septate, erect, simple or sparsely branched, 4–5 × 100–175μ; conidia borne singly on sharp pointed sterigmata, fusiform, typically 3-septate, simply curved or slightly sinuous; intercalary cells Prout's brown, densely granular; terminal cells lime green, less granular than intercalary cells; apical cell frequently longer and less acutely pointed than basal cell; spores occasionally

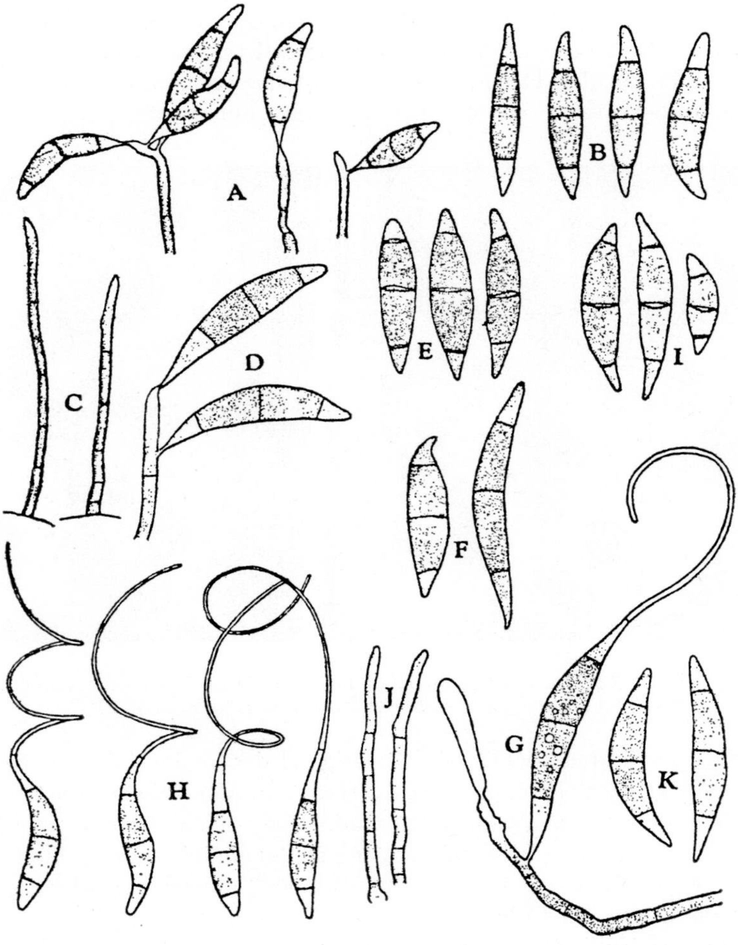

Fig. IV–4. Conidiophores and conidia of *Leptosphaeria salvinii* (A to F) and *Helminthosporium sigmoideum* var. *irregulare* (G to K). (Nakata & Kawamura, 1939.)

constricted at middle septum, 9·9–14·2×29–49μ, mostly 11–12·5×34–40μ (Fig. IV–4).

Ascigerous state (*L. salvinii*)—Perithecia dark, globose, embedded in outer tissues of sheath, 202–481μ, average diameter 381μ, beak rather short (30–70μ), frequently half the diameter of perithecium in width. Beak nonprotruding, tip flush with surface of outer epidermis of sheath, visible to naked eye; asci narrowly clavate, walls almost invisible and deliquescing by the time spores mature, short stalked, 90–128×12–14μ, mostly 103–125×13·5μ; ascospores biseriate, normally eight in ascus (rarely only four), three-septate when mature, usually somewhat constricted at septa, particularly at middle septum, brown, two end cells usually lighter in color and contents less granular than middle cells, fusiform, somewhat curved, 38–53×7–8μ, mostly 44×8μ (Fig. VI–10, F).

For *Helminthosporium sigmoideum* Cav. var. *irregulare* Cralley & Tullis: *Mycelium*—Hyphae white to olivaceous, septate, profusely branched, 2–5μ in diameter. In culture, aerial mycelium usually scant, submerged mycelium dark. On host, mycelium scanty. Numerous appressoria produced at times.

Sclerotial state—Sclerotia very numerous, irregular in outline, black, surface rough, 90–119×268–342μ (Fig. IV–3).

Conidial state—Conidiophores dark colored, septate, erect, simple, 4–5× 75–200μ; conidia borne singly on sharp pointed sterigmata, fusiform, typically three-septate, frequently with germ tubes (appendages) 2 or 3 times the length of the spore on spores still attached to the conidiophore, intercalary cells densely granular, terminal cells less granular than intercalary cells, 9–12× 41–58μ (Fig. V–4).

The major differences in morphology and cultural characteristics of the two organisms are that *H. sigmoideum* produces spherical, shiny, smooth, slightly larger sclerotia, mostly on the surface of the agar or plant tissue while *H. sigmoideum* var. *irregulare* has irregular, dull, rough, smaller sclerotia, mostly embedded in the agar or plant tissue. The former has more aerial mycelium and produces pigments in the medium but not the latter. The conidia of *H. sigmoideum* var. *irregulare* often have a long appendage or beak or germ tube and are not always 4-celled (Fig. IV–3, 4).

Taxonomy. Placing the conidial states of the two fungi in *Helminthosporium* Link ex Fr. has long been considered illogical. Hara (1936) created a new genus *Nakataea* and called the two organisms *N. sigmoideum* (Cav.) Hara and *N. irregulare* Hara. Subramanian (1956) created another new genus, *Vakrabeeja*, for *V. sigmoidea* (Cav.) Subram. Shoemaker (1959) advocated the splitting of the genus *Helminthosporium* (for which he preferred the spelling *Helmisporium*) and made a new combination for *V. sigmoidea* (Cav.) Subram. var. *irregulare* (Cralley and Tullis) Shoemaker. However, these names have not yet been widely adopted.

Physiology. The minimal, optimal and maximal temperatures for mycelial growth of *L. salvinii* were found to be 11–15°C, 27·5–30°C and 35°C, respectively. Best growth occurred at pH 4·05–6·1 (Anon., 1930; Ono & Suzuki, 1960).

Conidial germination occurs at 16–38°C, the optimum being 28°C, and at pH 3·5–8, optimum 5 (Togashi, 1955).

Appressoria are formed at 10–40°C, optimum 25°C (Ono & Suzuki, 1960). Hsieh (1966) reported that both appressoria and infection cushions are formed at 8–32°C, optimum 24–28°C.

The nutritional requirements of the two fungi have been studied by Misawa & Kato (1955a), Nonaka (1959), Sethunathan (1963, 1964), Hsieh (1966) and Jain (1967). The fungi use both organic and inorganic nitrogen including NH_4, NO_3, aspartic, glutamic and other amino acids and amines. NH_4 is absorbed before NO_3 and NO_2 is not used. Maltose and soluble starch are the best carbon sources while alcohols and organic acids are not utilized. Growth in mixed carbon sources is either additive (sucrose+maltose), synergistic (maltose+dulcitol) or inhibitory (maltose+rhamnose, sucrose+dulcitol). Jain (1967) found differential responses to carbon sources in *L. salvinii* and *H. sigmoideum* var. *irregulare*. For instance, sucrose supported good growth of both species while maltose supported good growth of the former but poor of the latter, and sorbitol, the reverse. Sadasivan & Subramanian (1954) considered the fungus heterotrophic for thiamine but autotrophic for biotin. Misawa & Kato (1955b) reported, however, that both are essential for growth. Sethunathan (1963, 1964) reported that at pH 4–8, unbuffered medium was favourable for development, buffering reducing growth and preventing sclerotia formation. The formation of sclerotia is often, but not always, correlated with mycelial growth, and is affected by carbon, nitrogen and other nutrient sources. Misawa & Kato (1960) reported revival of sclerotial production by adding rice plant decoction to culture media.

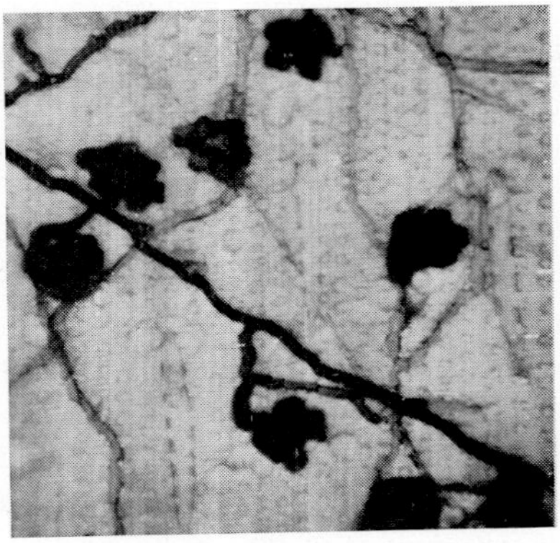

Fig. IV–5. Appressoria of *Helminthosporium sigmoideum* var. *irregulare* (× 400). (Hsieh, 1966.)

L. salvinii produces a yellow, orange or red pigment in the medium, while *H. sigmoideum* var. *irregulare* does not. The intensity of the pigment depends upon the conditions under which it is grown, being more intense at high temperature (25–30°C), high sugar content, pH 6–7·5, etc.

Sclerotia produce conidia while floating on water, on moist cotton, sand, or filter paper, but not when attached to rice stems or leaf sheaths. A sclerotium may germinate to produce conidia several times.

H. sigmoideum var. *irregulare* produces appressoria prior to infection, while *L. salvinii* produces both appressoria and infection cushions (Ono, 1953) (Figs. IV–5, 6). Appressoria are formed abundantly on leaf surfaces of many species of plants, but not on the surface of dead materials such as filter paper or cotton. Infection cushions are formed only on green stems and leaf sheaths

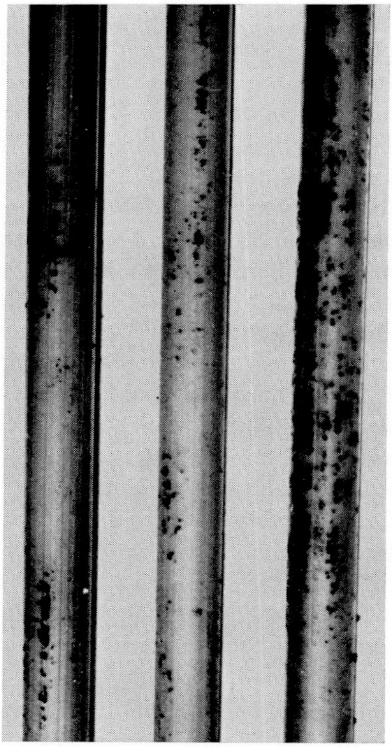

Fig. IV–6. Infection cushions of *Helminthosporium sigmoideum* on naked stems (× 2). (Hsieh, 1966.)

of rice and several other species of grass but not on dead tissues of these plants. The majority (85–90%) of the infection cushions are formed on the line of cells with stomata on leaves or leaf sheaths; this is perhaps due to stimulation by volatile substances from the stomata. The formation of appressoria and infection cushions is increased by high levels of nitrogen in the soil and is also affected by other environmental conditions (Ono, 1960; Hsieh, 1966).

Disease Cycle

The sclerotia of *L. salvinii* can survive the winter and other unfavourable conditions for a long time. Park & Bertus (1932) found them viable after

190 days on air-dry soil in the laboratory, 133 days buried in moist paddy soil, 319 days submerged in tap water, 525 days in a corked specimen tube or exposed to the sun or air-dried soil for 133 hours over a period of 26 days with temperatures from 47° to 60·8°C. Nisikado & Hirata (1937) found that sclerotia remained viable for 3 years at 20°C, 10–13 months at 25°C and 4 months at 35°C; submerged in water, even at 30°C, they survived for a year. Tullis & Cralley (1941) reported that when straw containing sclerotia was buried 4–6 inches deep in soil in Arkansas, a small percentage of viable sclerotia could be recovered after as long as 6 years.

In the rice field, the sclerotia are mostly distributed in the upper 2–3 inches of soil (Kawai, 1955). These sclerotia float on the surface of the water during ploughing, puddling, weeding and other field operations and the number of sclerotia near the leaf sheaths increases until the last weeding. These floating sclerotia easily come in contact with rice leaf sheaths and germinate to form appressoria or infection cushions, thereby causing infection. *L. salvinii* usually produces fewer appressoria than *H. sigmoideum* var. *irregulare*.

Pathogenicity and mode of infection have not been studied critically. Shaw (1913), Butler (1918), Park & Bertus (1932) and Tullis (1933b) successfully inoculated the fungi on to seedlings grown in test tubes, using young mycelium or suspensions of conidia or sclerotia. On older plants under more natural conditions, however, Park & Bertus (1932) reported that at the age of 6 weeks, it required 6 months for the outer sheaths to be killed (or perhaps die naturally). Su (1931) inoculated plants by placing sclerotia on the surface of the water, but produced no symptoms. The fungus was found in abundance on healthy mature plants and was also found floating in the water without causing any disease. Mundkur (1935) repeatedly inoculated plants in the field over a period of 3 years but was unable to induce any symptoms of stem rot. Luthra & Sattar (1936) obtained infection by sowing rice seeds or transplanting seedlings into soil in which there was diseased rice stubble. Kawamura (1941) successfully inoculated plants in the field by floating sclerotia on water and suggested the method for large scale field inoculation.

Recently, Hsieh (1966) reported the significance of the presence of wounds in the infection and development of the disease. The results of his inoculation experiments, which involved stems with or without wounds and with or without leaf sheaths, using mycelium mats of the two organisms, are shown in Table IV–1.

When the plants are not wounded, the stem rot fungi cause very low percentages of infection and lesions are of limited size even on leaf sheaths. If the plants are wounded, on the other hand, infection takes place readily and the leaf sheaths and internodes rot completely in susceptible varieties in 10 days. Wounding of the plants caused by lodging, insect attack or other factors directly affects the incidence of the disease in the field. Thus Ike & Watanabe (1953) reported that artificial lodging increases disease incidence and Kobari (1961) reported a positive correlation between stem borer incidence and stem rot. Our observations show that numerous appressoria and sclerotia may be found on plants and dead leaf sheaths in the field, but there are nevertheless few infections and few diseased plants. The stem rot fungi may therefore be considered chiefly as aggressive wound parasites.

Once infected, the tissues eventually die; when rotted numerous sclerotia are produced on them. These sclerotia produce conidia in the field. Nonaka & Yoshii (1956) found that the conidia were mostly produced during the latter part of the growing season. The air-borne conidia were mostly caught by spore traps placed at 10 cm above the ground, relatively few at 30 cm. Inoue (1961), however, considered these conidia to be relatively unimportant in causing infection. The sclerotia are carried by irrigation water from one field to another.

TABLE IV–1

PERCENTAGE OF INFECTION AND LESION DEVELOPMENT ON WOUNDED AND UNWOUNDED STEMS, WITH AND WITHOUT LEAF SHEATH, INOCULATED WITH STEM ROT FUNGI. (Hsieh, 1966).

TREATMENT	H. SIGMOIDEUM Sheathed stem		H. SIGMOIDEUM Unsheathed stem	H. SIGMOIDEUM VAR. IRREGULARE Sheathed stem		H. SIGMOIDEUM VAR. IRREGULARE Unsheathed stem
	On leaf sheath	On stem		On leaf sheath	On stem	
Not Wounded						
% infected[a]	12·0	0	2·5	13·5	0	2·0
Lesion development	5·4[b]	0	5·8[b]	5·6[b]	0	5·4[b]
Wounded						
% infected[a]	98·5	73·5	95·0	96·0	68·5	94·5
Lesion development	123·2[b]	36·3[c]	43·5[b]	115·6[b]	33·4[c]	39·6[b]

[a] Results from 400 tillers
[b] Length of lesion in mm, average of 200 tillers
[c] Disease index, average of 200 tillers

Misawa & Kato (1962) found the parenchymatous tissue in stem lesions to be disorganized, the lignified tissue (vascular bundles) becoming separated from the epidermis. Pectic substances in the middle lamella showed a change in staining reaction and sometimes had been digested, and viscosity decreased, suggesting that the pathogens showed enzymic action. They also found that pectic enzymes and cellulases were produced on a synthetic medium.

EFFECT OF ENVIRONMENTAL CONDITIONS

The nutritional conditions under which rice plants are grown have been reported to affect disease development. In experiments both in the glasshouse and in the field, Cralley (1939) found seedlings grown in culture solutions with high nitrogen to be very highly susceptible, while low nitrogen reduced susceptibility to a minimum, though the plants developed poorly. High potassium very greatly diminished the harmful effect of excessive nitrogen and plants developed well. Field test results were similar to those in the glasshouse. The effects of phosphorus application were similar to those of nitrogen but much less marked. Nitrogen and phosphorus together increased both disease incidence and yield more than did either nutrient alone. When sufficient potash was applied, together with nitrogen and phosphate, a low level of disease was maintained. The beneficial effect of potash in reducing the disease has also been reported by others (Adair & Cralley, 1950; Okamoto, 1950b; Orsenigo, 1955; Baldacci, 1955; Ono & Suzuki, 1960; Templeton, 1962).

Addition of sodium silicate to the culture solution (250–500 ppm) or to the soil (500 ppm) decreased the severity of the disease (Nonaka, Iwata & Yoshii, 1958). They found that the addition of silica resulted in a decrease in soluble nitrogen and an increase in the carbohydrate-soluble nitrogen ratio, and this was believed to be the reason for the increased resistance of the rice.

Removal of leaves increases stem rot severity, but removal of lower leaf sheaths decreases it (Nonaka & Yoshii, 1956, 1958).

Ike & Watanabe (1953) reported that lodging in the early as well as the late stages of growth of the rice plant accelerated the damage caused by the disease in certain varieties, and that varieties normally damaged by lodging generally show low resistance to stem rot under normal conditions.

Kobari (1961) found that disease incidence was two or three times higher in plants infested by stem borers than in plants free from these insects.

The age of the plant was found to be correlated with susceptibility in the later stages of growth (Nonaka & Yoshii, 1958a; Hsieh, 1966).

Varietal Resistance

Differences in varietal reaction to the disease have been observed by many workers. Tisdale (1921) found Early Prolific to be very susceptible, while Japanese varieties were apparently less susceptible in U.S.A. Hector (1931) stated that the Bengal variety Dudshar was apparently immune. Park & Bertus (1932) observed that Hmawbi 37 was very severely attacked while Hondurawela was free from the disease. Reyes (1929) found that the varieties Ramay, Elonelon, Arabon, Kinaturay and some others had promising degrees of resistance in the Philippines. He later (1936) developed the variety Raminad Str. 3, from a cross between Ramay (resistant) and Inadhica (susceptible) which was resistant to stem rot. Cralley (1936) tested 125 varieties for 4 years and reported that short-grain early maturing varieties are less susceptible and that the degree of resistance is a varietal character, only indirectly related to time of maturity. Among the more resistant varieties are: Bozu, Onsen, Akage, Spain Jap., Mubo Aikoku 22, Nakate Shinriki, Iyo Benkei, Aikoku, Mubo Aikoku, Mizuno Nishiki, Colusa, Butte, Asahi Mochi, Kameji and Fukui Shioda. Early Prolific, Early Blue Rose, Storm Proof and Nira are very susceptible. Ono (1949) also tested various groups of rice varieties and found considerable differences in reaction. Adair & Cralley (1950) evaluated the reaction of some commercial varieties to various diseases including stem rot and found no variety to be highly resistant, although Caloro, Asahi, Zenith, Arkrose and Magnolia were among the less susceptible ones. Goto & Fukatsu (1954) reported earlier maturing varieties to be more susceptible than later ones. Variety Gimbozu is remarkably resistant, strains of Shinriki and Sembon are usually resistant, while Aikoku, Kameji and Asahi are susceptible. Ono & Suzuki (1960) inoculated both *L. salvinii* and *H. sigmoideum* var. *irregulare* on a number of varieties and found some varieties to be more resistant or susceptible to one fungus than to the other. Hsieh (1966) and IRRI (1966) tested many varieties with similar results.

In the past, varietal resistance was tested either in a naturally infested field or by artificial inoculation with sclerotia, or by a combination of both methods. In view of the recent information that the presence of wounds on plants greatly affects infection by the pathogens and development of the disease, it appears that such testing methods may not produce uniform results. Besides, the necessary experiments usually take the entire growing season. A more efficient testing method has therefore been suggested (IRRI, 1966). The method uses 10 cut stems each about 30 cm long and taken from the lower part of a plant of each variety at the time of flowering. The leaf sheaths are removed and the internodes wounded by needle punctures. A small piece of the fungus from an agar culture is placed on the wounds and held in position by small pieces of scotch tape. The entire cut stems are covered by a plastic bag, placed over a pan of shallow water to maintain high humidity, and held at 28°C. Reading of reactions is made 10 days after inoculation. On resistant varieties small lesions (discoloration) appear around the wounds, whereas on susceptible varieties the entire internode rots and numerous sclerotia form inside it. Intermediate varieties produce various lengths of lesion, indicating different degrees of resistance. The method measures only the true resistance of a variety, as shown by lesion development. If the plants are more resistant to lodging or to wounding by natural causes or if there are more leaf sheaths protecting the stem, the plants will suffer less than the method indicates.

There are several scales for classifying degrees of resistance to stem rot. Cralley (1936) classified plants into four groups: (1) with diseased sheath only; (2) with mildly diseased culm; (3) with moderately diseased culm; and (4) with severely diseased culm. The disease index of a variety was calculated by adding together the number of plants in each group times the index number of each group, 1, 2, 3 and 4 respectively. Finally the varieties were classified into 3 groups, resistant, moderate and susceptible. Ono (1949), Yoshii, Koba & Watanabe (1949), Goto & Fukatsu (1954) and Inoue (1961) have used a more or less similar classification of plants as follows: (1) with lesion on leaf sheath; (2) with sclerotia in sheath; (3) with lesion on culm; and (4) with sclerotia in culm. Each class is given an index number from 1 to 10 (by Ono), 1 to 200 (Yoshii *et al.*), 5 to 57 (Inoue) and a disease index calculated. In the case of the cut stem wound inoculation method suggested by IRRI, the disease scale is divided into 10 units (0 and 1–9). The average lesion length in cm more or less directly fits into this scale, as on the most susceptible varieties the lesions reach about 9 cm (IRRI, 1967).

Very little is known about the pathogenic strains and variability which may exist in the stem rot fungi, as in most pathogenic fungi. The nature of resistance to stem rot is also obscure except that, as reported by Nonaka & Yoshii (1958b), carbohydrates decreased and nitrogen compounds increased in infected culms. Watanabe (1952) found a high correlation between the lowering of the physiological functions of the culm base and damage caused by the disease.

HOST RANGE

Tullis & Cralley (1933) reported that *L. salvinii* occurs on *Zizaniopsis miliacea* and *Echinochloa colonum* Link. Hsieh (1966) made wound inoculations on 39 species of grass weeds and found that *Eleusine indica* (L.) Gaertn., *Leptochloa chinensis* (L.) Nees and *Setaria pallide-fusca* Stapf & Hubb. were infected by both stem rot fungi. In Japan the disease has been found on other plants in nature and the fungi may invade several gramineous and cyperaceous plants when artificially inoculated on to them (Nakata & Kawamura, 1939). In the Philippines, preliminary experiments at IRRI (1966) showed that an organism found on *Zizania latifolia*, very similar to the stem rot fungi, failed to infect 8 rice varieties by wound inoculation.

CONTROL

Burning of the rice stubble after harvest has been suggested by many workers since the early days of stem rot investigations. Paracer & Luthra (1944) reported excellent results in the Punjab, India and the method has also been recommended in Ceylon, Philippines, U.S.A. (Tullis, Wood & Carter, 1933) and other areas.

Draining of the water and allowing the soil to crack before irrigating again has been reported to be useful and has been practised in some areas; but Cralley & Adair (1943) found that it did not help in increasing the yield.

The proper use of fertilizers will help in reducing the damage. As mentioned above, excess nitrogen tends to increase disease incidence and potassium tends to reduce the damage.

Fungicides have been tried in Japan (Hori & Iizuka, 1951; Shioyama, 1964) but are not generally recommended.

The use of resistant, nonlodging varieties is perhaps the only satisfactory basic approach to the solution of the problem.

LITERATURE CITED

ADAIR, C. R. & CRALLEY, E. M. 1950. 1949 rice yield and disease control tests. *Rep. Ser. Ark. agric. Exp. Stn* 15, 20 pp.

ANON. 1930. Plant Pathology. *Bull. Ark. agric. Exp. Stn* 257 (Ann. Rep. 42): 72–85.

BALDACCI, E. 1955. Il potassio e le malattie crittogamiche del riso. *Kalium-Symposium, 1955*: 471–483. (*Rev. Appl. Mycol.* **36**: 723, 1957.)

BRYCE, G. 1920. Annexure I. Report on the work of the Botanical and Mycological Division. *Rep. Dep. Agric. Ceylon, 1920*: C13–C15.

BUTLER, E. J. 1918. Fungi and disease in plants, pp. 230–232. Calcutta & Simla, Thacker, Spink & Co.

CATTANEO, A. 1876. Sullo *Sclerotium oryzae*, nuova parassità vegetale, che ha devastato nel corrente anno molte risaje di Lombardia e del Novarese. *Rendic. R. Ist. Lombardo Sci. Let.*, Ser 2, 9: 801–807. (*J. agric. Res.* **51**: 348, 1935.)

CAVARA, F. 1889. Matériaux de mycologie Lombarde. *Rev. Mycol.* **11**: 189. (*J. agric. Res.* **47**: 686, 1933.)

CHAUHAN, L. S., VERMA, S. C. & BAJPAI, G. K. 1968. Assessment of losses due to stem rot of rice caused by *Sclerotium oryzae*. *Pl. Dis. Reptr* **52**: 963–965.

CORBETTA, G. 1953. Ricerche e esperienze sulle malattie del riso (*Oryza sativa* L.). VIII. La malattia del riso da *Sclerotium oryzae* Catt. (=*Helminthosporium sigmoideum* Cav.= *Leptosphaeria salvinii* Catt.). *Phytopath. Z.* **20**: 260–296. (*Rev. appl. Mycol.* **33**: 445, 1954.)

CRALLEY, E. M. 1936. Resistance of rice varieties to stem rot. *Bull. Ark. agric. Exp. Stn* 329, 31 pp.

CRALLEY, E. M. 1939. Effects of fertilizers on stem rot of rice. *Bull. Ark. agric. Exp. Stn* 383, 17 pp.

CRALLEY, E. M. & ADAIR, C. R. 1943. Effect of irrigation treatments on stem rot severity, plant development, yield and quality of rice. *J. Am. Soc. Agron.* **35**: 499–507.

CRALLEY, E. M. & TULLIS, E. C. 1934. Rice disease investigations—stem rot. *Bull. Ark. agric. Exp. Stn.* 312 (Ann. Rep. 46): 52–53.

CRALLEY, E. M. & TULLIS, E. C. 1935. A comparison of *Leptosphaeria salvinii* and *Helminthosporium sigmoideum irregulare*. *J. agric. Res.* **51**: 341–348.

GOTO, K. & FUKATSU, R. 1954. Studies on the stem rot of rice plant. I. Varietal resistance and seasonal development. *Bull. Div. Pl. Breed. Cult. Tokai-Kinki natn. agric. Exp. Stn* 1: 27–39. [Jap. Engl. summ.]

HARA, K. 1936. Pathogenic fungi of Japan. Tokyo. [Jap.]

HECTOR, G. P. 1931. Annual report of the First Economic Botanist to the Government of Bengal for the year 1930–31. *Rep. Dep. Agric. Bengal, 1930–31*: 34–44. (*Rev. appl. Mycol.* **11**: 158, 1932.)

HERNANDEZ, A. 1923. Report of Plant Disease Section. *Rep. Philippine Bur. Agric.* **23**: 159–172.

HORI, M. & IIZUKA, K. 1951. Control effect of ceresan against rice stem rot. *Nogyo Gijitsu* **6** (9): 35–37. [Jap.]

HSIEH, S. P. Y. 1966. Stem rot of rice in the Philippines. *M.S. thesis, Univ. Philipp. Coll. Agric.*, 78 pp.

IKE, T. & WATANABE, S. 1953. Influence of lodging on the damage of stem rot disease in rice varieties. *Proc. Crop Sci. Soc. Japan* **21**: 247–248. [Jap. Engl. summ.]

INOUE, Y. 1961. On aspects of damage of rice plant judged from biological studies on the attack of rice stem rot disease caused by *Helminthosporium sigmoideum* Cav. *Spec. Bull. 1st Agron. Div. Tokai-Kinki natn. agric. Exp. Stn* 3, 118 pp. [Jap. Engl. summ.]

IRRI (International Rice Research Institute). Annual Reports for 1965, 1966 and 1967.

JAIN, S. S. 1967. The influence of carbon nutrition on the growth of *Sclerotium oryzae* and *S. oryzae* var. *irregulare* Roger. *Bull. natn. Inst. Sci. India* 35: 89–93.

KAWAI, I. 1955. Distribution of sclerotia of *Helminthosporium sigmoideum* in the paddy field. *Jubilee Publ. commem. 60th Birthdays of Profs. Y. Tochinai and T. Fukushi*: 255–258. Sapporo, Japan. [Jap. Engl. summ.]

KAWAMURA, E. 1941. A simple and successful method of the inoculation with *Sclerotium oryzae* on the rice plant. *Ann. phytopath. Soc. Japan* **11**: 20–21. [Jap.]

KOBARI, K. 1961. Relationship between stem rot disease and stemborer of rice. Abs. in *Ann. phytopath. Soc. Japan* **26**: 238 [Jap.]

LO, T. C. & Hsieh, S. P. Y. 1964. Studies on the stem rot of rice plant. I. Distribution and economic significance of the disease. *Pl. Prot. Bull., Taiwan* **6**: 121–135.

LUTHRA, J. C. & SATTAR, A. 1936. Some studies on the sclerotial disease of rice (*Sclerotium oryzae* Catt.) in the Punjab. *Indian J. agric. Sci.* **6**: 973–984. (*Rev. appl. Mycol.* **16**: 123, 1937.)

MISAWA, T. & KATO, S. 1955a. Physiology of the causal fungi of stem rot of rice plant—on the nitrogen metabolism. *Ann. phytopath. Soc. Japan* **19**: 125–128. [Jap. Engl. summ.]

MISAWA, T. & KATO, S. 1955b. Physiology of the causal fungi of stem rot of rice plant. II. On the growth factors. *Ann. phytopath. Soc. Japan* **20**: 65–70. [Jap. Engl. summ.]

MISAWA, T. & KATO, S. 1960. On the influence of RNA upon the sclerotium formation of stem rot fungus of rice plant. *Ann. phytopath. Soc. Japan* **25**: 75–79. [Jap. Engl. summ.]

MISAWA, T. & KATO, S. 1962. On the lodging of the rice plant caused by stem rot. *Ann. phytopath. Soc. Japan* **27**: 102–108. [Jap. Engl. summ.]

MIYAKE, I. 1910. Studien über die Pilze der Reispflanze in Japan. *J. Coll. Agric., Tokyo* **2**: 237–276.

MUNDKUR, B. B. 1935. Parasitism of *Sclerotium oryzae* Catt. *Indian J. agric. Sci.* **5**: 393–414.

NAKATA, K. & KAWAMURA, E. 1939. Studies on sclerotial diseases of rice. *Bureau Agric., Minist. Agric. For., Japan, Agric. Exp. Stn Records* 139, 175 pp. [Jap.]

NISIKADO, Y. & HIRATA, K. 1937. Studies on the longevitiy of sclerotia of certain fungi under controlled environmental factors. *Ber. Ohara Inst. landw. Forsch.* **7**: 535–547.

NONAKA, F. 1956. On the severity of rice stem rot when the plants were treated with leaf excision combined with varied quantities of sclerotial inocula of the causal fungus *Helminthosporium sigmoideum*. *Sci. Bull. Fac. Agric. Kyushu Univ.* **15**: 431–434. [Jap. Engl. summ.]

NONAKA, F. 1959. Studies on the nutritional physiology of rice stem fungi, *Leptosphaeria salvinii* and *Helminthosporium sigmoideum* var. *irregulare*. II. On the carbon and nitrogen sources. *Sci. Bull. Fac. Agric. Kyushu Univ.* **17**: 17–26. [Jap. Engl. summ.]

NONAKA, F. & IWATA, T. 1958. On the influence of sheath stripping on the severity of rice stem rot. (*Leptosphaeria salvinii* Catt.) of lower part of rice plant. *Sci. Bull. Fac. Agric. Kyushu Univ.* **16**: 473–479. [Jap. Engl. summ.]

NONAKA, F. & YOSHII, H. 1958a. Relation between the maturity of rice plant and the severity of stem rot caused by *Leptosphaeria salvinii* Catt. I. Seasonal observations on the severity and culm invasion of the causal fungus. *Sci. Bull. Fac. Agric. Kyushu Univ.* **16**: 439–445. [Jap. Engl. summ.]

NONAKA, F. & YOSHII, H. 1958b. Carbohydrate and nitrogen contents of the lower parts of rice culms affected with stem rot fungus (*Leptosphaeria salvinii* Catt.). *Sci. Bull. Fac. Agric. Kyushu Univ.* **16**: 459–463. [Jap. Engl. summ.]

NONAKA, F., IWATA, T. & YOSHII, H. 1958. Effect of silicic acid on the severity of rice stem rot caused by *Leptosphaeria salvinii* Catt. *Sci. Bull. Fac. Agric. Kyushu Univ.* **16**: 447–454. [Jap. Engl. summ.]

OKAMOTO, H. 1950. Relation of amount and periods of potassium fertilization to the stem rot disease of rice plants. Abs. in *Ann. phytopath. Soc. Japan* **14**: 110. [Jap.]

ONO, K. 1949. Varietal resistance to stem rot in rice. *Ann. phytopath. Soc. Japan* **13**: 14–18. [Jap.]

ONO, K. 1953. On the portions of infection cushion formation of stem rot fungus *Leptosphaeria salvinii* Catt. Abs. in *Ann. phytopath. Soc. Japan.* **17**: 179. [Jap.]

ONO, K. & SUZUKI, H. 1960. Studies on mechanism of infection and ecology of blast and stem rot of rice plant. *Spec. Rep. Forecast. Disease Insect Pest.* 4: 94–152. Minist. Agric. For. Fish., Japan. [Jap. Engl. summ.]

ORSENIGO, M. 1955. Effetto della nutrizione potassica del riso sulla gravità del mal dello sclerozio (*Leptosphaeria salvinii* Catt.), sull'elmintosporiosi (*Helminthosporium oryzae* Breda de Haan) e sulla malattia detta ' white tip ' (*Aphelenchoides oryzae* Yokoo), sia singolarmente che in combinazione sulle medesime piante. *Ann. Fac. Agric.* (*Publ. Univ. S. Cuore*, N.S. 51): 8–21. (*Rev. appl. Mycol.* **35**: 715, 1956.)

PARACER, C. S. & LUTHRA, J. C. 1944. Further studies on the stem rot disease of rice caused by *Sclerotium oryzae* Catt. in the Punjab. *Indian J. agric. Sci.* **19**: 44–48.

PARK, M. & BERTUS, L. S. 1932. Sclerotial diseases of rice in Ceylon. 2. *Sclerotium oryzae* Catt. *Ceylon J. Sci.*, Sect. A, **11**: 343–359.

PARK, M. & BERTUS, L. S. 1934. Sclerotial diseases of rice in Ceylon. 4. *Sclerotium oryzae* A strain. *Ceylon J. Sci.*, Sect. A, **12**: 11–23. (*Rev. appl. Mycol.* **15**: 114, 1936.)

REYES, G. M. 1929. A preliminary report on the stem rot of rice. *Philipp. agric. Rev.* **22**: 313–331.

REYES, G. M. 1936. Rice hybrids versus stem rot disease. *Philipp. J. Agric.* **7**: 413–418.

RHIND, D. 1924. Report of the mycologist, Burma, for the period ending 30th June, 1924, 6 pp. (*Rev. appl. Mycol.* **4**: 259, 1925.)

SADASIVAN, T. S. & SUBRAMANIAN, C. V. 1954. Studies on the growth requirements of Indian fungi. *Trans. Br. mycol. Soc.* **37**: 426–430.

SAKURAI, M. 1917. On *Sclerotium* diseases of rice plant. *Publ. Ehime Agric. Exp. Stn* 1, 60 pp. (*Bot. Abstr.* **10**: 195, 1922.)

SEETHUNATHAN, N. 1963. Studies on *Sclerotium oryzae*. *Riso* **12** (4): 21–27.

SEETHUNATHAN, N. 1964. Studies on *Sclerotium oryzae*. II. Nitrogen utilization. *Phytopath. Z.* **50**: 33–42.

SHAW, F. J. F. 1913. A sclerotial disease of rice. *Mem. Dep. Agric. India*, Bot. Ser. 6: 11–23.

SHIOYAMA, O., KURONO, H., MURATA, K. & MATSUMOTO, S. 1964. Fungicidal effect of organoarsenic compounds against rice stem rot fungus. *Noyaku Seisan Gijutsu* **11**: 8–12. [Jap.]

SHOEMAKER, R. A. 1959. Nomenclature of *Drechslera* and *Bipolaris*, grass parasites segregated from *Helminthosporium*. *Can. J. Bot.* **37**: 879–887.

SU, M. T. 1931. Report of the mycologist, Burma, Mandalay, for the year ending 31st March, 1931.

SUBRAMANIAN, C. V. 1956. Hyphomycetes II. *J. Indian bot. Soc.* **35**: 446–494.

TEMPLETON, G. E. 1962. Research on rice diseases in Arkansas. *Rice J.* **65** (7): 28.

TISDALE, W. H. 1921. Two sclerotial diseases of rice. *J. agric. Res.* **21**: 649–658.

TOGASHI, K. 1955. On the spore-formation from sclerotia and spore-germination of *Helminthosporium sigmoideum* Cav. var. *irregulare* Cralley et Tullis. *Bull. Fac. Agric. Niigata Univ.* **7**: 34–45. [Jap. Engl. summ.]

TULLIS, E. C. 1932. *Helminthosporium sigmoideum*, the conidial stage of *Sclerotium oryzae*. Abs. in *Phytopathology* **22**: 28.

TULLIS, E. C. 1933a. *Leptosphaeria salvinii*, the ascigerous stage of *Sclerotium oryzae*. Abs. in *Phytopathology* **23**: 35.

TULLIS, E. C. 1933b. *Leptosphaeria salvinii*, the ascigerous stage of *Helminthosporium sigmoideum* and *Sclerotium oryzae*. *J. agric. Res.* **47**: 675–686.

TULLIS, E. C. & CRALLEY, E. M. 1933. Laboratory and field studies on the development and control of stem rot of rice. *Bull. Ark. agric. Exp. Stn* 295: 3–22.

TULLIS, E. C. & CRALLEY E. M. 1941. Longevity of sclerotia of the stem rot fungus, *Leptosphaeria salvinii*. *Phytopathology* **31**: 279–281.

VINCENS, F. 1921. Maladies du riz. Rapport sommaire sur les travaux effectués au laboratoire de phytopathologie de l'Institut Scientifique de l'Indochine du 1er janvier 1919 au 1er juillet 1921, 19 pp. Saigon, Imprimerie Commerciale. (*Rev. appl. Mycol.* **1**: 157, 1922.)

WATANABE, B. 1952. The development of stem rot disease of rice plant and the physiological properties of the basal part of stem. *Bull. Kyushu agric. Exp. Stn* **1**: 271–276. [Jap. Engl. summ.]

YOSHII, H., KOBA, S. & WATANABE, B. 1949. On the scale for estimating degree of damage in rice by stem rot. *Ann. phytopath. Soc. Japan* **13**: 19–22. [Jap.]

BAKANAE DISEASE AND FOOT ROT

HISTORY

This disease is said to have been known in Japan since 1828 (Ito & Kimura, 1931). It was first described by Hori (1898), who wrongly identified the causal organism as *Fusarium heterosporum* Nees. Fujikuro found the perfect state which was described as *Lisea fujikuroi* by Sawada (1917). The fungus was later put in the genus *Gibberella* under the name of *G. fujikuroi* (Sawada) Ito (Ito & Kimura, 1931).

Kurosawa (1926) demonstrated the hypertrophic or ' bakanae ' effect of the fungus on its hosts. This unique phenomenon attracted the attention of biochemists and plant physiologists and led to the isolation of ' gibberellin ' and other growth regulators by Yabuta, Sumiki & Hayashi in 1935. This discovery has led more recently to further advances in the study of growth substances. The disease was described and called foot rot in India by Thomas (1931, 1933).

A great deal of work has been done on the disease in Japan since 1898, notably by Ito & Kimura, Nisikado, and Seto, on gibberellin and related compounds by Yabuta and his associates, and on the synthesis of gibberellin-like substances by Mori and his associates.

DISTRIBUTION AND DAMAGE

The disease is widely distributed in all rice growing areas. It has been called ' white stalk ' in China, ' palay lalake ' (man rice) in the Philippines, and also ' man rice ' in Guyana. Countries reporting its presence include Australia, Guyana, Cameroon, Ceylon, China, the former French Equatorial Africa,

India, Italy, Ivory Coast, Japan, Kenya, Nigeria, Surinam, Tanzania, Thailand, Trinidad, Uganda, U.S.A., Venezuela, Vietnam and others.

Estimates of losses due to the disease are few. Ito & Kimura (1931) reported up to 20% loss in Hokkaido. Pavgi *et al.* (1964) stated that losses of 15% occurred in eastern districts of Uttar Pradesh, India, and Kanjanasoon (1965) found 3·7–14·7% loss in Northern and Central Thailand. The disease is commonly seen in many parts of Southeast Asia but the percentage of infection is usually small.

SYMPTOMS

The most conspicuous and common symptoms are the bakanae symptoms, i.e. the abnormal elongation of the plant. Such symptoms may be observed in the seedbed as well as in the field. Infected seedlings in the seedbed are up to several inches taller than normal plants, and are thin and yellowish-green. Looked at from leaf-tip level, these affected seedlings stand out very conspicuously here and there, scattered throughout the field (Fig. IV–7). Severely diseased seedlings die before transplanting, and those that survive may die after transplanting. Not all infected seedlings have the bakanae symptoms; sometimes they are stunted or they may appear normal. The development of the bakanae symptoms depends on the strain of the causal organism involved and on environmental conditions such as temperature and moisture, which will be discussed below.

Fig. IV–7. Bakanae disease. Elongation symptoms in field. (Courtesy of Dr. Kanjanasoon.)

In crops reaching maturity the infected plants show tall lanky tillers bearing pale green flag leaves which are conspicuous above the general level of the crop. With sunshine and when a gentle breeze blows over the fields, these diseased plants are easily identified. Infected plants usually have only a small number of tillers and the leaves dry up one after another from below and die in a few weeks. Occasionally, infected plants survive until maturity but bear only empty panicles. While the infected plants are dying, a white or pink growth of the pathogenic fungus may be noticed on their lower parts. This growth consists of a thick mat of mycelium with large numbers of conidia and extends upwards after the plants are dead.

In India the infected plants also develop adventitious roots from the lower nodes (Thomas, 1931). Imura (1940) in Japan also reported that the leaf angles with the stem are wider in diseased than in healthy plants.

Causal Organism

Nomenclature. The perithecial state of *Fusarium moniliforme* Sheld. was first described as *Lisea fujikuroi* by Sawada (1917). It was later transferred to *Gibberella* as *G. fujikuroi* (Sawada) Ito (Ito & Kimura, 1931). In 1924 Wineland described *G. moniliformis* and proposed this name for the perithecial state of *F. moniliforme*. This later name is, however, invalid according to the International Code of Botanical Nomenclature but is still used by some pathologists (Snyder & Hansen, 1945; Snyder & Toussoun, 1965). *G. fujikuroi* occurs on a wide variety of host plants besides rice, causing seedling blights, stem rots and fruit decay.

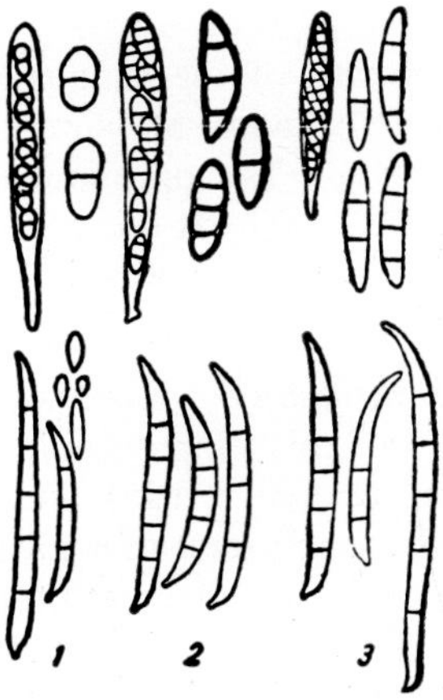

Fig. IV–8, Asci and ascospores (above) and conidia (below) of *Gibberella fujikuroi* (1), *G. baccata* (2) and *G. zeae* (3). (After Wei, 1957 redrawn from Wollenweber, 1931).

Morphology. The following description of the fungus from rice is adapted from Wollenweber & Reinking (1935); Perithecia dark blue, spherical to ovate, somewhat roughened outside, $250–330 \times 220–280$ $(190–390 \times 160–420)\mu$; asci cylindrical piston-shaped, flattened above, $90–102 \times 7–9$ $(66–129 \times 7–14)\mu$, 4-, 6-, seldom 8-spored, monostichous or indistinctly distichous; spores one-septate, about $15 \times 5{\cdot}2\mu$ (mostly $14–18 \times 4{\cdot}4–7\mu$), occasionally (in one-spored asci) larger $(27–45 \times 6–7\mu)$ (Fig. IV–8).

Microconidia are more or less agglutinated in chains and remain joined, or are cut off in false heads, later scattered in clear yellowish to rosy-white aerial mycelium as a dull, colourless powder, 1- or 2-celled, fusiform-ovate; macroconidia delicate, awl-shaped, slightly sickle-shaped or almost straight, narrowed at both ends, occasionally somewhat bent into a hook at the apex, distinctly or slightly foot-celled at the base, scattered, formed in sporodochia or pionnotes, in the mass clear, buff, or salmon-orange, when dry carrot-red or cinnamon-brown or rather pale, 3- to 5-, rarely 6- to 7-septate, measuring as follows:

0-septate $8{\cdot}4 \times 2{\cdot}4$, mostly $5–12 \times 2–3$ ($4–18 \times 1{\cdot}5–4$)μ
1-septate $17 \times 2{\cdot}9$, mostly $12–22 \times 2{\cdot}2–3{\cdot}5$ ($9–30 \times 2–5$)μ
3-septate 36×3, mostly $32–50 \times 2{\cdot}7–3{\cdot}5$ ($20–60 \times 2–4{\cdot}5$)μ
5-septate $49 \times 3{\cdot}1$, mostly $41–63 \times 2{\cdot}7–4$ ($37–70 \times 2–4{\cdot}5$)μ
7-septate $66 \times 3{\cdot}5$, mostly $61–82 \times 2{\cdot}7–4{\cdot}2$ ($58–90 \times 2{\cdot}5–4{\cdot}5$)μ

Chlamydospores absent. Sclerotia dark blue, spherical, 80×100μ; may or may not be present. Stroma more or less plectenchymatous, yellowish, brownish, violet etc.

The fungus is variable in septation of the conidia, and in the formation of microconidia, macroconidia, sclerotia, stromata etc.

Physiology. The fungus is easily cultured in various media; Richards' and Knop's solutions were commonly used by early workers. The effect of temperature on the mycelial growth of the fungus has been studied by several workers (Kurosawa, 1929; Nisikado, Matsumoto & Yamauti, 1933; Seto, 1935; Hemmi & Aoyagi, 1941). The optimum is about 27–30°C, maximum 36–40°C, and minimum 7–8°C. There are some variations in strains, and reports by various workers differ. Yogeswari (1948) reported that trace elements, such as boron, zinc and manganese, increase the growth of the fungus.

Metabolic products and variability. Kurosawa (1926) first demonstrated that the filtrate of a culture of this fungus could produce bakanae symptoms on rice seedlings. Hemmi & Seto (1928) and Seto (1932) confirmed the phenomenon but noticed that the filtrates of some isolates, grown under similar conditions, could not produce the bakanae symptoms. Nisikado & Matsumoto (1933a) reported that among 66 strains of *G. fujikuroi* obtained from rice and 5 strains of *G. moniliforme* var. *majus*, there were marked differences in pathogenicity as indicated by the degree of overgrowth. Studying a number of isolates, Chin (1940) found that some caused dwarfing, some caused elongation and others had no effect on the size of rice seedlings, so confirming the earlier work of Seto (1932).

The conditions under which the fungus is grown also affect the nature of the filtrate. Ito & Shimada (1931) reported that production of the growth-promoting substance is inhibited by the omission of KH_2 or PO_4 or $MgSO_4$ from the culture solution. They found the substance to be thermostable, and neither enzymic nor volatile. Seto (1932) found that, in Knop's solution, increasing the sugar content resulted in the filtrate having a stunting effect on rice seedlings. Kurosawa (1932) found potassium in the medium to be essential for the formation of the growth-promoting substance. Imura (1937) studied 14 isolates of the fungus as regards their ability to ferment alcohol and their pathogenicity and found these characteristics not to be correlated.

Viswanath-Reddy (1965) produced both stimulating (bakanae) and suppressive growth effects on given varieties by different inoculum potentials.

Tochinai & Ishizuka (1934) found that the presence of bakanae symptoms is limited to the actual period of contact with the toxin. When rice seedlings showing bakanae symptoms were transferred to an uninfected area, the plants recovered.

Yabuta & Hayashi (1934, 1939a, b) isolated two principal substances from the filtrate, fusaric acid and gibberellin. Yabuta, Sumiki & Uno (1939) made further studies of the conditions favourable for the production of gibberellin and fusaric acid. Glycerol was found to be a good source of carbon for gibberellin production while glucose was best for fusaric acid. The optimum pH for production of gibberellin was 3·4 while for fusaric acid it was 9. The proportion of the compounds produced differed with different fungus isolates. Stoll (1954) reported that fusaric acid and other compounds are produced most abundantly at 33°C in Richards' agar. Biological studies of the two substances showed the fusaric acid to cause stunting and gibberellin to cause elongation of many species of plants besides rice. Since the production of the two substances differs according to the strain of the fungus, as well as growth conditions and the composition of the medium in which it grows, the elongation and stunting effect of the filtrates described above is well explained. In further studies on the chemical structure of gibberellin, Yabuta *et al.* (1941a, b, c) separated gibberellin A, B, C, dihydro-gibberellin A and other substances in crystalline form. A series of studies on gibberellin-like substances has also been made by K. Mori and his associates since 1960 (Biochemical studies on ' bakanae ' fungus, in *Agricultural and biological Chemistry, Japan*).

Disease Cycle

The disease is seed-borne. Hemmi, Seto & Ikeya (1931) found that seeds are infected at the flowering stage. When severely infected, the kernel develops a reddish discoloration due to the presence of conidia of the pathogen. More often the whole seed is discoloured. The fungus may be isolated even from seeds which appear to be healthy, if they are collected from an infested field. Such seeds when germinated give rise to seedlings with bakanae symptoms, whereas the reddish-coloured seeds produce stunted seedlings. It appears, therefore, that whether overgrowth or stunting occurs may be determined by the degree of infection of the seed. Seto (1937) determined, by spraying a conidial suspension on to rice plants, that the most favourable stage of development for seed infection to take place is at the time of flowering. Infection continues for three weeks thereafter at a reduced level. The fungus also infects the branches of the panicle.

Seto (1937) reported that the fungus can infect seedlings at an early stage of their development. It becomes systemic, growing up within the plant but generally not penetrating the floral parts. Rajagopalan & Bhuvaneswari (1964) found that sowing ungerminated seeds in infested soil resulted in rapid progression of the disease and a high percentage of mortality; in contrast, only quite mild disease symptoms resulted when germinated seeds were sown. The first 72 hours after germination begins appears to be the critical period for disease development, largely because of the copious exudation by the germinating seeds of amino acids and sugars, which act as rich energy substrates for the fungus. Kanjanasoon (1965) obtained similar results, finding a higher percentage of infection when dry seeds were sown than when pre-soaked or germinated seeds were sown.

Nisikado & Kimura (1941) found microconidia and mycelium of the pathogen concentrated in the vascular bundles, especially the large pitted vessels and lacunae of the xylem. The phloem and parenchyma were not invaded to any extent. The pathogen was found to be discontinuously distributed, being present at one point, absent for an interval of 2 or 3 nodes, then present again.

The fungus survives the winter (or summer in the tropics) in infected seeds and other parts of diseased plants. Wollenweber & Reinking (1935) stated that the fungus survives for at least three years under dry conditions in a building, apparently not so long in the open, and not at all in stable manure. Iguchi (1964) also reported that the fungus may remain alive until the next season in Japan if it is kept indoors. Kanjanasoon (1965) found the fungus viable in seeds and infected plants after 4–10 months at room temperature and after more than 3 years in cold storage at 7°C.

Seto (1933a) stated that the disease can also be soil-borne, although it is generally regarded as being mainly carried in the seed. Kanjanasoon (1965) in Thailand showed that artificially inoculated soil caused 93% infection immediately after inoculation, infection decreasing with the passage of time until only 0·7% infection occurred after 90 days and no infection resulted after 180 days. This indicates that the fungus does not survive in soil for long in the tropics.

The fungus produces conidia on diseased plant parts and these are easily disseminated by wind and water to cause new infections. The copious production of conidia on diseased or dead culms in the field during the latter part of the growing season coincides with flowering and maturation of the crop, and so conidia are able to infect or contaminate the seeds. Kanjanasoon (1965) found 1–31·2% seedling infection in healthy-looking seeds from diseased fields.

Effect of Environmental Conditions

It is generally observed that bakanae plants are few or are not observed at all when the temperature is low. Seto (1933b, 1935) experimented with the effect of soil temperature on bakanae symptoms and found that a temperature of 35°C is most favourable for seedling growth and also for infection. At a soil temperature of 25°C, bakanae plants could still be found but at 20°C they failed to appear. However, the fungus was isolated from these outwardly healthy-looking plants. The optimum temperature for the growth of the fungus is about 27–30°C while the optimum temperature for infection is 35°C. Seto thus deduced that the effect of soil temperature is on the rice seedlings rather than on the fungus.

Seto (1933a, 1935) also found that elongated seedlings develop only under damp soil conditions; arrested growth and stunting occur in dry soil.

Thomas (1937) in India showed that the application of nitrogen to the soil stimulated the development of the disease and the effect was not modified by the addition of potassium or phosphorus. The growth of the fungus in culture was also stimulated by adding ammonium sulphate or asparagin to the medium. It was therefore suggested that the increase of disease in the field with the addition of nitrogen was due to increased growth of the fungus rather than to an increase in susceptibility of the host.

Varietal Resistance and Host Range

Ito & Kimura (1931) found that in Japan the varieties Shiroka, Akage No. 3 and Kairyo-Mochi No. 1 are resistant while Bozu, Chinko-Bozu and Iburiwase are susceptible. Thomas (1933) tested 41 varieties in India and found resistant varieties such as Wateribune had 1·13% disease, Aryan 1·85%, and GEB 24 4·83%, while susceptible varieties such as Co 13 had 95·56%, Tinnevelly Anarkomban 95·12%, and Gobikar 93·51%. Other varieties were intermediate in reaction. Reyes (1939) reported in the Philippines that Macan Bino, Apostal, Macan Binan and Guinangang strain 1 possess some degree of resistance while Macan 1, Macan Tago, Ramay and Magcumpal are susceptible. Hashioka (1952) tested 200 varieties and reported that varieties from the temperate regions are less affected than those from the tropics. Rajagopalan (1961) found 6 resistant varieties among the 20 which he tested in India. The varieties Lemello and Roverbella have been found to be resistant in Italy (Anon., 1959).

Kanjanasoon (1965) observed that 72 hours after inoculation of seeds with a spore mass, the mycelium of the fungus grows into a dense mass in between the crown and the leaf sheath surrounding the crown. The mycelium then invades the tissue of the crown of a susceptible variety, but in the case of a resistant variety, the fungus cannot grow well on the surface of the crown or the surrounding leaf sheath and cannot form a mycelial mass dense enough to penetrate the host tissue; and the initial amount of mycelium decreases rapidly.

Nisikado (1931, 1932) and Nisikado & Matsumoto (1933b) have shown that the fungus can also cause the bakanae phenomenon in maize, barley, sugarcane, sorghum and *Panicum miliaceum*.

Control

Seed treatment with organic mercury compounds, either in dust or liquid form, is effective in controlling this disease. Hoshino (1955) investigated the effect of the concentration of two groups, chloride and acetate, of these compounds and the duration of soaking as regards their fungicidal and phytotoxic action. At a low concentration (0·1% solution), which is commonly used in Japan, soaking the seed for 16–24 hours is recommended. Temperatures below 10°C should be avoided. The time of soaking may be shortened to 2 hours by increasing the concentration of the organic mercury compound to 0·25%, but at this concentration a long period of soaking (3 hours or more) is phytotoxic, particularly with the acetate group. A seed dressing in the form of a dust is easy to use; usually 1 g is used for each pound of seed.

LITERATURE CITED

ANON. 1959. Nuove razze di Riso. *Prog. agric.* **5**: 1414–1415. (*Rev. appl. Mycol.* **39**: 469.)

CHIN, W. F. 1940. A preliminary study on the physiological differentiation of *Fusarium fujikuroi* (Saw.) Wr. *Nanking J.* **9**: 305–321.

HASHIOKA, Y. 1952. Varietal resistance of rice to the ' Bakanae ' disease. (Studies on pathological breeding of rice V.) *Jap. J. Breed.* **1**: 167–171. [Jap. Engl. summ.]

HEMMI, T. & SETO, F. 1928. Experiments relating to stimulative action by the causal fungus of the ' bakanae ' disease of rice. (Preliminary report.) *Proc. imp. Acad. Japan* **4**: 181–184.

HEMMI, T., SETO, F. & IKEYA, J. 1931. Studies on the ' bakanae ' disease of the rice plant. II. On the infection of rice by *Lisea fujikuroi* Sawada and *Gibberella saubinetii* (Mont.) Sacc. in the flowering period. *Forschn. Geb. PflKrankh., Kyoto* **1**: 99–110. [Jap. Engl. summ.]

HEMMI, T. & AOYAGI, Z. 1941. Ecological studies on important fungi pathogenic to the crops in the Far East. I. Germination of the macroconidia of *Gibberella fujikuroi* in relation to some environmental factors. *Ann. phytopath. Soc. Japan* **11**: 66–80. [Jap. Engl. summ.]

HORI, S. 1898. Researches on ' bakanae ' disease of the rice plant. *Nojishikenjyo Seiseki* **12**: 110–119. [Jap.]

HOSHINO, Y. 1955. A study on the rice-seed disinfection with special reference to the Bakanae-disease. *Jubilee Publ. Commem. 60th Birthdays Profs. Y. Tochinai and T. Fukushi*: 290–299. Sapporo, Japan. [Jap. Engl. summ.]

IGUCHI, S. 1964. Overwintering of bakanae disease fungus. *Ann. Rep. Soc. Pl. Prot. North Japan* **15**: 39. [Jap.]

IMURA, J. 1937. On the alcoholic fermentation of *Gibberella fujikuroi*, the causal fungus of the ' bakanae ' disease of the rice plant and its relation to pathogenicity. *Forschn. Geb. PflKrankh., Kyoto* **3**: 289–309. [Jap. Engl. summ.]

IMURA, J. 1940. On the angles between blades and culms in the accelerated rice seedlings caused by *Gibberella fujikuroi. Ann. phytopath. Soc. Japan* **10**: 45–48. [Jap.]

ITO, S. & KIMURA, J. 1931. Studies on the ' bakanae ' disease of the rice plant. *Rep. Hokkaido agric. Exp. Stn* 27: 1–95+5. [Jap. Engl. summ.]

ITO, S. & SHIMADA, S. 1931. On the nature of the growth promoting substance excreted by the ' bakanae ' fungus. *Ann. phytopath. Soc. Japan* **2**: 322–338.

KANJANASOON, P. 1965. Studies on the bakanae disease of rice in Thailand. Doc. Agric. Thesis, Tokyo University, Japan.

KUROSAWA, E. 1926. Experimental studies on the filtrate of the causal fungus of the ' bakanae ' disease of the rice plant. (Preliminary report.) *Trans. nat. Hist. Soc. Formosa* **16**: 213–227. (Engl. abs. in *Jap. J. Bot.* **3**: 91, 1927.)

KUROSAWA, E. 1929. On the cultural characters of the ' bakanae ' disease fungi on various nutrient media and the temperature of their development. *Rep. nat. Hist. Soc. Formosa* **19**: 150–179. [Jap.]

KUROSAWA, E. 1932. On certain experimental results concerning the over-elongation phenomenon of rice plants which owe to the filtrate got from the culture solution of the ' bakanae ' fungi. *Rep. Taiwan nat. Hist. Soc.* **22**: 198–201. [Jap.] (Abs. in *Jap. J. Bot.* **6**: 72–73, 1933.)

NISIKADO, Y. 1931. Vergleichende Untersuchungen über die durch *Lisea fujikuroi* Saw. und *Gibberella moniliformis* (Sh.) Winel. verursachten Gramineenkrankheiten. (Vorläufige Mitteilungen.) *Ber. Ohara Inst. Landw. Forsch.* **5**: 87–106.

NISIKADO, Y. 1932. Uber zwei wirtschaftlich wichtige, parasitare Gramineenpilze: *Lisea fujikuroi* Sawada und *Gibberella moniliformis* Wineland. *Z. ParasitKde* **4**: 285–300.

NISIKADO, Y. & MATSUMOTO, H. 1933a. Studies on the physiological specialization of *Gibberella fujikuroi*, the causal fungus of the rice ' bakanae ' disease. *Trans. Tottori Soc. agric. Sci.* **4**: 200–211. [Jap. Engl. summ.]

NISIKADO, Y. & MATSUMOTO, H. 1933b. Weitere vergleichende Untersuchungen über die durch *Lisea fujikuroi* Sawada und *Gibberella moniliformis* (Sh.) Wineland verursachten Gramineenkrankheiten. *Ber. Ohara Inst. Landw. Forsch.* **5**: 481–500. (Abs. in *Jap. J. Bot.* **6**: 79–80, 1933.)

NISIKADO, Y., MATSUMOTO, H. & YAMAUTI, K. 1933. Zur Kenntnis der physiologischen Differenzierung der Fusariumarten. I; II. *Landw. Studien* **20**: 310–345; 346–375. (Abs. in *Jap. J. Bot.* **6**: 110, 1933.)

NISIKADO, Y. & KIMURA, K. 1941. A contribution to the pathological anatomy of rice plants affected by *Gibberella fujikuroi* (Saw.) Wollenweber. I. *Ber. Ohara Inst. Landw. Forsch.* **8**: 421–426.

PAVGI, M. S. & SINGH, J. 1964. Bakanae and foot rot of rice in Uttar Pradesh, India. *Pl. Dis. Reptr* **48**: 340–342.

RAJAGOPALAN, K. 1961. Screening of rice varieties for resistance to foot-rot disease. *Curr. Sci.* **30**: 145–147.

RAJAGOPALAN, K. & BHUVANESWARI, K. 1964. Effect of germination of seeds and host exudations during germination on foot rot disease of rice. *Phytopath. Z.* **50**: 221–226.

REYES, G. M. 1939. Rice diseases and methods of control. *Philipp. J. Agric.* **10**: 419–436.

SAWADA, K. 1917. Beitrage über Formosas-Pilze no. 14. *Trans. nat. Hist. Soc. Formosa* **31**: 31–133. [Jap.]

SETO, F. 1932. Experimentelle Untersuchungen über die hemmende und die beschleunigende Wirkung des Erregers der sogenannten ' Bakanae '-Krankheit, *Lisea fujikuroi* Sawada, auf das Wachstum der Reiskeimlinge. *Mem. Coll. Agric. Kyoto imp. Univ.* 18: 1–23.

SETO, F. 1933a. Untersuchungen über die ' Bakanae '-Krankheit der Reispflanze. III. Uber die Beziehungen zwischen der Bodenfeuchtigkeit und dem Krankheitsbefall durch Bodeninfektion. *Forschn Geb. PflKrankh., Kyoto* **2**: 125–137. [Jap. Germ. summ.]

SETO, F. 1933b. Untersuchungen über die ' Bakanae '-Krankheit der Reispflanze. IV. Uber die Beziehungen zwischen der Bodentemperaturen und dem Krankheitsbefall bei Bodeninfektion. *Forschn Geb. PflKrankh., Kyoto* **2**: 138–153. [Jap. Germ. summ.]

SETO, F. 1935. Beitrage zur Kenntnis der ' Bakanae '-Krankheit der Reispflanze. *Mem. Coll. Agric. Kyoto Univ.* 36: 1–81.

SETO, F. 1937. Studies on the ' bakanae ' disease of the rice plant. V. On the mode of infection of rice by *Gibberella fujikuroi* (Saw.) Wr. during and after the flowering period and its relation to the occurrence of the so-called ' Bakanae ' seedlings. *Forschn Geb. PflKrankh., Kyoto* **3**: 43–57. [Jap. Engl. summ.]

SNYDER, W. C. & HANSEN, H. N. 1945. The species concept in *Fusarium* with reference to discolor and other sections. *Am. J. Bot.* **32**: 657–666.

SNYDER, W. C. & TOUSSOUN, T. A. 1965. Current status of taxonomy in *Fusarium* species and their perfect stages. *Phytopathology* **55**: 833–837.

STOLL, C. 1954. Uber Stoffwechsel und biologisch wirksame Stoffe von *Gibberella fujikuroi* (Saw.) Woll., dem Erreger der Bakanaekrankheit. *Phytopath. Z.* **22**: 233–274. (*Rev. appl. Mycol.* **34**: 251.)

THOMAS, K. M. 1931. A new paddy disease in Madras. *Madras agric. J.* **19**: 34–36.

THOMAS, K. M. 1933. The ' foot rot ' of paddy and its control. *Madras agric. J.* **21**: 263–272.

THOMAS, K. M. 1937. Administrative report of the Government Mycologist for the year 1936–1937. In *Admin. Reps. Agric. Chem. Entom. Mycol., Madras, 1936–37*: 35–51.

TOCHINAI, Y. & ISHIZUKA, K. 1934. The after effect of the fungus filtrate of *Gibberella fujikuroi* on rice plants. *Trans. Sapporo nat. Hist. Soc.* **13**: 148–152.

VISWANATH-REDDY, M. 1965. Inoculum potential and foot-rot of rice (*Oryza sativa* L.). *Phytopath. Z.* **53**: 197–200.

WOLLENWEBER, H. W. 1931. Fusarium-Monographie. Fungi parasitici et saprophytici *Z. PflKrankh.* **3**: 269–516.

WOLLENWEBER, H. S. & REINKING, O. 1935. Die Fusarien. viii+355 pp., Berlin, Paul Parey.

YABUTA, T., KOBE, K. & HAYASHI, T. 1934. Biochemical studies of the ' bakanae ' fungus of rice. I. Fusarinic acid, a new product of the ' Bakanae ' fungus. *J. agric. Chem. Soc. Japan* **10**: 1059–1068. (Abs. in *Chem. Abstr.* **29**: 1132, 1935.)

YABUTA, T. & HAYSHI, T. 1939a. Biochemical studies on the ' bakanae ' fungus of rice. II. Isolation of ' gibberellin ', the active principle which makes the rice seedlings grow slenderly. *J. agric. Chem. Soc. Japan.* **15**: 257–266. [Jap.]

YABUTA, T. & HAYASHI, T. 1939b. Biochemical studies on the ' bakanae ' fungus of rice. III. Studies on physiological action of gibberellin on the plant. *J. agric. Chem. Soc. Japan* **15**: 403–413. [Jap.]

YABUTA, T., SUMIKI, Y. & UNO, S. 1939. Biochemical studies on the ' bakanae ' fungus of rice. IV. The cultural conditions for producing gibberellin or fusaric acid. *J. agric. Chem. Soc. Japan* **15**: 1209–1220. [Jap.]

YABUTA, T., SUMIKI, Y., ASO, K., TAMURA, T., IGARASHI, H. & TAMARI, K. 1941a. Biochemical studies on the ' bakanae ' fungus of rice. X. Chemical structure of gibberellin (1). *J. agric. Chem. Soc. Japan* **17**: 721–730.

YABUTA, T., SUMIKI, Y., ASO, K., TAMURA, T., IGARASHI, H. & TAMARI, K. 1941b. Biochemical studies of the ' bakanae ' fungus of rice. XI. Chemical structure of gibberellin (2). *J. agric. Chem. Soc. Japan* **17**: 894–900.

YABUTA, T., SUMIKI, Y., ASO, K., TAMURA, T., IGARASHI, H. & TAMARI, K. 1941c. Biochemical studies of the ' bakanae ' fungus of rice. XII. Chemical structure of gibberellin (3). *J. agric. Chem. Soc. Japan* **17**: 975–984.

YOGESWARI, L. 1948. Trace element nutrition of fungi. I. The effect of boron, zinc and manganese on *Fusarium* species. *Proc. Indian Acad. Sci.*, Sect. B, **28**: 177–201.

SHEATH BLIGHT

Miyake (1910) first described this disease in Japan and named the causal organism *Sclerotium irregulare*. Sawada (1912) later found the fungus to be identical with *Hypochnus sasakii*, described earlier by Shirai (1906). Reinking (1918) and Palo (1926) found a very similar disease in the Philippines, which they considered to be due to a fungus of the *Rhizoctonia solani* group. Park & Bertus (1932) reported the disease in Ceylon and referred the organism to *Rhizoctonia solani* Kuhn. Wei (1934) found the disease in China, and it has been seen in many countries in Asia. It was for some time considered to be a disease of the Orient, but recently it has also been reported from Brazil, Surinam, Venezuela and Madagascar. The sheath spot disease described from U.S.A. by Ryker & Gooch (1938) has very similar symptoms, but is due to a related but separate species, *R. oryzae* Ryker & Gooch.

Extensive studies of sheath blight have been made in Japan, particularly in recent years, emphasis being placed on chemical control.

In Japan, about 120,000 to 190,000 hectares have been reported to be infected and a loss of 24,000 to 38,000 tons of rice each year has been estimated in recent years (National Institute of Agricultural Sciences, Japan, 1954). Mizuta (1956) estimated that a 20% reduction in yield may be incurred if the disease develops up to the flag leaf. Wei (1934) reported a high percentage of infected plants in Southeast China. The disease is very common in the Philippines, where it causes severe damage in certain fields in some seasons.

The economic importance of the disease is at present tending to increase because more fertilizers are being used and the new high-yielding varieties which have large numbers of tillers increase the humidity of the plant layer.

Kazaka (1970) reviewed the disease and its control in Japan.

SYMPTOMS AND LOSSES

The disease causes spots on the leaf sheath. The spots are at first ellipsoid or ovoid, about 10 mm long, and greenish-grey. They enlarge and may reach

2 or 3 cm in length and are somewhat irregular in outline. The centre of the spot becomes greyish-white, with a brown margin. Sclerotia are formed on or near these spots, but are easily detached. The size and colour of spots and the formation of sclerotia depend upon environmental conditions. Under humid conditions the mycelium of the fungus may grow over the surface of the leaf sheaths and can spread a considerable distance (several cm) in 24 hours (Plate VIII).

In the field the spots are usually first observed near the waterline. When conditions are favourable to the pathogen, they are later formed also on the upper leaf sheaths and on the leaf blades. The presence of several large spots on a leaf sheath usually causes the death of the whole leaf, and in severe cases all the leaves of a plant may be blighted in this way. It is not unusual in the tropics to find most of the leaves of an affected rice plant killed by the fungus.

Hori (1969) reported 25% loss in yield when the disease reaches the uppermost flag leaves. There is a close correlation between the percentage of diseased hills in a field and the yield loss. Fungicidal treatment increased the yield by 1·6% when 5% of the hills were infected, by 6·4–7·1% when 50% were infected, and by 8·9–10·1% when 100% were infected.

THE CAUSAL ORGANISM

Nomenclature. There is considerable confusion concerning the name of the sheath blight fungus. Japanese workers have referred it to *Hypochnus sasakii* Shirai, a fungus first found by Sasaki on leaves of camphor trees. In Ceylon and the Philippines, it has been referred to *Rhizoctonia solani* Kuhn. Gadd & Bertus (1928) regarded the fungus as *Corticium vagum* Berk. & Curt. and Park & Bertus (1932) as the imperfect state (*Rhizoctonia solani*) of *Corticium solani* (Prill. & Delacr.) Bourd. & Galz.

Endo (1927) examined a specimen of the fungus from the Philippines, supplied by Palo, and found it to be identical with *H. sasakii*, and to differ from *Rhizoctonia solani* in some aspects. Matsumoto, Yamamoto & Hirane (1932, 1933) and Matsumoto (1934) made extensive studies of strains of *H. sasakii* from various hosts in Taiwan and compared them with a strain of *R. solani* from cotton in India and one from potato in Germany. They concluded that the morphological differences in the perfect state are not sufficient to warrant the classification of *H. sasakii* as a distinct species, but that in vegetative characters it could not be included in *Corticium vagum*. Matsumoto considered the name *C. sasakii* (Shirai) Matsumoto the most acceptable. Ryker & Gooch (1938) studied cultures from China and the Philippines and considered them to be large sclerotial strains of *Rhizoctonia solani*.

Rogers (1943) established the genus *Pellicularia* and the species *P. filamentosa* (Pat.) Rogers to include *Corticium solani* (Prill. & Delacr.) Bourd. & Galz. The rice fungus has also been called *P. filamentosa* (Pat.) Rogers f. *sasakii* and *P. sasakii* (Shirai) S. Ito. However, Venkatarayan (1949) and Talbot (1965) considered the genus *Pellicularia* Cooke as a *nomen confusum* which should be rejected. The perfect state of *R. solani* has also been called *Thanatephorus cucumeris* (Frank) Donk. The classification of the genus *Rhizoctonia* is still in a state of confusion, and recent studies on *R. solani* have shown that it is heterogenous and variable; strains have pathogenic (host) specificity on the one hand and genetic variability on the other. A thorough study of this group of fungi is needed. The rice fungus is here referred to as *Corticium sasakii* (Shirai) Matsumoto.

Morphology. The fungus grows readily on various common media. The hyphae are colourless when young, becoming yellowish brown when older, 8–12μ in diameter, with infrequent septations. On host tissues or on the test tube wall in culture, the hyphae sometimes give rise to short, swollen, much branched cells. These cells may possibly be associated with the infection process or with the formation of the sporogenous state. Sclerotia are superficial, more or less globose but flattened below, white when young, becoming brown or dark brown. Individual sclerotia measure up to 5 mm but may unite to form a larger mass. Measurements for both hyphae and sclerotia are greater when the fungus is grown in culture than on the natural host tissues.

The perfect state as described by Sawada (1912) and Matsumoto *et al.* (1932) has the measurements: basidia 10–15×7–9μ (10–16×8–9μ according to Matsumoto *et al.*); sterigmata 4·5–7×2–3μ (5–8×2·2–2·7μ), numbering 2–4; basidiospores 8–11×5–6·5μ (6–10×4–7μ) (Fig. IV–9).

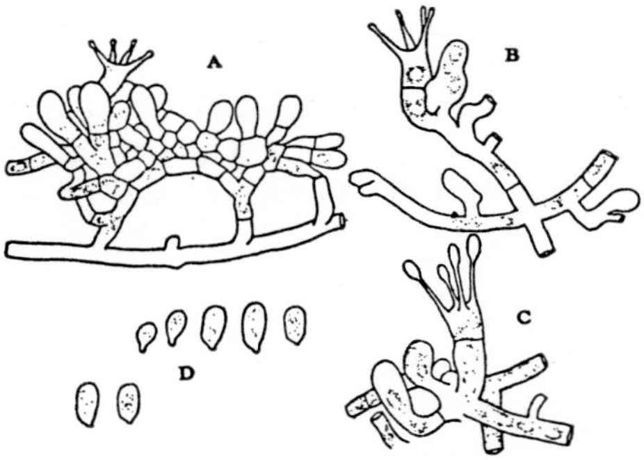

Fig. IV–9. Basidia and basidiospores of *Corticium sasakii.* (Nakata & Kawamura, 1939.)

About 40 isolates from rice collected in the Philippines were found to vary in size and number of sclerotia, colour of mycelium, and in other characters when grown in culture.

Hashioka & Makino (1969) attempted to separate various species by the number of nuclei in hyphal cells. They found that *Corticium sasakii* has mostly 6–8 nuclei, *C. solani* 4–9, *Rhizoctonia oryzae* 3–5, *C. gramineum* 1–3, and *C. rolfsii* more than 10.

Physiology. Hemmi & Yokogi (1927) reported the optimum temperature for mycelial growth to be 30°C, maximum 40–42°C, with little or no growth occurring at 10°C. Matsumoto, Yamamoto & Hirane (1933) indicated that the optimum temperature range was 28–31°C. Endo (1935a) determined the minimum, optimum and maximum pH for growth to be 2·5, 5·4–6·7 and 7·8 respectively. Kurodani, Yokogi & Yamamoto (1959) found that 2,4-D stimulated growth and increased pathogenicity. Hsu & Dough (1964) tested the effects of several chemical elements on the growth of the fungus. Hashioka & Makino (1969) reported that *Rhizoctonia oryzae* has an optimal temperature between 25 and 32°C and still grows well at 32–35°C; the figures for *Corticium sasakii* are 25–30°C and 20–25°C respectively, and for *C. solani* 25–27°C and 27–30°C respectively. *C. gramineum* has an optimal temperature between 20 and 25°C. Tu (1969) and Santos (1970), however, found varying thermal responses among many isolates of *C. sasakii.* Some have a wide optimal temperature range (24–36°C), others narrower (28–36°C, 24–32°C, and 28°C).

Hemmi & Endo (1931) found that sclerotia were formed most abundantly in the light and that their formation was also accelerated by a sudden fall in temperature. They (1928) devised a staining method to test the viability of the sclerotia, using a 1–2% aqueous solution of eosin, acid fuchsin, and other materials. The dead sclerotia were easily stained but not those which were alive.

Inoue & Uchino (1963) reported that no sclerotia were formed on media containing ammonium sulphate and peptone as nitrogen sources. The sclerotia germinate over a temperature range of 16–30°C, with an optimum at 28–30°C. High relative humidity, above 95–96%, is required for germination.

Santos (1970) found that the size and number of sclerotia formed on agar plates were affected by the carbon and nitrogen sources present in the medium, particularly the latter, irrespective of the isolates tested. Medium containing proline produces many (average 37 per petri dish) very large sclerotia (3·5×3·1 mm); few (3–8) large (2·5×2·2 mm) sclerotia are produced on isoleucine, threonine, leucine and valine media; abundant (162–244) moderately sized sclerotia are found on arginine, serine, glatamine, alanine and aspartic acid media. Tryptophan histidine and lysine media produce few very small sclerotia or none at all (Fig. IV–10).

The basidial state may be formed on steamed soil as reported by Stretton *et al.* (1964) or soil extract agar (Papavizas, 1965) for *Rhizoctonia.*

Chen (1958) isolated *p*-hydroxyphenylacetic acid from the fungus and Yoshimura (1955) studied its ability to decompose cellulose.

Variability. Chien & Chung (1963) studied 300 isolates from Taiwan, inoculated on 16 rice varieties. Based upon the degree of pathogenicity (number of leaves infected), they classified the 300 isolates into 7 cultural types and 6 physiologic races. They found correlations between cultural characteristics and races, for instance race 1 was of cultural type 3, race 2 of cultural type 6, etc. The susceptible and resistant reactions used to separate the races were not, however, clear-cut, many isolates having intermediate reactions. Tu (1967) also studied many strains from Taiwan and noted that strains with less aerial mycelium were usually more pathogenic. Akai, Ogura & Sato (1960) found, conversely, that strains with poor mycelial growth were less pathogenic. Some relationship has also been found between utilization of certain C and N sources and pathogenicity. Santos (1970) reported that cultures from media containing arginine, urea, threonine, glycine etc. are more virulent than those

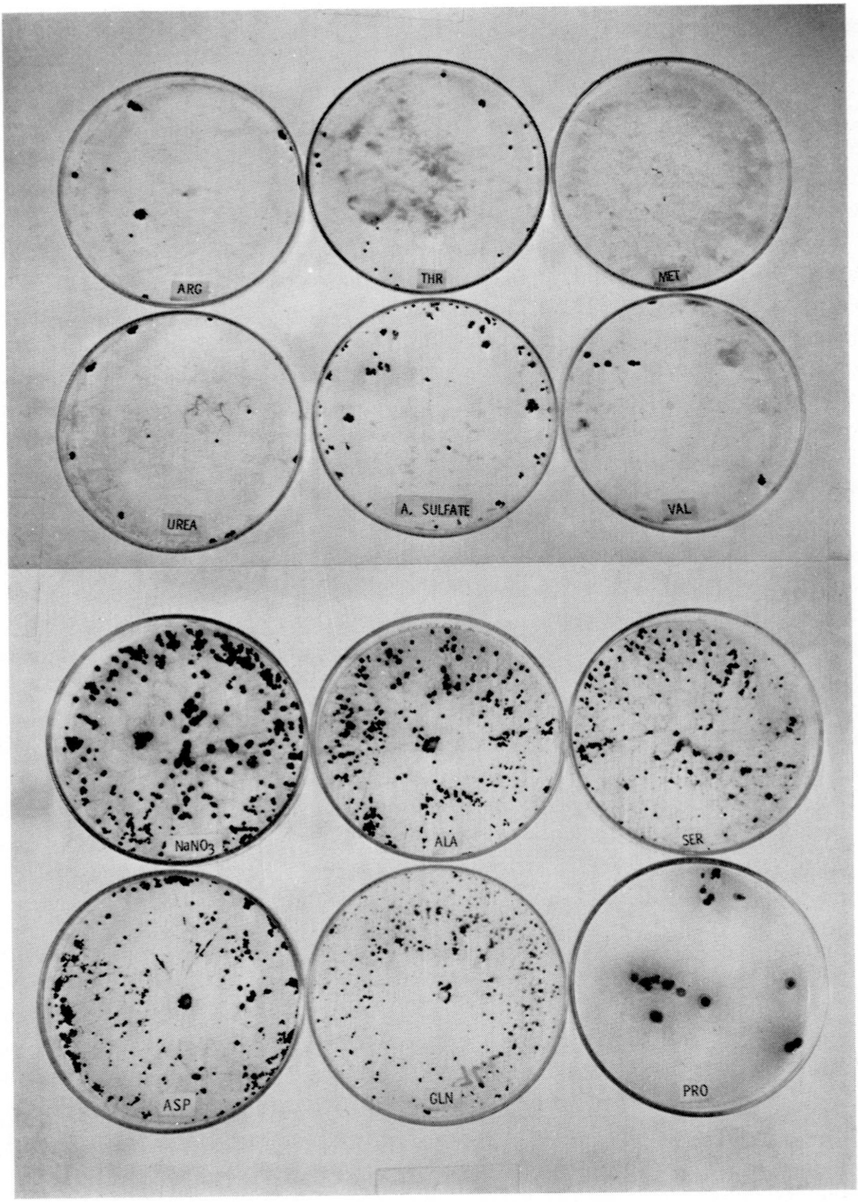

Fig. IV–10. Sclerotia production by *C. sasakii* on media with different sources of nitrogen. (Santos, 1970.)

from media containing serine, isoleucine, tryptophan, tyrosine, lysine or histidine, which have low pathogenicity.

Many isolates of *Rhizoctonia solani* from other crops have been shown to be pathogenic to the rice plant (Yokogi, 1927; Ryker, 1939; Sato & Shoji, 1957; Nonaka & Tanaka, 1964; etc.).

Hashioka & Makino (1970) from a brief survey reported 8 species causing sheath spots on rice: *Corticium sasakii*, *C. solani*, *Rhizoctonia oryzae*, *R. zeae*, *C. microsclerotia*, *C. rolfsii*, *C. gramineum* and *C.* sp.

Disease Cycle

Endo (1931) in Japan reported that the fungus is able to survive in soil over winter as sclerotia or as mycelium. The sclerotia lose their viability in dry soil after 21 months. Park & Bertus (1932) in Ceylon tested the survival of sclerotia under various conditions. At room temperature on dry or moist soil, they survived for at least 130 days, and for 224 days when submerged under 3 inches of tap water. Palo (1926) reported that the sclerotia may survive for several months in soil in the Philippines.

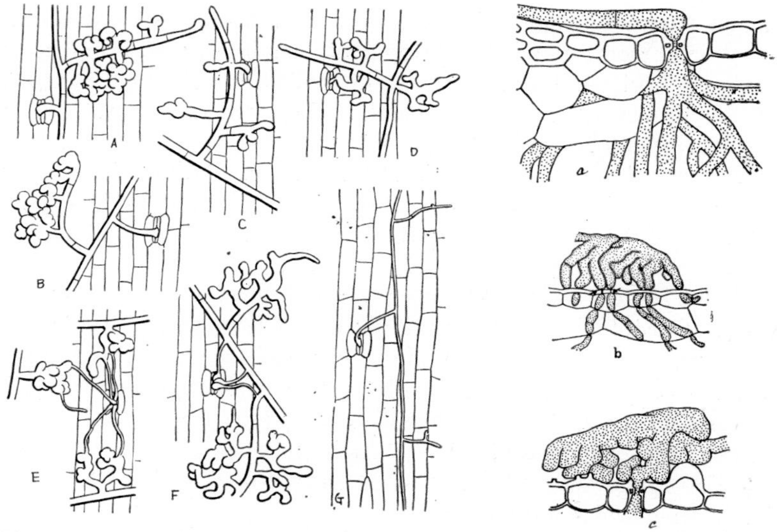

Fig. IV–11. Entrance of *Corticium sasakii* into host tissue. A to G through stomata (surface view), a and c (cross-section). b. through cuticle cells (cross-section). (Kozaka, 1961.)

The sclerotia float to the surface of the water during soil puddling, levelling, weeding and other operations of rice cultivation. They may be carried away or drift off and finally may come in contact with a rice plant, and so initiate infection. Soon after the primary lesions are formed, mycelium grows rapidly on the surface of the plant and inside its tissues, proceeding upwards as well as laterally and initiating secondary lesions. The mycelium is most active and infectious when the lesions are young. Only scanty mycelium develops on old bleached lesions and it is not highly infective.

The fungus enters the plant through the stomata or may penetrate directly through the cuticle, usually forming an infection cushion before penetration. It usually enters the leaf sheath from the inner surface, but may enter either surface of the leaf blade. Infection may occur in the temperature range 23–35°C, optimum 30–32°C. A high relative humidity, 96–97%, is required (Kozaka, 1965) (Fig. IV–11).

Endo (1930) and Hemmi & Endo (1933) recorded that at 32°C, infection took place in 18 hours, and at 28°C in 24 hours, with continuous wetting.

Effect of Environmental Conditions

Many workers have reported that the disease is especially destructive under conditions of high humidity and high temperature. While the temperature within the rice crop varies with that of the air temperature, humidity among the plants is greatly affected by the thickness of the stand, so that close planting and heavy applications of fertilizers, leading to thick growth, tend to increase disease incidence. For the same reason the disease is usually observed in the field after the plants have reached the maximum tillering stage, when high humidity tends to be maintained within the microclimate. Kozada (1961), who made extensive ecological studies in Japan, distinguished two phases of disease development; upward development, i.e. infection of the upper leaf sheaths, and horizontal development, i.e. infection of neighbouring stems. Under favourable conditions horizontal development is rapid. Upward development occurs only after the heading stage and under conditions favourable to the causal fungus. If conditions are not favourable after the heading stage, lesions are found only on the lower leaf sheaths. Endo (1935b) and Yoshimura (1955) both reported that sunlight inhibits and shading promotes infection.

The disease is more severe on plants grown in soil with high nitrogen (Kozaka, 1961, 1965; Chen, Chien & Hwang, 1961; Loo, Choun & Lee, 1963). Kozaka (1961) reported that the susceptibility of the leaf sheath is closely correlated with its nitrogen content, but not with its content of sugar or starch. Inoue & Uchino (1963) reported a higher disease incidence when a large amount of phosphate was added, and a lower incidence when much potash was used. Endo (1933) found that the addition of salt at 0·01–1% reduced infection, but this treatment also retarded plant growth.

The intensity of primary infection is closely related to the number of sclerotia in contact with the plant, but subsequent disease development is more affected by environmental conditions and the susceptibility of the host plant. It has been shown that plants become more susceptible as they grow older, and the environmental conditions during the later stages of growth greatly affect the incidence of and damage caused by the disease (Kozaka, 1961).

Varietal Resistance

Kozaka (1961) reported that the stems of rice plants as they approach the heading stage become more compact and are more subject to contact infection. Young leaf sheaths or blades when 2–3 weeks old are more resistant than when they are 5–6 weeks old. In older plants the encirclement of the leaf sheath around the stem becomes loose, which facilitates penetration of the mycelium to the inner side of the sheath.

Before the heading stage, the upper leaf sheaths and blades are more resistant than the lower ones, but after the heading stage the susceptibility of the upper parts increases with increasing plant age.

It has been generally observed that early varieties appear to suffer more than late ones, but Kozaka (1961) found no significant difference in varietal reaction among varieties tested when 40–50 days old. In Japan, late maturing varieties have more chance of escaping the disease than early varieties because of the lower temperature in autumn, when the upper leaves are not damaged. Tall varieties with few tillers usually have less infection than short varieties with many tillers.

Hashioka (1951a), in a field test in Japan, found varieties from India, Thailand, Burma, Europe and North America to be more resistant than local varieties. Tests recently made in Taiwan (Hsieh, Wu & Shian, 1965) in which were included many varieties grown at 4 localities and for several seasons, showed that there was wide variation in varietal reaction. Most of the varieties tested were susceptible or moderately susceptible, relatively few were in the resistant group, and none was immune. The tests also showed that *indica* varieties are more resistant than are *japonica* varieties, and that the incidence of sheath blight increases from north to south of Taiwan, as the temperature increases. A high level of soil nitrogen favoured the disease, and there was an apparent correlation between earliness and disease incidence. Kang, Lee & Kim (1965) also tested 155 varieties in Korea from 1962 to 1964 and reported that it is difficult to select varieties with resistance.

Yoshimura & Nishizawa (1954) studied various methods for testing varietal resistance, including: (1) field tests using artificial inoculation, (2) seedling tests, (3) inoculation at different stages, (4) tests in pots, and (5) sheath inoculation tests. They found: (1) field tests using artificial inoculation increased efficiency and the uniformity of infection; (2) seedling tests showed no differences in the percentage of stems becoming diseased; (3) the stage of maximum tillering is the optimum one for varietal testing; (4) for irrigated paddy field tests, cultures on sterilized straw inserted among the tillers of each hill for 2 months, or wrapping of the tillers after inoculation for one week, worked well; (5) the sheath inoculation method, similar to that used in blast disease, gave no indication of differences in resistance among the varieties tested.

Common inoculation methods used for testing varietal resistance, as well as for other experiments, are the use of naturally infected fields, with or without artificial inoculation, or the use of potted plants inoculated by insertion of the pathogen in culture among the stems, these being wrapped in paper to keep them moist. In either case, the plants have to be grown on until the early booting stage or beyond before disease incidence can be assessed, and they are therefore apt to be influenced by the usual variations in environmental conditions which occur naturally outdoors.

The different methods of inoculation were investigated at IRRI with the aim of finding a method which would save time and space. Among several methods tried, the seedling test method, similar to the procedures used in the rice blast nurseries, seems to work best. The seedlings, 3 weeks old, are inoculated with a culture of the fungus on rice straw by placing the straw on the soil surface at the base of the seedlings. The border rows are planted with a resistant variety (Kataktara DA-2). Readings are taken 2 weeks after inoculation. The results of tests of several hundreds of varieties are very similar to those obtained by Hsieh, Wu & Shian in lengthy field tests (1965).

Ono (1953) and Yoshimura (1954) suggested scales to estimate the severity of the disease or the degree of damage caused. Ono used the formula:

$$\text{Degree of damage} = \frac{0A + 10B + 15C + 25D + 40E}{N}$$

in which A is the number of plants without lesions, E is the number of plants in which all the sheaths are infected, and B, C, and D are intermediary. N is the total number of plants examined.

Yoshimura used another formula:

$$\text{Degree of severity } (\%) = \frac{3N_1 + 2N_2 + N_3 + 0N_4 \times 100}{3N}$$

in which N_4 is the number of tillers with the 4 uppermost sheaths healthy, N_1 is the number of tillers with all the 4 uppermost sheaths infected, and N_2 and N_3 are intermediate. N is the total number of tillers. At IRRI, in the seedling test mentioned above, a simple scale based on the percentage of sheaths infected is used. A scale of 0–9 is employed to simplify further the method of indicating the degree of resistance or susceptibility. On this scale 0 indicates immunity, 3 indicates that 21–30% of the sheaths are infected, etc.

Hashioka (1951) reported that resistance to sheath blight is inherited as a dominant character, and in crosses between resistant and susceptible varieties, the majority of the F_2 population were resistant. In some crosses a 3 : 1 ratio of resistant and susceptible plants was found.

Host Range

The fungus isolated from rice has been shown to infect many other plants, and similar fungi isolated from other crops are able to infect rice (Yokogi, 1927; Tervet, 1937; Ryker, 1939; Sato & Shoji, 1957). Nakata & Kawamura (1939) gave an indication of the wide range of hosts of the fungus from rice. Kozaka (1965) stated that plants of 188 species in 32 families may be infected by the rice fungus.

Chemical Control

Control of sheath blight by fungicides has been attempted for many years in Japan. Copper and organic mercury compounds have been used in the past, but recently organic arsine compounds have been shown to be more effective. Organic tin fungicides have also been tried (Tamura, 1965). One of the most extensive and comprehensive studies in this field was that made by Kozaka (1961) and his co-workers. They developed an efficient method of testing the effectiveness of various fungicides by the use of detached leaflets of broad bean in petri dishes. They reported the following results. (1) The three commercial inorganic copper fungicides tested have remarkable preventive and good residual action, but have little effect on mycelial growth and cannot inhibit the enlargement of lesions when applied after infection. (2) The 13 organic mercury compounds tested have two types of action. Those in Group I, which have methyl or ethyl radicals, have remarkable eradicative action and inhibit mycelial growth, but show little or no residual action. Those in Group II, with a phenyl radical, are less effective both in inhibiting mycelial growth and in protective action and have no effect on disease development after the appearance of lesions. (3) The 10 organic arsine compounds tested were the most effective in inhibiting mycelial growth, infection and enlargement of lesions. Particularly effective were methylarsine sulphide and urbacid (methylarsine bisdimethyl dithiocarbamate); they are effective in concentrations of about 50 ppm and only two applications, one at the first appearance of lesions and another at the booting stage, are required. (4) Several organic sulphur compounds were found to be ineffective. (5) The organic arsines are often phytotoxic, but toxicity can be reduced by adding a small amount of iron, e.g. as ferric chloride or ferric sulphate. Organic arsines linked with iron such as ferric-methylarsenate give good control and have apparently no injurious effects.

Methods of applying the fungicides have also been studied, including the use of various equipment (Sekizawa, Hashimoto & Ito, 1962; Sekizawa, Ito & Sakurai, 1962) and application of the chemicals in the irrigation water (Hashioka & Makino, 1961).

The use of antibiotics has also been investigated (Yamamoto, Iwasa, Shibata, Mizuno & Miyake, 1965). Recently it has been reported that an antibiotic, polyoxin, is almost as effective as the organo-arsenicals and is not phytotoxic.

PCP (pentachlorophenol), used for weed control in rice fields, has also been found useful in controlling sheath blight as a side effect (Ono & Iwata, 1961; Takasu & Nishimura, 1962; Inoue & Uchino, 1963; Endo, Shinohara & Hara, 1965).

LITERATURE CITED

AKAI, S., OGURA, H. & SATO, T. 1960. Studies on *Pellicularia filamentosa* (Pat.) Rogers. I. On the relation between the pathogenicity and some characters on culture media. *Ann. phytopath. Soc. Japan* **25**: 125–130. [Jap. Engl. summ.]

CHEN, Y. S. 1958. Metabolic products of *Hypochnus sasakii* Shirai. Isolation of *p*-hydroxyphenylacetic acid and its physiological activity. *Bull. agric. Chem. Soc. Japan* **22**: 136–142.

CHIEN, C. C. & CHUNG, S. C. 1963. Physiologic races of *Pellicularia sasakii* in Taiwan. *Agric. Res., Taiwan* **12** (2): 1–6.

ENDO, S. 1927. Ueber die Sclerotienkrankheit der Reispflanzen in den Philippinen. *Pl. Prot., Tokyo* **14**: 283–288. [Jap.] (Abs. in *Jap. J. Bot.* **3**: 77, 1927.)

ENDO, S. 1930. On the influence of the temperature upon the development of *Hypochnus*. *Ann. phytopath. Soc. Japan* **2**: 280–283. [Jap.]

ENDO, S. 1931. Studies on *Sclerotium* diseases of the rice plant. V. Ability of overwintering of certain important fungi causing *Sclerotium* disease of the rice plant and their resistance to dry conditions. *Forschn Geb. PflKrankh., Tokyo* **1**: 149–167. [Jap. Engl. summ.]

ENDO, S. 1933. Influence of salt on the pathogenicity of *Hypochnus sasakii* Shirai. *Trans. Tottori Soc. agric. Sci.* **4**: 362–367. (*Rev. appl. Mycol.* **14**: 120.)

ENDO, S. 1935a. On the influence of hydrogen-ion concentration on the mycelial growth of the causal fungi of sclerotial diseases of the rice plant. *Bull. Miyazaki Coll. Agric. For.* **8**: 1–11.

ENDO, S. 1935b. Effect of sunlight on the infection of the rice plant by *Hypochnus sasakii* Shirai. *Bull. Miyazaki Coll. Agric. For.* **8**: 75–78. (*Rev. appl. Mycol.* **15**: 48.)

ENDO, S., SHINOHARA, M. & HARA, S. 1965. Effects of PCP-Na herbicide on sheath blight of rice plants. I. The inhibitive mechanism of the PCP-Na herbicide to the occurrence of the disease. II. Residual effectiveness of PCP-Na herbicide on *Pellicularia sasakii* (Shirai) S. Ito in irrigation. *Bull. Coll. Agric. vet. Med. Nihon Univ.* **20**: 1–9; 11–17.

GADD, C. H. & BERTUS, L. S. 1928. *Corticium vagum* B. & C.—cause of a disease of *Vigna oligosperma* and other plants in Ceylon. *Ann. R. bot. Gard. Peradeniya* **11**: 27–49.

HASHIOKA, Y. 1951a. Varietal resistance of rice to sheath blight and the sclerotial diseases. (Studies on pathological breeding of rice, IV). *Jap. J. Breed.* **1**: 21–26. [Jap. Engl. summ.]

HASHIOKA, Y. 1951b. Inheritance of resistance to sheath blight in rice varieties. *Ann. phytopath. Soc. Japan* **15**: 98–99. [Jap.]

HASHIOKA, Y. & MAKINO, M. 1961. Examination of method of chemical application in irrigation water in paddy field against sheath blight disease of rice plant. *Proc. Kansai Pl. Prot. Soc.* **3**: 25–29. [Jap.]

HASHIOKA, Y. & MAKINO, M. 1969. *Rhizoctonia* group causing the rice sheath spots in the temperate and tropical regions, with special reference to *Pellicularia sasakii* and *Rhizoctonia oryzae*. *Res. Bull. Fac. Agric. Gifu Univ.* **28**: 51–63.

HEMMI, T. & ENDO, S. 1928. On a staining method for testing the viability of sclerotia of fungi. *Mem. Coll. Agric. Kyoto imp. Univ.* 7: 39–49. (*Rev. appl. Mycol.* **8**: 263.)

HEMMI, T. & ENDO, S. 1931. Studies on the *Sclerotium* diseases of the rice plant. III. Some experiments on the sclerotial formation and the pathogenicity of certain fungi causing sclerotium diseases of the rice plant. VI. On the relation of temperature and period of continuous wetting to the infection of rice plant by *Hypochnus sasakii* Shirai. *Forschn Geb. PflKrankh., Tokyo* **1**: 111–125; **2**: 202–218. [Jap. Engl. summ.] (*Rev. appl. Mycol.* **11**: 538; **13**: 124.)

HEMMI, T. & YOKOGI, K. 1927. Studies on *Sclerotium* diseases of the rice plant. I. *Agriculture Hort., Tokyo* **2**: 955–1094. [Jap. Engl. summ.] (*Rev. appl. Mycol.* **7**: 266.)

HORI, M. 1969. On forecasting the damage due to sheath blight of rice plants and the critical point for judging the necessity of chemical control of the disease. *Rev. Pl. Prot. Res.* **2**: 70–73.

HSIEH, Y. T., WU, Y. L. & SHIAU, K. N. 1965. Screening for sheath blight resistance in rice varieties. Kaohsiung Distr. agric. imp. Stn, Taiwan, 92 pp. [Chin. Engl. summ.]

HSU, S. C. & DOUGH, T. C. 1964. Effect of chemical elements on the growth of crop pathogens. II. Effect of Hg-, Ni-, Sn-, Pb-, Fe- and Al- salts on rice blast and sheath spot fungi. *Jnl. Taiwan agric. Res.* **3** (3): 38–43.

IKATA, S. & HITOMI, T. 1930. On the mode of primary infection through sclerotia and field observations on the basidiospore formation in the earliest stage of monko-disease (*Hypochnus sasakii* Shirai) of rice plants. *J. Pl. Prot., Tokyo* **17**: 17–28. [Jap.] (*Rev. appl. Mycol.* **10**: 55.)

INOUE, Y. & UCHINO, K. 1963. Studies on sheath blight of rice plant caused by *Pellicularia sasakii* (Shirai) S. Ito. I. Ecology of damage and chemical control. *Minist. Agric. For. Fish. Res. Council, Japan appointed expt. No. 4 at Yamaguchi Agric. Exp. Stn,* 136 pp. [Jap. Engl. summ.]

KANG, I. M., LEE, E. K., LEE, S. C. & LIM, Y. K. 1965. Studies on the resistance of rice variety of sheath blight (*Corticium sasakii*) in field. *Res. Reps Korea, Office of Rural Development* **8**: 235–241.

KOZAKA, T. 1961. Ecological studies on sheath blight of rice plant caused by *Pellicularia sasakii* (Shirai) S. Ito, and its chemical control. *Chugoku agric. Res.* **20**: 1–133. [Jap. Engl. summ.]

KOZAKA, T. 1965. Ecology of *Pellicularia* sheath blight of rice plant and its c mical control. *Ann. phytopath. Soc. Japan* **31** (Commem. Issue): 179–185. [Jap.]

KOZAKA, T. 1970. *Pellicularia* sheath blight of rice plants and its control. *Jap. agric. Res. Q.* **5**: 12–16.

KURODANI, K., YOKOGI, K. & YAMAMOTO, M. 1959. On the effect of 2,4-dichlorophenoxy-acetic acid on the mycelial growth of *Hypochnus sasakii* Shirai. *Forschn Geb. PflKrankh., Tokyo* **6**: 132–135. [Jap. Engl. summ.]

LOO, E. P., CHOUN, C. Q. & LEE, D. C. 1963. Studies on *Rhizoctonia* blight of rice. *Acta phytophylac. sin.* **2**: 431–440.

MATSUMOTO, T. 1934. Some remarks on the taxonomy of the fungus *Hypochnus sasakii* Shirai. *Trans. Sapporo nat. Hist. Soc.* **13**: 115–120.

MATSUMOTO, T. & HIRANE, S. 1933. Physiology and parasitism of the fungi generally referred to as *Hypochnus sasakii* Shirai. III. Histological studies in the infection by the fungus. *J. Soc. trop. Agric., Taiwan* **5**: 367–373.

MATSUMOTO, T., YAMAMOTO, W. & HIRANE, S. 1932. Physiology and parasitology of the fungi generally referred to as *Hypochnus sasakii* Shirai. I. Differentiation of the strains by means of hyphal fusion and culture in differential media. *J. Soc. trop. Agric., Taiwan* **4**: 370–388.

MATSUMOTO, T., YAMAMOTO, W. & HIRANE, S. 1933. Physiology and parasitism of the fungi generally referred to as *Hypochnus sasakii* Shirai. II. Temperature and humidity relations. *J. Soc. trop. Agric., Taiwan* **5**: 332–345.

MIYAKE, I. 1910. Studien über die pilze der Reispflanze in Japan. *J. Coll. Agric., Tokyo* **2**: 237–276.

MIZUTA, H. 1956. On the relation between yield and inoculation times of sheath-blight, *Corticium sasakii* in the earlier planted paddy rice. *Ass. Pl. Prot. Kyushu* **2**: 100–102. [Jap.]

NAKATA, K. & KAWAMURA, E. 1939. Studies on sclerotial diseases in rice. *Bureau Agric. Minist. Agric. For. Japan, Agric. Exp. Stn Records* 139, 176 pp. [Jap.]

NATIONAL INSTITUTE OF AGRICULTURAL SCIENCES, JAPAN. 1954. Insects and diseases of rice plants in Japan. Paper presented at the 4th Session, Int. Rice Commn, Tokyo, 33 pp.

NONAKA, F. & TANAKA, K. 1964. On the relation between the growth types of *Rhizoctonia solani* Kühn and their infectivity to the sheath of rice plant and to the seedlings of several other crops. *Agric. Bull. Saga Univ.* **19**: 9–24. [Jap. Engl. summ.]

ONO, K. 1953. Damage of sheath blight to rice. *J. Pl. Prot., Tokyo* **7**: 99–103. [Jap.]

ONO, K. & IWATA, K. 1961. Control of sheath blight disease of rice plant by PCP. *Nogyo Oyobi Engei* **36**: 71–74. [Jap.]

PALO, M. A. 1926. *Rhizoctonia* disease of rice. I. A study of the disease and of the influence of certain conditions upon the viability of the sclerotial bodies of the causal fungus. *Philipp. Agric.* **15**: 361–375.

PAPAVIZAS, G. C. 1965. Comparative studies of single-basidiospore isolates of *Pellicularia filamentosa* and *Pellicularia praticola*. *Mycologia* **57**: 91–103.

PARK, M. & BERTUS, L. S. 1932. Sclerotial diseases of rice in Ceylon. I. *Rhizoctonia solani* Kühn. *Ceylon J. Sci.*, Sect. A, **11**: 319–331.

REINKING, O. A. 1918. Philippine economic plant diseases. *Philipp. J. Sci.*, Sect. A, **13**: 165–274.

ROGERS, D. P. 1943. The genus *Pellicularia* (Thelephoraceae). *Farlowia* **1**: 95–118.

RYKER, T. C. 1939. The *Rhizoctonia* disease of Bermuda grass, sugarcane, rice and other grasses in Louisiana. *Proc. Sixth Congr. int. Soc. Sugarcane. Techn., 1938*: 198–201.

RYKER, T. C. & GOOCH, F. S. 1938. *Rhizoctonia* sheath spot of rice. *Phytopathology* **28**: 233–246.

SANTOS, L. G. 1970. Studies on the morphology, physiology and pathogenicity of *Corticium sasakii* (Shirai) Matsumoto with special reference on the effects of nitrogen and carbon sources. M.S. thesis, Univ. Philipp. Coll. Agric.

SATO, K. & SHOJI, T. S. 1957. Pathogenicity of *Rhizoctonia solani* Kühn isolated from gram-inaceous weeds in forestry nurseries. *Bull. For. Exp. Stn Meguro* 96: 89–104. [Jap. Engl. summ.]

SAWADA, K. 1912. On the 'shirakinibuyo' of camphor. *Spec. Bull. Formosan Agric. Exp. Stn* 4, 805 pp. [Jap.]

SEKIZAWA, H., HASHIMOTO, T. & ITO, S. 1962. Control effect of sheath blight disease of rice plant by mist spray. *Ann. Rep. Soc. Pl. Prot. North Japan* 13: 149–150. [Jap.]

SHIRAI, M. 1906. On *Hypochnus sasakii* n. sp. *Bot. Mag., Tokyo* **20**: 319–323.

STRETTON, H. M., MCKENZIE, A. R., BAKER, K. F. & FLENTJE, N. T. 1964. Formation of the basidial stage of some isolates of *Rhizoctonia*. *Phytopathology* **54**: 1093–1095.

TAKATSU, S. & NISHIMURA, J. 1962. Studies on the control of sheath blight of rice plant. 4. Effects of sodium salt of pentachlorophenol. *Bull. Hyogo Pref. agric. Exp. Stn* 10: 27–30. [Jap.]

TALBOT, P. H. B. 1965. Studies of '*Pellicularia*' and associated genera of Hymenomycetes. *Persoonia* **3**: 371–406.

TAMURA, H. 1965. Effect of organotin fungicides applied for the control of rice blast, and rice sheath blight. *Bull. natn. Inst. agric. Sci., Tokyo*, Ser. C, **19**: 47–79. [Jap. Engl. summ.]

TERVET, I. W. 1937. An experimental study of some fungi injurious to seedling flax. *Phytopathology* **27**: 531–546.

TU, J. C. 1967. Strain of *Pellicularia sasakii* isolated from rice in Taiwan. *Pl. Dis. Reptr* **51**: 682–684.

TU, J. C. 1968. Physiological specialization of strains of *Pellicularia sasakii* isolated from rice plants. *Pl. Dis. Reptr* **52**: 323–326.

VENKATARAYAN, S. V. 1949. The validity of the name *Pellicularia koleroga* Cooke. *Indian Phytopath.* **2**: 186–189.

WEI, C. T. 1934. *Rhizoctonia* sheath blight of rice. *Bull. Univ. Nanking Coll. Agric. For.* 15, 21 pp.

YAMAMOTO, H., IWATA, T., SHIBATA, M., MIZUNO, K. & MIYAKE, A. 1965. *Streptomyces multispiralis* nov. sp. and a new antibiotic, neohumidin, which controls sheath blight, *Pellicularia filamentosa sasakii* on rice. *Agric. Biol. Chem.* **29**: 360–368.

YOKOGI, K. 1927. On the *Hypochnus* disease of soybeans and its comparison with that of rice plants. *J. Pl. Prot., Tokyo* **14**: 12. (Abs. in *Jap. J. Bot.* **3**: 120–121, 1927.)

YOSHIMURA, S. 1954. On the scale for estimating degree of severity of sheath blight by *Hypochnus sasakii* Shirai in rice plant. *Ann. phytopath. Soc. Japan* **19**: 58–60. [Jap. Engl. summ.]

YOSHIMURA, S. 1955a. Nutritional and physiological studies on the rice sheath blight fungus (*Hypochnus sasakii* Shirai). I. On the cellulose decomposition by the rice sheath blight fungus. *Kyushu agric. Res.* **15**: 62–64. [Jap.]

YOSHIMURA, S. 1955b. On the effect of the shading upon the susceptibility of the rice plant to the sheath blight, *Hypochnus sasakii* Shirai. *Kyushu agric. Res.* **16**: 113. [Jap.]

YOSHIMURA, S. & NISHIZAWA, T. 1954. Studies on the method of testing varietal resistance of upland rice plants caused by sheath blight, *Hypochnus sasakii* Shirai. *Bull. Kyushu agric. Exp. Stn* **2**: 361–376. [Jap. Engl. summ.]

SHEATH SPOT

Ryker & Gooch (1938) made an extensive study of this disease, which was first described by Tullis (1934), who considered that it was caused by a species of *Trichoderma*. It occurs in U.S.A., in Louisiana, Arkansas and Texas, but is not considered important. It has also been reported from Japan and Indo-china (Vietnam).

The disease produces spots on the leaf sheath which very much resemble those of sheath blight, but are usually smaller and on the basal part of the plant.

Ryker & Gooch (1938) considered that they had sufficient grounds to regard the fungus as a new species and named it *Rhizoctonia oryzae*. The most obvious differences between this fungus and *R. solani* are found in culture. The sclerotial masses of *R. oryzae* are of indefinite size and shape, and are frequently formed above the main hyphal branches, giving a crow's foot pattern. The colour of the sclerotial masses varies somewhat with the substrate, but is generally a shade of salmon. The sclerotia of *R. solani* are large and round and more regular in shape, and generally brown. The sclerotial masses of *R. oryzae* are not formed on the host.

The mycelium is superficial or submerged in culture, 6–10μ in width, branching at an acute angle with a slight constriction and with a septum a short distance from the point of constriction. The hyphae intertwine and anastomose forming masses of sclerotial cells or pseudo-sclerotia of various sizes and shape (roughly 0·2–1·3×0·5 mm). The optimum temperature for growth is about 32°C; there is slight growth at 10°C, and fair growth at 35°C. The fungus also infects *Fimbristylis miliacea* Vahl, *Echinochloa crus-galli* (L.) Beauv. subsp. *edulis* Honda and other grasses but not chilli, eggplant, tomato and bean.

Two or more related fungi also cause similar spots on rice. Ryker & Gooch (1938) reported that *R. zeae* Voorhees, common on maize, also causes sheath spots on rice, and Roger (1941–42) reported it as attacking rice in Vietnam. The fungus is characterized by its small sclerotia, 0·5–1 mm in diameter, submerged or superficial in culture, and 0·1–0·5 mm in diameter, mostly superficial, on the host plant (maize). They are brown to dark brown at maturity, usually single, sometimes conglomerated, and are produced abundantly in culture. The hyphae are hyaline at first, reddish-brown in old cultures, salmon-pink on the host, later becoming dull grey, multiseptate, constricted at the septa.

Park & Bertus (1934) described a fungus which they called *Rhizoctonia solani* B strain, and which they regarded as a weak parasite of rice. Mundkur (1935) showed this fungus to be identical with *R. microsclerotia* Matz. Rogers (1943) included both in *Pellicularia filamentosa* but Padwick (1950) regarded them as separate species. This organism is also characterized by its small sclerotia, 230–780μ in diameter (mean 472μ). They are brown or purple brown, dull not shiny, superficial, roughly spherical, and loosely attached to the vegetative hyphae. Hyphae measure up to 8μ in diameter, and are septate, branching freely at right angles, with a constriction at the junction.

LITERATURE CITED

MUNDKUR, B. B. 1935. Parasitism of *Sclerotium oryzae* Catt. *Indian J. agric. Sci.* **5**: 393–414.

PADWICK, G. W. 1950. Manual of rice diseases. 198 pp., Kew, Commonwealth Mycological Institute.

PARK, M. & BERTUS, L. S. 1934. Sclerotial diseases of rice in Ceylon. V. *Rhizoctonia solani* B strain. *Ceylon J. Sci.*, Sect. A, **12**: 25–36.

ROGER, L. 1941–42. Les champignons à sclérotes parasites du riz. *Bull. econ. Indochine, 1941* (6) & *1942* (1–5), 302 pp.

ROGERS, D. P. 1943. The genus *Pellicularia* (Thelephoraceae). *Farlowia* **1**: 95–118.

RYKER, T. C. & GOOCH, F. S. 1938. *Rhizoctonia* sheath spot of rice. *Phytopathology* **28**: 233–246.

TULLIS, E. C. 1934. *Trichoderma* sheath spot of rice. *Phytopathology* **24**: 1374–1377.

OTHER SCLEROTIAL DISEASES

Several other species of sclerotium-forming fungi besides those discussed above are associated with leaf sheath diseases. Most of them have a low level of pathogenicity and others are perhaps only saprophytes. Those commonly found in Japanese literature are: *Sclerotium oryzae-sativae* Sawada, *S. fumigatum* Nakata ex Hara, *S. hydrophilum* Sacc., and *S. oryzicola* Nakata & Kawamura. Nakata & Kawamura (1939) studied several species in detail. A number of them have been reported from China (Wei, 1957) and Vietnam (Roger, 1941–42) but they are little known in other parts of the Asian tropics.

Sclerotium oryzae-sativae was first described by Sawada (1922) in Taiwan, but has since also been reported from Japan, Vietnam, and East, Northwest and Southwest China. The fungus causes lesions similar to those of sheath blight but they are smaller (0·5–1 cm) with distinct brown margins. Several lesions may occur together. It also attacks the culm which turns brown and may lodge and die.

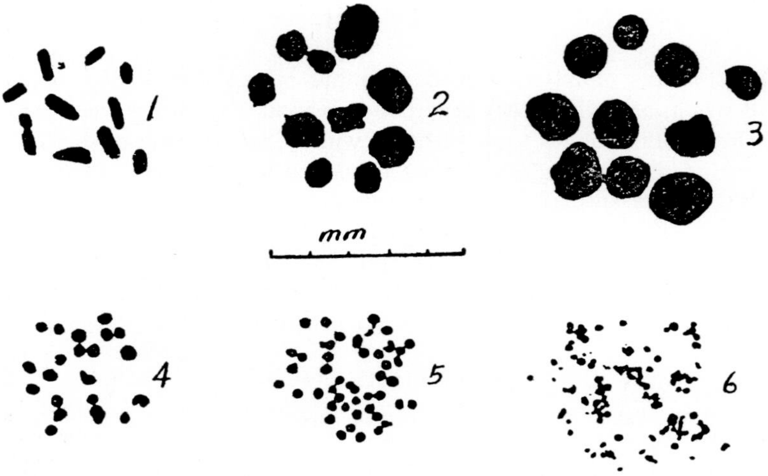

Fig. IV–12. Sclerotia of *Sclerotium oryzae-sativae* (1), *S. fumigatum* (2), *Corticium sasakii* (3), *S. hydrophilum* (4), *S. oryzae* (*Leptosphaeria salvinii*) (5), *Helminthosporium sigmoideum* var. *irregulare* (6). (Nakata & Kawamura, 1939.)

The sclerotia formed inside the hollow stem or in large cavities in the leaf sheath are pure brown or brownish-grey with a dull surface. They are covered with mycelium or are warty, very polymorphous, often oblong-cylindrical in outline, rarely round or regularly ovoid, often in pairs. Their dimensions are very variable because of their aggregation, 1620–3000 × 1020–2400μ, mostly 2250 × 1700μ, more regular and isolated in culture, 346–703μ for the spherical forms, 377–1126 × 289–904μ for the ovoid forms. The sclerotia are composed of swollen mycelium, 14–24μ in diameter, and no distinct outer and inner layers are visible in cross section (Figs. IV–12, 13). The mycelium is hyaline, even when old, loose or a little aggregated, much branched, variable in diameter, 4·2–6·3μ (av. 5·09μ), with moniliform cells 23–32·8 × 7·1–10·3μ. The optimum temperature for mycelial growth and sclerotia formation is 32°C, minimum 10°C and maximum 40–42°C. The optimum pH is 4·2, minimum and maximum being 2·7 and 8·2 respectively.

This fungus has also been found on *Juncellus serotinus* Clarke, *Zizania latifolia* Turcz and *Oryza cubensis* in nature.

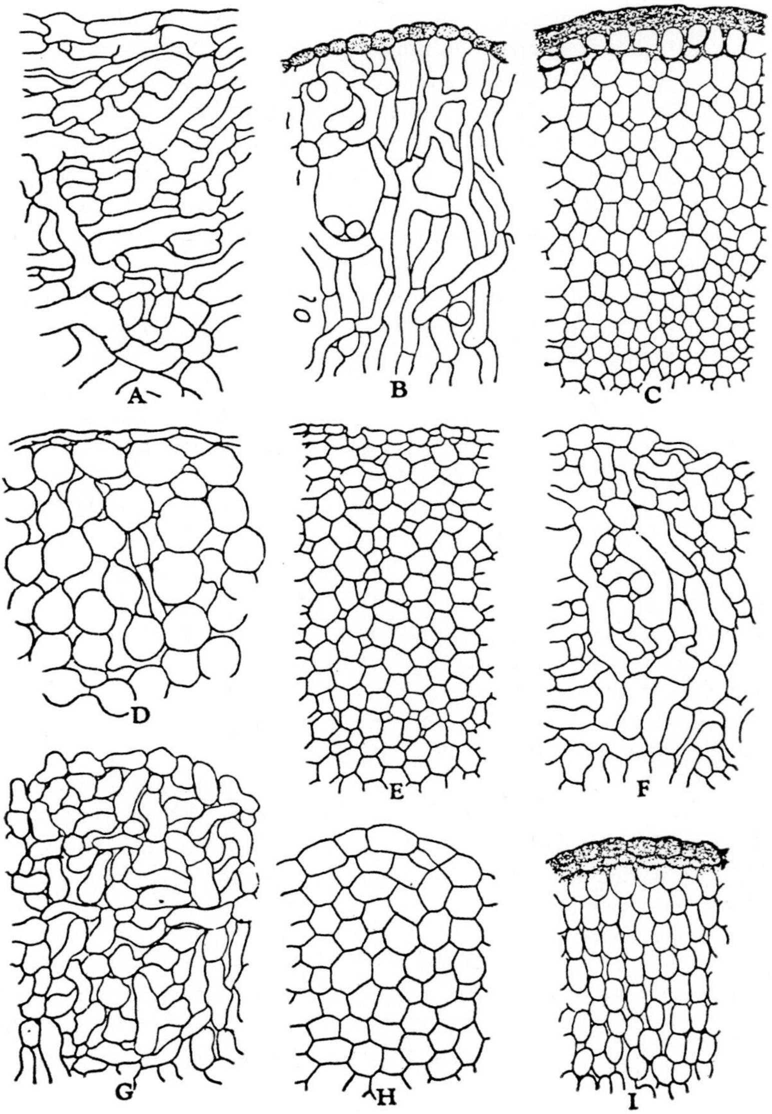

Fig. IV–13. Cross-section of the sclerotia of: A. *Corticium sasakii*. B. *Sclerotium hydrophilum*. C. *Helminthosporium sigmoideum*. D. *S. oryzae-sativae*. E. *H. sigmoideum* var. *irregulare*. F. *Rhizoctonia oryzae*. G. *S. fumigatum*. H. *S. oryzicola*. I. *Helicoceras oryzae*. (Nakata & Kawamura, 1939.)

Sclerotium fumigatum was described from Japan and has also been reported from Vietnam. It causes reddish brown lesions on the leaf sheath but not to such an extent as to cause lodging.

The sclerotia are more or less globose or oblong, often aggregated, surface rough, flat at the bottom, white, becoming grey or brownish grey at maturity, 0·3–1·5 mm, sometimes up to 2·5 mm, composed of uniform, dense mycelium, light yellowish brown, 5·5–9·9μ in diameter (Figs. IV–12, 13).

The mycelium is hyaline, irregularly branched, 5–6μ in diameter, growing inside the leaf sheath tissue. Optimum temperature for growth is 28–30°C, maximum 34–40°C. It grows at pH 5–9·8, optimum 5·5.

The fungus has also been found to infect *Coix agrostis* Loup and barnyard grass.

Sclerotium hydrophilum was first reported as *S. sphaeroides* Nakata from Japan but was found to be identical with *S. hydrophilum* after a comparative study. It causes indistinct yellowish lesions on the outer leaf sheaths, and is believed to be only a weak parasite on rice. In addition to Japan, it has been reported from Eastern China, U.S.A. and Bulgaria.

The sclerotia are globose or pear-shaped, 315–681 × 290–664μ, white, becoming yellowish brown and finally black; the surface is rough. In cross section, the sclerotium is composed of two layers, the dark brown outer layer consisting of cells 4–14 × 2–8μ, and the hyaline or light-yellow inner layer consisting of loose mycelium 3–6μ in diameter (Figs. IV–12, 13). Nakata & Kawamura (1939) distinguished this fungus from *Rhizoctonia microsclerotia* Matz by this two-layered composition of the sclerotia.

The fungus grows best at 30°C, with little or no growth below 15°C or above 39°C.

In nature it also attacks *Juncellus serotinus*, *Zizania latifolia* and *Digitaria sanguinalis* Scop.

Sclerotium oryzicola was reported from Japan to cause brown, irregular lesions, without a defined margin, on the sheath.

Sclerotia are formed inside the leaf sheath tissue, and are globose or irregular, dark reddish brown, surface rough, small, 70–100μ, composed of uniform, thin-walled, yellowish mycelium (Fig. IV–13).

Besides the above four species, Nakata & Kawamura (1939) reported three unnamed species, nos. 11, 12, 13, on the leaf sheath and dead straw of rice. *Sclerotium delphinii* Welch, *S. coffeicola* Stahel and *S. japonicum* Endo & Hidaka, which were reported from other host plants, were found to be able to infect rice by artificial inoculation.

LITERATURE CITED

Nakata, K. & Kawamura, E. 1939. Studies on sclerotial diseases of rice. I. Kinds of diseases and their organisms. *Bureau Agric., Minist. Agric. For. Japan, Agric. Exp. Stn Records* 139, 176 pp. [Jap.]

Padwick, G. W. 1950. Manual of rice diseases. 198 pp., Kew, Commonwealth Mycological Institute.

Roger, L. 1941–42. Les champignons à sclérotes parasites du riz. *Bull. écon. Indochine, 1941* (6) & *1942* (1–5), 302 pp.

Sawada, K. 1922. Descriptive catalogue of Formosan fungi II. *Rep. Govt Res. Inst., Dep. Agric. Formosa* (*Taiwan*) 2: 171–173. [Jap.]

Wei, C. T. 1957. Manual of rice pathogens. 267 pp., Peiping, Sci. Press. [Chin.]

SHEATH NET-BLOTCH

Sheath net-blotch disease was described by Matsuura (1942), who reported it as being caused by an unidentified species of *Cylindrocladium*. Aoyagi (1962) identified the fungus as *C. scoparium* Morgan, a species which also infects many other plants. Fujikawa *et al.* (1954–66) made a series of studies on the disease. It has not been reported on rice outside Japan.

SYMPTOMS

The disease is usually found on the leaf sheath near the water surface. It appears as small, yellow, water-soaked spots, gradually enlarging to 1–2 cm, oblong or elliptical in outline. The surface of the lesion becomes reticulated from the appearance of numerous dark brown lines which give it a net-like appearance. The cells of the infected part of the sheath often become whitish, contrasting with the brown net. Infected leaves turn yellow and die.

THE ORGANISM

The fungus sporulates on the lesions. The conidiophores branch dichotomously 2–3 times, terminating in phialides. The primary, secondary, tertiary and ultimate branches measure $27–39 \times 5–6\mu$, $18–21 \times 4–5\mu$, $10–12 \times 4\mu$ and $8–10 \times 4\mu$, respectively. The main axis is club-shaped and its apical swollen

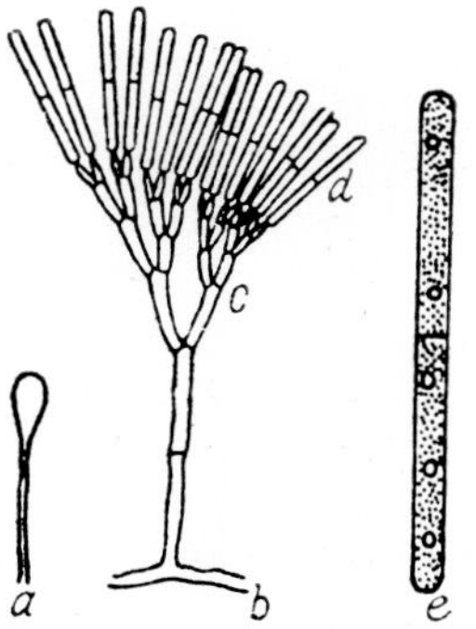

Fig. IV–14. *Cylindrocladium scoparium* Morgan a, tip of the main axis of conidiophore; b, mycelium; c, dichotomously branched conidiophore; d, e, conidia. (Aoyagi, 1962.)

part is $20–22 \times 7–11\mu$. Conidia are cylindric with rounded ends, $41–75 \times 2{\cdot}9–5{\cdot}8\mu$, hyaline and usually 1-septate. It grows well on many agar media made with plant decoctions, including PDA, and also grows quite well on many synthetic media (Fujikawa, Tomiku & Okatome, 1966) (Fig. IV–14).

The fungus attacks several diverse groups of plants, including other gramineous plants, legumes, cucurbits, and conifer seedlings (Fujikawa, 1954; Aoyagi, 1962).

Disease Cycle and Varietal Resistance

Fujikawa (1954) found that *C. scoparium* survived the winter on diseased straw in the field or when kept indoors.

Fujikawa *et al.* (1962) tested 30 rice varieties and found that Victory (glutinous), Norin Glutinous No. 1 and Norin No. 24 were highly resistant.

LITERATURE CITED

Aoyagi, K. 1962. Species name of the sheath net-blotch fungus of rice plants. *Ann. phytopath. Soc. Japan* **27**: 147–150. [Jap. Engl. summ.]

Fujikawa, T. 1954. Studies on the sheath-net blotch of rice plant. 2. Survival and host plants of the pathogen. *Kyushu agric. Res.* **3**: 107–108. [Jap.]

Fujikawa, T., Tomiku, T. & Okatome, Z. 1962. Varietal difference in resistance against sheath net-blotch in rice plants. *Nogyo Oyobi Engei* **37**: 395–396. [Jap.]

Fujikawa, T., Tomiku, T. & Okatome, Z. 1966. Studies on the sheath net-blotch of rice plant. XI. *Kyushu agric. Res.* **28**: 101. [Jap.]

Matsuura, I. 1942. New diseases of rice and lupin. *J. Ent. Path.* **29**: 286–293. [Jap.]

Fig. IV–15. Sheath rot symptoms.

SHEATH ROT

Sheath rot was first described by Sawada (1922) from Taiwan and the causal fungus was named *Acrocylindrium oryzae* Sawada. It is present in Japan and is very common in Southeast Asia. Chen (1957) reported 3–20% damage, and it may sometimes be as much as 85% in Taiwan (Ono, 1953).

SYMPTOMS

The rot occurs on the uppermost leaf sheaths enclosing the young panicles. The lesions start as oblong or somewhat irregular spots, 0·5–1·5 cm long, with brown margins and grey centres, or they may be greyish brown throughout. They enlarge and often coalesce and may cover most of the leaf sheath. The young panicles remain within the sheath or only partially emerge. An abundant whitish powdery growth may be found inside affected sheaths and young panicles are rotted (Fig. IV–15).

THE ORGANISM

Mycelium white, sparsely branched, septate, 1·5–2μ in diameter. Conidiophores arising from the mycelium, slightly thicker than the vegetative hyphae, branched once or twice, each time with 3–4 branches in a whorl. The main

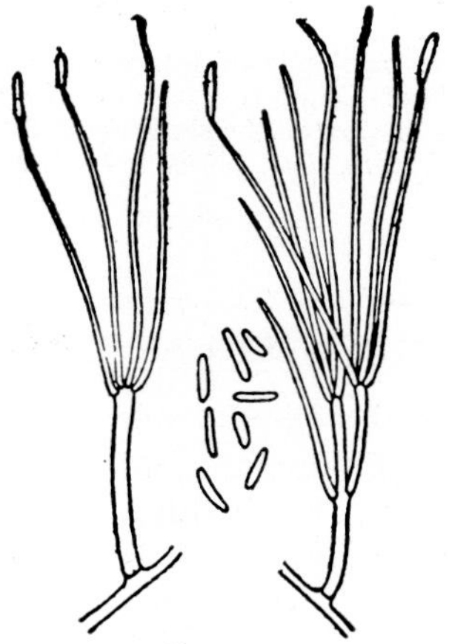

Fig. IV–16. *Acrocylindrium oryzae.* Conidiophores and conidia. (After Wei, 1957 redrawn from Sawada, 1922.)

axis 15–22 × 2–2·5μ, terminal branches tapering towards the tips, 23–45μ long, 1·5μ at the base. Conidia borne simply on the tip, produced consecutively, hyaline, smooth, single-celled, cylindrical, 4–9 × 1–2·5μ. Tasugi & Ikeda (1956) gave the conidial measurements as 2·1–8·5 × 0·5–1·6μ from the host plant, 1·8–13 × 1–1·6μ from culture (Fig. IV–16).

The fungus grows best at 30–31°C, little at 37°C, not at all at 13°C and is killed after 5 minutes at 50°C (Kawamura, 1940). Tasugi & Ikeda (1956) reported good growth at 20–28°C; conidia germinated at 23–26°C. Chen

(1957) reported that isolates of the fungus differ in their response to temperature, pH, and carbon and nitrogen sources, as well as in pathogenicity. It was found unable to infect other members of the Gramineae tested.

Kawamura (1946) and Tasugi & Ikeda (1956) artificially inoculated the fungus on to rice and established its pathogenicity. Wounding of the plants facilitated infection. Examining infected plants in Southeast Asia, it was found that almost all these plants were also infested by stemborers or had other injuries on the stems below. Chen & Chien (1964) found 56–57% of the diseased tillers to be infested by stemborers and 7–21% to be infected by yellow dwarf virus. This seems to suggest that the fungus tends to attack the leaf sheaths enclosing the young panicles when there is an injury which retards the emergence of the panicles.

Chen & Chien (1964) observed some varietal differences in susceptibility to the disease in the field in Taiwan.

LITERATURE CITED

CHEN, C. C. & CHIEN, C. C. 1964. Some observations on the outbreak of rice sheath rot disease. *Jnl Taiwan agric. Res.* **13**: 39–45. [Chin. Engl. summ.]

CHEN, M. J. 1957. Studies on sheath rot of rice plant. *J. Agric. For., Taiwan* **6**: 84–102. [Chin. Engl. summ.]

KAWAMURA, E. 1940. Notes on the sheath rot of rice plant with special reference to its causal organism, *Acrocylindrium oryzae* Saw. *Ann. phytopath. Soc. Japan* **10**: 55–59. [Jap.]

SAWADA, K. 1922. Descriptive catalogue of Formosan Fungi. II. *Rep. Govt Res. Inst. Dep. Agric. Formosa* 2, 135 pp. [Jap.]

TASUGI, H. & IKEDA, Y. 1956. Studies on the sheath rot of rice plant caused by *Acrocylindrium oryzae* Sawada. *Bull. natn. Inst. agric. Sci., Tokyo*, Ser. C, **6**: 151–166. [Jap. Engl. summ.]

CROWN SHEATH ROT

The fungus which causes this disease was first collected by Baker (1916) on dead rice straw from the Philippines. Merchers (1925) briefly mentioned that an *Ophiobolus* sp. was associated with a foot rot and root rot of rice in Arkansas. Tullis (1933) first proved its pathogenicity and studied the disease in more detail. For some time it was known to cause disease only in U.S.A., but it has now been reported from Japan, India and several countries in Africa, including parts of the former French Equatorial Africa and French West Africa, Ivory Coast, Kenya, Madagascar, Nigeria and Tanzania. The disease is usually of minor importance but Saccas & Fernier (1954) reported serious damage to rice in Ubangui-Chari (Central African Republic). It was called black sheath rot by Tullis and Arkansas foot rot or brown sheath rot by Padwick (1950).

SYMPTOMS

The disease is usually observed late in the season, just before harvest. The affected plants are characterized by a brown discoloration of the sheaths from the crown of the plants to the water line or for a considerable distance above. In the discoloured areas, the perithecia with their protruding beaks are found.

The number of perithecia produced on a single plant varies from a few to several hundred.

In the earlier stages of infection, dark reddish-brown mycelial mats as much as 1·5 cm long and as wide as the sheath may be found on the inner surface of the diseased sheaths. As the lesions become older, the mycelial mats increase in length, and when the sheaths become heavily infected the leaf blades above die. At maturity, the straw has a dull brownish cast.

Invasion of the culm proper by the fungus also occurs but perithecia are not commonly produced on the culm under field conditions.

Infected plants have a reduced number of tillers and frequently only one panicle per plant is produced.

The disease resembles the ‘ take-all ’ disease of wheat caused by *Ophiobolus graminis* Sacc. which has also been reported on rice. Both *Ophiobolus* spp. invade the leaf sheath, kill the crown tissue, form a mycelial mat beneath the leaf sheath, produce perithecia on the host tissue, inhibit tillering and cause premature ripening.

Tullis & Adair (1947) reported that the disease causes lodging and incomplete filling of grains. Veeraraghavan (1962) in India reported incomplete emergence of the panicles and chaffiness of the grains.

THE ORGANISM

Saccardo (1916) described *Ophiobolus oryzinus* from the Philippines. Perithecia are globose, loosely gregarious, subcutaneous, finally erumpent with a more or less prominent ostiole, 300–350μ in diameter. The perithecia walls

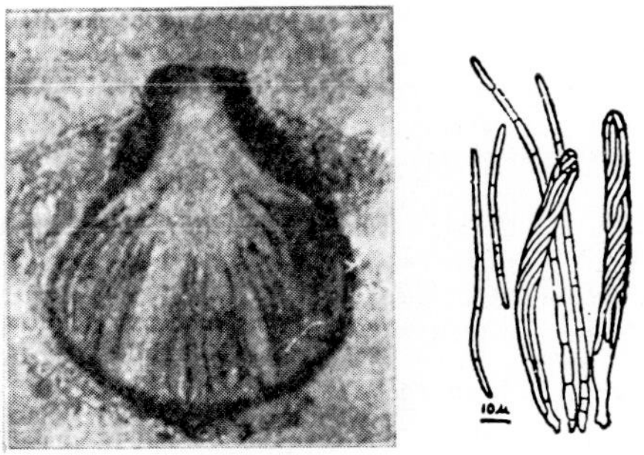

Fig. IV–17. *Ophiobolus oryzinus*, perithecium, asci and ascospores. (After Wei, 1957.)

are loosely cellular, ochraceous under the microscope, denser and more deeply coloured towards the apex and becoming reddish about the ostiole. The asci are cylindrical, rounded at the apex, with a short, narrow stipe, 7–11 × 95–110μ. The ascospores are 3–5 septate, contain many minute guttules, and are faintly coloured greenish-yellow.

Tullis (1933) studied collections from Arkansas and compared them with others from different areas: he found the fungus to be closest to *O. oryzinus*. He gave the measurements of the perithecia as 187–375μ in diameter; beaks hairy in culture; asci 96–111 × 12·5–14·5μ and ascospores 79–112 × 2–3μ (mostly 86–111μ long). Saccas & Fernier (1954) found the perithecia to be 155–415μ (mostly 325μ) in diameter; asci 72–123 × 10·5–14·5μ; ascospores 70–115 × 3·4–4·6μ (Fig. IV–17).

The ascospores are forcibly ejected from the perithecia if moistened after being dried for a time. They germinate from the two end cells and grow well on solid media. Tullis (1933) demonstrated that they are homothallic, perithecia being produced by single spore lines.

Disease Cycle

The fungus infects the plant through the basal leaves or the outer leaf sheaths by direct penetration or by forming appressoria before penetration. After penetrating the outer sheath, the mycelium spreads to form a mycelial mat between the sheaths. Finally the culm may be attacked. The air chambers of the leaf sheath are filled with balls of hyphae, which mature into perithecia.

A report from Nigeria (Harris *et al.*, 1962) states that the disease may be transmitted by seed, infected seeds producing 1–2% apparently diseased seedlings. The variety Oshodi was most susceptible while BG 79 and Agbede 140/54 were least susceptible. Tullis & Adair (1947) reported that Nira, Bluebonnet, Cody, Asahi, Fortuna, Kamrose, Teaxa Patna and Hill Patna were more resistant than other varieties.

LITERATURE CITED

Baker, C. F. 1916. Additional notes on Philippine plant diseases. *Philipp. Agric. For* **5**: 73–78.

Harris, E., Pyne, C. T. & Fatmi, A. S. 1962. Plant pathology. Rice. *Rep. Res. Spec. Serv., Minist. Agric. N. Nigeria, 1961–62*: 16.

Melchers, L. E. 1925. Diseases of cereal and forage crops in the United States in 1924. *Pl. Dis. Reptr* Suppl. 40: 107–188.

Padwick, G. W. 1950. Manual of rice diseases. 198 pp., Kew, Commonwealth Mycological Institute.

Saccas, A. M. & Fernier, H. 1954. Une grave maladie du riz due à *Ophiobolus oryzinus* Sacc. *Agron. trop.* **9**: 7–20. (*Rev. appl. Mycol.* **33**: 444.)

Saccardo, P. A. 1916. Notae mycologicae XX. *Nuovo Giorn. Bot. ital.*, N.S., **23**: 203.

Tullis, E. C. 1933. *Ophiobolus oryzinus* the cause of a rice disease in Arkansas. *J. agric. Res.* **46**: 799–806.

Tullis, E. C. & Adair, C. R. 1947. Black sheath rot of rice (*Ophiobolus oryzinus*) caused lodging of rice in Arkansas and Texas in 1947. *Pl. Dis. Reptr* **31**: 468.

Veeraraghavan, J. 1962. Occurrence of ‘black sheath rot’ disease of rice caused by *Ophiobolus oryzinus* Sacc. in India. *Indian Phytopath.* **15**: 14–17.

SHEATH BLOTCH

Sheath blotch, caused by *Pyrenochaeta oryzae* Shirai ex Miyake, was first observed in Japan in 1910. It has since been reported from Burma, China, India, Malaysia and Sierra Leone and is also present in the Philippines and Thailand. It usually attacks the leaf sheath but occasionally also the leaf blade and glumes. The blotches are oblong, about one inch long, and brownish

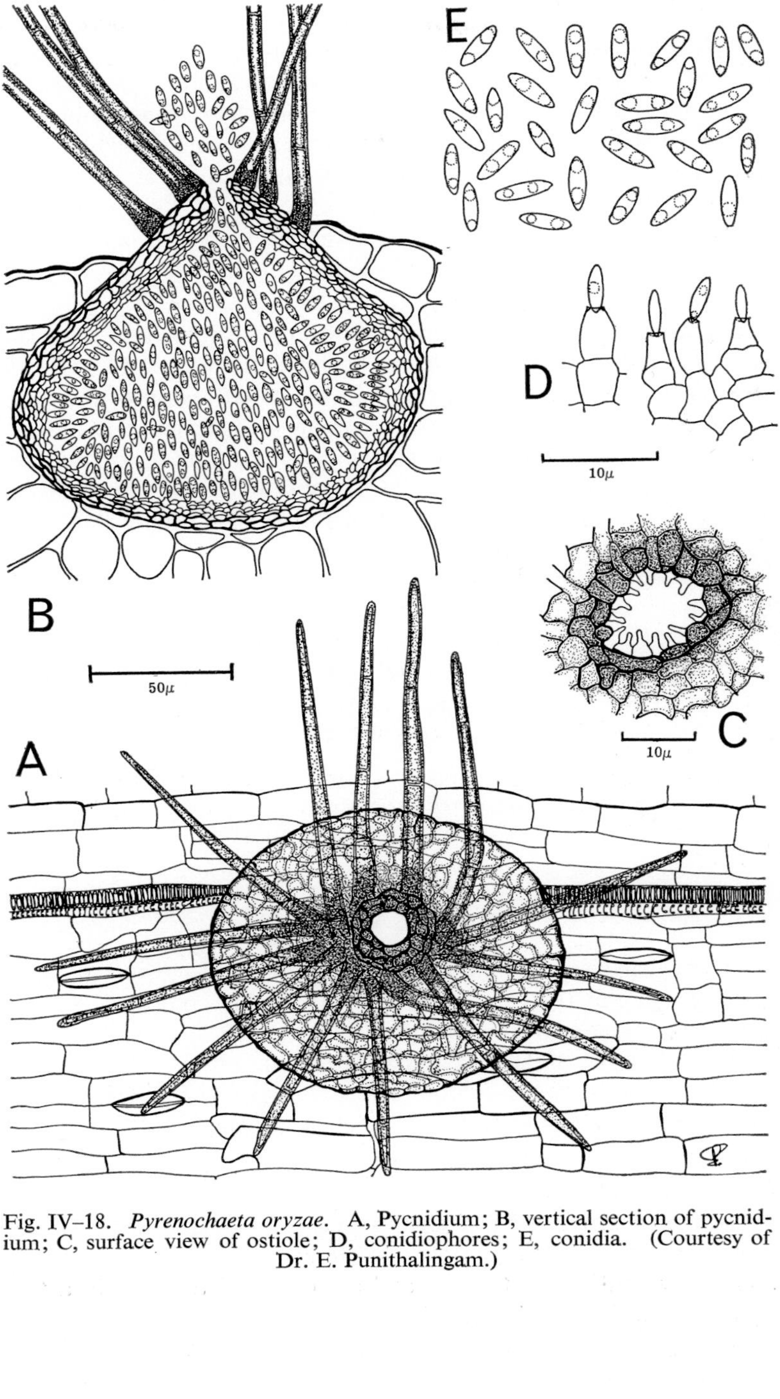

Fig. IV–18. *Pyrenochaeta oryzae*. A, Pycnidium; B, vertical section of pycnidium; C, surface view of ostiole; D, conidiophores; E, conidia. (Courtesy of Dr. E. Punithalingam.)

at first. Gradually the centre becomes grey or greyish brown and is dotted with the black fruiting bodies of the fungus. The margin remains brown.

The description of the fungus by Miyake (1910) may be translated as follows: Pycnidia immersed in the tissue with slightly projecting ostiole, dark brown, ellipsoid, 200μ in diameter, 120μ high, ostioles surrounded by 6–20 blackish, multiseptate setae 60–140×4–5μ, rigid when moist, but opening out in a star-shaped manner when dry; ostiole about 40μ broad outside, about 12μ inside; spores emitted in a tendril, fusoid, 4–6×1·5–2μ, guttulate at both ends (Fig. IV–18).

Hara (1918) described another species on rice leaves, *P. nipponica* Hara. It differs from *P. oryzae* in its smaller pycnidia (100–150μ), setae (45–75×2·5–3μ) and spores (3–4×1–1·5μ).

Recently, Mori, Makino & Osawa (1964) reported another *Pyrenochaeta* without giving it a name, but considered it different from both *P. oryzae* and *P. nipponica*. It attacks the leaf sheath and the collar region of the leaf when artificially inoculated. The dark brown, narrow and irregular lesions at first appear near the edges of the sheath and then surround the sheath and become yellowish brown. In many cases, the lesions appear on the site of oviposition of leafhoppers, and the fungus thus appears to be a wound parasite. The pycnidia are globose or elliptical, dark brown, 69–176×69–137μ. Several black, septate bristles are produced around the opening of the pycnidia. The pycnospores are hyaline, oval or elliptical, 2·5–5·4×1·2–2·7μ.

LITERATURE CITED

HARA, K. 1918. Diseases of the rice plant. 170 pp., Tokyo, Japan. [Jap.]

MIYAKE, I. 1910. Studien über die Pilze der Reispflanze in Japan. *J. Coll. Agric. Tokyo* **2**: 237–276.

MORI, K., MAKINO, T. & OSAWA, T. 1964. A leaf sheath disease of rice plant caused by *Pyrenochaeta* sp. *Bull. Shizuoka Pref. agric. Exp. Stn* 9: 25–31. [Jap. Engl. summ.]

WITCH WEED

Striga, the witch weed, is one of the few parasitic flowering plants found affecting crops in the tropics. Van Hall (1917) first reported *Striga lutea* Lour. as causing damage to rice plants in Sumatra. It has also been reported from Africa (Saunders, 1930) and India (Uttaman, 1950a). Another species, *S. hermonthica* Benth., was reported from tropical Africa (Chevalier, 1929) and found to parasitize maize and rice in Kenya (McDonald, 1928).

It attacks the roots of rice plants, which become stunted and take on a generally unhealthy appearance. Young plants suffer more than older ones.

Striga is an erect herb, 1–2 feet high, with a stout quadrangular stem. The leaves are green, undivided, lanceolate, 1–3 inches long and up to $\frac{1}{2}$ inch broad. The flowers are red to pink and are collected together in a terminal head about 6 inches long. The seeds are borne in a capsule and are so very minute that single seeds or groups of a few are hard to see with the naked eye.

Striga seeds germinate only when in contact with host roots, and Uttaman (1950b) reported that the roots exude a stimulating substance necessary for germination. The roots of *Striga* do not seem to function as normal roots and are apparently used to help the plant in securing as many connections as possible with the roots of the host plant and send haustoria into them. The haustorial branching vessels are comparable to the rootlets of a normal plant. The haustoria enter the host as far as the endodermis but the vessels reach the stelar region and establish parasitic connections with either the phloem or the xylem. In *in vitro* experiments, Uttaman (1950c) found the root extract of *Striga* to be toxic to rice plants and able to kill young rice seedlings. Older seedlings are able to resist the toxin better. This partly explains why an early sown rice crop is better able to resist the attack of *Striga* than a late one.

Being an annual plant, there is no means of propagation except by seed. It is important, therefore, that the parasite be destroyed before it bears seed so as to prevent further damage.

Striga has several other gramineous hosts, including maize, sorghum and *Eleusine coracana*.

LITERATURE CITED

CHEVALIER, A. 1929. Sur une Scrophularies (*Striga hermonthica*) parasite des céréales en Afrique tropicale. *C.R. Acad. Sci., Paris* **189**: 1308–1310.

McDONALD, J. 1928. Diseases of maize and notes on a parasitic maize weed in Kenya. *Bull. Dep. Agric. Kenya* 20: 1–7.

PALM, B. T. & HEUSSER, C. 1924. *Striga lutea* als Schädigen des Reises auf Sumatra. *Z. PflKrankh.* **34**: 11–18.

SAUNDERS, A. R. 1930. The witch weed. *Dep. Agric. Union S. Africa* **5**: 267–271.

UTTAMAN, P. 1950a. Parasitism of *Striga lutea* (Lour.) on rice and methods to protect rice plant against *Striga*. *Madras agric. J.* **37**: 99–118.

UTTAMAN, P. 1950b. A study on the germination of *Striga* seed and the mechanism and nature of parasitism of *Striga lutea* (Lour.) on rice. *Proc. Indian Acad. Sci.*, Sect. B, **32**: 133–142.

UTTAMAN, P. 1950c. An investigation into the nature of injury sustained by rice plant on account of parasitism by *Striga lutea* (Lour.). *Proc. Indian Acad. Sci.*, Sect. B, **32**: 222–226.

VAN HALL, C. J. J. 1917. Zicktenen plazen der cultuurgewassen in Nederlandsch-Indie in 1916. *Meded. Lab. Plantenziek.* 29: 1–37.

PART V

FUNGUS DISEASES — SEEDLING DISEASES

SEEDLING DAMPING OFF

Sawada (1912) first described this disease from Taiwan, and it has also been reported from Japan. Ito & Nagai (1931) stated that farmers in Hokkaido had to resow their seeds when there was a severe outbreak of the disease. It has been troublesome in California where rice seeds are sown in water. Damping off has not been recorded as being a problem in rice in the tropics.

Symptoms

Evidence of the occurrence of the disease is seen in the presence of a cottony growth of the hyphae of the parasitic fungi on the surface of the seeds in the nursery. The hyphae grow out frequently from a slit in the seed-coat opened or caused by breaking at germination or threshing. In a few days the hyphal growth may cover the entire seed and form a cottony mass. When severely

Fig. V–1. Seedling damping off. (After Wei, 1957.)

attacked, the seeds cannot germinate (Fig. V–1). If the seedlings are affected after the plumules have already grown to 0·5–1 cm in length, the growth of the seedlings is seriously checked and damping off often takes place. The progress of the symptoms is influenced by weather and the vigour of the seedlings.

The Causal Organisms

The disease is caused by many species of water moulds. Sawada (1912) isolated *Achlya prolifera* (Nees) de Bary. Ito & Nagai (1931) obtained 12 fungus species but not *A. prolifera*. Wei (1957) compiled the following list of species on rice:

Saprolegnia anisospora de Bary

Achlya racemosa Hildebrand, *A. americana* Humphrey, *A. oryzae* Ito & Nagai, *A. flagellata* Coker, *A. flagellata* Coker var. *yezoensis* Ito & Nagai and *A. megasperma* Humphrey

Dictyuchus monosporus Leitgeb, *D. anomalus* Nagai and *D. sterilis* Coker

Pythiogeton ramosum v. Minden, *P. uniforme* A. Lund and *P. dichotomum* Tokunaga

Pythium monospermum Pringsh., *P. oryzae* Ito & Tokunaga, *P. diclinum* Tokunaga, *P. graminicola* Subramaniam, *P. arrhenomanes* Drechsler, *P. aphanidermatum* (Edson) Fitzpatrick, *P. helicum* T. Ito, *P. nagaii* Ito & Tokunaga, *P. proliferum* Schenk, *P. rostratum* Butler, *P. debaryanum* Hesse, *P. echinocarpum* Ito & Tokunaga and *P.* sp.

Pythiomorpha miyabeana Ito & Nagai and *P. oryzae* Ito & Nagai

Webster *et al.* (1970) reported that the major species in California are *Pythium* spp. and *Achlya klebsiana* Pieters.

The pathogenicity of these species has not been studied in detail, but some have been observed to be obviously more pathogenic than others. Ito & Nagai (1931) reported that infection failed to take place at 10–12°C, but occurred readily at 18–25°C. *Achlya oryzae*, *Pythiomorpha miyabeana* and *P. oryzae* were able to cause infection at 32–35°C. These organisms are water-borne and are disseminated by irrigation water.

Effect of Environmental Conditions

The disease usually develops when the seedlings lack vigour. This is often brought about by cool weather, which retards germination and growth. Seeds which are mechanically injured or only partly filled are most subject to infection. Webster *et al.* (1970) found that *Pythium* spp. are more pathogenic at lower temperatures (25°C) while *Achlya klebsiana* has equal pathogenicity between 20 and 30°C. They also found that seed treatment, prior to soaking, with captan, thiram and difolatan controls the disease effectively.

LITERATURE CITED

Coker, W. C. 1923. The Saprolegniaceae with notes on other water moulds. 201 pp., Univ. North Carolina Press.

Hemmi, T. & Abe, T. 1928. An outline of the investigation on the seedling rot of rice caused by a water-mold, *Achlya prolifera* Nees. *Jap. J. Bot.* **4**: 113–123.

Ito, S. & Nagai, M. 1931. On the rot disease of the seeds and seedlings of rice plants caused by aquatic fungi. *J. Fac. Agric. Hokkaido imp. Univ.* **32**: 45–69.

Matthews, V. D. 1931. Studies on the genus *Pythium*. 136 pp., Univ. North Carolina Press.

Nagai, M. 1931. Studies on the Japanese Saprolegniaceae. *J. Fac. Agric. Hokkaido imp. Univ.* **32**: 1–32.

Padwick, G. W. 1950. Manual of rice diseases. 198 pp., Kew, Commonwealth Mycological Institute.

Sawada, K. 1912. Studies on the rot disease of rice seedlings in Formosa. *Spec. Bull. agric. Exp. Stn Formosa* 3: 1–84. [Jap.]

Webster, D. H., Hall, C. M. & Brandon, D. M. 1970. Seedling disease and its control in California rice fields. *Rice J.* **73**: 14–17.

Wei, C. T. 1957. Manual of rice pathogens. 267 pp., Peiping, Sci. Press. [Chin.]

SEEDLING BLIGHT

Reinking (1918) appears to have been the first to describe this disease on rice, from the Philippines. Tisdale (1921) investigated it in U.S.A. and Thompson (1928) found it in Malaya. It has since been observed in many tropical countries.

The disease is liable to cause some damage in upland rice, and also occurs in lowland rice seedbeds which dry up at times. It causes pre-emergence seed rot and post-emergence seedling blight. The base of the stem and the roots of diseased plants are dark in colour and often have a frosty appearance due to the presence of the mycelium of the causal fungus. Sclerotia may be found attached to the roots or the base of diseased seedlings. Infected young seedlings first show retarded growth, followed by yellowing and withering of the leaves; they then die slowly. A variety of saprophytic fungi may be found on the dying or dead leaves. The white silky mycelium of the pathogen is sometimes observed on the base of mature rice plants growing in upland conditions. No conspicuous symptoms are observed, however, on these plants, except perhaps a few dead outer leaf sheaths.

The causal organism was first named *Sclerotium rolfsii* Sacc. It has also been called *S. zeylanicum*, *Rhizoctonia destruens*, *Hypochnus centrifugus*, *S. delphinii*, *Corticium centrifugum*, *Corticium rolfsii* and *Botryobasidium rolfsii*. Aycock (1966) published a very extensive review on the organism and expressed the view that *Pellicularia rolfsii* West appears to be the most acceptable binomial. However, the generic name *Pellicularia* is objected to by several workers, and the name *Corticium rolfsii* Curzi is at present considered most acceptable.

The fungus has many strains and variable morphology. It grows luxuriantly on many culture media, producing white aerial mycelium and abundant sclerotia. The mycelium usually has numerous clamp connections but the frequency of these varies with the strain or biotype. Sclerotia are spherical to ellipsoid; they were described by Saccardo as being 0·5–0·8 mm in diameter, but they vary according to the strain and under different nutritional and other environmental conditions. In some cases they are reported to have reached 3×5 mm; they are at first white, later becoming tan to brown. Hymenia, which appear infrequently, are at first white, becoming yellow to buff. Basidia are *c.* $7 \cdot 9 \times 4$–5μ, with 2–4 sterigmata. Basidiospores range from *c.* $3 \cdot 5$–5×6–7μ, these measurements being variable in different strains. Wu (1968) noted that the pigmentation of sclerotia is related to oxygen rather than to light or darkness, as pigmentation occurs in the light and in the dark but no sclerotia turn brown under anaerobic conditions.

The genetic variability of *C. rolfsii* has been shown by West (1947) and Lyle (1953), by their studies of single basidiospore cultures (Aycock, 1966). From these studies, it seems reasonable to include various forms, such as *Corticium centrifugum*, *S. delphinii* etc., which have been considered separate species by some authors, within the single species *C. rolfsii*.

Not many studies have been made on the fungus on rice, although it has been extensively investigated on other hosts. Tisdale (1921) found that isolates of

C. rolfsii from wheat and rice caused marked reduction in germination of rice on inoculation; isolates from soybean caused less reduction, and one from tall oat grass only a very slight reduction. The number of plants surviving and showing chlorotic symptoms was largest with the rice and wheat isolates, less with the tall oat grass, and only very small with the soybean isolates. Bertus (1929) in Ceylon reported that only the outer leaf sheaths were affected, when rice seedlings were inoculated with mycelium. Weerapat & Schroeder (1966) reported that among the strains isolated from rice, pepper (*Capsicum*), groundnut and cotton, the rice strain was most virulent on 11 rice varieties, the cotton strain least so. Severity increased with increasing levels of inoculum. Severe disease developed at 25–35°C, maximum development being at 30°C. Tisdale (1921) found that flooding with irrigation water completely checks further development of the disease.

From the studies of the organism on other crops, it has been found that the fungus utilizes numerous simple and complex carbohydrates. It grows in synthetic media but is apparently unable to synthesize thiamine. Growth occurs over a broad range of pH (2·1 to 8), but is best on acid media, the optimum being pH 3–5. The temperature range for growth is *c.* 10–40°C, optimum 25–35°C. Sclerotial formation is also greatest at or near the optimum temperature for growth (Aycock, 1966).

The fungus can penetrate uninjured host tissue, although mechanical or other injuries facilitate ingress of the fungus into the host. The mycelium is both intercellular and intracellular in the host tissue. In the field, a high incidence of disease is induced by a high soil moisture content, optimum temperature for growth of the fungus, adequate aeration of the soil, and a high content of soil organic matter. Jones *et al.* (1938) and Tullis (1940) have observed that seedling blight is common during periods of warm, moist weather.

Sclerotia are the principal means of overwintering, or of surviving the hot, dry summer in the tropics. They float, and doubtless the fungus is disseminated by irrigation or rain water.

The fungus has a very wide host range, infecting plants of nearly 100 families (Aycock, 1966).

Mendiola & Ocfemia (1926) noted that only 21 of 126 Philippine rice varieties tested remained free from infection in their experiments. Weerapat & Schroeder (1966) reported that the 100 rice varieties tested by them in Texas were all susceptible before emergence but that Milfer No. 2 and Taichung (Native) 1 were resistant after emergence.

In general, deep ploughing to cover crop debris, which removes from the surface soil the substrate necessary for saprophytic growth of the fungus, tends to reduce disease incidence. Seed treatment with chemicals should also be tried.

LITERATURE CITED

AYCOCK, R. 1966. Stem rot and other diseases caused by *Sclerotium rolfsii*. (The status of Rolf's fungus after 70 years). *Tech. Bull. N. Carol. agric. Exp. Stn* 174, 202 pp.

BERTUS, L. S. 1929. *Sclerotium rolfsii* Sacc. in Ceylon. *Ann. Rep. Bot. Gard., Peradeniya* 11: 173–187.

JONES, J. W., JENKINS, J. M., WYCHE, R. H. & NELSON, M. 1938. Rice culture in the Southern States. *Fmrs' Bull. U.S. Dep. Agric.* 1808, 29 pp.

LYLE, J. A. 1953. A comparative study of *Sclerotium rolfsii* Sacc. and *Sclerotium delphinii* Welch. Ph.D. Dissertation, Univ. Minnesota.

MENDIOLA, N. B. & OCFEMIA, G. O. 1926. The work of breeding disease resistant crop plants at the College of Agriculture at Los Baños. *Philipp. Agric.* **15**: 117–128.

REINKING, O. A. 1918. Philippine economic plant diseases. *Philipp. J. Sci.*, Sect. A, **13**: 225–228.

THOMPSON, A. 1928. Notes on *Sclerotium rolfsii* Sacc. in Malaya. *Malay. agric. J.* **16**: 48–58.

TISDALE, W. H. 1921. Two *Sclerotium* diseases of rice. *J. agric. Res.* **21**: 649–658.

WEERAPAT, P. & SCHROEDER, H. W. 1966. Effect of soil temperature on resistance of rice to seedling blight caused by *Sclerotium rolfsii*. *Phytopathology* **56**: 640–644.

WEST, E. 1947. *Sclerotium rolfsii* Sacc. and its perfect stage on climbing fig. *Phytopathology* **37**: 67–69.

WU, L. C. 1968. Maturation of sclerotia produced by *Sclerotium rolfsii*. Fifth International Congress Photobiology, Aug. 1968, New Hampshire, no. Bg-8. [Abstr.]

PART VI

FUNGUS DISEASES—DISEASES OF GRAIN AND INFLORESCENCE

FALSE SMUT (GREEN SMUT)

HISTORY

The fungus causing false smut was first described by Cooke (1878) and named *Ustilago virens*; the description was based upon a specimen from India. Material sent from Japan to Patouillard (1887) was named *Tilletia oryzae*; Brefeld (1895) transferred it to *Ustilaginoidea oryzae* (Pat.) Brefeld. Takahashi (1896) studied the fungus and called it *Ustilaginoidea virens* (Cke.) Tak. Omori (1896) still considered it as a true smut and called it *Sphacelotheca virens*. Sakurai discovered the perfect state in 1934 and named it *Claviceps virens* (Cke.) Sakurai ex Nakata (1934), but this name has not been taken up and the fungus is generally referred to as *Ustilaginoidea virens* (Cke.) Tak.

The fungus was also described at a very early date in Chinese literature, but without a scientific name. The presence of the disease was believed to be an indication of a year of good crops, as conditions favourable for the growth of the fungus, particularly high moisture or rainfall, are similar to those resulting in good growth of the rice crop.

The disease usually causes little damage except in restricted areas under special conditions. Reinking (1918) mentioned severe infection in the Philippines during damp weather and Seth (1945) reported an epidemic in Burma in 1935. It is, however, present in most of the major rice-growing areas, including Burma, Ceylon, China, India, Indonesia, Japan, Malaysia, Pakistan, Philippines, Taiwan, Thailand and Vietnam in Asia; Bolivia, Brazil, Colombia, Guyana, Mexico, Panama, Peru, Surinam, Trinidad, U.S.A. and Venezuela in America; Congo, Ghana, Guinea, Ivory Coast, Liberia, Madagascar, Mozambique, Nigeria, Rhodesia, Sierra Leone, Sudan, Tanzania and Zambia in Africa; and Italy, Fiji and Papua–New Guinea (see CMI Distribution Map 347).

SYMPTOMS

The fungus transforms individual grains of the panicle into greenish spore balls of a velvety appearance. The spore balls are small at first and are visible in between the glumes, growing gradually to reach 1 cm or more in diameter, and enclosing the floral parts. They are slightly flattened, smooth and yellow, and are covered by a membrane. The membrane bursts as the result of further growth and the colour of the ball becomes orange and later yellowish-green or greenish-black. At this stage, the surface of the ball cracks. When cut open, it is white in the centre, and consists of tightly woven mycelium together

PLATE VIII

Upper picture — *False smut*, showing infected kernels.

Lower left — *Sheath blight*, showing severely infected plants; lesions reach flag leaves which are completely blighted.

Lower right — *Sheath blight*, showing lesions and sclerotia formed on leaf sheath.

PLATE VIII

with the glumes and other tissues of the host. There are 3 outer layers at different stages of development. The innermost layer is yellowish, with radiating mycelium and spores in the process of formation. The next layer is orange-coloured and is composed of mycelium and spores. The outermost layer is green, and consists of mature spores together with remaining fragments of mycelium. The surface is covered with powdery dark green spores (Plate VIII).

Usually only a few grains in a panicle are infected, but several may sometimes occur.

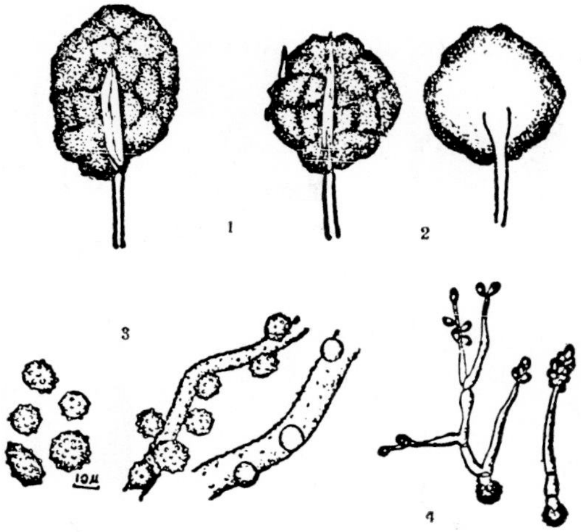

Fig. VI–1. *Ustilaginoidea virens.* 1, smut balls. 2, cross-section. 3, chlamydospores and their attachment to mycelium. 4, germinating chlamydospores and conidia. (After Brefeld, 1895.)

THE CAUSAL ORGANISM

Ustilaginoidea virens (Cke.) Tak. is the name generally accepted for the fungus as *Ustilago virens* antedates *Tilletia oryzae*.

Chlamydospores formed on the spore balls are borne laterally on minute sterigmata on radial hyphae, and are spherical to elliptical, warty, olivaceous, $3–5 \times 4–6\mu$; younger spores are smaller, paler, almost smooth. Chlamydospores germinate in culture by germ-tubes which become septate and form conidiophores bearing conidia at the tapering apex. These conidia are ovoid and very minute (Fig. VI–1).

Some of the green spore balls develop one to four sclerotia in the centre. These sclerotia overwinter in the field and produce stalked stromata the following summer or autumn. The stromata form a swelling at the tip of the stalk, are more or less globose, and contain perithecia round the periphery. Each flask-shaped perithecium contains about 300 asci. The asci are cylindrical with a hemispherical apical appendage, $180–220 \times 4\mu$, and 8-spored. Ascospores are hyaline, filiform, unicellular, $120–180 \times 0{\cdot}5–1\mu$, ($50–80 \times 0{\cdot}5–1{\cdot}5\mu$, Hashioka *et al.*, 1951) (Fig. VI–2). Ikegami (1963a) traced the development of the sclerotia and noticed that they are usually most numerous in the larger, greenish-black smut balls. Hashioka, Ikegami & Horin (1966) compared the fine structure of the chlamydospores of the fungus with that of true cereal smut spores, and found distinct differences in spore wall ornamentation.

Hashioka, Yoshino & Yamamoto (1951) reported that chlamydospores germinated in water to produce fine germ tubes bearing 1–3 conidia. There was more growth of germ tubes and a greater number of conidia when they were germinated in sugar solution or potato decoction than in water. The conidia germinated well in nutrient solution but seldom in water. Chlamydospore germination and mycelial growth are best at 28°C and slight at 12°C; no growth occurs at 36°C. Growth is best at pH 6·02–6·72; there is little or no growth at pH 2·77 and only slight growth at pH 9·05.

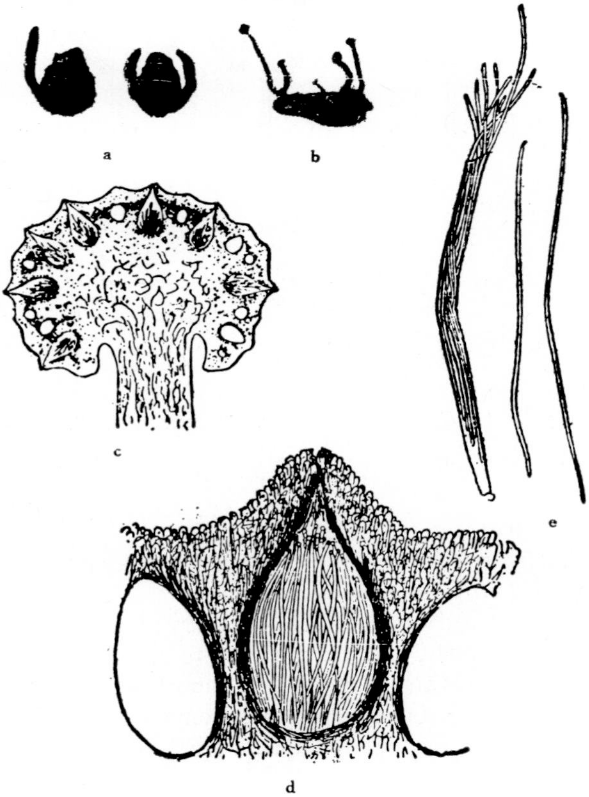

Fig. VI–2. *Ustilaginoidea virens*. a–b, germinating sclerotia. c, cross-section of stroma. d, perithecia. 3, ascus and ascospores. (After Nakata, 1934.)

Hashioka and his co-workers also produced perithecia and mature stromata from sclerotia kept for 4–5 weeks in moist sand at 24–30°C. In culture, chlamydospores are produced in 20–40 days on Saito's soy agar, rice decoction agar, and steamed rice. Lepori (1951) found that growth was better on malt agar and peptone glucose broth than on potato agar. Ikegami (1962b) studied the effects of different carbon and nitrogen sources on the growth of the fungus.

Yamashita (1965) detected alkaloids in infected plant parts and considered that the sclerotia of *U. virens* resemble the ergots formed by certain other fungi.

Disease Cycle

In the temperate regions, the fungus survives the winter by means of sclerotia as well as chlamydospores. It is believed that primary infections are initiated mainly by ascospores produced from the sclerotia. Chlamydospores play an important role in secondary infection, which is a major part of the disease cycle.

Hisada (1936), from field observations, suspected that infections occur when rice plants are at the booting stage. As a result of histological studies, Raychaudhuri (1946) described two types of infection. One type takes place at a very early stage of flowering, when the ovary is destroyed, but the style, stigmas, and anther lobes remain intact and are ultimately buried in the spore mass. The second type takes place when the grain is mature. Spores accumulate on the glumes, absorb moisture, swell, and force the lemma and palea apart. The fungus finally contacts the endosperm and the growth is greatly accelerated. Ultimately the whole grain is replaced and enveloped by the fungus.

Yoshino & Yamamoto (1952) and more recently Ikegami (1960) successfully inoculated rice plants with chlamydospores and ascospores (Ikegami) by injecting a spore suspension into the leaf sheath enclosing the young panicle. They inferred that infection occurs at the rather short period just before heading. Hashioka, Yoshino & Yamamoto (1951) observed that 99·6% of the smut balls contained intact anthers, indicating that most infections take place just before flowering.

Ikegami (1962a, 1963b) also inoculated coleoptiles of germinating seeds with spore suspensions and found that the fungus infected the host tissues. Later hyphae could be found in the growing points of the tillers until the late tillering stage, but the fungus was not seen at the ear primordium.

Chlamydospores are air-borne, but do not free very easily from the smut balls because of the presence of a sticky material. Sreeramulu & Vittal (1966) found the maximum number of spores in the air at the time of heading of the rice plants. On normal dry days, a diurnal periodicity occurred with a maximum number of spores at 22.00 hours; rainfall reduced the number of spores trapped. The maximum number of spores recorded was 328,200/m^3 of air.

Effect of Environmental Conditions

It is generally believed that high moisture favours the development of false smut. Ikegami (1960) concluded that plants grown under conditions of high fertility, favourable for the vegetative growth of rice, were more susceptible to the disease. Other workers in Japan have expressed a similar view. Rao (1964) also reported that weather and nutritional conditions have some effect on the incidence of the disease.

Host Range and Varietal Resistance

Haskell & Diehl (1929) reported that a species of *Ustilaginoidea* on maize tassels from Louisiana and Panama was similar morphologically to *U. virens* on rice, but it is not known if the fungus from maize could attack rice. Rao & Reddy (1955) found *U. virens* on *Oryza officinalis* Wall. in India. Veeraraghavan (1962) also reported the finding of the fungus on wild species of *Oryza*.

Little information is available on varietal resistance, but certain varieties have been noted to be more frequently attacked than others. Rao (1964) observed in India that 186 varieties remained free from infection among 297 examined in a single area, despite the occurrence of environmental conditions favouring disease development.

CONTROL

The disease usually does not warrant special control measures. Seed treatments were recommended some years ago but there is no sound scientific basis for recommending them. With the better understanding of the disease cycle which more recent work has provided, it is now possible to combat false smut by spraying or dusting with a fungicide a few days before heading (Hashioka, 1952).

LITERATURE CITED

BREFELD, O. 1895. *Ustilaginoidea oryzae* nov. genus auf *Oryza sativa. Untersuchungen aus dem Gesammtgebiete der Mykologie* 12: 194–202.

COOKE, M. C. 1878. Some extra-European fungi. *Grevillea* **7**: 13–15.

HASHIOKA, Y. 1952. Application of new fungicide in rice cultivation. *Agriculture Hort., Tokyo* **27**: 485–489. [Jap.]

HASHIOKA, Y., IKEGAMI, H. & HORINO, O. 1966. Fine structure of the rice false smut chlamydospores in comparison with that of cereal smut spores. *Res. Bull. Fac. Agric. Gifu Univ.* **22**: 40–49.

HASHIOKA, Y., YOSHINO, M. & YAMAMOTO, T. 1951. Physiology of the rice false smut, *Ustilaginoidea virens* (Cke.) Tak. *Res. Bull. Saitama agric. Exp. Stn* **2**: 1–20. [Jap. Engl. summ.]

HASKELL, H. R. & DIEHL, W. W. 1929. False smut of maize, *Ustilaginoidea. Phytopathology* **19**: 589–592.

HISADA, K. 1936. Infection stage of false smut of rice. *Ann. phytopath. Soc. Japan* **6**: 72–76.

IKEGAMI, H. 1960. Studies on the false smut of rice. IV. Infection of the false smut due to inoculation with chlamydospores and ascospores at the booting stage of rice plants. *Res. Bull. Fac. Agric. Gifu Univ.* **12**: 45–51.

IKEGAMI, H. 1962a. Studies on the false smut of rice. V. Seedling inoculation with the chlamydospores of the false smut fungus. *Ann. phytopath. Soc. Japan* **27**: 16–23. [Jap. Engl. summ.]

IKEGAMI, H. 1962b. Studies on the false smut of rice. VIII. Carbon and nitrogen sources of the false smut fungus. *Res. Bull. Fac. Agric. Gifu Univ.* **16**: 45–54. [Jap. Engl. summ.]

IKEGAMI, H. 1963a. Occurrence and development of sclerotia of the rice false smut fungus. *Res. Bull. Fac. Agric. Gifu Univ.* **18**: 47–53. (Studies IX.)

IKEGAMI, H. 1963b. Invasion of chlamydospores and hyphae of the false smut into rice plants. *Res. Bull. Fac. Agric. Gifu Univ.* **18**: 54–60. (Studies X.)

LEPORI, L. 1951. Alcune osservazioni sulla morfologia e biologia dell'*Ustilaginoidea virens* (Cke) Tak. su materiale thailandese. *Notiz. Mal. Painte* **16**: 40–46. [Engl. summ.]

NAKATA, K. 1934. Illustrated manual of diseases of crop plants, p. 19. Tokyo. [Jap.]

OMORI, J. 1896. Some remarks on Mr. Takahashi's paper on the identity of *Ustilago virens* Cooke and *Ustilaginoidea oryzae* Brefeld. *Bot. Mag., Tokyo* **10**: 29–31.

PATOUILLARD, N. 1887. Contributions à l'étude des champignons extra européens. *Bull. Soc. mycol. Fr.* **2**: 119–131.

RAO, G. P. & REDDY, V. T. C. 1955. Occurrence of *Ustilaginoidea virens* (Cke.) Tak. on *Oryza officinalis* Wall. *Indian Phytopath.* **8**: 72–73.

RAO, K. M. 1964. Environmental conditions and false smut incidence in rice. *Indian Phytopath.* **17**: 110–114.

RAYCHAUDHURI, S. P. 1946. Mode of infection of rice by *Ustilaginoidea virens* (Cke.) Tak. *J. Indian bot. Soc.* **25**: 145–150.

REINKING, O. A. 1918. Philippine economic plant diseases. *Philipp. J. Sci.*, Sect. A., **13**: 217–274.

SETH, L. N. 1945. Studies on the false smut disease of paddy caused by *Ustilaginoidea virens* (Cke.) Tak. *Indian J. agric. Sci.* **15**: 53–55.

SREERAMULU, T. & VITTAL, B. P. R. 1966. Periodicity in the air-borne spores of the rice false smut fungus, *Ustilaginoidea virens*. *Trans. Br. mycol. Soc.* **49**: 443–449.

TAKAHASHI, Y. 1896. On *Ustilago virens* Cooke and a new species of *Tilletia* parasitic on rice plants. *Bot. Mag., Tokyo* **10**: 16–20.

VEERARAGHAVAN, J. 1962. False smut disease caused by *Ustilaginoidea virens* (Cke.) Takahashi on certain wild species of *Oryza*. *Oryza* **1**: 86–87.

YAMASHITA, S. 1965. Detection of alkaloids in the unhulled rice infected by the false smut fungus of rice. *Ann. phytopath. Soc. Japan* **30**: 64–66. [Jap.]

YOSHINO, M. & YAMAMOTO, T. 1952. Pathogenicity of the chlamydospores of the rice false smut. *Agriculture Hort., Tokyo* **27**: 291–292. [Jap.]

KERNEL SMUT

The disease was described independently by Takahashi (1896) from Japan and by Anderson (1899) from U.S.A. Anderson identified the fungus as *Tilletia corona* Scrib., but this was erroneous, as pointed out by Earle (1899) and confirmed by Anderson (1902) himself. The name now generally accepted for the fungus is *T. barclayana* (Bref.) Sacc. & Syd. For a long time, it was thought to infect germinating rice grains, causing systemic infection as is the case in most *Tilletia* smuts, but later it was found to infect the opening flowers individually.

Besides Japan and U.S.A., the disease is known to occur in many countries, including Burma, China, India, Indonesia, Malaysia, Nepal, Pakistan, Philippines, Taiwan, Thailand and Vietnam in Asia; Guyana, Mexico, Trinidad and Venezuela in America; and Sierra Leone in Africa (see CMI Distribution Map 75).

Only a few grains in each panicle are affected and those affected often only partially, and the disease is usually considered to be unimportant. However, Su (1933) found in one season in Mandalay a loss of 2–5%. Fulton (1908) reported 3–4% infected grain in South Carolina, which caused serious damage to rice flour on account of the dark colour arising from the smutted grains. Reyes (1933) in the Philippines reported stunting of seedlings and a reduction in the number of tillers and yield when smutted seeds were planted. Recently, Templeton, Johnston & Henry (1960) observed increased severity in Arkansas.

SYMPTOMS

The disease is found in the field at the time of maturity of the rice crop. When the grains are examined, diseased ones show minute, black pustules or streaks bursting through the glumes. In a severe infection, a short beak-like or spur-like outgrowth is produced by the rupturing glumes (Fig. VI–3). This occurs because frequently only a part of the grain is affected and the unaffected portion causes the protruding or twisting of the grain. Sometimes the entire grain is replaced by a powdery, black mass of smut spores. This powdery, black mass scatters on to other grains or leaves and this is often the easiest way to detect the disease in the field. Sometimes the infected grains may be detected by their dull colour externally before the glumes are burst open by the masses of spores.

If not severely infected, the seeds may germinate but the seedlings are stunted (Reyes, 1933).

Fig. VI–3. Kernel smut. Above, infected grain. Below, partially smutted kernels. (Templeton, 1961, 1963a.)

THE CAUSAL ORGANISM

Morphology and nomenclature. The smut was originally called *Tilletia horrida* by Takahashi (1896) and he described the fungus as follows: 'Spore masses pulverulent, black, produced within the ovaries and remaining covered by the glumes. Spores globose, irregularly rounded, or sometimes broad-elliptical, the round ones 18·5–23·0μ in diameter and the elongated 22·5–26·0×18·0–22·0μ in size. Epispore deep olive brown, opaque, thickly covered with conspicuous spines. The spines hyaline or slightly coloured, pointed at the apex, irregularly polygonal at the base, more or less curved, 2·5–4·0μ in height, and 1·5–2·0μ apart at their free ends. Sporidia filiform or needle-shaped, curved in various ways, 10 to 12 in number and 38–53μ in length' (Fig. VI–4).

A number of observations and spore germination studies have been made since then (Butler, 1913; Reyes, 1933; Teng, 1931; Lin, 1936; Wei, 1934). There are several aspects in which this fungus has been regarded as differing from a typical *Tilletia*. It produces a large number of sporidia in a single whorl on the promycelium and the primary sporidia do not fuse. The spore often has a short hyaline apiculus and a broad, hyaline, warty outer coat of a gelatinous nature. The fungus infects only a few grains in a panicle and often only a part of the grains. For these reasons Padwick & Khan (1944) considered the smut to belong to the genus *Neovossia* and gave it the name *N. horrida* (Tak.) Padwick & Azmat. Khan. Fischer (1953), however, retained the name *T. horrida* Tak.

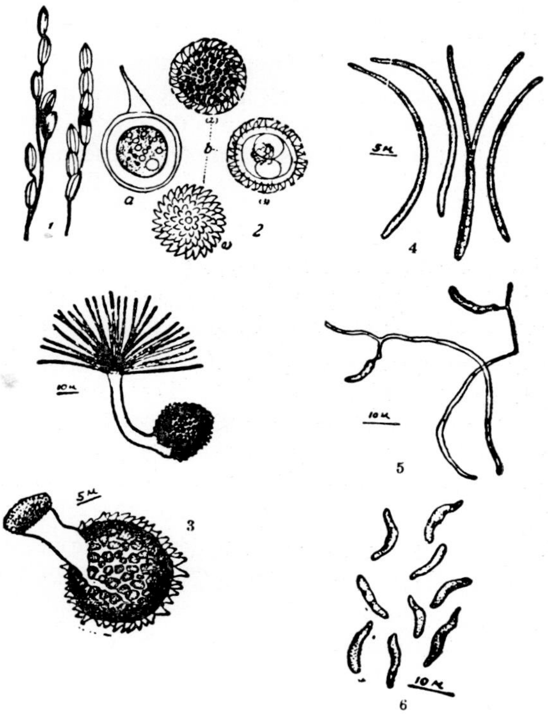

Fig. VI–4. *Tilletia barclayana*. 1, infected grains. 2, chlamydospores, different views. 3, germinating chlamydospores. 4, sporidia. 5, sporidia producing secondary sporidia. 6, secondary sporidia (C. K. Lin, 1936.)

Tullis & Johnston (1952) made further studies on spore germination and inoculated the rice kernel smut to several grass hosts (*Pennisetum* spp.) on which they found that it produced spores essentially identical with those of *Neovossia barclayana* Brefeld. They agreed with Padwick and Khan that the fungus belongs to *Neovossia* and further suggested that it is identical with *N. barclayana*, which has priority. They found that *N. horrida* had slightly larger and more conspicuous markings on the spores but regarded it as still within the limits of variation of a species. Their inoculation experiments were only partially successful, but they inoculated the sporidia of the rice fungus on to young seedlings of the 4 species of *Pennisetum* tested in petri dishes. Infection is now known to be not systemic and to take place in the flowers, not in seedlings. This leaves some room for doubt concerning the validity of the results.

Duran & Fischer (1961) agreed that the fungus on rice was the same as that found on grasses but regarded the mature spores to be more representative of the genus *Tilletia* than *Neovossia*. They therefore disposed it as *Tilletia barclayana* (Bref.) Sacc. & Syd., and that name is now generally used.

Physiology. Teng (1931) found that the chlamydospores germinated near the surface of water drops in a petri dish near a window, indicating the need for light and a good supply of oxygen. Lin (1936, 1955) further found that chlamydospores require a dormant period of about 5 months, besides light and oxygen. Germination, which was not affected by different nutrient solutions, was best at about 30°C. Exposing the chlamydospores directly to sunlight for 2 hours or to fluorescent light for 4–6 hours promotes subsequent germination in darkness.

Disease Cycle

Chlamydospores live for a year or more under normal conditions and have been found viable after 3 years in stored grains; they also survive passage through the digestive tracts of domestic animals.

Germinated chlamydospores produce many sporidia on the promycelium and as many as 50 or 60 have been observed. These sporidia further produce secondary sporidia in great numbers either directly or from mycelium produced by the primary sporidia. The secondary sporidia are sickle-shaped and are forcibly discharged, so that they are easily disseminated and so spread infection (Vanterpool, 1932).

The disease was at first thought to be systemic, as in the case of other diseases caused by species of *Tilletia*. Anderson (1899) and Butler (1913) found mycelium within the stem tissues and Butler believed that the fungus entered the host tissue while still below ground. Reyes (1933) planted smutted and healthy seeds of rice and found that many smutted seeds, but no healthy ones, gave rise to smutted plants. When Tullis & Johnson (1952) inoculated the germinating seeds of *Pennisetum* spp., a proportion of plants gave rise to smutted grains, while none occurred in the control. These tests did not, however, give adequate proof that the disease is systemic.

Chowdhury (1946), using a vacuum method, inoculated 97 rice panicles and found that 72 of them became infected. He concluded that the fungus causes local infection of florets by sporidia and does not infect the seedlings. This is supported by the fact that invariably only a few grains in a panicle are infected and often the grains are only partially infected. Recently, Templeton, Johnston & Henry (1960) and Templeton (1961) confirmed that the sporidia infect the opening flowers, which become smutted, but not the seedlings. There is no positive evidence that the disease is systemic.

Heavy application of nitrogenous fertilizers and the planting of late heading varieties increase the incidence of the disease (Templeton, 1963a).

Host Range and Varietal Resistance

Deighton (1955) found the fungus on wild rice, *Oryza barthii*, in Sierra Leone, and Thompson & Johnston (1953) reported it on *O. minuta* in Malaya. If it is accepted that the fungus on rice is *N. barclayana*, the host range also includes species of *Brachiaria*, *Digitaria*, *Eriochloa*, *Panicum* and *Pennisetum* (Duran & Fischer, 1961).

Reyes (1933) observed in the Philippines that several rice varieties, including Palagad, Sipot and Apostol, were susceptible whereas Elon-elon was apparently immune. Kameswar (1962) and Chauhan & Verma (1964) noticed that early maturing varieties are more susceptible than the late maturing ones in India. Templeton (1963b) mentioned that the variety Nova, as well as some others, is susceptible while Arkrose, Zenith and Saturn are resistant (Templeton, 1968) in U.S.A.

Control

The disease is not economically important and special control measures are normally unnecessary. Since the disease is not seed-borne and infection does not take place through germinating seeds as is usually the case in a bunt disease of cereals, the seed treatment recommended by earlier workers would have no value.

LITERATURE CITED

Anderson, A. P. 1899. A new *Tilletia* parasitic on *Oryza sativa* L. *Bot. Gaz.* **27**: 467–472.

Anderson, A. P. 1902. *Tilletia horrida* Tak. on rice plant in South Carolina. *Bull. Torrey bot. Club* **29**: 35–36.

Butler, E. J. 1913. Diseases of rice. *Bull. agric. Res. Inst., Pusa* 34, 37 pp.

Chauhan, L. S. & Verma, S. C. 1964. Bunt resistance of paddy varieties in Uttar Pradesh. *Sci. Cult.* **30**: 201.

Chowdhuri, S. 1946. Mode of transmission of the bunt of rice. *Curr. Sci.* **15**: 111.

Deighton, F. C. 1955. Plant pathology section. *Rep. Dep. Agric. Sierra Leone, 1953*: 33–34.

Duran, R. & Fischer, G. W. 1961. The genus *Tilletia*. 138 pp., Washington State University.

Earle, F. S. 1899. Letters. *Bot. Gaz.* **38**: 138.

Fischer, G. W. 1953. Manual of the North American smut fungi. 343 pp., New York, Ronald Press.

Fulton, H. R. 1908. Diseases affecting rice in Louisiana. *Bull. La agric. Exp. Stn* 105, 28 pp.

Kameswar Row, K. V. S. R. 1962. Incidence of ' bunt ' disease on rice. *Sci. Cult.* **28**: 534–535.

Lin, C. K. 1936. Factors affecting the germination of chlamydospores of *Tilletia horrida* Tak. *Bull. Coll. Agric. For. Univ. Nanking* **45**: 1–11.

Lin, C. K. 1955. Studies on the nature of light reaction in chlamydospore germination of rice kernel smut. *Acta phytopath. Sin.* **1**: 183–190. [Chin. Engl. summ.]

Padwick, G. W. & Azmatullah Khan. 1944. Notes on Indian fungi. II. *Mycol. Pap.* 10, 17 pp.

Reyes, G. M. 1933. The black smut or bunt of rice (*Oryza sativa* Linnaeus) in the Philippines. *Philipp. J. Agric.* **4**: 241–270.

Su, M. T. 1933. Report of the Mycologist, Burma, Mandalay for the year ending the 31st March 1933. 12 pp.

TAKAHASHI, Y. 1896. On *Ustilago virens* Cooke and a new species of *Tilletia* parasitic on rice plant. *Bot. Mag., Tokyo* **10**: 16–20.
TEMPLETON, G. E. 1961. Local infection of rice florets by the rice kernel smut organism, *Tilletia horrida. Phytopathology* **51**: 130–131.
TEMPLETON, G. E. 1963a. Kernel smut of rice as affected by nitrogen. *Arkans. Fm Res.* **12** (5): 12.
TEMPLETON, G. E. 1963b. Rice disease research in Arkansas. *Rice J.* **66**: 68, 70.
TEMPLETON, G. E. 1968. Rice disease studies—1968, in Arkansas Rice Research Station. *Rice J.* **72** (7): 41.
TEMPLETON, G. E., JOHNSTON, T. H. & HENRY, S. E. 1960. Kernel smut of rice. *Arkans. Fm Res.* **9** (6): 10.
TENG, S. C. 1931. Observations on the germination of the chlamydospores of *Tilletia horrida* Tak. *Contr. Biol. Lab. Sci. Soc. China*, Bot. Ser., **6**: 111–114.
THOMPSON, A. & JOHNSTON, A. 1953. A host list of plant diseases in Malaya. *Mycol. Pap.* 52, 38 pp.
TULLIS, E. C. & JOHNSON, A. G. 1952. Synonymy of *Tilletia horrida* and *Neovossia barclayana. Mycologia* **44**: 773–788.
VANTERPOOL, T. C. 1932. Cultural and inoculation methods with *Tilletia* species. *Science, N.Y.* **75**: 22–23.
WEI, C. T. 1934. Rice diseases. *Bull. Coll. Agric. For. Univ. Nanking*, N. S., 16, 39 pp. [Chin. Engl. summ.]

UDBATTA DISEASE

This disease was first described by Sydow (1914) from India, where the fungus seems to be common in several states. It was also found in Southwest China by Tai & Siang (1948). Infections of 9–11% of the panicles in Bombay, India, and 5–20% and occasionally 30% in Yunnan, China, have been observed.

SYMPTOMS

The symptoms are evident at the time of panicle emergence, the infected panicles while still inside the sheaths being matted together by the mycelium of the fungus. They emerge as single, small, cylindrical rods covered by white mycelium. In time they become hard and sclerotium-like, bearing many black dots. The infected plants are usually stunted. Sometimes the white mycelium and conidia form narrow stripes on the flag leaves along the veins before the panicles emerge. Usually all tillers of a plant are infected, indicating a systemic infection (Fig. VI–5).

THE CAUSAL ORGANISM

Ephelis oryzae was described by Sydow as having a hard, effuse, greyish-black stroma, occupying the whole length of the inflorescence and surrounding it entirely. Pycnidia immersed in the stroma, black, slightly convex, finally more or less exposed; spores needle-shaped, hyaline, $20–35 \times 1\mu$. Tai & Siang (1948) gave a description as follows: Sporodochia black, cupulate or convex, roundish, measuring 1–1·5 mm diameter; conidiophores branched, hyaline, $57–85 \times 0{\cdot}8–1{\cdot}4\mu$; conidia hyaline, acicular, aseptate straight or curved, $12–22\ (–40) \times 1{\cdot}2–1{\cdot}5\mu$ (Fig. VI–5).

Narasimhan & Thirumalachar (1943) found what they considered to be the perfect state on infected grains in sand culture and named it *Balansia oryzae* comb. nov. but without a proper diagnosis.

The spores germinate at 18–30°C, best at 26°C. They grow slowly but produce abundant conidia. They are resistant to dryness, and about one-third of spores air-dried for 162 days germinated (Tai & Siang, 1948).

Deighton (1937) reported a similar disease from Sierra Leone and referred it to *E. pallida* Pat., distinguished from *E. oryzae* by its smaller spores. He (1956) also reported the *Balansia* state on rice and several wild grasses.

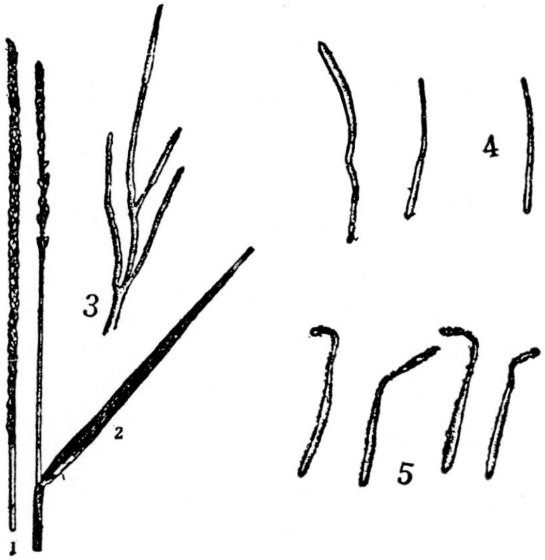

Fig. VI–5. *Ephelis oryzae.* 1–2, infected panicles and leaf. 3, conidiophore. 4, conidia. 5, germinating conidia. (Tao & Siang, 1948.)

Disease Cycle and Host Range

Tai & Siang (1948) believed that spores on infected kernels are able to overwinter. Mohanty (1964) made inoculation experiments and found that the fungus is internally seed-borne, infection being initiated at the time of emergence of the panicle. Experiments also showed that *E. oryzae* is not soil-borne and that diseased seed lots which were untreated produced more infected panicles than did those given hot water treatment.

It has been observed in several countries that some rice varieties are more often infected than others. Mohanty (1964) tested 26 varieties in India and found one (CB–11) to be highly resistant while 8 others were resistant.

The fungus has several grass hosts, including *Isachne elegans, Eragrostis tenuifolia, Arthraxon ciliaris* var. *coloratus, Saccolepis indica, Cynodon dactylon, Pennisetum* sp. and rye (Rao, Reddy & Reddy, 1959; Venkatarayan, 1937). Deighton (1956) reported *E. pallida* on 13 grass hosts, including *Oryza brachyantha.*

LITERATURE CITED

Deighton, F. C. 1937. Mycological Work. *Ann. Rep. Dep. Agric., Sierra Leone, 1936*: 44–46.

Deighton, F. C. 1951. Plant Pathology Section. *Rep. Dep. Agric., Sierra Leone, 1949*: 14–16.

DEIGHTON, F. C. 1956. Diseases of cultivated and other economic plants in Sierra Leone. 76 pp., Government of Sierra Leone.
MOHANTY, N. N. 1964. Studies on 'Udbatta' disease of rice. *Indian Phytopath.* **17**: 308–316.
NARASIMHAN, M. J. & THIRUMALACHAR, M. J. 1943. Preliminary note on the perfect stage of *Ephelis oryzae* Syd. (*Balansia oryzae* (Syd.) comb. nov.). *Curr. Sci.* **12**: 276.
RAO, P. G., REDDY, G. S. & REDDY, T. C. V. 1959. Four new hosts of *Ephelis*. *Sci. Cult.* **25**: 74–75.
SYDOW, H. 1914. Beiträge zur Kenntnis der Pilzflora des Südlichen Ostindiens. II. *Ann. mycol., Berl.* **12**: 484–490.
TAI, F. L. & SIANG, W. N. 1948. 'I-chu-hsiang' disease of rice caused by *Ephelis oryzae* Sydow in Yunnan. *Acta Agric.* **1**: 125–131.
VENKATARAYAN, S. V. 1937. Report of work done in Mycological Section during the year 1935–1936. *Adm. Rep. agric. Dep. Mysore, 1935–36*: 51–55.

BLACK KERNEL

Several species of *Curvularia* are found on rice grains causing discoloration (Boedijn, 1933; Bugnicourt, 1950; Grove & Skolko, 1945; Padwick, 1950; Wei, 1957), and a few of them mould the grain and may even cause leaf spots under certain conditions. Martin (1939, 1940) found that the infected grains produce black kernels after being polished. When severely infected, they may also cause seedling blight or weakening of the seedlings.

In isolating fungi associated with discoloured rice grains, *Curvularia* spp. often constitute a high percentage. Rao & Salam (1954) in India, for instance, reported 60% of the discoloured grains to be infected by species of *Curvularia*.

Ten or more species have been reported on rice from different parts of the world. Perhaps the most common are *C. lunata* (Wakker) Boedijn and *C. geniculata* (Tr. & Earle) Boedijn. Boedijn described the two species as follows (translated (Padwick, 1950) from the German):

C. lunata: Mycelium septate, much branched, subhyaline to light brown in the substratum, brown above it; single hyphae 2–5μ in diameter. Conidiophores dark brown, unbranched, septate, sometimes bent and knotted near the tip, 70–270 × 2–4μ. Conidia (? one or more) at the tip, in a whorl one over another or more or less spirally arranged; boat-shaped, rounded at the tip, mostly a little constricted at the base, with 3 septa, the second cell much larger and darker-coloured than the others, the conidium being bent at this cell; 19–30 × 8–16μ (mostly 23 × 11μ).

C. geniculata: Mycelium septate, abundantly branched, subhyaline to brown; single hyphae 2·5–7μ diameter. Conidiophores hyaline, septate, sometimes constricted at the base, lighter coloured at the tip; 340–900μ, at the base about 2–2·5μ, at the tip 3·5–5μ broad. Conidia arranged mostly in thick panicles, boat-shaped, asymmetrical or more or less strongly bent, with 4 septa; the third cell much larger and darker coloured than the others; 19–45 × 7–14μ (mostly 24 × 9μ).

The spore measurements vary considerably depending upon the substrate and according to different authors, and some of the species are difficult to separate from each other. Padwick (1950) considered *Brachysporium tomato*

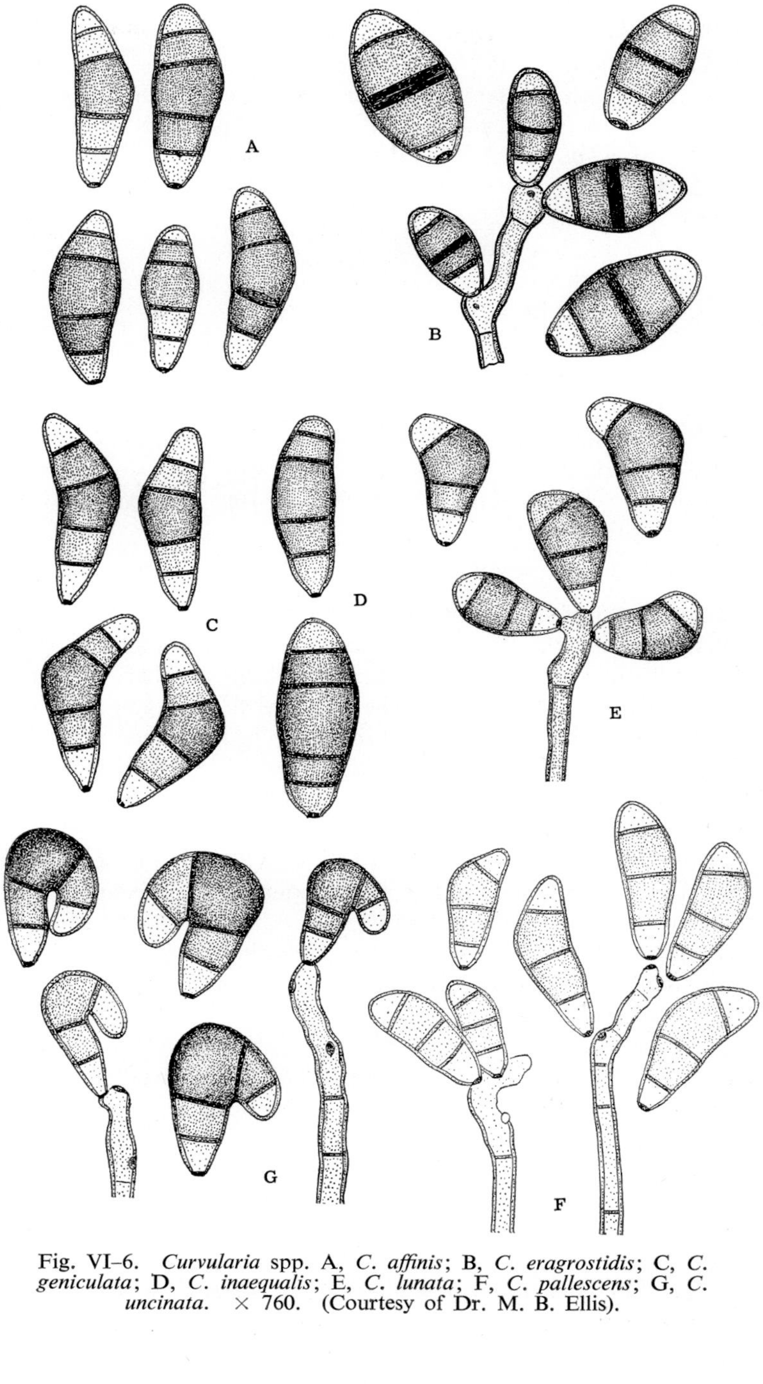

Fig. VI–6. *Curvularia* spp. A, *C. affinis*; B, *C. eragrostidis*; C, *C. geniculata*; D, *C. inaequalis*; E, *C. lunata*; F, *C. pallescens*; G, *C. uncinata*. × 760. (Courtesy of Dr. M. B. Ellis).

(Ell. & Barth.) Hiroe & Watanabe and *B. oryzae* Ito & Ishiyama to be very similar to *C. lunata*. Wei (1957) attempted to separate some of the species by the following key:

I. Conidia usually 3-septate
  A. Conidia with two larger and darker cells, not or slightly curved
    1. Conidia short and broad, barrel-shaped, 19–26×11–17μ—*C. maculans* (Bancroft) Boedijn
    2. Conidia long and narrow, cylindrical, 21–36×7–11μ—*C. pallescens* Boedijn
    3. Conidia long and broad, 26–43×11–26μ—*C. oryzae* Bugn.
  B. Conidia with second cell larger and darker, curved, 19–30×8–16μ—*C. lunata* (Wakker) Boedijn

II. Conidia usually 4-septate
  A. Conidia slightly curved
    1. Conidia narrower, 25–36×8–12μ—*C. affinis* Boedijn
    2. Conidia broader, 28–36×11–14μ—*C. geniculata* (Tr. & Earle) Boedijn
  B. Conidia curved, broader, 27–35×12–16μ—*C. inaequalis* (Shear) Boedijn

III. Conidia usually 3–4-septate, 17–38×8–14μ—*C. uncinata* Bugn.

The size and shape of conidia of some of the species are shown in Fig. VI–6.

Most of the species also live as saprophytes. In working with *C. lunata*, Martin (1939, 1940) succeeded in producing a large number of discoloured grains by inoculating the fungus at the time of flowering and by injecting inoculum into the seeds. He considered insects to be instrumental in spreading the disease, and rice straw stacks to be sources of infection in U.S.A. *C. lunata* has also been reported as infecting tomato and pepper fruits and *C. geniculata* as infecting cabbage, flax and pea.

C. tuberculata Jain (1962) was described recently on rice grains and *C. verruculosa* Tandon & Bilgrami has been reported as infecting leaves and spikelets of rice in India (Aulakh, 1966).

Ellis (1966) treated *C. maculans* as synonymous with *C. eragrostidis* (P. Henn.) Meyer and listed *C. intermedia* Boedijn (with its perfect state *Cochliobolus intermedius* Nelson) on rice.

The perfect states of *C. lunatus* and *C. geniculata* are *Cochliobolus lunatus* Nelson & Haasis (1964) and *Cochliobolus geniculatus* Nelson (1964) respectively. They were found to be heterothallic.

LITERATURE CITED

Aulakh, K. S. 1966. Rice, a new host of *Curvularia verruculosa*. *Pl. Dis. Reptr* **50**: 314–316.

Boedijn, K. B. 1933. Ueber einige phragmasporen Dematiazeen. *Bull. Jard. Bot. Buitenzorg*, Ser. 3, **13**: 120–134.

Bugnicourt, F. 1950. Les espèces du genre *Curvularia* isolées des semences de riz. *Rev. gén. Bot.* **57**: 65–77.

Ellis, M. B. 1966. Dematiaceous Hyphomycetes. VII: *Curvularia*, *Brachysporium* etc. *Mycol. Pap.* 106, 57 pp.

GROVE, J. W. & SKOLKO, A. J. 1945. Notes on seed-borne fungi. III. *Curvularia*. *Can. J. Res.*, Sect. C, **23**: 94–104.
JAIN, B. L. 1962. Two new species of *Curvularia*. *Trans. Br. mycol. Soc.* **45**: 539–544.
MARTIN, A. L. 1939. Possible cause of black kernels in rice. *Pl. Dis. Reptr* **23**: 247–249.
MARTIN, A. L. & ALTSTATT, G. E. 1940. Black kernel and white tip of rice. *Bull. Texas agric. Exp. Stn* 584, 14 pp.
NELSON, R. R. 1964. The perfect stage of *Curvularia geniculata*. *Mycologia* **56**: 777–778.
NELSON, R. R. & HAASIS, F. A. 1964. The perfect stage of *Curvularia lunata*. *Mycologia* **56**: 316–317.
PADWICK, G. W. 1950. Manual of rice diseases. 198 pp., Kew, Commonwealth Mycological Institute.
RAO, P. N. & SALAM, M. A. 1954. *Curvularia* species from discoloured grains from Hyderabad. *J. Indian bot. Soc.* **33**: 268–271.
WEI, C. T. 1957. Manual of rice pathogens. 267 pp., Peiping, Sci. Press. [Chin.]

MINUTE LEAF AND GRAIN SPOT

Species of *Nigrospora* are often found on various old or dead parts of rice plants throughout the world, mainly occurring as saprophytes. When rice plants are weakened by nutritional or climatic conditions or are suffering from other diseases or from insect attack, these fungi may affect the glumes, culms, leaves or various other parts of the rice plant. They have been reported to cause ear blight and blackening of ears under special conditions (Hopkins, 1950; Prasad *et al.*, 1960).

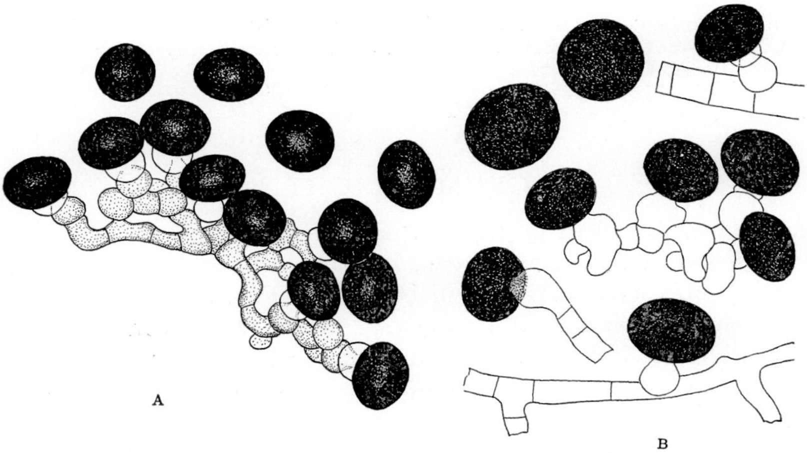

Fig. VI–7. A, *Nigrospora oryzae* (conidial state of *Khuskia oryzae*). B, *N. sphaerica*. × 650. (Courtesy of Dr. M. B. Ellis).

The characteristic symptoms are the presence of numerous minute, black pustules, less than 0·5 mm in diameter. These pustules are easily mistaken for pycnidia, but their somewhat powdery appearance may be seen under low magnification.

The fungus enters the host tissue through the stomata. The mycelium is restricted to an area along the outer layers of parenchyma, with very little tendency to penetrate deeper. It produces sporodochium-like bodies after rupturing the host tissue (Kar, 1966).

The conidiophores are short, simple, inflated below the tip and bear apically, single, one-celled, globose or subglobose, smooth, dark conidia (Fig. VI–7).

Four species have been reported on rice. They are separated mainly by shape and size of the conidia.

N. oryzae (Berk. & Br.) Petch—conidia tend to be subglobose, measuring *c.* 12–17 × 13–18 (13 × 15)µ.

N. sphaerica (Sacc.) Mason—conidia perfectly globose, very smooth, 16–18µ.

N. panici Zimm.—conidia globose or somewhat flattened, with a hyaline membranous cap over the upper part, 25–30 × 22–25µ.

N. padwickii Prasad, Agnihotri & Agarwal—conidia large, 37·9 × 34·8µ.

N. oryzae and *N. sphaerica* are the species most commonly reported on rice. The perfect state of *N. oryzae* was described by Hudson (1963) and named *Khuskia oryzae*. *N. padwickii* was described rather recently from India and *N. panici* is rarely reported on rice.

LITERATURE CITED

HOPKINS, J. C. F. 1950. A descriptive list of plant diseases in Southern Rhodesia and list of bacteria and fungi. *Mem Dep. Agric. S. Rhod.* 2, 106 pp.

HUDSON, H. J. 1963. The perfect state of *Nigrospora oryzae*. *Trans. Br. mycol. Soc.* **46**: 355–360.

KAR, A. K. 1966. Host-pathogen relationship in *Nigrospora* disease of rice. *Indian Phytopath.* **19**: 115–116.

PADWICK, G. W. 1950. Manual of rice diseases. 198 pp., Kew, Commonwealth Mycological Institute.

PRASAD, N., AGNIHOTRI, J. P. & AGARWAL, J. P. 1960. A new species of *Nigrospora* Zimm. on *Oryza sativa* L. *Curr. Sci.* **29**: 352–353.

WEI, C. T. 1957. Manual of rice pathogens. 267 pp., Peiping, Sci. Press. [Chin.]

GLUME BLIGHT

Glume blight is caused by *Phyllosticta glumarum* (Ell. & Tr.) Miyake. It was first found in U.S.A. and Japan, but has now also been reported from Brazil, Ceylon, China, territories in the former French West Africa, India, Sierra Leone, Tanzania and other areas. Under special conditions it may cause considerable losses (Inoue & Matsunaga, 1952). Chu (1933) reported a 25% loss in Chekiang, China, in 1932.

The fungus infects the glumes during the 2 or 3 weeks following emergence of the panicle. The lesions are at first small, oblong, and brown; they gradually enlarge, becoming whitish, and finally bear numerous small black dots, the pycnidia, in the centre. The spots may coalesce to become irregular in shape. When infection occurs early, no grain is formed. Late infections result in partly filled grains or the grains become discoloured and brittle.

The fungus was named *Phoma glumarum* by Ellis & Tracy in Ellis & Everhart (1888). Miyake (1910) transferred it to *Phyllosticta* because of the thin, membranaceous pycnidial wall and the production of definite and regular spots. Pycnidia, at first covered by the epidermis, become erumpent and finally superficial, scattered or gregarious, globose or subglobose, 48–133 × 40–95µ, with protruding ostioles; ostiole opening about 10–18µ in diameter. Pycnidial walls are dark brown, lighter below, yellowish brown at the base, consisting of 3 layers of cells, the outermost layer with thick-walled cells and the cells of the inner layer thin-walled. Conidia oblong to ovoid, smoky-hyaline, 3–6×2–3μ.

When moistened, the conidia are extruded from the pycnidia in tendrils. They germinate in water and infect the glumes. Damage from rain-storms and wounds facilitate infection.

Several related species have been reported on rice glumes, including *P. oryzina* (Sacc.) Padwick, *P. oryzicola* Hara (=*Trematosphaerella oryzae* (Miyake) Padwick) and *P. glumicola* (Speg.) Hara, but their parasitic nature on rice is not well established. In addition, several species of *Phyllosticta* and *Phoma* have been found on leaves and leaf sheaths. A key for separating 7 commonly occurring species of *Phyllosticta* is as follows (Wei, 1957):

I. Pycnidia superficial
  A. Conidia oblong, 3–6×2–3μ—*P. glumarum* (Ell. & Tr.) Miyake
  B. Conidia long-oblong, 4·8–6×1·8–2μ—*P. oryzina* (Sacc.) Padwick

II. Pycnidia partially buried in the tissue
  A. Conidia with a guttule at each end, 5–7×3–4μ—*P. oryzicola* Hara
  B. Conidia without guttule, 4–5×2μ—*P. glumicola* (Speg.) Hara

III. Pycnidia buried in the tissue
  A. Conidia elliptical, 7·5–10×3–4μ—*P. miyakei* Syd.
  B. Conidia cylindrical, 3–4×1–1·5μ—*P. miurai* Miyake
  C. Conidia ovoid, 2·5–3·5×1·8–2·5μ—*P. oryzae* (Cke. & Mass.) Miyake

LITERATURE CITED

CHU, H. T. 1933. Observations on the physiological characters of *Phoma glumarum*, the causal fungus of grain blight of rice plant. *Yearb. Bur. Ent. Hangchow, China, 1932*: 192–198. [Chin. Engl. summ.]

ELLIS, J. B. & EVERHART, B. M. 1888. New fungi from various localities. *J. Mycol.* **4**: 121–124.

INOUE, Y. & MATSUNAGA, M. 1952. On the glume blight of the rice plant. *Kyushu agric. Res.* **9**: 13–14. [Jap.]

MIYAKE, I. 1910. Studien über die Pilze der Reispflanze in Japan. *J. Coll. Agric., Tokio* **2**: 237–276.

PADWICK, G. W. 1950. Manual of rice diseases. 198 pp., Kew, Commonwealth Mycological Institute.

WEI, C. T. 1957. Manual of rice pathogens. 267 pp., Peiping, Science Press. [Chin.]

SCAB

The fungus which causes scab disease of wheat also infects rice and produces similar symptoms. It was reported on rice in Italy by Cattaneo (1877) (as *Botryosphaeria saubinetii*) and in Japan by Miyake (1910) (as *Gibberella saubinetii*) on glumes and leaves. It has since been recorded in several other countries, including Brazil, China, India and Uganda. The disease does not usually cause heavy damage but it may be severe in conditions favourable for disease development, such as high humidity.

The fungus causes lesions on or discoloration of the glumes, which are at first white, and later yellow, salmon or carmine. Sometimes the whole grain is affected. These lesions bear sporodochia and masses of conidia. Infected grains are light, shrunken and brittle, and often do not germinate; when they do germinate they give rise to diseased seedlings. The fungus may also attack the nodes, which become blackened and disintegrate, the stems wilting and breaking (Kasai, 1923).

Scab is caused by *Fusarium graminearum* Schwabe, perithecial state *Gibberella zeae* (Schw.) Petch. There has in the past been considerable confusion regarding the name of the perithecial state and it has often been referred to as *Gibberella saubinetii* (Mont.) Sacc. This is incorrect as this name is a synonym of *G. cyanogena* (Desm.) Sacc. *G. rosea* (Link) Snyder & Hansen f. *cerealis* Snyder & Hansen (1945) and the conidial state name *Fusarium roseum* f. *cerealis* Snyder & Hansen (1945), both invalid names, are occasionally used for this species.

Wollenweber & Reinking (1935) described the fungus under the name *F. graminearum* as follows (Padwick, 1950):

Fertile layer of conidial stage variously coloured, white to rose, golden yellow, ochre (becoming blue-violet on addition of ammonia) or carmine-purple-red, partly plectenchymatous, spreading, and more or less covered with fleecy aerial mycelium partly restricted, erumpent, sclerotial, resembling in appearance the ochre to bright orange-red layers of pionnotal conidia masses or (more rarely) sporodochia with which it is covered. Conidia sometimes compressed as in *Fusarium culmorum*, sometimes longer as in its varieties, spindle- to sickle-shaped, moderately curved, tapering at both ends, with conically pointed or notched apex and foot-celled base, 3–5-septate, more rarely 1–2- or 6–9-septate;

3-septate	41×4·3	mostly	30–47×3·5–5 (25–66×3–6)μ
5-septate	51×4·9	,,	41–60×4·3–5·5 (28–72×3·2–6)μ
7-septate	73×5·4	,,	61–82×4·5–6·5 (50–88×4–7)μ
9-septate	80×5·5	,,	61–96×4·5–7 (55–106×4–8)μ

Chlamydospores absent or very sparsely distributed.

They described the perithecia under the name *Gibberella saubinetii* as follows: Perithecia blue-black, oval or round, with or without a conical ostiole, surrounded at the top by a ring composed of a large-celled outgrowth of the peridium, which in itself is rather thick, many-layered, and of plectenchymatous structure, 200×170 (150–300×100–250)μ; asci 37–84×8–15μ; ascospores in one or indistinctly two rows, spindle-shaped, slightly curved to almost straight, blunt to conical at both ends, 3-septate, 22·7×4, mostly 18–27×3·4–5 (16–33×3–6)μ, more rarely 1-septate and 19×3·5 (14–24×2·5–5)μ, occasionally 4-septate, in isabella-coloured masses, at first moist, then powdery and dry after emerging through the opening of the perithecium.

Hemmi, Seto & Ikeya (1931) found that this fungus and *Gibberella fujikuroi* infected the seeds in the flowering period, less readily in the milky stage and afterwards. It is also known that the fungus can be soil-borne.

Ikeya (1933) compared isolates from rice with those from wheat, and found that they seemed to be morphologically and physiologically alike. The optimum temperature for mycelial growth of both was about 28°C. Chung *et al.* (1964) found that an isolate from wheat infected rice and other plants and caused post-emergence blight in rice.

Several other species of *Fusarium* have been reported on rice grains. Bugnicourt (1952a, b) found *F. moniliforme* Sheld. var. *minus* Wollen., *F. decemcellulare* Brick. (=*Calonectria rigidiuscula* (Berk. & Br.) Sacc.), *F. kuhnii* (Fuckel) Sacc., *F. annulatum* Bugn., and *F. gibbosum* App. & Wollen. emend. Bilai in material from the Pacific Islands. Saccas (1950) reported *F. nivale* (Fr.) Ces. var. *oryzae* Saccas to cause considerable damage to rice panicles in French Equatorial Africa.

Other species of *Fusarium* have been reported on straw and other parts of rice, including *F. merismoides* Cda, *F. lateritium* Nees (=*Gibberella baccata* (Wallr.) Sacc.), *F. avenaceum* (Fr.) Sacc., *F. culmorum* (W. G. Smith) Sacc., *F. heterosporum* Nees (=*Gibberella cyanea* (Sollm.) Wollen.), *F. oxysporum* Schl. f. *vasinfectum* (Atk.) Snyder & Hansen, and *F. reticulatum*. They may live on rice as saprophytes or weak parasites.

LITERATURE CITED

BUGNICOURT, F. 1952a. Note sur la mycoflore des semences de riz dans les territoires du Pacifique Sud. *Rev. Mycol.* **17**, Suppl. colon. 1: 26–29.

BUGNICOURT, F. 1952b. Une espèce fusarienne nouvelle, parasite du riz. *Rev. gén. Bot.* **59**: 12–18.

CATTANEO, A. 1877. Contributo allo studio dei miceti che nascono sulle pianticelle di riso. *Archivio trienn. Labor. Bot.critt., Pavia* **2-3**: 117–128.

CHUNG, H. W., CHUNG, H. S. & CHUNG, B. J. 1964. Studies on pathogencity of wheat scab fungus (*Gibberella zeae*) to various crop seedlings. *J. Pl. Prot. Korea* **3**: 21–25.

HEMMI, T., SETO, F. & IKEYA, J. 1931. Studies on the ' bakanae ' disease of rice plant. II. On the infection of rice by *Lisea fujikuroi* Sawada and *Gibberella saubinetii* (Mont.) Sacc. in the flowering period. *Forschn Geb. PflKrankh., Kyoto* **1**: 99–110. [Jap. Engl. summ.]

IKEYA, J. 1933. On a disease of the rice plant caused by *Gibberella saubinetii* (Mont.) Sacc. *Forschn Geb. PflKrankh., Kyoto* **2**: 292–313. [Jap. Engl. summ.]

KASAI, M. 1923. Cultural studies with *Gibberella saubinetii* (Mont.) Sacc. which is parasitic on the rice plant. *Ber. Ohara Inst. landw. Forsch.* **2**: 259–272.

MIYAKE, I. 1910. Studien über die Pilze der Reispflanze in Japan. *J. Coll. Agric., Tokyo* **2**: 237–276.

PADWICK, G. W. 1950. Manual of rice diseases. 198 pp., Kew, Commonwealth Mycological Institute.

SACCAS, A. 1950. Un *Fusarium* parasite des panicules de riz. *Rev. Bot. appl.* **30**: 483–500. (*Rev. appl. Mycol.* **30**: 429–430.)

SNYDER, W. C. & HANSEN, H. N. 1945. The species concept in *Fusarium* with reference to Discolor and other sections. *Am. J. Bot.* **32**: 657–666.

WEI, C. T. 1957. Manual of rice pathogens. 267 pp., Peiping, Sci. Press. [Chin.]

WOLLENWEBER, W. H. & REINKING, O. A. 1935. Die Fusarien. 355 pp., Berlin, Paul Parey.

RED BLOTCH OF GRAINS

Ito & Iwadare (1934) reported that two species of *Epicoccum*, *E. oryzae* and *E. neglectum*, caused red blotch disease of rice grains in Japan. The disease occurs when the rice panicles fall on the ground before or after maturity. Infected seeds fail to germinate and damage can be severe at times. The fungi have been reported from Brazil, Italy, Portugal, Turkey, Uganda, U.S.A., and other areas. They are often found on weakened or entirely dead plant parts and are not considered as virulent parasites. Another species, *E. purpurascens*, is also found on dead rice plant tissue.

E. oryzae Ito & Iwadare was described as follows: spots red to rose, irregular; hyphae branched, hyaline, at length olivaceous, septate, 3·7–6·2μ broad; sporodochia globose or subglobose, dark, dot-like, 42–210μ and in masses many times that diameter; conidiophore 2·5–7·5μ long, at first yellow, finally olivaceous; conidia globose, subglobose to piriform, granular-verrucose, 1–5-celled, olivaceous, 9·9–23·1 × 6·6–16·5μ.

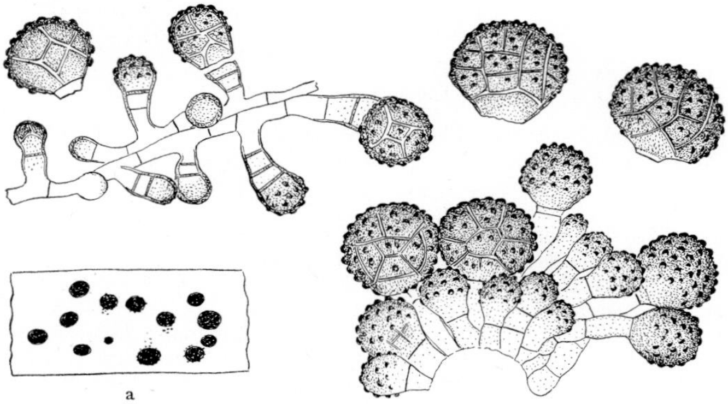

Fig. VI–8. *Epicoccum purpurascens*. a, habit sketch; other figures × 650. (Courtesy of Dr. M. B. Ellis).

E. neglectum Desmazières has spherical, slightly rough, reticulate conidia about 17μ.

E. purpurascens Ehrenberg ex Schlecht. also has reticulate conidia, 16–22μ in diameter (Fig. VI–8).

Ito & Iwadare (1934) reported that both *E. oryzae* and *E. neglectum* produce pinkish-red pigments on various standard media. They distinguished 3 strains of *E. neglectum* based upon temperature reaction in respect to pigmentation of media, and infection. Optimum temperature range for infection is 14–25°C, varying somewhat among the species and strains.

E. neglectum has also been found on wheat, oats, maize and buckwheat, but it cannot infect the healthy leaves or stems of any of these hosts (Ito & Iwadare, 1934).

In the opinion of some taxonomists the 3 species of *Epicoccum* mentioned above are all to be regarded as *E. purpurascens*.

LITERATURE CITED

Ito, S. & Iwadare, S. 1934. Studies on the red blotch of rice grains. *Rep. Hokkaido agric. Exp. Stn* 31: 1–85. [Jap. Engl. summ.]

SPECKLED BLOTCH

Several species of *Septoria* are found on rice. *S. oryzae* Cattaneo is reported to infect leaves, leaf sheaths, and glumes; *S. miyakei* Sacc. & Trav. to infect glumes; and *S. poae* Cattaneo to infect leaves, causing whitish to greyish spots. Haskell (1962) reporting *S. oryzae* on rice glumes in Florida called the disease speckled blotch. These species of *Septoria* have been reported from Italy, Japan, Brazil and U.S.A. Other species are also found which are more or less saprophytic on dying or dead plant parts.

S. oryzae has innate-protruding pycnidia, black, numerous, in clusters, arranged in parallel between the veins, 80–120μ in diameter. Conidia cylindrical to long-elliptical, straight or slightly curved, round at both ends, hyaline and not septate at first, later becoming yellow to yellowish brown and 1–3 septate, 15–23 × 2–3·5 (21 × 3)μ.

S. miyakei has immersed pycnidia, black, ellipsoid with a papillate ostiole, 140–150μ in diameter, 100–110μ high, conidia long with rounded or constricted ends, hyaline, often curved, nonseptate, 30–40 × 2·5–3·2μ.

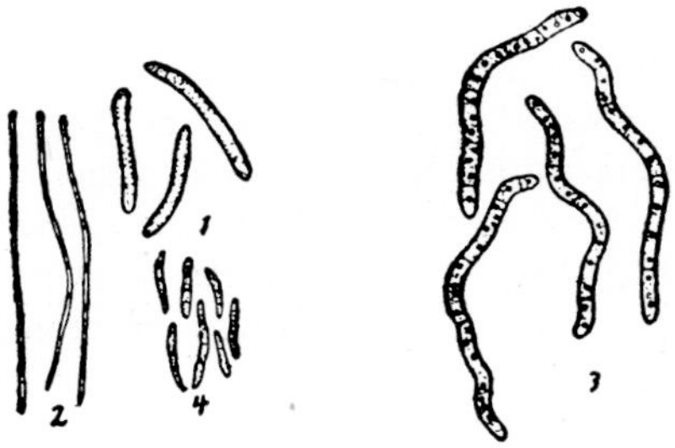

Fig. VI–9. Conidia of (1) *Septoria miyakei*, (2) *S. oryzicola*, (3) *S. curvula*, and (4) *S. oryzae*. (Wei, 1957).

S. poae has fleshy, white pycnidia, on roughly square, black spots, conidia very slender, 50μ long, yellowish–hyaline. Miyake (1910) pointed out that the white pycnidia of this species indicate that it does not belong to the Sphaerioidaceae, but to the Nectrioidaceae, and is close to the genus *Trichocrea*.

The species of *Septoria* found on rice may be separated by the appearance of the conidia (Fig. VI–9) and by the following key (Wei, 1957).

I. Pycnidia black
  - A. Conidia hyaline
    - 1. Conidia single-celled (nonseptate)
      - a. Conidia short and broad, 30–40 × 2·5–3·2μ—*S. miyakei* Sacc. & Trav.
      - b. Conidia long and narrow, 50–80 × 0·5–0·8μ—*S. oryzicola* Hara
    - 2. Conidia many-celled (septate), 50–80 × 2·5–3μ—*S. curvula* Miyake

B. Conidia coloured
  1. Conidia cylindrical
    a. Conidia 4-celled, 21×3μ—*S. oryzae* Cattaneo
    b. Conidia 3–6-celled, 20–30×3μ—*S. oryzae* var. *brasiliensis* Speg.
  2. Conidia elliptical to round, worm-like, 30–45×2μ—*S. dehaani* Hara

II. Pycnidia white—*S. poae* Cattaneo

LITERATURE CITED

HASKELL, R. J. 1926. Diseases of cereals and forage crops in the United States in 1925. *Pl. Dis. Reptr*, Suppl. 48: 301–381.
MIYAKE, I. 1910. Studien über die Pilze der Reispflanze in Japan. *J. Coll. Agric., Tokyo* **2**: 237–276.
WEI, C. T. 1957. Manual of rice pathogens. 267 pp., Peiping, Sci. Press. [Chin.]

OTHER DISEASES ON FOLIAGE AND GLUMES

Besides those discussed above, many other fungi are found on leaves and glumes of rice plants. Most of them are found on weakened plants or moribund plant parts and some are probably only saprophytic. Occasionally, however, some of these fungi have been reported to cause diseases in healthy tissues, and under very humid conditions several of them may develop together to cause a composite disease called false blast. The more commonly occurring fungi are briefly mentioned below:

1. *Mycosphaerella* Johanson

M. oryzae (Catt.) Miyake has been found in Japan, Italy, Portugal and Austria, causing brown spots on leaves. *M. tassiana* (de Not.) Johans. has been reported from Japan and China, causing whitish spots with a dark brown border on the tips and margins of leaves. *M. tulasnei* (Jancz.) Lindau has an imperfect state *Cladosporium herbarum* (Pers.) Link ex S. F. Gray which causes grey mould disease of old leaves under humid conditions. It has been found on rice in several areas, including China, Japan, India, Austria and North America. It is also found on other crops in many parts of the world. *M. hondai* Miyake has been reported from China, Japan and Korea on leaves which become pale and covered with perithecia, appearing as black dots. *M. malinverniana* (Catt.) Miyake is found on dying leaves in Japan, India and Italy. It has been suspected to be the perfect state of *Pyricularia oryzae* Catt. but no positive proof of this has been found. *M. graminicola* Fuckel and *M. usteriana* (Speg.) Hara are found on leaves (Padwick, 1950; Wei, 1957).

The following key has been constructed as an aid in distinguishing between species of *Mycosphaerella* (Wei, 1957):

I. Spore more or less fusoid
  A. Spores not constricted at septum, 14–15×4–4·2μ—*M. oryzae*
  B. Spores constricted at septum, 17–26×5–9μ—*M. tassiana*

II. Spores more or less ovoid or oblong
   A. Spores cylindrical or oblong
      1. Spores cylindrical, slightly tapering at the ends, 11–29×4–9μ—*M. tulasnei*
      2. Spores oblong, slightly tapering at the ends, 10–12×2–2·5μ—*M. graminicola*
   B. Spores ovoid
      1. Spores septate at the middle, 10–12×3μ, hyaline—*M. usteriana*
      2. Spores septate slightly below the middle
         a. Spores 10–12×3–4·5μ, greenish—*M. hondai*
         b. Spores 20×10μ, hyaline—*M. malinverniana*

2. *Sphaerulina* Sacc.

S. miyakei Hara causes leaf spotting in Japan. The spots are small and dark brown at first, gradually enlarging, becoming irregular and coalescing. Finally they are greyish with a brown border and bear black dots in the centre. The upper portion of the leaf may die as a result of attack. *S. oryzina* Hara has been shown to be the perfect state of *Cercospora oryzae* Miyake, the cause of narrow brown leaf spot. *S. oryzae* Miyake is found on dead leaves (Padwick, 1950; Wei, 1957).

S. miyakei: Perithecia scattered or gregarious, covered by the epidermis then erumpent, globose or subglobose, 100–150μ in diameter, membranaceous, dark brown, ostiole papilliform; asci clavate, cylindrical or oblong, rounded at the apex, with a short stipe, 8–spored, 50–75×10–15μ; spores cylindrical or elliptical, 3-septate, hyaline, 20–27×5–7μ (Fig. VI–10B).

S. oryzae has elliptical spores, 15–18×3–5μ, at first 1-septate. *S. oryzina* has also elliptical spores but larger, 20–23×4–5μ (Fig. VI–10A).

3. *Trematosphaerella* Kirschst.

T. oryzae (Miyake) Padwick was reported to be very harmful in Japan. The affected leaves gradually become pale, beginning at the tip and edges but without the formation of definite spots, and eventually no green part of the leaf is left. On the glumes it forms brownish-black spots which later become pale. The grains are not formed or are incompletely filled. The fungus has also been reported from China, India and the Philippines. *T. cattanei* (Thuemen) Padwick, reported from China and Austria, causes greyish spots on old plant parts. *T. elongata* (Hara) Wei was reported on glumes in Japan (Padwick, 1950; Wei, 1957).

T. oryzae: Perithecia immersed with slightly papillate, erumpent, oval to ellipsoid ostiole, dark brown, 70–125μ broad, 90–125μ high; asci cylindrical, tapering downward, 35–55×7–9μ, thin-walled, 8–spored; spores in 2 rows, fusoid, usually curved, 4-celled, slightly constricted at the septa, 16–23×4–5μ, dark yellow.

T. elongata has oblong-cylindrical spores, 16–23·5×5–6μ. *T. cattanei* also has fusoid spores but they are larger, 22–28×4μ (Fig. VI–10G).

4. *Metasphaeria* Sacc.

M. albescens Thuemen has been reported from Burma, China, Japan and Italy, causing dark brown leaf spots which quickly enlarge to become yellowish brown, finally greyish white and covered with small black dots. It is not considered as a virulent parasite and is often found together with *M. oryzae* (Catt.) Sacc. on glumes. *M. oryzae* (Catt.) Sacc. is reported from Italy. It attacks all plant parts, which later become covered with black dots. Infected plants may be reduced in size and grains may not mature normally. *M. cattanei* Sacc. and *M. oryzae-sativae* Hara are weak parasites (Padwick, 1950; Wei, 1957).

M. albescens was described as follows: Perithecia compressed-globose, black, covered by the epidermis, 112–178 × 93–140μ, with slightly protruding ostiole; asci fusiform, narrower at both ends, short-pedicelled to almost sessile, with obtuse apex, 70–85 × 16–22μ, 8-spored; paraphyses very few, branched; ascospores oblong-fusoid, 3–5-septate, bow-shaped, hyaline, 18–24 × 5–6μ (Fig. VI–10C).

Wei (1957) observed conidia in culture. They are hyaline, allantoid to spindle-shaped, usually 1-septate, 6–15 × 3–4μ. He also mentioned that specimens from Burma and China have smaller asci, 40–65 × 12·5–20μ and 40–61 × 12–15μ, respectively.

The four species of *Metasphaeria* may be separated by the following key (Wei, 1957):

I. Spores oblong or cylindrical
   A. Spores 15–18 × 2·5–3·5μ—*M. oryzae-sativae*
   B. Spores 27 × 6μ—*M. cattanei*

II. Spores fusiform
   A. Spores 3–5 septate—*M. albescens*
   B. Spores 5 (4–6) septate—*M. oryzae*

5. *Melanomma* Nits. ex Fuckel

M. glumarum Miyake is suspected to be the perfect state of *Phyllosticta glumarum*, the cause of glume blight. It attacks the glumes, causing dark brown spots which become bleached later and on which are produced many black dots. When attack is severe no grain is formed; light attacks reduce the size of the grain and the glumes are discoloured. *M. glumarum* has been reported from China, Japan, India and the Philippines.

M. glumarum: Perithecia sparse, globose to ellipsoid, black, with an ostiole at the tip, about 150μ in diameter and somewhat more in height; asci cylindrical with very short pedicel, usually slightly curved, 8-spored, 70–90 × 10μ; spores in two rows, fusoid, usually somewhat curved, dark, 24–30 × 4–5μ, 3-septate, with one or two guttules in each cell. Paraphyses not observed (Fig. VI–10D).

M. oryzae Hara was found on dead plants in Japan, and has smaller spores, 10–12 × 2·5–3μ, yellowish brown.

Luc (1953) reported *M. glumarum* f. *africana* on glumes from the Ivory Coast. It differs from *M. glumarum* in having smaller and narrower asci (40–50 × 5–7μ) and shorter ascospores (15–19 × 3–4·5μ).

6. *Leptosphaeria* Ces. & de Not.

L. michotii (West.) Sacc. has been reported to attack leaves, culms and glumes in Japan and Korea. Diseased areas turn whitish. *L. oryzicola* Hara causes zonate spots on leaves in Japan. *L. culmicola* (Fr.) Karst., *L. inecola* Hara and *L. culmifraga* (Fr.) Ces. & de Not. are found on dead leaves and glumes. *L. salvinii* Catt. causes the stem rot which is discussed separately. These species may be distinguished by the following key (Wei, 1957) (Fig. VI–10F).

I. Spores 3-celled—*L. michotii*

II. Spores 4-celled, elliptical
  A. On leaves, spores 17–20×4–4·5μ—*L. oryzicola*
  B. On leaves, sheaths and culms, spores 20–28×9μ—*L. salvinii*

III. Spores 6-celled
  A. Spores cylindrical oblong, 24–27×5–6μ—*L. culmicola*
  B. Spores elliptical, 34–45×4–6μ—*L. inecola*

IV. Spores 8–10-celled—*L. culmifraga*

7. *Sphaeropsis* Sacc.

Sphaeropsis oryzae (Catt.) Sacc. was reported from Italy and Brazil to cause premature drying of leaves and partially filled grains. *S. vaginarum* (Catt.) Sacc. was recorded on leaf sheaths from Italy and Japan. *S. japonicum* Miyake was found on glumes in Japan. The three species differ in the size of their dark conidia. *S. oryzae* has globose conidia, 14μ in diameter; *S. vaginarum* has ovoid to pyriform conidia, 15×9μ; while the conidia of *S. japonicum* are ellipsoid, cylindrical or irregular, 12–17×4–6μ (Padwick, 1950; Wei, 1957).

8. *Coniothyrium* Sacc.

Several species of *Coniothyrium* have been reported on rice. *C. japonicum* Miyake, *C. brevisporum* Miyake and *C. anomalum* Miyake produce brown spots on the tips and margins of rice leaves, the spots later becoming bleached to whitish lesions. These and other species are separated by the shape and size of their conidia (Wei, 1957) (Fig. VI–10I).

I. Conidia cylindrical to oblong
  A. Conidia 6–9×2–3μ—*C. japonicum*
  B. Conidia 11–13×5–6μ—*C. oryzae* Cav.

II. Conidia oblong to short-ellipsoid or ovoid
  A. Conidia ellipsoid, 10–12×2·5–3μ—*C. inevorum* Hara
  B. Conidia oblong to ellipsoid, 3–3·5×2–2·5μ—*C. oryzivorum* Hara
  C. Conidia oblong or ovoid, 4–5×2–3μ—*C. brevisporum*
  D. Conidia oblong or ovoid, 6–7·5×2–3μ—*C. anomalum*

9. *Diplodia* Fr. and *Diplodiella* (Karst.) Sacc.

Diplodia oryzae Miyake has been reported from Japan, India and South Africa on rice leaves and glumes. Pycnidia are produced under the epidermis, 82–137μ in diameter; conidia ellipsoid, oblong to cylindrical, 7–9×2–3μ.

Diplodiella oryzae Miyake was found on leaves and glumes from Japan and Yunnan, China. Pycnidia are superficial, 120–180μ high and 120–220μ in diameter; conidia elliptical, 9–13 × 2·2–3·5μ (Fig. VI–10H) (Padwick, 1950; Wei, 1957).

10. *Phaeoseptoria* Speg.

Phaeoseptoria oryzae Miyake infects rice leaves and glumes. The infected area is discoloured but no definite spot is formed. It has been reported from Japan and Yunnan, China. *P. japonica* Hara was reported from Japan to cause greyish discoloration of the glumes. *P. oryzae* has 4–6-septate conidia, 30–45 × 2·5–3μ. The conidia of *P. japonica* are 3-septate, 20–30 × 2·5–3μ. Another species found in Yunnan by Tai, but not named, has 3-septate conidia, measuring 14–19 × 3–3·6μ (Wei, 1957).

11. *Pestalotia* de Not.

Pestalotia kawakami Sawada was reported to infect the leaves of upland rice in Taiwan. It produces large irregular spots with dark brown margins. The centre of the spot is usually bleached to a straw or grey colour and bears small, black dots. Another species known to occur only on dead straw and leaves is *P. oryzae* Hara. The former has smaller conidia, 20–24 × 4–5μ with hairs 5–10μ long; the conidia of the latter are 25–35 × 7–10μ, with hairs 20–45μ long (Padwick, 1950; Wei, 1957).

12. *Oospora* Wallr.

Two species of *Oospora* occur on rice, *O. oryzae* Ferraris and *O. oryzetorum* Sacc. The former was reported from Italy and Japan on glumes and leaves, producing whitish, irregular spots. *O. oryzetorum* was reported from the Philippines, infecting glumes of maturing panicles with a whitish powdery bloom. The grains fail to fill. *O. oryzae* has elliptic or ovoid conidia, 3·5–6 × 2–2·5μ. *O. oryzetorum* has globose conidia or nearly so, 2·5 × 3·5μ (Padwick, 1950; Wei, 1957).

13. *Cladosporium* Link ex Fr.

C. miyakei Sacc. & Trott. has been reported in Vietnam to cause spots on the leaves, which dry out. The foliage ultimately turns yellow and is covered with the greenish, powdery conidia of the fungus. In Ceylon it is referred to as wet weather mould. Recently, it has also been reported from India.

C. miyakei: Mycelium superficial and creeping, with black dots; conidiophores dark, of various lengths, usually 45–70 × 4–5μ, septate, alternately denticulate at the tip; spores dark, mostly 2-celled, often 1–4-celled, of variable size, 7–20 × 4–6μ, constricted at the septa (Wei, 1957).

C. herbarum is a common saprophyte, the perfect state of which is *Mycosphaerella tulasnei* (Jancz.) Lindau. *C. oryzae* Sacc. & Sydow was reported from Italy on rotten culms. Because of variation in spore size and shape, these species are difficult to separate. However, *C. oryzae* has discoid stromata and *C. miyakei* has denticulate conidiophores (Padwick, 1950; Wei, 1957).

14. *Helminthosporium* Link ex Fr.

H. rostratum Drechs. was reported to cause leaf spots on rice in West Bengal (Chattopadhyay & Dasgupta, 1959). The spots are small, oval, delimited by

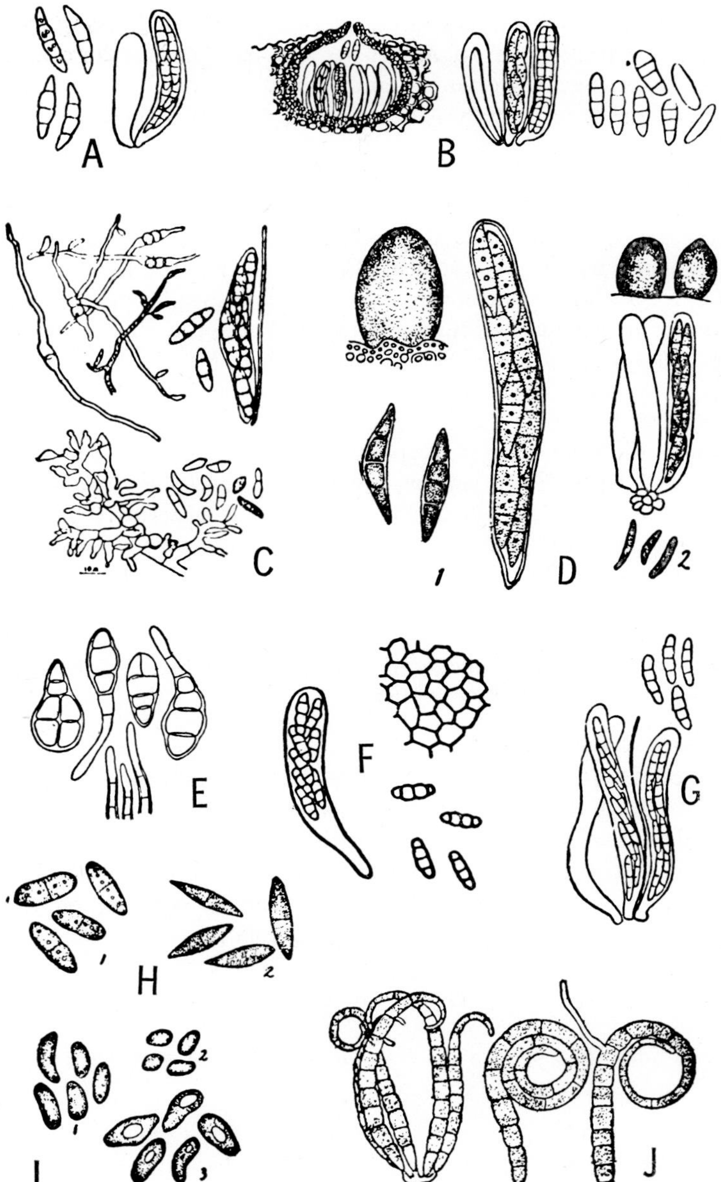

Fig. VI–10. A, *Sphaerulina oryzina*, asci and ascospores. B, *Sphaerulina miyakei*, perithecium, asci and ascospores. C, *Metasphaeria albescens*, ascus, ascospores and germinating ascospores. D, *Melanomma glumarum* (1) and *M. oryzae* (2), perithecia, ascus, and ascospores. E, *Alternaria oryzae*, conidiophores and conidia. F, *Leptosphaeria salvinii*, ascus and ascospores. G, *Trematosphaerella elongata*, asci and ascospores. H, *Diplodia oryzae* (1) and *Diplodiella oryzae* (2), conidia. I, *Coniothyrium japonicum* (1), *C. brevisporum* (2) and *C. anomale* (3), conidia. J, *Helicoceras oryzae*, conidiophore and conidia. (Rearranged from Wei, 1957).

veins, faint brown to straw-coloured, about 2–3×1–1·5 mm, often occurring together with the brown spots caused by *H. oryzae* but differing from them. The fungus was found first on *Eragrostis* and later on sorghum and several other grass hosts as well as on maize and wheat.

The rice fungus from India was described as follows: Conidiophores emerge through stomata or between epidermal cells, singly or 2–4 in a group, dark olivaceous, closely septate near apex and with swollen basal cell, 262·8–1044× 5·8–9·6μ (av. 482·5×6·8μ) for isolate 1 and up to 570μ (av. 312×6·5μ) for isolate 2, shorter on artificial media; conidia dark olivaceous, straight, oblanceolate to frequently curved, widest near the base to the middle, usually tapering from base to apex which is rounded; the basal and distal septa are thicker and more prominent; isolate 2 strongly rostrate in the apical end; 11·5–149×7·7–15·3μ (av. 66·8×10·9μ), average number of septa 5–8.

Besides *H. oryzae* (brown spot), and *H. sigmoideum* and *H. sigmoideum* var. *irregulare* (stem rot), a few other species have also been reported from rice. *H. tetramera* McKinney, a weak parasite on glumes, was found in China, India and Costa Rica. Some authors consider this species as *Curvularia spicifera* (Bainier) Boedijn (*Cochliobolus spicifer* Nelson). Bugnicourt (1955) described two new species, *H. hawaiiense* and *H. australiense*, from seeds from the South Pacific region. The former has also been reported from Madras, India.

H. tetramera has 4-celled conidia at maturity, 20·4–40·8×8·5–20·4μ, mostly 30·6–34×10·2–13·6μ. Both *H. hawaiiense* and *H. australiense* have narrower conidia, 6–10μ in diameter. In the former the conidia are 19–51× 6·1–9·9μ, in the latter 15–39×6·6–10μ.

15. *Alternaria* Nees ex Wallr.

Hara stated that *A. oryzae* Hara causes mouse-grey spots on glumes, but others have expressed the opinion that it is only a weak parasite. Growth on the kernels is dark and sooty, and the kernels fail to fill. Mycelium branched, yellowish brown, 4–4·4μ in diameter; conidiophores filiform, perpendicular to the mycelium, straight or somewhat geniculate, septate, 66–100×4–5μ, dark brown; conidia obclavate, pyriform, elliptic, or fusiform, 3–7-septate, constricted at the septa, with several longitudinal septa, muriform, 22–52× 9–15μ, dark brown, finely dotted on the surface (Padwick, 1950; Wei, 1957) (Fig. VI–10E).

Various unidentified species of *Alternaria* have been reported frequently on rice seed from various parts of the world.

16. *Helicoceras* Linder

H. oryzae Linder & Tullis attacks the leaf sheaths, causing brown irregular spots, and also the glumes, causing discoloration. It was reported from U.S.A. and Japan some years ago, and more recently has also been noted from China, India, Thailand and Vietnam.

The fungus forms minute sclerotia inside the infected leaf sheath, black on the surface, olivaceous inside. It has been considered as one of the sclerotial diseases of rice in Japan. Conidiophores erect, simple, 0–5-septate, 24–120× 3·6–4·8μ; conidia 3–6-acropleurogenous, cylindrical to long-ellipsoid, attenuated and rounded at both ends, twisted in one or two convolutions, 1–14-septate, constricted at the septa, spiny, olivaceous, $16 \cdot 8–117 \cdot 6 \times 7 \cdot 2–18\mu$ (Padwick, 1950; Wei, 1957) (Fig. VI–10J).

LITERATURE CITED

Bugnicourt, F. 1955. Deux espèces nouvelles d'*Helminthosporium* isolées de semences de riz. *Rev. gén. Bot.* **62**:238–243.

Chattopadhyay, S. B. & Dasgupta, C. 1959. *Helminthosporium rostratum* Drechs. on rice in India. *Pl. Dis. Reptr* **43**:1241–1244.

Hara, K. 1959. Monograph of rice diseases. 3rd ed. 139 pp., Tokyo, Japan Kokodo. [Jap.]

Luc, M. 1953. Champignons graminicoles de Côte d'Ivoire. I. Pyrénomycètes. *Rev. Mycol.* **18**, Suppl. colon. 1: 1–37.

Miyake, I. 1910. Studien über die Pilze der Reispflanze in Japan. *J. Coll. Agric., Tokyo* **2**: 237–276.

Padwick, G. W. 1950. Manual of rice diseases. 198 pp., Kew, Commonwealth Mycological Institute.

Wei, C. T. 1957. Manual of rice pathogens. 267 pp., Peiping, Sci. Press. [Chin.]

GRAIN DISCOLORATION

Rice grains may be infected by various organisms before or after harvest, causing discoloration, the extent of which varies according to season and locality.

The discoloration may appear externally on the glumes or internally on the kernels, or both. On the glumes, the symptoms vary, depending upon the organism involved and the degree of infection. Sometimes there are distinct black dots, usually caused by fungus fruiting bodies or other structures, on normal or bleached areas of the glumes. In other cases there are brown or blackish blotches, which may be small flecks, or large enough to cover the entire glumes. Often the lesions are pale or greyish in the centre with a dark brown margin. The kernels internally, and sometimes also externally, are stained with various distinct colours. *Monascus purpureus* Went and *Oospora* sp. cause a red colour (Uyeda, 1901; Aoi, 1921). (*M. purpureus* has been cultured on rice kernels and used as food and food colouring material in China for many centuries.) *Wolkia decolorans* (Van der Wolk) Ramsbottom is the cause of yellow grain (Van der Wolk, 1913). *Penicillium puberulum* Bain. stains the grain orange (Schroeder, 1963). *Xanthomonas itoana* (Tochinai) Dowson causes black rot. A pink colour may be caused by *Fusarium* and green, blue and yellow by *Aspergillus* and *Penicillium* spp.

The Organisms

A large number of fungi and bacteria are associated with the discoloration of rice grains. They may be divided into two major groups. One group consists of field fungi which are more or less parasitic and infect the grains before harvest; the other contains the storage moulds which usually are saprophytes and develop after harvest. Most of the former group have been discussed separately in preceding sections. The more commonly found are:

Cochliobolus miyabeanus (*Helminthosporium oryzae*) and other *Helminthosporium* spp., *Pyricularia oryzae, Alternaria padwickii, Gibberella fujikuroi, G. rosea* f. *cerealis, Nigrospora* spp., *Epicoccum* spp., *Curvularia* spp., *Phyllosticta glumarum, Alternaria* spp. and *Helicoceras oryzae.*

The prevalence of these organisms varies greatly. Tullis (1936) isolated many fungi from discoloured grains in U.S.A. The most common were *Alternaria* spp. and *Curvularia lunata*, followed by *Trichoconis padwickii* and *Phoma* spp., more rarely *Helminthosporium oryzae* and *Fusarium* spp., and occasionally *Cladosporium herbarum, Penicillium* sp. and *Helicoceras oryzae.* Many isolates were unidentifiable. Baldacci & Picco (1948) in Italy reported that 50–55% of the grains which they studied bore spots. After incubation in a moist chamber, 23·4% were identified as *Cochliobolus miyabeanus*, 13·4% *Alternaria* sp., 4·9% *Epicoccum purpurascens*, 4·9% *Penicillium* sp., 1·2% *Fusarium* sp., 1·2% *Cephalosporium* sp. and 19·7% sterile spp. In 30·8% no fungus was found. Padmanabhan (1949) in India found *Trichoconis padwickii* to be predominant over *Curvularia lunata, Cochliobolus miyabeanus* and *Nigrospora* sp. Teunisson (1954) reported that in a sample of grain giving a mould count of 90,000/g before storage, 40% were *Fusarium*, 25% *Penicillium* spp., and 22% the *Aspergillus flavus-oryzae* group and others. Rao & Salam (1954) found that 60% of discoloured grains in Hyderabad, India yielded five species of *Curvularia.* Chevaugeon & Ravisé (1957) reported from Senegal that under certain weather conditions over 263 species of organisms were detected on grains, the most important being *Rhizopus stolonifer, Helminthosporium oryzae, Curvularia maculans, Fusarium nivale* and *Phyllosticta glumarum.* Johnston (1958) in Malaysia found 9·1% of the seeds examined to be infected with *Helminthosporium oryzae*, 7·9% *Trichoconis padwickii* and 14·6% *Fusarium* spp.; also present were species of *Aspergillus, Penicillium* and *Curvularia.* Baldacci & Corbetta (1964) recorded 31·8% bacteria, 30·2% *Epicoccum* sp., 19% *Alternaria* sp., 9·2% *Helminthosporium oryzae*, 2·6–0·42% *Penicillium* sp., *Fusarium* sp., *Curvularia* sp. and *Pyricularia oryzae.*

Most of the storage moulds are species of *Aspergillus* and *Penicillium. Absidia, Mucor* and *Rhizopus* spp. are also found, and occasionally species of *Chaetomium, Dematium, Monilia, Oidium, Streptomyces, Syncephalastrum* and *Verticillium* have been reported. Del Prado & Christensen (1952) found that *Aspergillus glaucus, A. niger, A. terreus* and *Penicillium* sp. were most common on samples from Louisiana and *Fusarium* sp., *A. niger* and *Hormodendrum* sp. from Surinam. Naito (1953) identified 12 species of *Penicillium* and 8 species of *Aspergillus* and other moulds in Japan. Iizuka (1957, 1958) studied the storage moulds of Burma and Thailand and found that they were mostly species of *Aspergillus, Penicillium* and *Streptomyces.* Hirayama & Udagawa (1957, 1958) reported 25 species of *Penicillium* and 28 species of *Aspergillus* from rice imported into Japan from Southeast Asia. Udagawa (1959) further reported 14 species of *Penicillium* and 1 species of *Aspergillus.* Among the species recorded, *P. rugulosum, P. citrinum, P. cyclopium, P. implicatum, P. chrysogenum, P. adametzi, P. funiculosum, A. flavus-oryzae, A.*

candidus, *A. glaucus* and *A. versicolor* were common. Hyun (1963) in Korea reported that *A. repens*, *A. restrictus* and *A. flavus* were the dominant species.

Methods of Detection

Detecting and identifying organisms associated with discoloured grains is not always easy. Culture conditions employed may favour some organisms but not others; and some fungi are more competitive in culture than others because of their faster or more abundant growth. Surface sterilization has advantages and disadvantages. Many fungi do not sporulate in culture. Bacteria associated with the grains have usually not been identified. Neergaard & Saad (1962) developed routine methods for seed health testing, employing the blotter method and the agar plating method simultaneously. The blotter method involves keeping the seeds in moist petri dishes at 22°C for 6 days. In the agar plating method the seeds are kept at 28°C for 8 days. They found that 0·1% 2,4–D inhibits seed germination but not fungal growth. They also found that among 5 kinds of fungi involved in their experiments, *Epicoccum* became dominant at 15–22°C, *Helminthosporium* and *Alternaria* at a higher temperature; *Pyricularia* and *Cladosporium* were least competitive. Larter (1956) in Malaysia reported tests of seeds from a field heavily infested by blast disease but no *Pyricularia oryzae* was isolated, although other fungi were. Tanaka & Inagaki (1957) added 7·5% NaCl to the media used in their studies for isolating *Penicillium citrinum*, *P. islandicum* and *P. citreoviride*.

Damage

These micro-organisms have various effects on the grains. When infection is by field fungi, the main effects are reduced viability and grain quality; seedling blights and other diseases may occur when the seeds are planted. When infection is by storage moulds, besides a reduction in viability and grain quality, there may also be production of toxins. Uraguchi (1942) reported a toxin formed in yellowed rice due to *Penicillium* sp. Iizuka (1958) also found a toxin which was poisonous to mice. Tran-Vy *et al.* (1958), however, reported that rice infected by *P. citrinum* was not toxic to mice. In view of present knowledge on the occurrence of mycotoxins in fungus-infected cereal grains, the toxic aspect of fungus-infected rice grains should be further explored (Kinosita & Shikata, 1965; Miyake & Saito, 1965).

Effect of Environmental Conditions

Infection of grains by field fungi is greatly affected by climatic conditions, particularly moisture. It takes place from some considerable time before the grains mature up to the period just before storage. Tisdale (1922) stated that stackburn disease is likely to develop in a wet season when the rice remains in the field and the stacks do not dry out. Martin (1936) recorded severe infection of *Curvularia lunata* during a wet season. Ito & Iwadare (1934) found red blotch of grains (*Epicoccum* spp.) to be favoured by lodging of the plants, when the ears touch the ground. Kimura (1937) found that inoculation before the flowering period resulted in most discoloration by *Phyllosticta glumarum*, *Cochliobolus miyabeanus*, *Alternaria* sp., *Epicoccum hyalopes*,

Gibberella fujikuroi and other fungi. Much of the rice in Asia does not get threshed immediately after cutting, but is piled or stacked for some time in the field. Moist weather at this period may induce discoloration of large numbers of grains.

In storage, relative humidity (R.H.) and temperature greatly affect the development of saprophytic moulds. Ghosh (1951) found that mould development depends more on the R.H. of the storage atmosphere than the moisture content of the grains, development being fairly rapid on all samples exposed to 65–100% R.H. He also found that species of *Aspergillus* grow best at 30–35°C, *Penicillium* at 25°C, *Mucor* at 30°C, *Fusarium* at 35°C, and *Cladosporium* and *Alternaria* at 22–25°C. Naito (1953) found heavier infection by *Penicillium*, *Aspergillus* and other fungi at 20°C than at 10° or 30°C, and a moisture content of 18–20% was most conducive to mould growth. Houston *et al.* (1957) reported that an increase in mould population occurred at 60–100°F in all cases where the moisture content was above 14%. The increase at 60°F was small, but at 90° and 100°F it became so rapid that the sample with 15–16% moisture quickly became a mouldy mass. The moulds hold moisture, and this in turn hastens deterioration of the rice. Schroeder & Sorensen (1961) found that during storage, field fungi decreased while storage moulds increased. This increase was accompanied by more grain discoloration particularly at low rates of aeration. Christensen (1969) reported that in seeds of varieties Nato and Bluebonnet stored at 17–20% moisture content at 12 and 27°C, *Aspergillus glaucus*, *A. candidus*, *A. flavus-oryzae* and *Penicillium* sp. increased with time. No other storage fungi increased appreciably, even when conditions favoured them, due to competition from other organisms.

Injuries from insects in the field (Su, 1938) or during storage (Hyun, 1963) and from wind and rainstorms may also increase grain discoloration.

Control

Control or prevention of mould infection of grains depends upon both pre- and post-harvest measures, including those taken during cutting, threshing and drying. Control of the major field fungi has been discussed above. Conditions for storage of rice in U.S.A. have been reviewed by Dachtler (1959). Low moisture, aeration and avoidance of high temperatures are essential. Grains with *c.* 13·5–14% moisture are generally satisfactory for storage (Del Prado & Christensen, 1952; Houston *et al.*, 1957; Dachtler, 1959); higher moisture content together with high temperature are most conducive to mould development. Schroeder (1964) reported that the use of infrared drying and additional treatment with sodium propionate (5000 ppm) reduced infection by field fungi.

Naito (1953) studied the osmotic pressure (O.P.) relationship between the grain and storage moulds. He found the O.P. of moulds to vary from 19·6 to 134·8 atmospheric pressures and that the fungi will grow on rice when their O.P is about five times higher than the O.P. of the seed. When the grains contain 14·5% moisture, the O.P. is so high that the moulds will not grow on them.

LITERATURE CITED

AOI, K. 1921. Studies on the reddish discoloration of polished rice grains. *Bull. imp. cent. agric. Exp. Stn Japan* **45**: 29–69. [Jap.]

BALDACCI, E. & CORBETTA, G. 1964. Ricerche sulla microflora delle cariossidi di riso dopo conversazione in magazzino e in condizioni sperimentali. *Il Riso* **13**: 79–88. (*Rev. appl. Mycol.* **44**, 2788.)

BALDACCI, E. & PICCO, D. 1948. Osservazioni sulle malattie del riso durante gli anni 1946 e 1947. *Risicolt.* **36**: 73–77, 113–117.

CHEVAUGEON, J. & RAVISE, A. 1957. Régime de l'eau et maladies parasitaires du riz en Afrique Occidentale. *J. Agric. trop. Bot. appl.* **4**: 143–151.

CHRISTENSEN, C. M. 1969. Influence of moisture content, temperature, and time of storage upon invasion of rough rice by storage fungi. *Phytopathology* **59**: 145–148.

DACHTLER, W. C. 1959. Research on conditioning and storage of rough and milled rice. *Agric. Res. Serv. USDA ARS* 20–27, 55 pp.

DEL PRADO, F. A. & CHRISTENSEN, C. M. 1952. Grain storage studies. XII. The fungus flora of stored rice seed. *Cereal Chem.* **29**: 456–462.

GHOSH, J. J. 1951. The effect of environmental factors on fungal deterioration of stored rice grains. *Sci. Cult.* **17**: 42–43.

HIRAYAMA, S. & UDAGAWA, S. 1957. Taxonomic studies of fungi on stored rice grains. I. *Penicillium* group (Penicillia and related genera) 1. *Bull. Fac. Agric. Mie Univ.* **14**: 21–41.

HIRAYAMA, S. & UDAGAWA, S. 1958. Taxonomic studies of fungi on stored rice grains. II. *Aspergillus* group. *Bull. Fac. Agric. Mie Univ.* **16**: 7–28.

HOUSTON, D. F., STRAKA, R. P., HUNTER, I. R., ROBERTS, R. L. & KESTER, E. B. 1957. Changes in rough rice of different moisture content during storage at controlled temperatures. *Cereal Chem.* **34**: 444–456.

HYUN, J. S. 1963. Development of storage fungi in polished rice infested with rice weevil, *Sitophilus oryzae* L. *Seoul Univ. J.*, Ser. B, **13** (8): 77–86.

IIZUKA, H. 1957. Studies on the microorganisms found in Thai rice and Burma rice. Part I. On the microflora of Thai rice. *J. gen. appl. Microbiol.* **3**: 146–161.

IIZUKA, H. 1958. Studies on the microorganisms found on Thai rice and Burma rice. Part II. On the microflora of Burma rice. *J. gen. appl. Microbiol.* **4**: 108–119.

ITO, S. & IWADARE, S. 1934. Studies on the red blotch of rice grains. *Rep. Hokkaido Agric. Exp. Stn* 31: 1–84. [Jap. Engl. summ.]

JOHNSTON, A. 1958. Fungicidal treatment of padi seed. *Malay. agric. J.* **41**: 282–289.

KIMURA, K. 1937. On the relation of fungi to discoloured rice seeds. *Forschn Geb. PflKrankh.* **3**: 209–233. [Jap. Engl. summ.]

KINOSITA, R. & SHIKATA, T. 1965. On toxic moldy rice. In *Mycotoxins in Foodstuffs*: 111–132. Cambridge, Mass., MIT Press.

LARTER, L. N. H. 1956. Notes on current investigations (research), April to June, 1956. *Malay. agric. J.* **39**: 220–227.

MARTYN, E. B. 1936. Report on the Botanical and Mycological Division for the year 1935. *Div. Rep. Dep. Agric. Br. Guiana, 1935*: 89–92.

MIYAKE, M. & SAITO, M. 1965. Liver injury and liver tumors induced by toxins of *Penicillium islandicum* Sopp growing on yellowed rice. In *Mycotoxins in Foodstuffs*: 133–146. Cambridge, Mass., MIT Press.

NAITO, H. 1953. Studies on mold microflora in stored rice grains, with special reference to osmotic pressure. *Ann. phytopath. Soc. Japan* **18**: 41–45. [Jap. Engl. summ.]

NEERGAARD, P. & SAAD, A. 1962. Seed health testing of rice; a contribution to development of laboratory routine testing methods. *Indian Phytopath.* **15**: 85–111.

PADMANABHAN, S. Y. 1949. Occurrence of fungi inside rice kernels. *Curr. Sci.* **18**: 442–443.

RAO, P. N. & SALAM, M. A. 1954. *Curvularia* species from discoloured grains from Hyderabad-DN. *J. Indian bot. Soc.* **33**: 268–271.

SCHROEDER, H. W. 1963. Orange stain, a storage disease of rice caused by *Penicillium puberulum*. *Phytopathology* **53**: 843–845.

SCHROEDER, H. W. 1964. Sodium propionate and infrared drying for control of fungi infecting rough rice (*Oryza sativa*). *Phytopathology* **54**: 858–862.

SCHROEDER, H. W. & SORENSON, J. W. 1961. Mold development in rough rice as affected by aeration during storage. *Rice J.* **64**: 6, 8–10, 12, 21–23.

SU, M. T. 1938. Report of the mycologist, Burma, Mandalay for the year ending 31st March 1938. 10 pp.

TANAKA, Y. & INAGAKI, N. 1957. Studies on the technique for the isolation for the presence of rice grain fungi. I. *Bull. natn. Hyg. Lab. Tokyo* **75**: 443–459.

TEUNISSON, DOROTHEA J. 1954. Influence of storage without aeration on the microbial population of rough rice. *Cereal Chem.* **31**: 462-474.

TISDALE, W. H. 1922. Seedling blight and stack-burn of rice and hot water seed treatment. *Bull. U.S. Dep. Agric.* 1116.

TRAN-VY, TRUONG-VAN-CHOM & BUI-DUY-TAM. 1958. Experimental researches on the toxicity of rice infected by *P. citrinum* Thom. *Proc. symp. on phytochemistry, Kuala Lumpur, 1957*: 178–180.

TULLIS, E. C. 1936. Fungi isolated from discolored rice kernels. *Tech. Bull. U.S. Dep. Agric.* 540, 12 pp.

UDAGAWA, S. 1959. Taxonomic studies of fungi on stored rice grains. III. *Penicillium* group (Penicillia and related genera) 2. *J. agric. Sci., Tokyo* **5**: 5–21. [Jap.]

URAGUCHI, I. 1942. Pharmacological studies on a toxin formed in the yellowed rice (*Penicillium* sp.). Rep. 1. [Jap.]

UYEDA, Y. 1901. Ueber den ' Benikoji-pilz ' aus Formosa. *Bot. Mag., Tokyo* **15**: 160–163.

WOLK, P. C. VAN DER. 1913. *Protascus colorans*, a new genus and a new species of the Protascineae group, the source of ' yellow grains ' in rice. *Myc. Centralbl.* **3**: 153–157.

PART VII

DISEASES CAUSED BY NEMATODES

WHITE TIP

History

Kakuta (1915) first found this disease in Kyushu, Japan, and called it black grain disease. A similar disease on Italian millet was found by Nakano (1916), also in Kyushu. Yoshii (1944) called it nematode heart blight, studied the course of infection, and found that seed treatment controlled the disease. He called the nematode *Aphelenchoides oryzae* Yokoo, and a detailed description was given by Yokoo in 1948. Yoshii & Yamamoto (1950), Goto & Fukatsu (1952, 1956), Fukano & Yokoyama (1951) and Nishizawa (1953a, b), and others studied various aspects of the disease and it became well known.

In U.S.A., the disease was first noticed in 1935 (Todd & Atkins, 1958), but was at first attributed to iron or magnesium deficiency. Cralley (1949), however, reported it to be caused by a nematode similar to that occurring in Japan. Allen (1952) found *Aphelenchoides oryzae* Yokoo to be identical to *A. besseyi* Christie, 1942, and *A. besseyi* is the name now in general use. Todd & Atkins (1958, 1959) studied in detail the culture of the nematode, infection, seed treatment with chemicals and hot water, and varietal resistance.

Distribution and Damage

White tip has been recorded in many countries. Besides Japan and U.S.A. as mentioned above, it has been reported from Australia, Ceylon, Cuba, El Salvador, India, Indonesia, Italy, Mexico, East Pakistan, Philippines, Thailand etc. (Feakin, 1970). Since 1964 it has also been reported from Cameroon, Kenya, Malagasy, Nigeria, Senegal, Sierra Leone and many other countries in Africa (Barat *et al.*, 1969; Huu-Hai-Vuong, 1969). It has been known since 1929 in U.S.S.R. (Huu-Hai-Vuong, 1969).

Yoshii & Yamamoto (1950, I) estimated 10–35% losses in yield from the disease. Atkins & Todd (1959) reported 40–50% reduction in grain yield when the nematode was artificially inoculated on to susceptible varieties. During a three-year period they estimated 17% loss on susceptible varieties and 7% on resistant varieties. Hung (1959) in Taiwan estimated 29–46% yield loss in 10 varieties surveyed. The disease in Japan is less damaging than previously and is gradually diminishing because of extensive seed treatment programmes. In many tropical countries very little attention has yet been paid to the disease.

Symptoms

The common name ' white tip ' is based on the characteristic symptoms caused on the leaf tips, which become chlorotic or white for a distance of up to 5 cm. These white areas later become brownish and tattered (Fig. VII–1). Diseased plants are stunted, lack vigour, and produce small panicles. Both the length of the panicles and the number of spikelets are reduced, the reduction being most evident in the terminal portion of the panicles where lemmas and paleas are often absent. Affected panicles show high sterility, distorted glumes, and small and distorted kernels.

In the field, the white tip symptoms usually become evident at the beginning of the elongation period of the plants. The upper leaves, particularly the flag leaf, of severely diseased plants are markedly twisted, to such a degree that emergence of the panicles from the boot is incomplete. The panicles of severely diseased tillers mature late. Infected leaves are darker green than normal, and diseased plants give rise to tillers at high nodes.

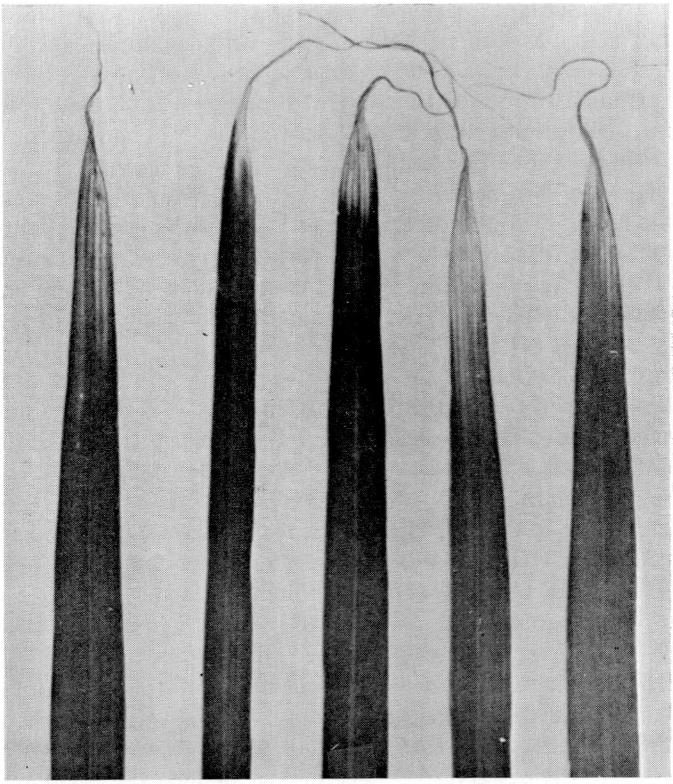

Fig. VII–1. Symptoms of white tip disease on leaves.

Not all infected plants show symptoms. Yoshii & Yamamoto (1950, I). estimated that three quarters of the infected tillers did not show symptoms Atkins & Todd (1959) reported that the more resistant varieties rarely showed foliar or panicle symptoms, although they contained nematodes and their yields were reduced. It should be noted that some insect injuries during the tillering stage may result in white tips or white bands on the leaves, but in most cases leaf perforations accompany these injuries.

The Causal Organism

Taxonomy and morphology. Allen (1952) studied the morphology of the bud and leaf nematodes and demonstrated that the white tip nematode, referred to as *Aphelenchoides oryzae* Yokoo, 1948 is identical to *A. besseyi* Christie, 1942. His description of the species is as follows (Fig. VII–2):

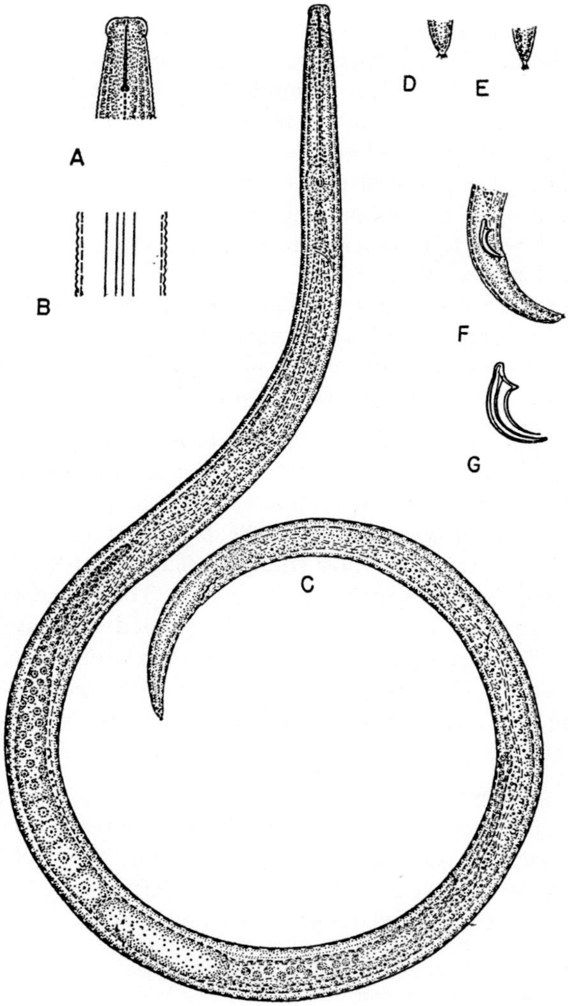

Fig. VII–2. *Aphelenchoides besseyi.* A - Head; × 1000. B - Lateral field; × 1000. C - Female; × 500. D and E - Female tail terminus; × 1000. F - Male tail; G - Spicule; × 1000. (After Allen, 1952).

Females*: L, 0·62–0·88 mm; a, 38–58; b, 9–12; c, 15–20; V,$^{43\text{-}33**}$ 66–72,$^{4\text{-}8***}$

Males*: L, 0·44–0·72 mm; a, 36–47; b, 9–11; c, 14–19; T, 50–65.

Female (Neotype): L, 0·77 mm; a, 50; b, 10; c, 19; V, 70%; ovary 33%; posterior uterine branch 6% (3 times body width). Body slender. Cuticle marked by fine striae. Lateral field occupying one-fourth of body diameter, marked by four incisures. Lip region expanded, wider than neck at base of lips. Lips without annulation. Six-radial head sclerotization delicate. Cheilorhabdions near oral aperture moderately sclerotized, and appearing as dark cuticularized pieces. Spear 10μ long with moderately well-developed knobs. Median oesophageal bulb well developed. Nerve ring one body width behind median bulb. Excretory pore located anterior to nerve ring. Oesophageal gland extending 5 body widths behind median bulb, joining oesophagus immediately behind median bulb. Intestine joining the oesophagus as a slender tube immediately behind median bulb. Ovary relatively short. Oocytes not arranged in tandem, several in a cross section. Posterior uterine branch short, narrow, usually not containing spermatozoa. Tail tapering conoid. Terminus armed with four mucronate points. Mucrons usually divergent, with star-shaped appearance.

Male: Male tail curvature about 180 degrees when relaxed by gentle heat. Three pairs of ventro-submedian papillae, the anterior pair being adanal. Spicules ventrally curved. The ventral piece with a moderate ventral process at the distal end. Terminus armed with four variable mucronate points.

Culture. Iyatomi & Nishizawa (1954) reported that *A. besseyi* multiplied on eight species of fungi, *Alternaria citri*, *A. kikuchiana*, *A. brassicae*, *Pyricularia oryzae*, *Cochliobolus miyabeanus*, *Helminthosporium sigmoideum* (*Leptosphaeria salvinii*), *Colletotrichum lagenarium* and *Phytophthora* sp. Among these, the three species of *Alternaria* were most suitable. The nematode failed to multiply on *Corticium sasakii* and *Fusarium bulbigenum*. Todd & Atkins (1958) tried to culture the nematode on unhulled rice inoculated with surface disinfected diseased seed. The nematode multiplied together with contaminating fungi, including species of *Alternaria*, *Curvularia*, *Helminthosporium* and *Fusarium*, but failed to multiply in a fungus-free medium. In both cases the nematode from culture produced white tip symptoms when inoculated on to rice seedlings.

Physiology. Tamura & Kegasawa (1957) determined the temperature at which *A. besseyi* would leave rice seed soaking in water. The nematodes left infected seeds soaked in water at temperatures of 20°C, 25°C and 30°C, swimming away soonest at 30°C. At 35°C no nematodes left the seeds. After soaking for 150 hours at 20°C, 35% of the nematodes remained in the seeds, 53% of

* Extremes of measurements—L=length; a=length divided by greatest width; b=length divided by length of oesophagus; c=length divided by length of tail; V=position of the vulva as a percentage of body length measured from the anterior end; T=end of testis as a percentage of body length measured from the cloaca.

** Position of the anterior end of the ovary as a percentage of body length measured from the vulva.

*** Position of the posterior end of the uterine branch as a percentage of body length measured from the vulva.

these being dead; at 25°C, 22% remained and 67% of these were dead; at 30°C, 18% remained and 91% of these were dead; and at 35°C, all were dead. It is concluded that the optimum temperature for nematodes to leave the seeds is 20°C, but many remain behind. Yoshii & Yamamoto (1950) noticed that the nematodes move vigorously in water at 20–50°C and longevitiy in water is 30 days.

To examine the nematode in infected material, several dehulled seeds are usually soaked in 50 ml warm water for 12–24 hours; or folded young leaves or meristematic tissues are cut into small pieces and soaked in warm water. Some nematodes will come out a few hours after soaking begins but the majority emerge only after 12–24 hours.

Tikhonova (1966) recorded the influence of air temperature and relative humidity on the life cycle of the nematode and reported that it produces 10 generations each season in the Tadzhik SSR, 13 generations in the Uzbek SSR and 8 generations in the Kazakh SSR. The rate of reproduction would be higher in tropical countries.

Huu-Hai-Vuong (1969) observed that the optimum temperature for the development of the nematode is in the region of 28°C, while marginal temperatures are 13° and 42°C. The life cycle is from 3 to 6 days at 25° and 31·3°C. It ranges from 9 to 24 days at temperatures of 20·6–14·7°C. The lethal temperature is 42°C for 16 hours or 44°C for 4 hours. For normal development, the nematode needs an atmospheric humidity of at least 70%; saturation encourages its movement. In dry grains, the nematode becomes anabiotic.

Disease Cycle

The nematode survives on infected seed for a long time. Cralley (1949) found viable nematodes on rice seed eight months after harvest. Yoshii & Yamamoto (1950, II) found living nematodes on unhulled grains which had been stored for three years. But observations indicated that the nematode is unlikely to survive over winter outdoors in Japan. Todd (1952) and Atkins (1958) found that the nematodes remained viable on dry seeds up to 23 or 24 months.

Yoshii & Yamamoto (1950, II) found that *A. besseyi* is an ectoparasite, occurring within the young folded leaves in the early stages, entering the spikelets after ear formation, and mostly occurring within the glumes when the grains are mature. After threshing, very few nematodes were found on the straw, except some on the empty spikelets which remained attached. Todd (1950), using staining techniques, also showed that the nematode is entirely external. Todd & Atkins (1958) found the nematode mostly on the inner surface of the hulls, with some on the kernels, around the young leaves and flowers, and in washings from leaf sheaths, but not within the tissues.

Goto & Fukatsu (1952) studied the number and distribution of the nematodes at various stages in the growth of infected rice plants. They found a few nematodes remaining on the seed in germinating seeds. At the seedling stage, just before transplanting, no nematodes could be found. During the tillering stage they were found in the cavities above the growing tip of the rudimentary culm, or on the young leaf surrounded by the innermost leaf sheath. As the plant grew, the number of nematodes increased and they were found in the larger young leaves. When the young ear was formed, they tended to assemble in the ear while some entered the preflowering glumes. At the time of heading, most of them were on the outside of the glumes, but during flowering the number of nematodes outside the glumes suddenly decreased, suggesting entrance into the glumes. Small numbers remain inside the sheath of the flag leaf after flowering and later they may be rarely found outside the glumes. More nematodes were found in plants with many white tips, but healthy looking plants in infected fields also contained nematodes. The main stems are more frequently invaded than are other tillers, and they contain more nematodes. In a panicle, more nematodes are found towards the middle, slightly less in the terminal part. Larger size grains also have more nematodes and empty grains contain markedly fewer.

Yoshii & Yamamoto (1950, III) reported that infection from the soil is very rare in the field; during 1944–48, only 5 out of 2900 plants grown from clean seed in infested soil showed disease symptoms, and only in one year, 1948. However, when healthy and diseased seeds were sown in adjacent plots in the seedbed, 22·9–68·6% of the former were infected. Irrigation water containing nematodes, and diseased husks applied to the seedbed also facilitate infection of healthy seeds. When healthy seedlings were transplanted together with diseased ones in the same hill, it was found that they became infected. Healthy seedlings were also infected when transplanted in alternate rows with diseased seedlings. All these experiments indicate the migration of the nematode in the field, causing infection of germinating seeds or transplanted seedlings. Cralley (1952) showed that the nematode is carried in the viable state in seed from infested plants and does not overwinter in soil.

Tamura & Kegasawa (1958) reported that the nematode did not attack the roots of seedlings.

Effect of Environmental Conditions

From pot experiments, Tamura & Kegasawa (1959) concluded that (1) the ability of the nematode to infect rice seedlings appeared to be less in wet soil than in flooded soil; (2) plants grown at a high temperature in the glasshouse had more injured stems than plants kept in the shade, but the number of nematodes was smaller in the grains from the former; (3) the effects of nitrogen, calcium silicate, and urea fertilizers on nematode injury were indistinct.

Tamura & Kegasawa (1958) also reported that movement of the nematode between hills was related to the amount of nitrogenous manure present. With double manuring their movement reached a maximum at the first-leaf stage of the seedlings, whereas with normal manuring, maximum movement did not take place until the third to fourth-leaf stage.

Nishizawa (1953b) reported that in potted plants inoculated with the husks of nematode-infected grains and also with sclerotia of the stem rot fungus, separately and together, stem rot infection was greater when inoculation was with the stem rot fungus alone.

VARIETAL RESISTANCE

Nishizawa (1953a), Nishizawa *et al.* (1953), Goto & Fukazu (1956) and others in Japan, and Atkins & Todd (1959) in U.S.A. studied varietal resistance to white tip disease. Nishizawa (1953a), summarizing the results of field experiments from 1949 to 1952, concluded that nine varieties, Norin No. 8, Tozan Nos. 36, 37, 38 and 58, Nankai No. 3, Asa-Hi (not Asahi) and Asahi No. 1 (from Fukuoka and Kagoshima) were symptomless; six varieties, Norin Nos. 6, 37 and 39, Norin-mochi No. 5, Saikai No. 37 and Nakate Ginbozu were resistant; while 24 other varieties were moderately susceptible and nine were highly susceptible. Resistance is inherited from variety Asa-Hi. There were more variations of symptoms from year to year in susceptible varieties than in resistant ones.

Goto & Fukatsu (1956) tested in the field 20 varieties inoculated with husks of diseased seeds and found that variety Tozan 38 showed no symptoms in 3 years, Norin No. 8, Norin-mochi No. 5 and Hatsushimo showed only slight symptoms, and Asahi No. 1, Aichi-Asahi and Kinai-Omachi No. 2 were very susceptible.

Atkins & Todd (1959) found varieties Early Prolific, Blue Rose, Caloro, Calrose, Colusa, Kamrose, Lacrosse, Arkrose and Zenith to be very susceptible, while Blue Bonnet, Blue Bonnet 50, Asahi, Century 52, Century Patna 231, Fortuna, Improved Blue Bonnet, Nira, Rexoro, Texas Patna, Toro and TP49 were resistant.

Hung (1959) found the Ponlai varieties Chianung 242, Kaohsiung 10 and Taichung 65 to be very susceptible in Taiwan but the leading native varieties did not show white tip symptoms.

Nishizawa *et al.* (1953) showed that the amount of inoculum (husks of diseased grains) affects the symptoms, including length of stem, weight of ear, weight of 1000 grains, etc. on the susceptible variety Zuiho, but not on the symptomless var. Nankai No. 3. Todd & Atkins (1958) succeeded in inoculating plants by injecting a suspension of nematodes into the inside of the leaf sheath.

Fukano & Yokoyama (1951) concluded that in testing varietal resistance both the number of infected stems and the degree of white tip should be considered because some varieties are infected without showing white tip symptoms.

Goto & Fukatsu (1956) in laboratory experiments showed that there is variation in the attractiveness of seedlings of different varieties to *A. besseyi*, the variations being strongly correlated with susceptibility in the field. The nematodes were attracted to young growing plant parts, to expressed juice and to an aqueous extract of germinating seeds. The reaction is chemotactic. They also found that the rate of multiplication of the nematode was lower in resistant than in susceptible varieties. The nematodes were also smaller in resistant varieties.

HOST RANGE

The nematode is known to occur on a rather wide range of hosts. Yoshii & Yamamoto (1950, II) reported that *Setaria viridis* (foxtail) may be infected by the nematode and *Panicum sanguinale* (crab grass) and *Cyperus iria* may be slightly infected, but not *Panicum* (*Echinochloa*) *crusgalli* (barnyard grass). They also showed by cross-inoculation that the 'ear-blight' disease of Italian millet (*Setaria italica*) is caused by the same nematode and that exceedingly large numbers of nematodes can be found in infected ears. Huu-Hai-Vuong (1969) detected the nematode on *Cyperus polystachyus* and *Imperata cylindrica.* It is also known to occur in millet, strawberry, Vanda orchids, chrysanthemum, Saintpaulia and other hosts.

Control

Since the nematode is seed-borne, seed treatment by both hot water and chemicals has been studied by several workers.

Cralley (1949) found that hot water treatment of seed for 15 minutes at 52–53°C reduced infection from 75% to less than 1% in glasshouse tests. He (1952) later tried to obtain seed free from infestation by presoaking in cool water for 8–12 hours, preheating in water at 55°C for 15 seconds, soaking for 15 minutes at 50°C and 5 minutes in cool water, and finally drying.

Yoshii & Yamamoto (1950, IV; 1951) first used a soak in water at 50–52°C for 5–10 minutes following soaking in water at below 20°C for 16–20 hours; later they suggested the immersion of dry seed in water at 56–57°C for 10–15 minutes, 60 days before sowing; this destroys all seed-borne pathogens. They reported that injury to the seeds occurs only at 60°C applied for more than 20 minutes.

Todd & Atkins (1959) presoaked the seed for 24 hours in cool water, then for 15 minutes at 51–53°C; this treatment proved to be nematicidal and not phytotoxic. Without presoaking, good results were obtained with 10–15 minutes immersion at 55–61°C, or with 15 minutes in water at 54°C.

Many chemicals have been tried for controlling the nematode. Yoshii & Yamamoto (1950, IV) sprayed nicotine sulphate, at 1:500 dilution, on to the panicles of infested plants at the heading stage, and reduced the number of nematodes on the panicles; but the treatment also reduced the yield of grain. Tullis (1951) fumigated variety Zenith with methyl bromide, 1·25 lb/1000 ft^3 for 6 hours, and killed the nematode. The treatment caused some injury to the more sensitive variety Blue Bonnet. Treatment was also successful if the seed was fumigated first with 1 lb/1000 ft^3 for 15 hours followed after several days aeration by a second treatment with 0·5 lb/1000 ft^3 for 15 hours.

Cralley & French (1952) tried seed treatment with (1) 25% Parathion dust or 50% Systox on Carbon dust at 2 oz/bushel; (2) soaking in 1:1000 mercuric chloride solution for 12 hours; (3) fumigation with methyl bromide; and (4) Aagrano dust, 2 oz/bushel. They found that these treatments resulted in significant reduction in the severity of the disease.

Nishizawa (1953c) reported that immersing the roots of infected seedlings for 24 hours in Folidol, an organo-phosphorus insecticide, resulted in heavier and taller plants, a greater number of ears and heavier grain. Gomi & Kogure (1956) also reported Folidol to be more effective in control of the nematode than hot water seed treatment.

Atkins & Todd (1952) and Todd & Atkins (1959) tested 50 chemicals in the laboratory and found that most of the fungicides, insecticides and nematicides tried were too injurious for field use against the nematode. The moisture content of seed during fumigation with methyl bromide greatly affects viability. Experimental nematicides N–244 and N–245 (10% dust) at 2–4 oz per bushel gave complete control in several tests without injury to the seed. However, later samples from batches treated on a production basis were ineffective.

Hirota & Yamada (1958) tested octylchlororesorcinol (OCR) and dialkyl sodium sulphosuccinates, using in addition surfactants of the alkyl groups. OCR was more effective with dibutyl and di-isoamyl ester surfactants in concentrations between 0·1% and 0·001%. Higher concentrations were less effective due to increased mobility of the nematode.

Tanaka (1959) tested rhodanate acetic esters (REE-200, REM-200, REB-200) by dipping infested seeds in 1:600 and 1:200 solutions of each chemical containing 20% of active ingredient. Best results were from 1:600 REE-200 for 24 hours, with 99% germination and no diseased plants. Fukano (1962) reported effective control with ethyl thiocyanoacetate. Laboratory and field tests showed that complete control of the disease may be obtained by soaking the seed for 24 hours in a 20% emulsion of the chemical at 1:200–1:400. Nakajima & Shimada (1962) reported that emulsions of methyl thiocyanoacetate (REM; Tanaka, 1959), ethyl thiocyanoacetate (REE, also called Sassen), and butyl thiocyanoacetate (REB) gave 100% control, without injury to the seeds, by dipping rice seeds in 1:100–1:500 concentrations of the active ingredient for 24 hours.

Certain cultural practices have also been recommended to reduce the damage from *A. besseyi*. Cralley (1949) recommended early planting (in April). Yoshii & Yamamoto (1951) reported that by advancing the time of sowing by 60 days, nematode infection could be avoided. Cralley (1956) found that seeding directly into water, instead of before flooding with water, resulted in only a small percentage of infected plants. Tamura & Kegasawa (1959, II) tried to devise a seedbed which would afford some measure of protection against the nematode. Unfortunately all these measures have their limitations for general application.

LITERATURE CITED

ALLEN, M. W. 1952. Taxonomic status of the bud and leaf nematodes related to '*Aphelenchoides fragariae*' (Ritzema Bos 1891). *Proc. Helminth. Soc. Washington* **19**: 108–120.

ATKINS, J. G. & TODD, E. H. 1952. Laboratory screening of chemicals for control of rice white tip. Abs. in *Phytopathology* **42**: 463.

ATKINS, J. G. & TODD, E. H. 1959. White tip disease of rice. III. Yield tests and varietal resistance. *Phytopathology* **49**: 189–191.

BARAT, H., DELASSUS, M. & HUU-HAI-VUONG. 1969. The geographical distribution of white tip disease of rice in tropical Africa and Madagascar. *In* Nematodes of tropical crops. *Tech. Commun. Commonw. Bureau Helminth.* 40: 269–273.

CRALLEY, E. M. 1949. White tip of rice. Abs. in *Phytopathology* **39**: 5.

CRALLEY, E. M. 1952. Control of white tip of rice. *Arkans. Fm Res.* **1** (1): 6.

CRALLEY, E. M. 1956. A new control measure for white tip. *Arkans. Fm Res.* **5** (4): 5.

CRALLEY, E. M. & FRENCH, R. G. 1952. Studies on the control of white tip of rice. Abs. in *Phytopathology* **42**: 6.

FEAKIN, SUSAN D. (Editor). 1970. Pest control in rice. *PANS Manual* 3: 99–107.

FUKANO, H. 1962. Ecological studies on white tip disease of rice plant caused by *Aphelenchoides besseyi* Christie and its control. *Bull. Fukuoka agric. Exp. Stn* 18, 108 pp. [Jap. Engl. summ.]

FUKANO, H. & YOKOYAMA, S. 1951. Studies on the white tip on rice plant, with special reference to the damage and varietal resistance. *Kyushu agric. Res.* **8**: 88–90. [Jap.]

GOMI, M. & KOGURE, M. 1956. Effect of Folidol on the control of the rice white tip nematode. *Proc. Kanto-Tosan Pl. Prot. Soc.* **3**: 20–21. [Jap.]

GOTO, K. & FUKATSU, R. 1952. Studies on the white tip of rice plants. II. Number and distribution of the nematode on the affected plants. *Ann. phytopath. Soc. Japan* **16**: 57–60. [Jap. Engl. summ.]

GOTO, K. & FUKATSU, R. 1956. Studies on the white tip of rice plant. III. Analysis of varietal resistance and its nature. *Bull. natn. Inst. agric. Sci.*, Ser. C, **6**: 123–149.

HIROTA, K. & YAMADA, T. 1958. Effects of anionic surfactants to the nematicide. Studies on supplements of pesticides XVI. *Botyu-Kagaku, Kyoto* **23**: 227–229. [Jap. Engl. summ.]

HUNG, Y. P. 1959. White tip disease of rice in Taiwan. *Pl. Prot. Bull., Taiwan* **1** (4): 1–4. [Chin.]

HUU-HAI-VUONG. 1969. The occurrence in Madagascar of the rice nematodes, *Aphelenchoides besseyi* and *Ditylenchus angustus*. *In* Nematodes of tropical crops. *Tech. Commun. Commonw. Bureau Helminth.* 40: 274–288.

IYATOMI, K. & NISHIZAWA, T. 1954. Artificial culture of the strawberry nematode, *Aphelenchoides fragariae*, and the rice white tip nematode, *Aphelenchoides besseyi*. *Jap. J. appl. Zool.* **19**: 8–15. [Jap.]

KAKUTA, T. 1915. On black grain disease of rice. *J. Pl. Prot., Tokyo* **2**: 214–218. [Jap.]

NAKAJIMA, M. & SHIMADA, K. 1962. The effect of thiocyanoacetate derivatives in controlling *Aphelenchoides besseyi* (white tip nematode of rice). *J. Tokyo Soc. Vet. Zoo. Tech. Sci.* **12**: 42–50. [Jap. Engl. summ.]

NAKANO, I. 1916. On the sterility of purple spikelet disease of Italian millet. *J. Pl. Prot., Tokyo* **3**: 33–36. [Jap.]

NISHIZAWA, T. 1953a. Studies on the varietal resistance of rice plant to the rice nematode disease 'Senchu Shingare Byo'. (VI). *Bull. Kyushu Agric. Exp. Stn* **1**: 339–349. [Jap. Engl. summ.]

NISHIZAWA, T. 1953b. On the relation between the rice nematode disease 'white tip' and the stem rot of rice plants. *Ann. phytopath. Soc. Japan* **17**: 137–140. [Jap. Engl. Summ.].

NISHIZAWA, T. 1953c. On the prevention of the rice nematode disease, 'Senchu Shingare Byo' by Folidol. *Botyu Kagaku* (Scientific Insect Control) **18**: 1–6. [Jap. Engl. summ.]

NISHIZAWA, T. & YAMAMOTO, S. 1951. Studies on the varietal resistance of rice plant to the rice nematode disease 'Senchu Shingare Byo'. II. A test of the leading varieties and part of breeding lines of rice plants in Kyushu. *Kyushu agric. Res.* **8**: 91–92. [Jap.]

NISHIZAWA, T., YAMOMOTO, S. & MIZUTA, H. 1953. Studies on the varietal resistance of rice plant to the rice nematode disease 'Senchu Shingare Byo' (VII). *Bull. Kyushu agric. Exp. Stn* **2**: 71–80. [Jap. Engl. summ.]

TAMURA, I. & KEGASAWA, K. 1957–59. Studies on the ecology of the rice nematode, *Aphelenchoides besseyi* Christie. I. On the swimming away of rice nematodes from the seeds soaked in water and its relation to the water temperatures. II. On the parasitic ability of rice nematode and their movement into hills. III. The injured features of the rice plant and the population density of nematodes found in the unhulled rice grain with special reference to the type of nursery bed. IV. The injurious features and population dynamics of nematodes in unhulled rice grain with special reference to the cultural environment of rice plant. *Jap. J. Ecol.* **7**: 111–114, 1957; **8**: 37–42, 1958; **9**: 1–4; 65–68, 1959. [Jap. Engl. summ.]

TANAKA, I. 1959. Rice seed treatment test with rhodanate-acetic esters for the control of white tip nematode. *Kyushu agric. Res.* **21**: 152–153. [Jap.]

TIKHONOVA, L. 1966. [Harmful parasite.] *Zashch. Rast. Vredit. Bolez.* 6: 18–19. [Russ.] (Abs. in *Helminth. Abstr.* **36**, 2525, 1967).

TODD, E. H. 1952. Further studies on the white tip disease of rice. *Proc. Ass. sth. agric. Wkrs* **49**: 141.

TODD, E. H. & ATKINS, J. G. 1958. White tip disease of rice. I. Symptoms, laboratory culture of nematodes, and pathogenicity tests. *Phytopathology* **48**: 632–637.

TODD, E. H. & ATKINS, J. G. 1959. White tip disease of rice. II. Seed treatment studies. *Phytopathology* **49**: 184–188.

TULLIS, E. C. 1951. Control of the seed-borne nematode of rice by fumigation with methyl bromide. *Prog. Rep. Texas agric. Exp. Stn* 1413, 4 pp.
YOKOO, T. 1948. '*Aphelenchoides oryzae*' Yokoo n. sp., a nematode parasitic to rice plant. *Ann. phytopath. Soc. Japan* **13**: 40–43. [Jap. Engl. summ.]
YOSHII, H. 1946. Nematode heart blight of rice. A preliminary report. *Agriculture Hort., Tokyo* **19**: 981–982. [Jap.]
YOSHII, H. & YAMAMOTO, S. 1950. A rice nematode disease 'Senchu Shingare Byo'. I. Symptoms and pathogenic nematode. II. Hibernation of *Aphelenchoides oryzae.* III. Infection course of the present disease. IV. Prevention of the present disease. *J. Fac. Agric. Kyushu Univ.* **9**: 209–222; 223–233; 289–292; 293–310.
YOSHII, H. & YAMAMOTO, S. 1951. On some methods for the control of rice nematode disease. *Sci. Bull. Fac. Agric. Kyushu Univ.* **12**: 123–131. [Jap. Engl. summ.]

STEM NEMATODE

HISTORY, DISTRIBUTION AND DAMAGE

The disease caused by this nematode was first discovered by Butler (1913) in eastern Bengal (now East Pakistan) where it was called 'ufra' or 'dak pora'. He made accurate observations and precise descriptions of the disease in Bengal and determined the cause to be a nematode which he called *Tylenchus angustus* Butler. Jack (1923) mentioned the nematode as a parasite of rice in Malaya. Seth (1939) found it in Burma where it was called 'akhet-pet'. Singh (1953) reported 'ufra' disease in most paddy tracts in Uttar Pradesh of India. Reyes & Palo (1956) observed a similar disease in the Philippines involving a species of *Ditylenchus*, and it was recently found again in Quezon Province of the Philippines. Sasser & Jenkins (1960) indicated its presence in Egypt. Hashioka (1963) discovered it in South Thailand and made transmission and other experiments. Dried specimens of the disease were received by IRRI in 1965 and 1966 from Madagascar, where it was reported to cause serious damage on the high plateaus (Huu-Hai-Vuong, 1969).

The disease has caused great damage in the Ganges delta of East Pakistan, a marshy area about 100 miles long and wide, with its centre at the junction of the Pabna and Meghna rivers. Butler estimated the losses in monetary terms as being 25–35 Rs. per acre; it is now difficult to estimate the amount of loss in yield of grain that this represents. Both Butler and Padwick (1950) observed that in some infested fields no grain harvest was expected and the crop was cut for cattle feed. Singh (1953) reported up to 50% damage in Uttar Pradesh, India. Hashioka (1963) estimated that 500 hectares of low-lying rice fields at Pathalung in South Thailand suffered 20–90% loss of yield.

SYMPTOMS

The symptoms of the disease, while most clear and evident in the panicle, may also be observed in the earlier stages of growth. In the seedling stage, when plants are artificially inoculated, a marked chlorosis of the leaves is followed by withering and death (Padwick, 1950). In Thailand, Hashioka (1963) described mosaic-like discolorations arranged in a splash pattern throughout the blade, becoming more evident with time. The entire leaf may become twisted or severely malformed. In some cases, the basal portion of the young leaves becomes wrinkled, following the whitish-green discoloration. Symptoms in the field, usually observed two months after planting, vary according to the activity of the nematodes. Discoloration and malformation of the leaves, as in seedlings, may be observed, but sometimes the symptoms are masked and the plants look healthy. At a later stage, a few scattered brown stains appear on the leaves and sheaths. Later the stains become darker and parts of the upper internodes of the stem turn dark brown.

Fig. VII–3. Panicles and leaves of rice affected by the stem nematode.

At the heading stage, the symptoms on the panicle vary considerably according to the degree and time of infection. The panicles are twisted or distorted in other ways, and bear sterile, partly filled or empty spikelets towards the base (Fig. VII–3). When the nematodes do not damage the panicle at an early stage in its formation, it emerges normally from the sheath, displaying the symptoms described above. If the nematode damage occurs early, the entire panicle remains enclosed in the leaf sheath. There are also intermediate forms in which panicles partly emerge and are weak and sterile. Butler (1913, 1919) reported the former type as ' swollen ' ufra and the latter as ' ripe ' ufra.

Padwick (1950) observed a strong tendency towards branching of the stem in the infected portions, and even the formation of two or three distorted ears surrounded by one sheath. Hashioka (1963) also reported that the infected stem might have 2–4 branches, while the ear of the main stem developed to a normal size.

Padwick (1950) mentioned that brown spots on the sheaths as described by Butler in connection with this disease might be caused by the fungus *Corticium sasakii* (Shirai) Matsumoto. In Thailand, Hashioka (1963) reported that such spots appear to be due to *A crocylindrium oryzae* Sawada.

THE CAUSAL ORGANISM

Butler (1913) called the nematode *Tylenchus angustus*. The name was subsequently changed to *Anguillulina angusta* (Butler) Goodey, 1932, and later to *Ditylenchus angustus* (Butler) Filipjev, 1936, which is the name now in common use. The description of the nematode by Goodey is as follows (Fig. VII–4):

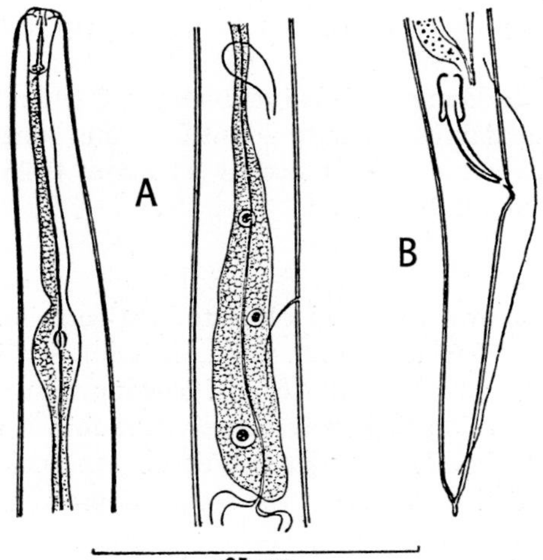

Fig. VII–4. *Ditylenchus angustus*. A, Œsophageal region, highly magnified. B, Male tail highly magnified, lateral view. (Goodey, 1932).

Morphology: Female: length, 0·7 mm to 1·23 mm; breadth, 0·015 mm to 0·022 mm; oesophagus, 0·14 mm to 0·15 mm; tail, 0·045 mm to 0·052 mm; stylet, 0·01 mm; a=58–36, b=8–7, c=20–17, V=80%.

Male: length, 0·6 mm to 1·1 mm; breadth, 0·014 mm to 0·019 mm; oesophagus, 0·13 mm to 0·14 mm; tail, 0·034 mm to 0·048 mm; stylet, 0·01 mm; spiculaes, 0·02 mm; gubernaculum, 0·008 mm; a=47–36, b=7–6, c=23–18.

In general structure and anatomy the adults are similar to *A. dipsaci*. Body very slender, tapering but little anteriorly and posteriorly. Cuticle with fine transverse striations about 1·5μ apart. Head like a flattened disc, showing the six radial ridges; no papillae seen on it. Stylet typical, with three rounded basal swellings. Oesophagus usual structure; median bulb ellipsoidal; neck region rather slender, posterior part rather elongated containing three uni-nucleate oesophageal gland cells. Ducts from latter typical. Intestine, rectum and anus normal.

Female: Ovary reaching almost to beginning of intestine, not reflexed but gradually widening backwards with ova in single file. At about 0·28 mm in front of the vulva a constriction in the wall of the genital tube is found leading to the receptaculum seminis, this measures about 0·14 mm long and is separated from the uterus by another constriction. There is a post-vulval uterine sac extending almost two-thirds of the distance from vulva to anus. Lips of vulva rounded and slightly protuberant. Tail almost cylindrical in shape but terminus suddenly tapering to a sharp pointed process.

Male: Gonad anterior, single, extending sometimes almost to oesophagus; not reflexed. Rest of gonad practically as in *A. dipsaci*. Spicules paired and shaped as in *A. dipsaci* with anterior third expanded and rather oblong in outline, shaft tapering gradually to pointed tips. Gubernaculum simple. Caudal alae arise a little in advance of heads of spicules and are inserted just short of the tip of the tail; they have faintly crenate edges. Tip of tail sharply pointed as in female.

Eggs. Butler gave 0·08 mm to 0·084 mm long by 0·016 mm to 0·02 mm wide.

Larvae. First stage larvae measure about 0·17 mm long and soon grow to 0·25 mm. Butler says an ecdysis occurs at about 0·25 mm and another, probably the last, at about 0·6 mm.

Disease Cycle

The nematode is known to be an obligate ecto-parasite. Infection occurs when seedlings are a few days old, when, if humidity is sufficient, active nematodes climb up and enter the tender growing tissues. They enter the inner portions by working their way between the folded leaves and sheaths. They never penetrate bodily through the tissue but suck the sap from the epidermal cells with their stylets. As the seedlings grow, the nematodes move up to the new tissues.

The eelworms are found mainly at the base of the peduncle, the stem just above the node, and within the glumes. They are most numerous when there is a protective covering formed by the leaf sheath and are most commonly found on protected young tissues, chiefly in the cavity of the leaf sheath covering the young folded leaves. Though at times the number of eelworms may be sufficient to form a cottony weft or gray coating on the surface, they are not highly gregarious in habit.

Reproduction only occurs on the rice plant. The number of generations which can occur in a season is uncertain, and the number of eggs laid by a female is also unknown (Padwick, 1950).

When the crop matures, the nematodes become inactive, each being tightly coiled in a circular fashion with the head at the centre. When placed in water they uncoil and become active and vigorously mobile, progressing with a wriggling or snake-like motion. On drying out, they again coil up, only to return to the active state when again moistened. At 31°C they become active much more quickly than at 16–19°C and retain their vigorous activity longer. They can move on a solid surface in an atmosphere with 85% or more relative humidity and free water is not an absolute necessity. The power to resume motility has been shown to remain under desiccation for over six months, and the power to uncoil for 15 months. Thus the nematodes may survive easily from one rice harvest to the next infective period. The period of survival is much reduced by immersion in water and the nematodes could certainly not survive under such conditions from one season to the next.

Hashioka (1963) reported that the coleoptiles of germinating seeds are readily infested by immersing them in a freshly prepared nematode suspension and incubating at 28–30°C for two to three days. In the tillering stage, inoculation can be made by injecting a suspension of nematodes into the sheath cavity. The disease gradually extends from the inoculated tiller to neighbouring tillers. The disease may be transmitted by close leaf contact under humid conditions, and Hashioka (1963) also demonstrated possible transmission by rain splash or irrigation water. Transmission through soil and seed is unlikely.

Conditions for Development

The rice stem nematode is restricted to certain tropical areas where the climate and cultural conditions are suitable for its development. In eastern Bengal, the humidity is high during the growing and ripening season and rainfall is negligible during the winter and spring months. The prevailing climatic conditions are the same for the southern part of Thailand. In both areas, large amounts of infested plant material are left in the field after harvest. Under such conditions the nematodes are dormant during the dry season and become active in the wet season when rice is planted. Since the nematode is an obligate parasite and active only in moist conditions, any alteration of such conditions would limit the development of the disease. In Bengal, the disease is prevalent during the wet season of July to December. The *aman* crop (winter rice), which is sown between March and May and harvested in November and December, is infested. The *aus* (autumn rice) is less affected because it matures faster and is harvested before the nematodes reach their maximum population. The *boro* rice, which grows from January to April or May, almost escapes infestation due to the prevailing dry weather and lower temperatures.

Varietal Resistance and Host Range

Using the coleoptile inoculation method, Hashioka (1963) found that of five Thai varieties tested, Khao-tah-oo was the most resistant (42·9% infection) and Khao-tah-haeng 17 the most susceptible (80·9% infection). Among eight other varieties tested, Ginga with 17·8% infection and Norin No. 22 with 30% were the most resistant while Tadukan with 74·7% was the most susceptible. The *japonica* varieties seemed to be more resistant, but this was partly due to the fact that the coleoptiles of the *japonica* varieties were at an earlier stage of development at the time of inoculation. The coleoptile inoculation method used consisted of pouring 8 ml of nematode suspension (5–30 nematodes/ml) on to the coleoptiles of 3–5-day-old germinated seeds in petri dishes. Padwick (1950) stated that every class of rice and every variety tested in Bengal was susceptible to ufra disease.

Hashioka (1963) reported that in inoculation experiments *D. angustus* infected several species of *Oryza*, including *O. sativa* var. *fatua*, *O. glaberrima*, *O. cubensis*, *O. officinalis*, *O. meyriana*, *O. latifolia*, *O. eichingeri*, *O. alta* and *O. minuta*; but it could not attack maize or several species of grasses tested. It was concluded that only *Oryza* spp. serve as hosts of the nematode.

The nematode is believed to be incapable of feeding upon decomposing vegetable matter and it is concluded that it is an obligate and highly specialized parasite (Padwick, 1950).

Control

Not much is known on the control of the disease. Removal and destruction of the straw in infested fields to reduce the amount of inoculum would be helpful. If possible infested fields should be drained or flooded while rice is not growing in them, for long enough to kill the nematode; but both these operations would often be difficult in practice. Adjusting the cropping pattern so as to avoid severe damage might also be tried. The use of nematicides requires to be investigated.

LITERATURE CITED

Butler, E. J. 1913. Ufra disease of rice. *Agric. J. India* **8**: 205–220.

Butler, E. J. 1919. The rice worm (*Tylenchus angustus*) and its control. *Mem. Dep. Agric. India*, Bot. Ser. 10: 1–37.

Filipjev, I. N. 1936. On the classification of the Tylenchineae. *Proc. helminth. Soc. Wash.* **3**: 80–82.

Goodey, T. 1932. The genus *Anguillulina* Grev. and v. Ben., 1959, vel Tylenchus Bastian, 1865. *J. Helminth.* **10**: 75–180.

Hashioka, Y. 1963. The rice stem nematode *Ditylenchus angustus* in Thailand. *Pl. Prot. Bull. F.A.O.* **11**: 97–102.

Huu-Hai-Vuong. 1969. The occurrence in Madagascar of the rice nematodes, *Aphelenchoides besseyi* and *Ditylenchus angustus*. *In* Nematodes of tropical crops. *Tech. Commun. Commonw. Bureau Helminth.* 40: 274–288.

Jack, H. W. 1923. Rice in Malaya. *Malay. agric. J.* **11**: 103–119, 139–169.

Padwick, G. W. 1950. Manual of rice diseases. 198 pp., Kew, Commonwealth Mycological Institute.

Reyes, G. M. & Palo, A. V. 1956. Nematode disease of rice. *Araneta J. Agric.* **3**: 72–77.

Sasser, J. N. & Jenkins, W. R. 1960. Nematology, fundamentals and recent advances with emphasis on plant parasitic and soil forms. Univ. North Carolina Press.

Seth, L. N. 1939. Report of the mycologist, Burma, for the year ending 31st March 1939. 6 pp., Rangoon, Supt Govt Printing and Stationery.

Singh, B. 1953. Some important diseases of paddy. *Agric. Anim. Husb. India* **3** (10–12): 27–30.

ROOT NEMATODE

History and Distribution

This nematode was first reported from Indonesia by Van Breda de Haan (1902) in connection with the ' mentek ' disease of rice, and was named *Tylenchus oryzae* Van Breda de Haan. Van der Vecht & Bergman (1952) made extensive studies on the nematode and its relation to mentek. They failed to prove that mentek is due to the nematode but found that the nematodes injured the roots and caused growth retardation. It was reported from Japan by Imamura (1931) as *Tylenchus apapillatus* Imamura and is presently known to occur in rice fields throughout Japan. Atkins, Fielding & Hollis (1955) reported it in Texas and Louisiana, U.S.A. Thorne (1961) collected the nematode in the islands of Java, Bali and Sumatra in Indonesia, and in Thailand and the Philippines, and Chantanao (1962) found it to be widespread in Thailand. It has also been reported from Ceylon, East Pakistan, India, Madagascar, Malaysia, Nigeria, Sierra Leone, Venezuela and other areas, and is apparently widely distributed in all rice-growing regions of the world.

Symptoms and Damage

There is no specific symptom of the disease on the above-ground parts except a gradual retardation of growth. Van der Vecht & Bergman (1952) observed the nematode penetrating into the roots of healthy rice plants, feeding on the parenchyma tissues, and multiplying in them, the root cortex becoming discoloured. When the soil was inoculated before rice seedlings were transplanted, the rate of tillering was reduced considerably, sometimes as much as 50–60%. Studies in Japan (Kawashima & Hori, 1962; Genma, Shibuya & Kikuchi, 1964; Kawashima & Fujinuma, 1965; Kurokawa & Nakamura, 1965) showed that the height, tillering ability, and weight of foliage and roots of infected plants were reduced. A high correlation was found between frequency of infection and brown discoloration of the root tissues. Root rot of rice induced by physiological factors (autumn decline) becomes more serious when this nematode is present.

Application of nematicides resulted in crop yield increases of 10–40%. These increases do not, however, necessarily indicate damage due to nematodes alone, as will be discussed below.

The Causal Organism

Van Breda de Haan (1902) first described a nematode on rice roots in Java as *Tylenchus oryzae*, but provided only a meagre description. Goodey (1936) described and illustrated the nematode in detail and renamed it *Anguillulina oryzae*. Thorne (1949) established the genus *Radopholus* and named the rice root nematode *R. oryzae*. Luc & Goodey (1962) differentiated the genus *Hirschmannia* from *Radopholus* and called the nematode *H. oryzae*. Later (1963) they found that the generic name *Hirschmannia* was preoccupied and a new generic name *Hirschmanniella* was used. The nematode was called *Hirschmanniella oryzae* (Breda de Haan, 1902) Luc & Goodey, 1936. Sher (1968) recognized the name *Tylenchus oryzae* Soltwedel, 1889 for this nematode which thus became *Hirschmanniella oryzae* (Soltwedel, 1889) Luc & Goodey, 1963.

Sher (1968) revised the genus *Hirschmanniella* and gives the following description for the species:

Measurement: ♀: L=1·14–1·63; a=50–67; b=8·8–12·1; b^1=4·5–7·2; c=15–19; c^1=4·3–5·5; V=50–55; stylet=16–19μ; m=47–50; O=15–19.

♂: L=1·01–1·40; a=52–61; b=9·1–11·3; b^1=4·6–5·7; c=16–18; c^1=4·1–5·4; stylet=16–18; m=47–50; O=13–18; gubernaculum=7–9μ; spicules=18–26.

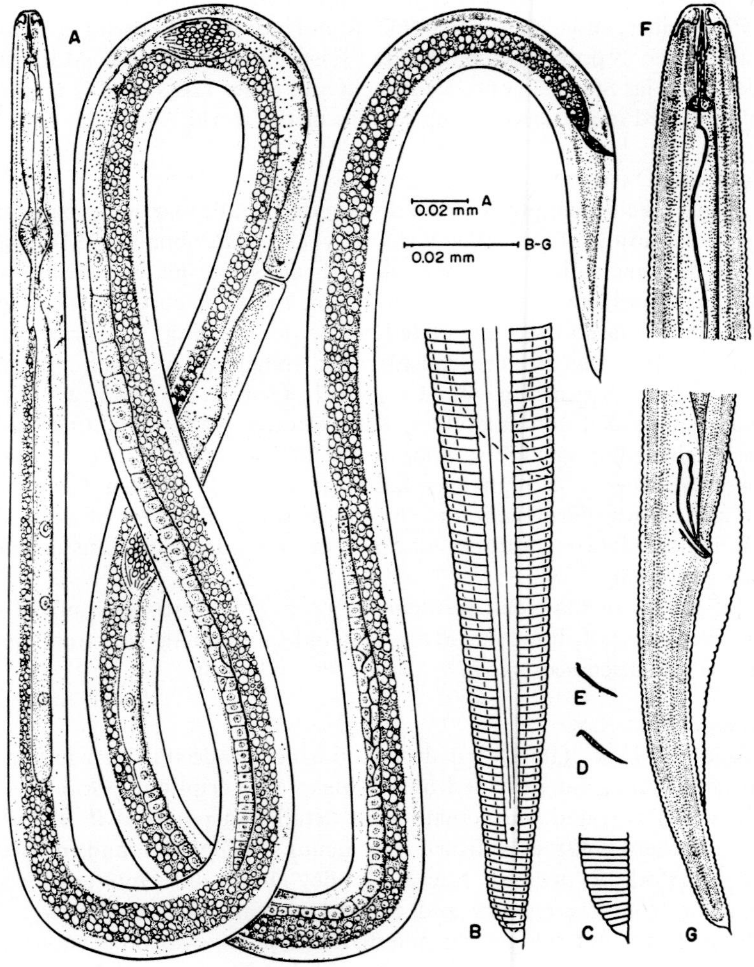

Fig. VII–5. *Hirschmanniella oryzae*. A, Female. B, Female, posterior region. C, Female, tail terminus. D, E, Gubernacula. F, Male, anterior end. G, Male, posterior region. (Sher, 1968).

Female: Lip region flattened, rounded edges, 3 to 4 annules. Stylet knobs rounded, sometimes slightly sloping anterior surfaces. Intestine not overlapping rectum. Lateral field usually not areolated; occasionally incomplete areolation, usually in posterior region. Tail terminus with distinct ventral mucro, sometimes a fine pointed ventral projection (Fig. VII–5).

Male: Similar to female except for sexual differences and occasionally areolation of the lateral field is less complete.

There are several species of *Hirschmanniella* which associate with rice roots. Taylor (1969) stated that until evidence that rice is parasitized by many species of *Hirschmanniella* became available, there was a tendency for authors to group all *Hirschmanniella* species found in rice roots and other hosts under the name *H. oryzae*, even though they differed considerably from Goodey's description. Taylor *et al.* (1966) surveyed rice nematodes in Thailand and found at least six species in the material collected, none of which fitted Goodey's description. Sher (1968) distinguished two species from *H. oryzae* as described by Thorne (1961), the small one as *H. oryzae*, as described above, and the large one as *H. thornei* sp. nov. Sher (1968) also reported *H. imamuri* sp. nov. on rice roots and soil from Tokyo, Japan, *H. caudacrena* sp. nov. from rice soil from Louisiana and *H. belli* sp. nov. from rice soil from California.

Another species of *Hirschmanniella*, *H. spinicaudata* (Sch. Stek.) Luc & Goodey, 1963, which was also reported under the name *Radopholus lavabri* Luc, 1957, is said to be parasitic on rice, but not much is known as to its pathogenicity. *H. mucronatus* (Das, 1960) Luc & Goodey, 1963, is reported from paddy fields in India, the Philippines, Thailand and East Pakistan.

Disease Cycle

Van der Vecht & Bergman (1952) reported that the adult nematodes of both sexes enter young roots through the epidermis at some distance from the tip and move in both directions so that in older roots they may exist anywhere between the base and the tip. The minimum time of development from egg to adult is at least one month and the multiplication factor per generation may be as great as 13. The nematodes are still alive in the soil after 10 weeks in the absence of host plants.

In Japan, the nematode is more frequently found in damp paddy fields than in well-drained soil and high correlation is found between the pH of the paddy soil and the nematode population. Infection frequency on the rice roots depends greatly on the number of nematodes overwintering in the field. Usually they overwinter in dead rice roots in the larval or adult form, but in swampy fields hibernation in the egg state frequently takes place. They seldom survive through the winter in well-drained paddy fields because of the dryness of the soil in early spring (Kawashima, 1962).

The root nematode has a rather wide host range in the tropics (Van der Vecht & Bergman, 1952; Chantanao, 1962). This fact suggests that other hosts are capable of carrying the nematodes over between rice crops.

The application of nitrogenous fertilizer seems to increase the nematode population, while potash fertilizer or compost maintains a low population level throughout the season. In Japan a comparatively high grain yield and a low population of nematodes have been reported in paddy fields where an annual application of either calcium silicate or compost has been made (Ichinohe, 1966).

Van der Vecht & Bergman (1952) in Indonesia reported that the rice variety Gendjah Ratji contained a small number of nematodes per plant in comparison to the large number present in Tjina, though both varieties received the same inoculum. This suggests the possibility of developing resistant varieties.

Kawashima (1963) in Japan made an estimation of the susceptibility to the nematode of 17 common varieties, based on the number of nematodes extracted from the roots, but no significant differences were found.

Van der Vecht & Bergman (1952) reported that the nematode lives and reproduces in the roots of more than 20 species of plants, mostly members of the Cyperaceae and Gramineae. Kawashima (1963) listed 25 plant species in 8 families as host plants including the lotus (*Nelumbo nucifera* Gaertn.).

Chemical Control

Yamashita first reported in Japan that soil treatment with D-D, EDB and Vapam in rice fields resulted in a considerable increase in yield of naked barley and rice (Ichinohe, 1966). Kawashima & Hori (1962a) used D-D, EDB and DBCP for fumigation in rice fields with a varying water content. The D-D treatment in a well-drained field resulted in a 38% increase in rice yield. In a moderately drained field, it is suggested that D-D should be injected before ploughing. The effect in a badly drained field was very slight. The beneficial effect of D-D remained until the second crop, in which there was a 27% increase in yield. Since then many tests have been made in Japan (Ichinohe, 1966) and D-D treatments have been shown to increase rice yields most effectively. DBCP has less effect and EDB treatment induces abnormal plant growth. Taylor *et al.* (1966) in Thailand have also reported better growth and yield of rice in soil treated with nematicides.

The increase in yield resulting from the application of nematicides is not necessarily due only to the control of nematodes. Hollis *et al.* (1959) concluded that fumigants stimulated plant growth by hygienic and nutritive effects. The hygienic effects of soil fumigation did not result from elimination of the nematodes alone, in this case *Tylenchorhynchus martini*, or a single complex of nematodes and other microorganisms, but from suppression of unknown injurious and competitive soil factors. A report from Rothamsted Experimental Station (1965) stated that fumigant application causes a temporary mineralization of soil nitrogen in the form of readily available ammonia. This was thought to result from the decomposition, by the surviving population, of the bodies of microorganisms killed by the fumigant. Experiments in Japan (Kureha *et al.*, 1965; Yoshida & Suzuki, 1965) also demonstrated that when D-D is used, the application of nitrogen fertilizer to rice plants should be reduced by 40% or 60% to avoid excess fertilization which induces lodging. Higher yields were thus obtained.

LITERATURE CITED

ATKINS, J. G., FIELD, M. J. & HOLLIS, J. P. 1955. Parastitic or suspected plant parasitic nematodes found in rice soil from Texas and Louisiana. *Pl. Dis. Reptr* **39**: 221–222.

CHANTANAO, A. 1962. Nematodes of rice and other plants in Thailand. *Tech. Bull. Dep. Ent. Pl. Path. Kasetsart Univ.* 1: 1–15.

DAS, V. M. 1960. Studies on the nematode parasites of plants in Hyderabad (Andhra Pradesh, India). *Z. ParasitKde* **19**: 553–605.

GENMA, T., SHIBUYA, T. & KIKUCHI, S. I. 1964. Investigation on nematode parasitism in roots of rice plant and its injuries. *J. Yamagata Agric. For. Soc.* **22**: 15–20. [Jap. Engl. summ.]

GOODEY, T. 1936. On *Anguillulina oryzae* (Van Breda de Haan 1902) Goodey, 1932, a nematode parasite of the roots of rice, *Oryza sativa* L. *J. Helminth.* **14**: 107–112.

HOLLIS, J. P., WHITLOCK, L. S., ATKINS, J. G. & FIELDING, M. J. 1959. Relations between nematodes, fumigation and fertilization in rice culture. *Pl. Dis. Reptr* **43**: 33–40.

ICHINOHE, M. 1966. Rice-infesting nematodes in the Pacific Basin. Paper presented at the Div. Meeting on Pl. Prot., 11th Pac. Sci. Congr., Tokyo, pp. 31–46.

IMAMURA, S. 1931. Nematodes in paddy fields, with notes on their population before and after irrigation. *J. Coll. Agric. imp. Univ. Tokyo* **11**: 193–240.

KAWASHIMA, K. 1962. Ecology of '*Radopholus oryzae*' attacking rice roots. *J. Pl. Prot., Tokyo* **16**: 57–59. [Jap.]

KAWASHIMA, K. 1963. Investigations on *Hirschmannia oryzae*. I. Varietal susceptibility to the nematode. II. Susceptibilities of weeds to the nematode. III. Durable effect of fungicides. *Ann. Rep. Soc. Pl. Prot. N. Japan* 14: 111; 112–113; 158–159. [Jap.]

KAWASHIMA, K. & FUJINUMA, T. 1965. On the injury to the rice plant caused by rice root nematode (*Hirschmanniella oryzae*): injury to the rice seedlings. *Bull. Fukushima Pref. agric. Exp. Stn* **1**: 57–64. [Jap. Engl. summ.]

KAWASHIMA, K. & HORI, T. 1962. Damage of rice seedlings by *Radopholus oryzae. Ann. Rep. Soc. Pl. Prot. N. Japan* 13: 120–121. [Jap.]

KAWASHIMA, K. & HORI, T. 1962a. An investigation on control of *Radopholus oryzae. Ann. Rep. Soc. Pl. Prot. N. Japan* 13: 174–175.

KUREHA, Y., MATSUNO, T. & NAKAMURA, T. 1965. Effect of nematicide (D-D) on rice root nematodes and its influence on the growth of rice plant. *Proc. Kanto-Tosan Pl. Prot. Soc.* **12**: 96–97. [Jap.]

KUROKAWA, S. & NAKAMURA, R. 1965. Damage of rice plant caused by rice root nematode. *Proc. Ass. Pl. Prot. Hokuriku* **13**: 92–94. [Jap.]

LUC, M. 1957. *Radopholus lavabri* n. sp. (Nematoda: Tylenchidae) parasite du riz au Cameroun Français. *Nematologica* **2**: 144–148.

LUC, M. & GOODEY, J. B. 1962. *Hirschmannia* n. g. differentiated from *Radopholus* Thorne, 1949. (Nematoda: Tylenchoidea). *Nematologica* **7** 197–202.

LUC, M. & GOODEY, J. B. 1963. *Hirschmanniella* nom. nov. for *Hirschmannia. Nematologica* **9**: 471.

ROTHAMSTED EXPERIMENTAL STATION. 1965. Pedology Dept, Soil Chemistry, Partial sterilization of soil. *Rep. for 1964*: 79–80.

SHER, S. A. 1968. Revision of the genus *Hirschmanniella* Luc & Goodey, 1963 (Nematoda – Tylenchoidea). *Nematologica* **14**: 243–275.

TAYLOR, A. L. 1969. Nematode parasites of rice. *In* Nematodes of tropical crops. *Tech. Commun. Commonw. Bureau Helminth.* 40: 264–268.

TAYLOR, A. L., KAOSIRI, T., SITTAICHI, T. & BUANGSUWON, D. 1966. Experiments on the effect of nematodes on the growth and yield of rice in Thailand. *Pl. Prot. Bull. F.A.O.* **14**: 17–23.

THORNE, G. 1949. On the classification of Tylenchida, new order (Nematoda, Phasmidia). *Proc. helminth. Soc. Wash.* **16**: 37–73.

THORNE, G. 1961. Principles of nematology. 553 pp., New York, McGraw Hill Co.

VAN BREDA DE HAAN, J. 1902. Een aaltjes-ziekte der rijst, ' omo mentèk ' of ' omo bambang '. *Meded. Lands Plant.* 53: 1–65.

VAN DER VECHT, J. & BERGMAN, H. H. 1952. Studies on the nematode '*Radopholus oryzae*' (Van Breda de Haan) Thorne and its influence on the growth of the rice plant. *Contr. gen. agric. Res. Stn Bogor* 131, 82 pp.

YOSHIDA, T. & SUZUKI, T. 1965. Chemical control for rice root nematode. *Proc. Kanto-Tosan Pl. Prot. Soc.* **12**: 98. [Jap.]

ROOT-KNOT NEMATODE

History and Distribution

Tullis (1934) was the first to notice a minor rice disease due to the root-knot nematode, in Arkansas, U.S.A. The nematode was then referred to as *Heterodera marioni* (Cornu) Goodey, now referred to *Meloidogyne.* Steiner (1934) also found it in a glasshouse bed previously used for a tobacco crop. Ichinohe (1955) reported the disease on an upland rice variety in Chiba, Japan. The nematode was referred to as *Meloidogyne incognita* var. *acrita* Chitwood, 1949. Van der Linde (1956) reported from South Africa that *Meloidogyne javanica* (Treub, 1885) Chitwood, 1949, *M. incognita* var. *acrita* and *M. arenaria* subsp. *thamesi* Chitwood, 1952, penetrate and reproduce in rice but not *M. hapla* Chitwood, 1949. Kanjanasoon (1964) reported that about 2000 hectares in northern Thailand, where tobacco, soybean and vegetables are grown in rotation with rice, were generally infested; and Hashioka (1963) found it in the deep water floating rice region of Thailand. The nematodes in Northern Thailand were referred to as *Meloidogyne javanica* and those in the deep water area as *M. exigua.* Israel, Rao & Rao (1964) reported *M. incognita* var. *acrita* on rice in India but the nematode was later identified as *M. graminicola* by Golden & Birchfield (1968), who also reported that the root-knot nematode in Laos was *M. graminicola.* Root-knot specimens have also been received by the writer from East Pakistan.

Symptoms and Damage

The characteristic symptoms are the enlargement of and formation of knots on the roots of rice seedlings. Growth of the root tips is retarded or stopped. Usually no distinct symptoms are visible on the foliage, but in very severe cases the plants are stunted and the leaves turn yellowish, the lower ones drying up.

The damage caused is usually slight and of a temporary nature. The nematode becomes inactive after long flooding in the paddy field, and flooding may therefore serve as a measure of control for this nematode as it has for some others (Hollis, 1966).

Patnaik (1969) observed that *M. graminicola* requires a minimum of 41 hours for the larvae to fix their heads inside the apical meristem of the root. Hypertrophy and hyperplasia start in cortical cells, forming a gall within 72 hours. Giant cells are formed 4 days after invasion. At the highest level of inoculum, i.e. 16 egg masses per plant, each containing an average of 770 eggs, discoloration of seedlings commences 10–12 days after inoculation, followed by tip drying and bronzing, which proceeds downwards and from the margin towards the midrib of the leaf blade. Newly emerged leaves appear distorted and crinkled along the margins. The symptoms are prominent on plants up to 50 days old. As the plant age advances, these symptoms become less and less pronounced. New tillers of infected plants are reduced in height. Severely infected plants show early flowering and poor setting of grains. Rice plants under normal conditions, however, could support a population of

1785 galls without any visible symptoms. As many as 62 nematodes are found inside a gall with 45 ovipositing females.

ORGANISMS, HOST RANGE AND VARIETAL RESISTANCE

As indicated above, several species of *Meloidogyne* may cause root-knot of rice, the presence of the nematodes being related to the other crops grown in rotation with rice. *M. incognita* var. *acrita* is known as the sweet potato root-knot nematode and also produces root-knot on maize, wheat, barley, red pepper, watermelon, fig etc. but not on groundnut and cotton (Ichinohe, 1955). Kanjanasoon (1964) found that *M. exigua* infected several species of grasses, sedges etc. in Thailand, including *Paspalum scrobiculatum*, *Panicum repens*, *Fuirena glomerata*, *Cyperus procerus*, *C. pulcherimus*, *Fimbristylis miliacea*, *Eleocharis* sp. and *Jussiaea linifolia*. These rice field weeds apparently serve as hosts when rice is not present.

Israel, Rao & Rao (1964) reported that rice varieties differ in susceptibility to *M. incognita* var. *acrita* in India. Variety Ch. 47 had an average of only 0·7 galls per plant while HR–19 had 53·5 galls; others were intermediate in susceptibility. Rao *et al.* (1969) reported that though many rice varieties can be infested by inoculation, only a few allow complete development of *M. graminicola*. Varieties such as TKM 6 and Patni 6 were found to be resistant, and BM 1514, IR52–18–2, IR8–118–1 and IR34–17–2 to be less susceptible than the others tested.

As mentioned above, Golden & Birchfield (1968) identified the rice root-knot nematode from Laos and India as *M. graminicola*. Many rice varieties were reported as being infected, but their relative resistance is not known. *M. graminicola* has a wide range of hosts. Originally found on *Echinochloa colonum*, it also infects oats, bushbean and several species of grass, but not soybean, sweet potato, cotton, maize, pepper, tomato, watermelon and several other crops.

LITERATURE CITED

GOLDEN, A. M. & BIRCHFIELD, W. 1968. Rice root-knot nematode (*Meloidogyne graminicola*) as a new pest of rice. *Pl. Dis. Reptr* **52**: 423.

HASHIOKA, Y. 1963. Report to the Government of Thailand on blast and other diseases of rice in Thailand. *F.A.O. E.P.T.A. Rep.* 1734, 41 pp.

HOLLIS, J. P. 1966. Nematodes in Louisiana rice field, nature and significance of population control by flooding. Paper presented at 11th meeting of FAO-IRC Working Party on Rice Production and Protection. Lake Charles, La, U.S.A., 20 pp. (Mimeographed.)

ICHINOHE, M. 1955. Two species of the root-knot nematodes in Japan. *Jap. J. appl. Zool.* **20**: 75–82. [Jap. Engl. summ.]

ISRAEL, P., RAO, Y. S. & RAO, V. N. 1964. Rice nematode—host and parasite relationship. Paper presented at 10th meeting of FAO-IRC Working Party on Rice Production and Protection, Manila, Philippines, 3 pp. (Mimeographed.)

ISRAEL, P., RAO, Y. S. & RAO, V. N. 1966. Rice parasitic nematodes. Paper presented at 11th meeting of FAO-IRC Working Party on Rice Production and Protection, Lake Charles, La, U.S.A., 6 pp. (Mimeographed.)

KANJANASOON, P. 1964. Rice root-knot nematodes and their host plants. Paper presented at 10th meeting of FAO-IRC Working Party on Rice Production and Protection, Manila, Philippines, 2 pp. (Mimeographed.)

PATNAIK, N. C. 1969. Pathogenicity of *Meloidogyne graminicola* (Golden & Birchfield, 1965) in rice. *All India Nematology Symposium, New Delhi, 1969*: 12 (Abs.).

RAO, Y. S., BISWAS, H., PANDA, M., RAO, P. R. V. J. & RAO, V. N. 1969. Screening rice varieties for their reaction to root and root-knot nematodes. *All India Nematology Symposium, New Delhi, 1969*: 59 (Abs.).
STEINER, G. 1934. Root-knot and other nematodes attacking rice and some associated weeds. *Phytopathology* **24**: 916–928.
TULLIS, E. C. 1934. The root-knot nematode of rice. *Phytopathology* **24**: 938–942.
VAN DER LINDE, J. 1956. The *Meloidogyne* problem in South Africa. *Nematologica* **1**: 177–183.

RICE CYST NEMATODE

Okada (1955) reported and briefly described an unnamed species of *Heterodera* which had been found to attack upland rice in Japan. Later, Luc & Berdon Brizuela (1961) described *Heterodera oryzae*, a new species parasitic on lowland rice in the Ivory Coast. The species is characterized by its dark brown to black cysts, 0·31–0·81 × 0·22–0·68 mm (mean 0·57 × 0·45 mm), lemon-shaped, with a subcrystalline layer and oriented in zig-zag lines but with no trace of regular transverse markings. The vulval cone is prominent and the cone top is of the ambifenestrate type with bullae. The second stage larvae measure 0·37–0·50 mm (mean 0·44 mm). Their stylet is 19·5–22μ long. These larvae are characterized mainly by their lateral field with only three lines and by the acutely pointed tail with a hyaline portion much longer than the stylet (35–45μ). Males are rare and resemble those of *H. schachtii*.

Berdon Brizuela & Merny (1964) observed the life cycle of the nematode. The fourth moult in the male occurs on the 16th day after the larva has penetrated into the root. Egg formation can be observed between the 20th and 25th day and the first second-stage larvae begin to move out between the 26th and the 30th day. Several generations may occur in the same rice crop. In infested roots, areas of disintegration of the cell walls have been observed within the stele. No true giant cells were observed but tyloses in the vessels were noted and also the formation of a tissue around the female body, tending to solate it from the surrounding cortex tissues.

Watanabe *et al.* (1963; *cf.* Ichinohe, 1966) found the cyst nematode to be the cause of the failure of successive cropping of upland rice in Japan. The nematode disturbed the nutrition and physiology of the plants, and affected plants bore few root hairs, were stunted during the seedling stage, and in severe cases died, the leaves showing severe chlorosis. Mature plants had yellowish leaves and their growth was retarded. Number of tillers and grain weight were reduced. They reported that *Panicum* (*Echinochloa*) *crus-galli* and *P. crus-galli* var. *frumentaceum* are also hosts of the nematode and noted that no rice variety is highly resistant. A one or two year rotation with non-host plants, or fallowing, reduces the population of the nematode in the upper layers of the soil. The rice crop following the rotation was good, but the population of nematodes built up again by harvest time. D-D and EDB soil treatment significantly reduced the population, the low population level being maintained by a high dosage of D-D (4 ml/30 × 30 cm) but not by a low dosage (Oda *et al.*, 1964).

LITERATURE CITED

BERDON BRIZUELA, R. & MERNY, G. 1964. Biologie d'*Heterodera oryzae* Luc & Berdon 1961. I. Cycle du parasite et réactions histologiques de l'hôte. *Revue Path. vég. Ent. agric. Fr.* **43**: 43–53.

ICHINOHE, M. 1966. Rice-infesting nematodes in the Pacific basin. Paper presented at Div. Meet. Pl. Prot. 11th Pacific Sci. Congr., Tokyo, pp. 31–46. (Mimeographed.)

LUC, M. & BERDON BRIZUELA, R. 1961. *Heterodera oryzae* n. sp. (Nematoda-Tylenchoidea) parasite du riz en Cote D'Ivoire. *Nematologica* **6**: 272–279.

ODA, K., HOSHINO, M., TAKITA, Y., YANAKA, S., KUMAZAWA, T., TANAKA, S. & KEGASAWA, K. 1964. Studies on rice cyst nematodes. IV. Damage analysis; V. Effect of nematicides; VI. Effect of crop rotation. *Proc. Kanto-Tosan Pl. Prot. Soc.* **11**: 109–111. [Jap.]

OKADA, T. 1955. On morphological characters of the upland rice nematode, *Heterodera* sp. *Abs. paper Ann. Meet. appl. Zool. & Entom. Japan 1955*: 11. [Jap.]

RICE STUNT NEMATODE AND OTHER MIGRATORY PARASITIC NEMATODES

The stunt nematode *Tylenchorhynchus martini* was described by Fielding (1956) as a new species, which was found to be widespread in the sugarcane and rice soils of Louisiana and in the rice soils of Texas, U.S.A. It was reported from East Pakistan by Timm & Ameen (1960) and from India by Israel, Rao & Rao (1966). Fielding described the nematode as follows:

Female: total length=0·75 mm; a=31, b=5; c=13·8; $V={}^{26}54^{21}$.

The lip region bears three annules and is set off by a slight constriction. Cuticle marked by conspicuous annules which average 1·9μ apart at mid-body, and are slightly smaller near the head. The annules on the tail are variable in size. Female terminus blunt, tail shape distinctive. Phasmids prominent, located slightly anterior to middle of tail. Cervical papillae (deirids) not observed. Lateral fields marked by four lines, anteriorly beginning in region of buccal stylet knobs, and ending near caudal terminus, and are 1/4 as wide as body width at mid-body. Annulation interrupted at lateral fields. Buccal stylet about 19μ long, slender, with strong basal knobs. Dorsal oesophageal gland opening near base of spear. Median oesophageal bulb ovoid, with a conspicuous crescent-shaped valve. Nerve ring encircling isthmus of oesophagus near the middle. Basal portion of oesophagus forming a distinct bulb bearing three gland nuclei. Excretory pore located near the proximal end of the basal oesophageal bulb. Hemizuonid located two body annules anterior to excretory pore.

Vulva a transverse slit. Vagina extending inward 1/3 the distance across the body. Ovaries paired, outstretched, oogonia forming a single line. Eggs 1/2 as wide and twice as long as body width.

Males not observed. Among 80,000 specimens examined not a single male was seen.

Resembled *Tylenchorhynchus claytoni* Steiner 1937, but different because of simple body annulations, shape of female tail, and presence of a slight constriction setting off lip region.

Johnston (1958) found the optimum soil moisture for the survival of *T.*

martini was 40–60% field capacity while for other nematodes it was 75–100%. Johnston (1959) studying the effect of fatty acid mixtures on the nematodes, reported that when nematodes were submerged in concentrations of 10^{-2}, 10^{-3} or 10^{-4} M of butyric, propionic, acetic and formic acids in distilled water, separately or in combination, the effectiveness was related to molecular weight, butyric being most active. The effectiveness of acid mixtures was related to combined molecular weight.

Atkins & Fielding (1956) found that fumigation with methyl bromide gave better control of *T. martini* than did ethylene dibromide, D-D or Nemagon. Hollis *et al.* (1959) reported that soil fumigation with D-D and methyl bromide and fertilizer treatments of phosphorus, potash and nitrogen exerted independent additive effects on rice yields. The total fumigant effect was divided into nutritive and hygienic factors. Laboratory studies showed that *T. martini* and fungi in the rice fields were not sufficiently pathogenic to account for the full hygienic effects achieved by fumigation and they postulated the destruction by the fumigant of many unknown competitive factors.

Hollis (1966) reported that *T. martini* and *Hirschmaniella oryzae* are the most prevalent parasitic nematodes in the rice fields of Louisiana, but they cause little or no damage to rice. The normal practice of flooding exerts a high degree of control over nematode populations. The decline in the nematode population after flooding coincides with the rise of total hydrogen sulphide in the field; the nematodes are killed *in vitro* by hydrogen sulphide at concentrations of 10 ppm or higher. Another species of *Tylenchorhynchus*, *T. elegans*, was described by Siddiqi (1961) from rice soils in India.

Several other parasitic nematodes have been reported in rice. Imamura (1931) collected the ring nematode *Criconemoides komabaensis* (Imamura, 1931) Taylor, 1936, in Japan. Hollis *et al.* (1968) reported *C. onoensis* in rice soils of Louisiana causing stubby root, root-knot, stunting, chlorosis and yellow dwarf of rice seedlings and recorded a yield loss of 15% in 1967. Steiner (1934) reported that the root lesion nematode *Pratylenchus pratensis* (de Man, 1880) Filipjev, 1936, attacks rice. In Pakistan Timm (1956) reported on rice the ring nematode *Criconemoides rusticus* (Micolcetzky 1915) Taylor, 1936, the lance nematode *Hoplolaimus galeatus* (Cobb, 1913) Sher, 1961, and the spiral nematode *Helicotylenchus multicinctus* (Cobb, 1893) Golden, 1956. Luc (1959) found a root lesion nematode, *Pratylenchus brachyurus* (Godfrey, 1929) T. Goodey, 1951, in rice roots in Madagascar. *Hoplolaimus indicus* is a parasite of rice roots in India (Das & Rao, 1969).

Two migratory dorylaimid nematodes, *Xiphinema parasetariae* and *X. orbum*, have been reported from around rice roots in Guinea and India by Luc (1958) and Siddiqi (1964), respectively. The genus *Xiphinema* is especially important to agriculture because several species are known to be vectors of important soil-borne plant viruses.

LITERATURE CITED

Atkins, J. G. & Fielding, M. J. 1956. A preliminary report on the response of rice to soil fumigation for the control of stylet nematode, *Tylenchorhynchus martini*. *Pl. Dis. Reptr* **40**: 488–489.

DAS, P. K. & RAO, S. 1969. Life history and pathogenicity of *Hoplolaimus indicus* Sher in rice. *All Indian Nematology Symposium, New Delhi, 1969*: 12. (Abs.)

FIELDING, M. J. 1956. *Tylenchorhynchus martini*, a new nematode species found in the sugar cane and rice fields of Louisiana and Texas. *Proc. helminth. Soc. Wash.* **23**: 47–48.

HOLLIS, J. P. 1966. Nematodes in Louisiana rice field—nature and significance of population control by flooding. Paper presented at 11th meeting of FAO-IRC Working Party on Rice Production and Protection, Lake Charles, La, U.S.A. (Mimeographed.)

HOLLIS, J. P., EMBABI, M. S. & ALHASSAN, S. A. 1968. Ring nematode disease of rice in Louisiana. Abs. in *Phytopathology* **58**: 728–729.

HOLLIS, J. P., WHITLOCK, L. S., ATKINS, J. G. & FIELDING, M. J. 1959. Relations between nematodes, fumigation and fertilization in rice culture. *Pl. Dis. Reptr* **43**: 33–40.

IMAMURA, S. 1931. Nematodes in the paddy field, with notes on their population before and after irrigation. *J. Coll. Agric. imp. Univ. Tokyo* **11**: 193–240.

ISRAEL, P., RAO, Y. S. & RAO, V. N. 1966. Rice parasitic nematode. Paper presented at 11th meeting of FAO-IRC Working Party on Rice Production and Protection, Lake Charles, La, U.S.A. (Mimeographed.)

JOHNSTON, T. M. 1958. The effect of soil moisture on *Tylenchorhynchus martini* and other nematodes. *Proc. La Acad. Sci 1957* **20**: 55–60.

JOHNSTON, T. M. 1959. Effect of fatty acid mixtures on the rice stylet nematode (*Tylenchorhynchus martini* Fielding, 1956). *Nature, Lond.* **183**: 1392.

LUC, M. 1958. *Xiphinema* de l'ouest Africain: Description de cinq nouvelles espèces (Nematoda: Dorylaimidae). *Nematologica* **3**: 57–72.

LUC, M. 1959. Nématodes parasites ou soupçonnés de parasitisme envers les plantes de Madagascar. *Bull. Inst. Rech. agron. Madagascar* 3: 89–101.

SIDDIQI, M. R. 1961. Studies on *Tylenchorhynchus* spp. (Nematoda: Tylenchida) from India. *Zentbl. ParasitKunde* **21**: 46–64.

SIDDIQI, M. R. 1964. Three new species of *Dorylaimoides* Thorne & Swanger, 1936, with a description of *Xiphinema orbum* n. sp. (Nematoda: Dorylaimoidea). *Nematologica* **9**: 626–634.

STEINER, G. 1934. Root-knot and other nematodes attacking rice and some associated weeds. *Phytopathology* **24**: 916–928.

TIMM, R. W. 1956. Nematode parasites of rice in East Pakistan. *Pakistan Rev. Agric.* **2**: 115–118.

TIMM, R. W. & AMEEN, M. 1960. Rice nematodes in East Pakistan. Abs. in *Proc. 12th Pak. Sci. Conf. 1960* Pt 3, Sect. B: 25–26.

PART VIII

PHYSIOLOGICAL DISEASES

Physiological diseases in the broadest sense refer to all disorders or abnormalities brought about by non-parasitic causes such as low or high temperatures beyond the limits for normal growth of rice, deficiency or excess of nutrients in the soil and water, pH and Eh and other soil conditions which affect the availability and uptake of nutrients, toxic substances such as H_2S produced in soil, water stress and reduced light.

Takahashi (1960), in his review on physiological diseases of rice, concluded that, with a few exceptions, all physiological diseases are more or less related to the development of highly reduced conditions in the soil; this concept is somewhat narrow. Tanaka & Yoshida (1970) stated that the highly reduced conditions as expressed by Eh *per se* may not cause the disorder, but that some chemical changes taking place under such reduced conditions may bring about a deficiency or excess of nutrients, or inhibit the uptake of nutrients or water. More precisely, therefore, they are nutritional disorders. For instance, detailed studies on 'akiochi' disease in Japan have revealed that it is due to deficiencies in K, Si, Mn, Mg or other elements (Baba & Harada, 1954; Baba *et al.*, 1965).

The symptoms of several virus diseases are similar to those of physiological diseases. Some of the more important so-called physiological diseases mentioned by Takahashi in his review, such as 'penyakit merah' and 'mentek', are now known to be due to viruses. The suffocation disease, first thought to be physiological, is also mainly due to a virus. Several other diseases which have not been critically studied are called physiological diseases because their causes are obscure or unknown.

Most of the physiological diseases are the subjects of studies in plant physiology and soil chemistry. IRRI Technical Bulletin No. 10, entitled 'Nutritional disorders of rice in Asia' and prepared by Tanaka & Yoshida, has been published recently. Therefore, these diseases are only briefly mentioned here.

Papers reporting on damage caused by unfavourable climatic factors are relatively few, but injuries resulting from low temperatures have been reported from temperate regions such as North Japan (Hokkaido) where the early decline of the autumn temperature may cause sterility. Shimazaki *et al.* (1964) reported that this sterility was mainly due to the failure of anthers to dehisce and flowers to open, and also partly due to failure of fertilization, i.e. some pollen grains did not germinate and some embryos did not develop even when fertilized. Considerable work has been done on this subject in Japan (Ministry of Agric. and Forest., 1965). The temperature range for proper flowering and fertilization is generally said to be 15–30°C. Prolonged periods beyond this range cause varying degrees of sterility.

Injuries from low temperatures have also been reported from the highlands of Vietnam at the reproductive stage of crop growth and from East Pakistan at the vegetative stage, but no detailed studies have been made.

In Central Iraq, the writer observed high sterility (50–90%) in many of the hundreds of introduced rice varieties, presumably due to high air temperatures (reaching 50°C) which destroyed the pollen.

Nutritional disorders due to deficiency or excess of major and minor elements are common in rice plants. Tanaka & Yoshida (1970) briefly described the visual symptoms as follows:

1. *Nitrogen deficiency*—Whole plant yellow or only the youngest leaf green, leaves erect, plants stunted and number of tillers reduced.

2. *Phosphorus deficiency*—Leaves dark green, erect and narrow, plants stunted and number of tillers reduced.

3. *Potassium deficiency*—Leaves dark green with tips of lower leaves becoming yellow, leaves drooping, often with brown spots. Plants stunted and number of tillers slightly reduced.

4. *Sulphur deficiency*—Symptoms similar to, and difficult to distinguish from, those of nitrogen deficiency.

5. *Calcium deficiency*—Tips of leaves become white and in extreme cases growing points die, leaves remain green, root elongation retarded and the tips become brown.

6. *Magnesium deficiency*—Interveinal chlorosis in lower leaves.

7. *Iron deficiency*—Entire leaves become chlorotic or whitish.

8. *Manganese deficiency*—Interveinal chlorosis on youngest leaves, older leaves remain relatively yellowish green. The chlorotic streaks spread downward from the tips and later some dark brown necrotic spots appear on these leaves. Newly emerging leaves become short and narrow with severe chlorosis.

9. *Zinc deficiency*—The midribs of the youngest leaves, especially at the base, become chlorotic; more generally, brown blotches and streaks appear on lower leaves, followed by stunted growth.

10. *Boron deficiency*—Plants stunted, the emerging leaves show white tips and die in severe cases.

11. *Copper deficiency*—Leaves bluish green, becoming chlorotic near the tips, with the chlorosis developing downward along both sides of the midrib, followed by dark-brown necrosis of the tips. New emerging leaves fail to unroll and remain needle-like or, occasionally, the upper half of the leaf unrolls while the basal portions develop normally.

12. *Iron toxicity*—Tiny brown spots appear on lower leaves, starting from the tips; leaves remain green. In severe cases, entire leaves look purplish brown.

13. *High salt toxicity*—Stunted growth and reduced tillering, the leaf tips become whitish and frequently some parts become chlorotic.

14. *Low silicon content*—Leaves become soft and drooping.

The exact diagnosis of these disorders depends upon plant tissue and soil and water analyses or must be confirmed by foliar sprays of the chemicals which are suspected to be deficient. Tanaka & Yoshida (1970) gave the content of the nutritional elements in the rice plant below which deficiency symptoms may appear (Table VIII–1).

TABLE VIII–1

DEFICIENCY AND TOXICITY CRITICAL CONTENTS OF VARIOUS ELEMENTS IN THE RICE PLANT*
(Tanaka & Yoshida, 1970)

Element	Deficiency (D) or toxicity (T)	Critical content	Plant part analyzed	Growth stage**
N	D	2·5%	Leaf blade	Til
P	D	0·1%	Leaf blade	Til
	T	1·0%	Straw	Mat
K	D	1·0%	Straw	Mat
	D	1·0%	Leaf blade	Til
Ca	D	0·15%	Straw	Mat
Mg	D	0·10%	Straw	Mat
S	D	0·10%	Straw	Mat
Si	D	5·0%	Straw	Mat
Fe	D	70 ppm	Leaf blade	Til
	T	300 ppm	Leaf blade	Til
Zn	D	10 ppm	Shoot	Til
	T	1500 ppm	Straw	Mat
Mn	D	20 ppm	Shoot	Til
	T	>2500 ppm	Shoot	Til
B	D	3·4 ppm	Straw	Mat
	T	100 ppm	Straw	Mat
Cu	D	<6 ppm	Straw	Mat
	T	30 ppm	Straw	Mat
Al	T	300 ppm	Shoot	Til

* Figures for critical contents collected from references cited but adjusted to round figures.
** Mat—ma.urity. Til—tillering.

The measurement of soil pH usually provides useful clues in determining the deficiency and toxicity of the elements. When the pH is low, iron toxicity and phosphorus deficiency are likely to occur; when high, iron deficiency, zinc deficiency, phosphorus deficiency and high salt toxicity are likely to occur.

Tanaka & Yoshida (1970) also summarized the reported nutritional disorders of the rice plant and their possible causes (Table VIII–2). Among these diseases, akiochi, akagare and others which occur in Japan are well known and have been studied in detail. Straighthead occurs in U.S.A. and in other countries. The cause is still obscure but remedies are known. Zinc deficiency in India, Philippines and West Pakistan have recently been well documented. It is suspected that zinc deficiency may be found in many other areas. Bronzing in Ceylon has also been well studied. Mentek, penyakit merah and suffocation disease are now known to be caused by viruses. Other suspected nutritional disorders have not yet been critically studied. It is difficult to determine the causes of some of these diseases because of the complicated conditions in the soil. One also has to consider the possibility of virus infections, the symptoms of virus diseases often being similar to those of physiological diseases.

A few of the well known physiological diseases are briefly discussed below:

Akiochi (Autumn decline)

This disease has occurred widely in Japan, damaging 600,000 hectares or 20 % of the total rice acreage of the country. It occurs in the so-called ' degraded paddy soil ' which is well-drained and sandy, has a very low base exchange capacity, is deficient in Fe, Mn, Ca and K, and where fertilizers are easily leached; and also in poorly drained paddy soil which has a high water table, much organic matter, little active iron, and produces H_2S in the warm summer months (Baba & Harada, 1954).

TABLE VIII–2

LIST OF REPORTED NUTRITIONAL DISORDERS OF THE RICE PLANT
(Summary from Tanaka & Yoshida, 1970)

Country	Name of disorder	Possible causes
Burma	Amyit-Po	K deficiency?
	Myit-Po	P deficiency
	Yellow leaf	S deficiency
Ceylon	Bronzing	Fe toxicity
Colombia	Espiga Erecta	Similar to straighthead, unknown
Hungary	Brusone	Blast?, H_2S toxicity?
India	Khaira disease	Zn deficiency
	Bronzing	Fe, Mn, H_2S toxicity
	Yellowing	Unknown
Indonesia	Mentek	Virus disease
Japan	Akagare type I	K deficiency (Fe toxicity)
	Akagare II	Zn deficiency
	Akagare type III	I toxicity
	Akiochi	Hydrogen sulphide toxicity, Si, Mg, K deficiency
	Hideri Aodachi	Unknown
	Aogare	Imbalance of water economy due to excess N, K deficiency and accelerated transpiration
	Straighthead	Unknown
Korea	Aiochi (stifle)	Similar to Akiochi in Japan
Malaya	Penyakit Merah	
	brown spot type	Fe toxicity
	yellowing type	Virus disease
Pakistan	Pansukh	Unknown
	Hadda	Zn deficiency
Portugal	Branca	Cu deficiency? Similar to straighthead
Taiwan	Suffocation disease	Virus disease
U.S.A.	Straighthead	Unknown
	Alkali disease	Fe deficiency
Vietnam	Acid sulphate soil	Fe toxicity, P deficiency

The affected plants generally show normal growth in the early stages but begin to decline gradually just before heading. The typical symptoms are: (1) occurrence of *Helminthosporium* leaf spot, (2) early drying of lower leaves and (3) root injury (Baba *et al.*, 1965).

The causes of akiochi are: (1) malnutrition, due to deficiencies in K, Si, Mn and Mg, which is intensified by the addition of N because of the imbalance of the K/N or Si/N ratio; (2) nitrogen deficiency in the later stages of growth because of the low nutrient-holding capacity of sandy soil; (3) root injury and inhibition of nutrient absorption due to the presence of H_2S and organic acids; and (4) root injury and retarded absorption of Si and K due to high soil and water temperature and high light intensity (Baba *et al.*, 1965).

The remedies recommended for the disease include: (1) application of furnace slag containing Si, Ca, Mg and Mn; (2) split applications of nitrogenous fertilizers and potash; (3) use of fertilizers containing no SO_4 radical; (4) use of resistant varieties; and (5) improvement of soil. Varieties resistant to *Helminthosporium* brown spot are also said to be resistant to akiochi.

Akagare

Akagare occurs in Japan, and similar diseases also occur in other countries. Studies in Japan have distinguished three types of the disease according to the detailed symptoms present and the causal factors involved.

Type I occurs in ill-drained, sandy, lowland rice soils and ill-drained mucks or boggy soils. Generally at the tillering stage, the leaves first turn dark green, then small reddish-brown spots appear near the tips of the older leaves. The spots spread all over the leaves and the leaves die, starting from the tips. The roots turn light brown in mucks or boggy soils, and in many cases may darken or become rotten.

Type II occurs mostly in badly drained muck or boggy lowland soils. The midribs of the leaves turn yellow, and then reddish-brown spots appear around the discoloured part and increase until the whole leaf becomes reddish-brown. In severe cases the spots appear without any previous yellow discoloration. The roots of affected plants are reddish-brown and some of them darken and rot.

Type III occurs in reddish, heavy clay loam, volcanic-ash soils, or humic volcanic-ash soils newly converted from upland cultivation. All are extremely deficient in phosphorus and they are often acidic. Small brown spots appear first at the tips of older leaves and subsequently spread all over the surface, giving the leaves a yellow-brown or brown discoloration.

Type I is caused by potassium deficiency, and the symptoms are very similar to those of potassium deficiency observed in culture solution. This disorder can be prevented by the application of potassium fertilizers. Type II is caused by a combination of factors including (1) high organic matter content; (2) production of H_2S; (3) low iron content of the surface soil; (4) excessive ferrous iron; and (5) abnormal reducing power of the soil in addition to a poor natural supply of potassium. This type cannot be prevented by the mere application of potash. The major cause is the absorption of such harmful organic acids as butyric and acetic acids, and soluble ferrous iron and hydrogen sulphide. Type III is partly caused by phosphorus and potassium deficiencies. Application of phosphorus noticeably reduces the disease and application of potassium likewise reduces the disease to some extent. The application of lime intensifies the disease since it favours organic decomposition which lowers the pH (Baba *et al.*, 1965).

Recent studies (Tanaka *et al.*, 1969; Shiratori *et al.*, 1969; Watanabe & Tensho, 1970) have demonstrated more precisely that Type II is caused by zinc deficiency and Type III by iodine toxicity. Remedies may be applied accordingly.

Varieties differ in their reaction to akagare. Okamoto (1950) found that upland varieties are more susceptible while lowland varieties have varying susceptibility. Varieties resistant to potassium deficiency are said to be resistant to all three types.

General preventive measures include: (1) use of potassium fertilizer; (2) application of red-coloured hill soil and of fertilizer without a sulphate radical; (3) avoidance of the use of green manure; and (4) improved methods of cultivation and irrigation (Baba *et al.*, 1965).

Bronzing

Some 40,000 hectares of the ill-drained soils in the low wet zone of Ceylon are affected by this disease.

Many small, brownish spots appear on the tips of the older leaves, and then spread until almost the entire leaf surface, except the midrib, turns brown; this causes the leaves to die, starting from the tips. In severe cases, the brown discoloration appears even on the youngest open leaves. Ponnamperuma (1958) reported that the colour of affected leaves varies somewhat with varieties but that it is usually purplish or orange. Ota & Yamada (1962) distinguished two types of the disease: type I begins to occur 1 or 2 weeks after transplanting, while type II appears 1 or 2 months after transplanting.

The soil conditions which favour the development of the disease are: (1) low pH of dried soil, 4·3–4·9; (2) high content of iron oxide; (3) poor drainage; and (4) contiguity with ferruginous lateritic highland soil (Ponnampenama, 1958). According to Ponnamperuma *et al.* (1955, 1958) bronzing is assoruated with a high content of ferrous iron in the soil solution when the soil ciH is low. It may be corrected by adding lime to raise the pH. Ota & Yapmad (1962), however, observed bronzing where the pH was high and no bronzing when the free iron was high. Hydrogen sulphide has been suggested as a possible primary cause of bronzing (Yamada, 1959; Inada, 1966), and it does aggravate iron toxicity. A combination of high aluminium and low calcium contents has also been suggested as the cause of bronzing.

Liming is effective in correcting the disease, and the application of compost also reduces the incidence of the disease. Some local varieties in the wet zone of Ceylon show a degree of resistance to bronzing (Takahashi, 1960; Tanaka & Yoshida, 1970).

Straighthead

First described in U.S.A. (Tisdale & Jenkins, 1921) where it occurred on rice fields newly converted from virgin or grass upland, straighthead has also been reported from Japan, Portugal, Colombia and other countries. In U.S.A. several thousand hectares have been reported to be affected.

The panicles of affected plants remain upright at maturity because of high (or complete) sterility and as a consequence of their very light weight. The hulls, either the paleas or lemmas or both, are distorted, beak-shaped, reduced, or may be lacking. The pistil and stamens of affected flowers are generally absent. In severe cases, the entire panicles are reduced in size and emerge slowly or incompletely from the bootleaf. The affected plants sometimes remain in the vegetative phase, are dark green, and produce tillers at the lower nodes. Atkins *et al.* (1968) also reported that the germination of seeds harvested from straighthead fields is lower than normal, and that some seeds germinated with two or three coleoptiles.

Straighthead is often severe in slightly sandy clay soils which are not drained completely. The exact cause is not known, but it is believed that the disorder may be due to unfavourable soil conditions aggravated by prolonged flooding. Cu deficiency was suspected to be the cause of branca in Portugal (Cunha & Baptista, 1958).

Several varieties are known to be resistant to the disorder (Atkins *et al.*, 1956), and drainage just prior to the stem elongation stage has been found effective in controlling the disease (Cheaney, 1955).

LITERATURE CITED

ATKINS, J. G., BEACHELL, H. M. & CRANE, L. E. 1956. Reaction of rice varieties to straighthead. *Prog. Rep. Texas agric. Exp. Stn* 1865, 2 pp.

ATKINS, J. G., BOLLICH, C. N. & SCOTT, J. E. 1968. Reduction of rice yields and seed germination by straighthead. *Int. Rice Commn Newsl.* **17** (3): 45.

BABA, I. & Harada, T. 1954. Physiological diseases of rice plant in Japan. *Jap. J. Breed.* **4** (Studies, Rice Breeding): 101–150.

BABA, I., INADA, K. & TAJIMA, K. 1965. Mineral nutrition and the occurrence of physiological diseases. In *Proc. symp. on the mineral nutrition of the rice plant. IRRI, 1964*: 173–195. Baltimore, Maryland, Johns Hopkins Press.

CHEANEY, R. L. 1955. Effect of time of draining of rice on the prevention of straighthead. *Prog. Rep. Texas agric. Exp. Stn* 1774, 5 pp.

CUNHA, J. M. DE A. & BAPTISTA, J. E. 1958. Estudo da branca do arroz. I. Combate da doença. *Agron. lusit.* **20**: 17–64.

INADA, K. 1966. Studies on the bronzing disease of rice plant in Ceylon. I; II. *Trop. Agriculturist* **122**: 19–29; **125**: 31–46.

MINISTRY OF AGRICULTURE AND FORESTRY, JAPAN. 1955. Literature on cold-weather injuries of rice plant. Japan. Agric. Improv. Bureau. [Jap.]

OKAMOTO, H. 1950. Akagare disease of rice and its prevention. *Agriculture Hort., Tokyo* **25**: 437–439. [Jap.]

OTA, Y. & YAMADA, N. 1962. Physiological study on bronzing of rice plant in Ceylon. (Preliminary Report). *Proc. Crop Sci. Soc. Japan* **31**: 90–97.

PONNAMPERUMA, F. N. 1958. Lime as a remedy for physiological disease of rice associated with excess iron. *Int. Rice Commn Newsl.* **7** (1): 10–13.

PONNAMPERUMA, F. N., BRADFIELD, R. & PEECH, M. 1955. Physiological diseases of rice attributable to iron toxicity. *Nature, Lond.* **175**: 265.

SHIMAZAKI, Y., SATAKE, T., ITO, N., DOI, Y. & WATANABE, K. 1964. Studies of cool weather injuries of rice plant in northern part of Japan. III. Sterile spikelets in rice plants induced by low temperatures during the booting stage. IV. Effect of day and night temperature accompanied by shading treatment during booting stage upon induction of sterile spikelets in rice plant. *Res. Bull. Hokkaido natn. Agric. Exp. Stn* 83: 1–16. [Jap. Engl. transl. by Takamatsu & Imura.]

SHIRATORI, K., SUZUKI, T. & MIYOSHI, H. 1969. On a disease of rice plant similar to 'Akagare' occurring in paddy field where soil dredged from the Tone River is used. *Bull. Chiba agric. Exp. Stn* 9: 73–81.

TAKAHASHI, J. 1960. Review of investigations on physiological diseases of rice. I and II. *Int. Rice Commn Newsl.* **9** (1): 1–6; (2): 17–24.

TANAKA, A., SHIMONO, K. & ISHIZUKA, Y. 1969. On ' Akagare ' of lowland rice caused by zinc deficiency. *J. Sci. Soil Manure, Japan* **40**: 415.

TANAKA, A. & YOSHIDA, S. 1970. Nutritional disorders of rice in Asia. *Tech. Bull. int. Rice Res. Inst.* 10.

TISDALE, W. H. & JENKINS, J. M. 1921. Straighthead of rice and its control. *Fmrs' Bull. U.S. Dep. Agric.* 1212, 16 pp.

WATANABE, I. & TENSHO, K. 1970. Further study on iodine toxicity in relation to ' reclamation akagare ' disease of lowland rice. *Soil Sci. Plant Nutr.* **16**: 192–194.

YAMADA, N. 1959. Some aspects of the physiology of bronzing. *Int. Rice Commn Newsl.* **8** (3): 11–16.

INDEX

[Note: host names are in general those used by original authors; hosts should therefore be looked for under common and alternative Latin names]